Lecture Notes in Control and Information Sciences

Edited by A. V. Balakrishnan and M. Thoma

Vol. 1: Distributed Parameter Systems: Modelling and Identification
Proceedings of the IFIP Working Conference, Rome, Italy, June 21–26, 1976
Edited by A. Ruberti
V, 458 pages. 1978

Vol. 2: New Trends in Systems Analysis
International Symposium, Versailles, December 13–17, 1976
Edited by A. Bensoussan and J. L. Lions
VII, 759 pages. 1977

Vol. 3: Differential Games and Applications
Proceedings of a Workshop, Enschede, Netherlands, March 16–25, 1977
Edited by P. Hagedorn, H. W. Knobloch, and G. J. Olsder
XII, 236 pages. 1977

Vol. 4: M. A. Crane, A. J. Lemoine
An Introduction to the Regenerative Method for Simulation Analysis
VII, 111 pages. 1977

Vol. 5: David J. Clements, Brian D. O. Anderson
Singular Optimal Control: The Linear Quadratic Problem
V, 93 pages. 1978

Vol. 6: Optimization Techniques
Proceedings of the 8th IFIP Conference on Optimization Techniques, Würzburg, September 5–9, 1977
Part 1
Edited by J. Stoer
XIII, 528 pages. 1978

Vol. 7: Optimization Techniques
Proceedings of the 8th IFIP Conference on Optimization Techniques, Würzburg, September 5–9, 1977
Part 2
Edited by J. Stoer
XIII, 512 pages. 1978

Lecture Notes in Control and Information Sciences

Edited by A.V. Balakrishnan and M. Thoma
Series: IFIP TC7 Optimization Conferences

IFIP

7

Optimization Techniques

Proceedings of the
8th IFIP Conference on Optimization Techniques
Würzburg, September 5–9, 1977

Part 2

Edited by J. Stoer

Springer-Verlag Berlin Heidelberg GmbH 1978

Editor
Prof. Dr. J. Stoer
Institut für angewandte Mathematik und Statistik
Universität Würzburg
D-8700 Würzburg
Am Hubland

With 82 Figures

ISBN 978-3-540-08708-3 ISBN 978-3-540-35890-9 (eBook)
DOI 10.1007/978-3-540-35890-9

Originally published by Springer-Verlag Berlin Heidelberg New York in 1978.

PREFACE

These Proceedings contain most of the papers presented at the 8th IFIP Conference on Optimization Techniques held in Würzburg, September 5-9,1977.
The Conference was sponsored by the IFIP Technical Committee on System Modelling and Optimization (TC 7) with the cooperation of

- European Research Office (ERO), London
- Gesellschaft für Angewandte Mathematik und Mechanik (GAMM)
- Bayerisches Staatsministerium für Unterricht und Kultus
- Bundesministerium für Forschung und Technologie
- Deutsche Forschungsgemeinschaft.

The Conference was attended by 241 scientists from 28 countries. The program offered a broad view of optimization techniques currently in use and under investigation. Major emphasis was on recent advances in optimal control and mathematical programming and their application to modelling, identification and control of large systems, in particular, recent applications in areas such as biological, environmental and socio-economic systems.

The Proceedings are divided into two volumes: In the first are mainly collected the papers dealing with optimal control, in the second those dealing with mathematical programming and various application areas.

The international Program Committee of the Conference consisted of:

A.V. Balakrishnan (Chairman, USA), L.V. Kantorovich (USSR), W.J. Karplus (USA), R. Kluge (GDR), H.W. Knobloch (GER), J.L. Lions (France), G.I. Marchuk (USSR), C. Olech (Poland), L.S. Pontryagin (USSR), A. Prekopa (Hungary), E. Rofman (Argentina), A. Ruberti (Italy), B.F. de Veubeke† (Belgium), K. Yajima (Japan).

TABLE OF CONTENTS

MATHEMATICAL PROGRAMMING, THEORY

NONLINEAR AND STOCHASTIC PROGRAMMING

<u>INTEGER PROGRAMMING, NETWORKS</u>

PART 1

TABLE OF CONTENTS

INVITED SPEAKERS

ROUND TABLE DISCUSSION ON WORLD MODELS

COMPUTATIONAL TECHNIQUES IN OPTIMAL CONTROL

---

*paper not received

DIFFERENTIAL GAMES

OPTIMAL CONTROL OF PARTIAL DIFFERENTIAL EQUATIONS

SEMI-INFINITE PROGRAMMING: CONDITIONS OF OPTIMALITY AND APPLICATIONS

R. P. Hettich
Universität Bonn
Institut für Angewandte Mathematik
Wegelerstrasse 6
5300 Bonn/Germany

H. Th. Jongen
Department of Applied Mathematics
Twente University of Technology
P.O. Box 217
Enschede/The Netherlands

1. Introduction.

In this paper we consider the following problem:

Problem 1.1. Let $Y \subset R^m$ be a compact, $X^* \subset R^n$ an open set. Given functions $f: X^* \to R$, $g: X^* \times Y \to R$, and $e^j: X^* \to R$, $j = 1, \ldots, p$, let

$$X = \{x \in X^* | g(x,y) \leq 0,\ y \in Y,\ e^j(x) = 0,\ j = 1,\ldots,p\}. \quad (1.1)$$

Find $\bar{x} \in X$, such that

$$f(x) \leq f(\bar{x}) \quad (1.2)$$

for all $x \in X$ (global maximum) or at least for all $x \in X \cap U_{\bar{x}}$, $U_{\bar{x}}$ some neighborhood of $\bar{x}$ (local maximum).

A local or global maximum is called strict if equality in (1.2) implies $x = \bar{x}$. Lateron, the functions f, g and e^j will be assumed to have continuous derivatives up to order one or two.

We remark that rather general nonlinear Chebyshev approximation problems such as monotone, one-sided, and restricted range approximation as well as problems with interpolatory constraints (cf. [11]) can be stated as problems of the above type. Furthermore many problems from different areas are included, e.g. programming problems with restrictions depending on time or place. An example may be found in [1], where a problem of air pollution control is considered.

If Y is a finite set, Problem 1.1 is a usual (finite) programming problem. Otherwise, the problem is called semi-infinite. For finite problems, in [5] first and second order conditions are given being necessary or sufficient for a given point $\bar{x} \in X$ to be a local maximum. There are two types of conditions. A first one based on the first order condition given by John [6] for p = 0 and by Mangasarian and Fromovitz [8] for the general case (Y finite):

There are numbers $\lambda \geq 0$, $\lambda_y \geq 0$, $y \in Y$, and μ_j, $j = 1, \ldots, p$, not all equal to zero, such that

$$L_x(\lambda,\lambda_y,\mu_j,\bar{x}) = 0 \tag{1.3}$$

where

$$L(\lambda,\lambda_y,\mu_j,x) = \lambda f(x) - \sum_{y\in Y} \lambda_y g(x,y) - \sum_{j=1}^{p} \mu_j e^j(x) \tag{1.4}$$

and a second one based on the Kuhn-Tucker-condition [7], where λ may be taken equal to one if some additional condition (constraint qualification) holds.

In this paper we restrict ourselves to conditions, where λ is allowed to be zero. This is motivated by the following observations:

Conditions where $\lambda \neq 0$ are given in [13] and, for the finite case, in [9]. Checking the constraint qualifications assumed for these conditions to hold, requires rather complete information about X in a neighborhood of the point $\bar{x}$ under consideration, which will not be available in general.

While, for $\lambda = 0$, no information about the object function f is contained in (1.3), this is not the case in the second order conditions presented in this paper or in [5] for the finite case. For illustration consider the following example:

Let $n = 2$, $Y = \{1,2\}$, $g(x,1) = g^1(x) = x_2^2 - x_1^2$, $g(x,2) = g^2(x) = -x_1$, and $p = 0$. Then, $X = \{\begin{pmatrix} x_1 \\ x_2 \end{pmatrix} \mid x_1 \geq 0,\ |x_2| \leq x_1\}$. Take $\bar{x} = \begin{pmatrix} 0 \\ 0 \end{pmatrix}$. $g_x^1(\bar{x}) = \begin{pmatrix} 0 \\ 0 \end{pmatrix}$

shows that (1.3) holds with $\lambda = \lambda_2 = 0$, $\lambda_1 \neq 0$, for every function f. If not, by accident, $f_x(\bar{x}) = \binom{a}{0}$, $a \leq 0$, this choice is the only possibility to realize (1.3). However, in the second order conditions ([5]) the sign of $q(\xi) = \xi^T L_{xx}(\lambda, \lambda_y, \mu_j, \bar{x})\xi$ must be considered on the intersection P of the halfspaces given by $\xi^T f_x(\bar{x}) \geq 0$ and $\xi_1 \geq 0$, $q(\xi) < 0$, $\xi \in P \sim \{0\}$, being sufficient and $q(\xi) \leq 0, \xi \in P$, necessary for $\bar{x}$ to be a local maximum. Clearly, P depends on f. If $f_x(\bar{x}) \neq \binom{a}{0}$, $a \leq 0$, we have $(\lambda_1 = 1) q(\xi) = 2\xi_1^2 - 2\xi_2^2$. This is nonnegative if and only if $\xi \in X$. Therefore, the necessary condition requires that there is no direction ξ of increase, i.e. $\xi^T f_x(\bar{x}) > 0$, such that $\bar{x} + t\xi \in X$, t sufficiently small. But this is clearly equivalent to the Kuhn-Tucker -condition when X is described by $\bar{g}^1(x) = x_2 - x_1 \leq 0$, $\bar{g}^2(x) = -x_2 - x_1 \leq 0$, instead of g^1, g^2.

Apart from checking the optimality of a given point, there are other important applications of second order conditions:

If the test for the necessary condition has negative result, automatically a direction of increase is obtained. This may be used to "complete" a number of methods, for instance steepest descent methods [14], by an additional step which makes it possible to leave or avoid first order critical points not being maxima.

On the other hand, second order sufficient conditions are an important tool in proving convergence of a number of methods ([10],[12], [4]). Moreover, they are very useful in investigating the dependence of the solution on parameters ([10]). In the last section we will come back to this point.

2. First order conditions of optimality

In this section we assume the functions f, g, e^j, $j = 1, \ldots, p$ to be continuous and C^1-functions with respect to the variable x.

Let $\bar{x} \in X$ be a given point. Define

$$Y_0(\bar{x}) = \{y \in Y | g(\bar{x}, y) = 0\}. \tag{2.1}$$

We call $Y_0(\bar{x})$ the activity set. Let $Y_0(\bar{x})$ be given by means of some set K of indices, not necessarily countable

$$Y_0(\bar{x}) = \{\bar{y}^k | k \in K\}. \tag{2.2}$$

Furthermore we define

$$P = \{\xi \in R^n | \xi^T f_x(\bar{x}) \geq 0,\ -\xi^T g_x(\bar{x},\bar{y}^k) \geq 0,\ k \in K,\ \xi^T e_x^j(\bar{x}) = 0$$
$$j = 1, \ldots, p\},$$

$$P_0 = \{\xi \in P | \xi^T f_x(\bar{x}) = 0\}, \tag{2.3}$$

$$K_s = \{k \in K | \xi^T g_x(\bar{x},\bar{y}^k) = 0 \quad \text{for all } \xi \in P\}. \tag{2.4}$$

The inequalities $-\xi^T g_x(\bar{x},\bar{y}^k) \geq 0$, $k \in K_s$ are called singular.

Theorem 2.1. First order sufficient condition.

If $P = \{0\}$, then $\bar{x}$ is a strict local maximum.

Proof. Suppose that $\bar{x}$ is not a strict local maximum. Then there is a sequence of points $x^i \in X$, $x^i \to \bar{x}$, $x^i \neq \bar{x}$, such that

$$-[f(x^i) - f(\bar{x})] \leq 0, \tag{2.5}$$

$$g(x^i,\bar{y}^k) = g(x^i,\bar{y}^k) - g(\bar{x},\bar{y}^k) \leq 0,\ k \in K \tag{2.6}$$

$$e^j(x^i) - e^j(\bar{x}) = 0,\ j = 1, \ldots, p. \tag{2.7}$$

Let $x^i = x + t_i\xi_i$, $\xi_i \in R^n$, $||\xi_i|| = 1$, $t_i > 0$. It is no restriction to assume that ξ_i converge, $\xi_i \to \xi_0$, $||\xi_0|| = 1$.

Clearly, $t_i \to 0$. Formula (2.5) implies

$$-[f(x^i) - f(\bar{x})] = -t_i\xi_i^T f_x + o(t_i) \leq 0,$$

showing $-\xi_0^T f_x \leq 0$. Analogously (2.6) and (2.7) imply $\xi_0^T g_x^k \leq 0, k \in K$, $\xi_0^T e_x^j = 0$, $j = 1, \ldots, p$. Thus $\xi_0 \in P$, $\xi_0 \neq 0$, contrary to the assumption.

Remark 2.1. From the proof of Theorem 2.1 it is obvious that Theorem 2.1 remains to be true if Y is not compact.

Theorem 2.2. First order necessary condition.

Suppose $\bar{x}$ is a local maximum of Problem 1.1. Then there exist real numbers $\lambda \geq 0$, μ_j, $j = 1, \ldots, p$ and $\lambda_k \geq 0$, $k \in \bar{K}_s \subset K_s$, where $\bar{K}_s$ is a finite set, not all equal to zero, such that

$$\lambda f_x = \sum_{k \in \bar{K}_s} \lambda_k g_x^k + \sum_{j=1}^{p} \mu_j e_x^j . \tag{2.8}$$

Moreover, λ can be chosen unequal to zero if and only if $P = P_0$.

We note that, except for the last assertion, Theorem 2.2 is in fact the optimality condition of Fritz John [6]. The last assertion can be proved analogously to the proof of Theorem 4.2 in [5].

3. The Constraint -Reduction-Lemma.

In this section we assume the functions f, g, e^j, $j = 1, \ldots, p$ to be C^2-functions.

Let $\bar{x} \in X$. If Y is a finite set, then, by continuity, there exists a neighborhood $U_{\bar{x}}$ of $\bar{x}$ such that for all $x \in U_{\bar{x}} \cap X$ we have $Y_0(x) \subset Y_0(\bar{x})$. However this is no longer true if Y is an infinite subset of R^m. In fact this is a basic difference between finite and semi-infinite optimization problems and will play a main role in formulating optimality conditions of second order.

In order to control the behavior of the set $Y_0(x)$ depending on x we formulate two conditions, one describing the structure of the set Y and the other stating non-degeneracy of the activity set.

Condition A. There is given a finite set L of indices and C^2-functions $h^\ell: R^m \to R$, $\ell \in L$, such that

$$Y = \{y \in R^m | h^\ell(y) \leq 0, \ \ell \in L\}.$$

For $y \in Y$ let

$$L(y) = \{\ell \in L | h^\ell(y) = 0\}. \tag{3.1}$$

Then, for every $y \in Y$, the gradients $h_y^\ell(y)$, $\ell \in L(y)$ are linearly independent.

Given $\bar{x} \in X$. We note that every $\bar{y}^k \in Y_0(\bar{x})$ is a maximum of $g(\bar{x},y)$ with respect to Y. Condition B states that these maxima should be non

-degenerate. Consequently these maxima are isolated and from compactness of Y it follows that there are at most a finite number of them.

Condition B. Condition B holds at $\bar{x} \in X$ if for every $\bar{y}^k \in Y_0(\bar{x})$ the following is satisfied:

There exist real numbers $\bar{\chi}_{k\ell} > 0$, $\ell \in L(\bar{y}^k)$, such that

$$g_y^k - \sum_{\ell \in L(\bar{y}^k)} \bar{\chi}_{k\ell} h_y^{\ell k} = 0 \tag{3.2}$$

and

$$\mu^T M_k \mu \quad := \mu^T [g_{yy}^k - \sum_{\ell \in L(\bar{y}^k)} \bar{\chi}_{k\ell} h_{yy}^{\ell k}] \mu < 0 \tag{3.3}$$

for all $\mu \in H_k \sim \{0\}$,

$$H_k = \{\mu \in R^m | \mu^T h_y^{\ell k} = 0, \ \ell \in L(\bar{y}^k)\}. \tag{3.4}$$

Here, g_y^k, $h_y^{\ell k}$ etc. stand for $g_y(\bar{x},\bar{y}^k)$, $h_y^\ell(\bar{y}^k)$ etc.

Condition B states in particular that the maxima $\bar{y}^k$ of $g(\bar{x},y)$ are such that a second order sufficient condition holds (cf. [5]) together with strict complementary slackness with respect to h^ℓ and $\bar{\chi}_{k\ell}$.

Conditions A and B make it possible to describe the effect of a variation of $\bar{x}$ on the position of the maxima $\bar{y}^k$:

Note that $\bar{y}^k$ and $\bar{\chi}_{k\ell} > 0$, $\ell \in L(\bar{y}^k)$, are solutions of the system

$$\left.\begin{aligned} g_y(\bar{x},y^k) - \sum_{\ell \in L(\bar{y}^k)} \chi_{k\ell} h_y^\ell(y^k) &= 0 \\ -h^\ell(y^k) = 0, \ \ell \in L(\bar{y}^k)&. \end{aligned}\right\} \tag{3.5}$$

The Jacobian matrix of (3.5) w.r.t. y^k and $\chi_{k\ell}$ at $\bar{y}^k$, $\bar{\chi}_{k\ell}$ is given by

$$\begin{pmatrix} M_k & G_k \\ G_k^T & 0 \end{pmatrix} \tag{3.6}$$

with M_k given by (3.3) and

$$G_k = (\ \ldots\ldots, -h_y^{\ell k}, \ \ldots\ldots \ )_{\ell \in L(\bar{y}^k)} \ . \tag{3.7}$$

By Condition A G_k has full rank given by the number of indices in $L(\bar{y}^k)$. Suppose, the matrix (3.6) is singular. Then there are μ and ν, not both 0, such that

$$M_k\mu + G_k\nu = 0 \tag{3.8}$$

$$G_k^T\mu = 0 . \tag{3.9}$$

(3.9) implies $\mu = 0$ or $\mu \in H_k$ (cf. (3.4)). From $\mu = 0$ it would follow that $G_k\nu = 0$ and, because G_k has full rank, $\nu = 0$. Thus $\mu \neq 0$, $\mu \in H_k$. Premultiplying (3.8) by μ^T gives $\mu^T M_k \mu = 0$ contrary to Condition B. Thus (3.6) is non-singular. Application of the Implicit Function Theorem shows that there is a neighborhood $U_{\bar{x}} \subset X^*$ of $\bar{x}$ and for every $k \in K$ (finite set!) uniquely determined C^1-maps $y^k\colon U_{\bar{x}} \to R^m$, $\chi_{k\ell}\colon U_{\bar{x}} \to R$, $\ell \in L(\bar{y}^k)$ such that we have identically in $U_{\bar{x}}$:

$$\left.\begin{aligned} & g_y(x,y^k(x)) - \sum_{\ell \in L(\bar{y}^k)} \chi_{k\ell}(x) h_y^{\ell}(y^k(x)) \equiv 0 \\ & h_{\ell}(y^k(x)) \equiv 0, \ \ell \in L(\bar{y}^k) \end{aligned}\right\} . \tag{3.10}$$

From the foregoing and the compactness of Y it is easily seen that there is a neighborhood $V_{\bar{x}} \subset U_{\bar{x}}$ of $\bar{x}$ such that for every $x \in V_{\bar{x}}$ the (global) maxima of $g(x,y)$ w.r.t. Y are contained in the finite set $\{y^k(x),\ k \in K\}$. This leads to the following lemma.

Constraint-Reduction-Lemma. Let $\bar{x} \in X$ and suppose that Conditions A and B hold. Then there is a neighborhood $V_{\bar{x}}$ of $\bar{x}$, and uniquely defined C^1-maps $y^k\colon V_{\bar{x}} \to R^m$, $k \in K$ (finite) such that with $\phi^k(x) = g(x,y^k(x))$

$$X \cap V_{\bar{x}} = \{x \in V_{\bar{x}} \mid \phi^k(x) \le 0,\ k \in K,\ e^j(x) = 0,\ j = 1,\dots,p\}.$$

Furthermore the composite functions ϕ^k are C^2-functions.

Proof. It remains to prove that ϕ^k are C^2. From (3.10) it follows

$$\phi^k(x) = g(x,y^k(x)) \equiv g(x,y^k(x)) - \sum_{\ell \in L(\bar{y}^k)} \chi_{k\ell}(x)\, h^{\ell}(y^k(x)).$$

Consequently, $\phi^k \in C^2$ follows from

$$\phi_x^k = g_x + [y_x^k]^T \underbrace{[g_y^k - \sum_{\ell \in L(\bar{y}^k)} \chi_{k\ell} h_y^{\ell k}]}_{\equiv 0} - \sum_{\ell \in L(\bar{y}^k)} \chi_{k\ell,x} \underbrace{h^{\ell}(y^k(x))}_{\equiv 0}$$

and the fact that g is a C^2-function and y^k a C^1-map.

4. Second order conditions of optimality

From the results of Section 3 we conclude that $\bar{x}$ is a (strict) local maximum for Problem 1.1 if and only if it is a (strict) local maximum for the following finite programming problem:

Problem 4.1. Let $V_{\bar{x}}$, $\phi^k(x)$, $k \in K$, be given as in the Constraint-Reduction-Lemma. Define

$$\hat{X} = \{x \in V_{\bar{x}} | \phi^k(x) \le 0,\ k \in K,\ e^j(x) = 0,\ j = 1,\dots,p\}. \tag{4.1}$$

Find $\hat{x} \in \hat{X}$ such that

$$f(x) \le f(\hat{x}),\ x \in \hat{X}\ .$$

Let

$$\hat{P} = \{\xi \in R^n | \xi^T f_x(\hat{x}) \ge 0,\ -\xi^T \phi_x^k(\hat{x}) \ge 0,\ k \in K,\ \xi^T e_x^j(\hat{x}) = 0,\ j = 1,\dots,p\}, \tag{4.2}$$

$$\hat{P}_0 = \{\xi \in \hat{P} | \xi^T f_x(\hat{x}) = 0\}, \tag{4.3}$$

$$\hat{K}_s = \{k \in K | \xi^T \phi_x^k(\hat{x}) = 0 \quad \text{for all } \xi \in \hat{P}\}, \tag{4.4}$$

and

$$q(\lambda,\lambda_k,\mu_j,\xi) = \xi^T(\lambda f_{xx}(\hat{x}) - \sum_{k \in K_s} \lambda_k \phi_{xx}^k(\hat{x}) - \sum_{j=1}^{p} \mu_j e_{xx}^j(\hat{x}))\xi. \tag{4.5}$$

In [5] it is shown that the following are necessary (sufficient) conditions for a (strict) local maximum for Problem 4.1:
Let $\hat{x}$ be a local maximum for Problem 4.1. Then, for every $\xi \in \hat{P}$, there are numbers $\lambda \ge 0$, $\lambda_k \ge 0$, $k \in \hat{K}_s, \mu_j$, not all zero, such that

$$\lambda f_x(\hat{x}) = \sum_{\lambda \in \hat{K}_s} \lambda_k \phi_x^k(\hat{x}) + \sum_{j=1}^{p} \mu_j e_x^j(\hat{x}) \tag{4.6}$$

and $q(\lambda,\lambda_k,\mu_j,\xi) \le 0$. Moreover, $\hat{P} \ne \hat{P}_0$ implies $\lambda = 0$.

On the other hand, if, for every $\xi \in \hat{P}$, there exist λ, λ_k, μ_j as above such that (4.6) holds and, if $\xi \ne 0$, $q(\lambda,\lambda_k,\mu_j,\xi) < 0$, then $\hat{x}$ is a strict local maximum for Problem 4.1.

We compute $\phi_x^k = \phi_x^k(\bar{x})$ and $\xi^T \phi_{xx}^k \xi = \xi^T \phi_{xx}^k(\bar{x})\xi$, with $\bar{x}$ a given point where Conditions A and B (cf. Section 3) and the condition of Theorem 2.2 hold. From the proof of the Constraint-Reduction-Lemma we have

$$\phi_x^k(x) = g_x(x, y^k(x)). \tag{4.7}$$

Because of (4.7) we have $P = \hat{P}$, $P_O = \hat{P}_O$, $K_s = \hat{K}_s$. Let $\xi \in P$. Then

$$\xi^T \phi_{xx}^k \xi = \xi^T g_{xx}^k \xi + \xi^T g_{xy}^k y_x^k \xi = \xi^T g_{xx}^k \xi - (-(g_{yx}^k \xi)^T) y_x^k \xi. \tag{4.8}$$

The vector $\pi_k = y_x^k \xi$ is the directional derivative of $y^k(x)$ in the direction ξ for $x = \bar{x}$. For $x = \bar{x} + t\xi$ we find from (3.5) and (3.6)

$$\begin{pmatrix} M_k & G_k \\ G_k^T & 0 \end{pmatrix} \begin{pmatrix} \pi_k \\ \dot{\chi}_{k\ell} \end{pmatrix} = \begin{pmatrix} -g_{xy}^k \xi \\ 0 \end{pmatrix}, \tag{4.9}$$

where $\dot{\chi}_{k\ell} = \frac{d}{dt}\chi_{k\ell}(\bar{x}+t\xi)\big|_{t=0}$ (shortly viewed as a $|L(\bar{y}^k)|$-vector).

Formulae (4.8), (4.9) imply

$$\xi^T \phi_{xx}^k \xi = \xi^T g_x^k \xi - \pi_k^T M_k \pi_k - \dot{\chi}_{k\ell}^T G_k^T \pi_k = \xi^T g_{xx}^k \xi - \pi_k^T M_k \pi_k. \tag{4.10}$$

Substituting (4.7) and (4.10) in the above conditions for Problem 4.1 we obtain the following conditions for Problem 1.1.

<u>Assumption 4.1</u>. f, g and e^j, $j = 1,\dots,p$, are twice continuously differentiable w.r.t. all variables. Conditions A and B hold for $x = \bar{x}$.

<u>Theorem 4.1</u>. <u>Second order necessary condition</u>.

Let $\bar{x} \in X$ be a local maximum for Problem 1.1 and suppose that Assumption 4.1 holds. Then, for every $\xi \in P$, there exist $\lambda \geq 0$, $\lambda_k \geq 0$, $k \in K_s$, μ_j, not all equal to zero, such that (2.8) holds and

$$q(\lambda, \lambda_k, \mu_j, \xi) = \xi^T (\lambda f_{xx} - \sum_{k \in K_s} \lambda_k g_{xx}^k - \sum_{j=1}^{p} \mu_j e_{xx}^j) \xi + \sum_{k \in K_s} \lambda_k \pi_k^T M_k \pi_k \leq 0. \tag{4.11}$$

Moreover, if $P \neq P_O$, then λ is always equal to zero.

<u>Theorem 4.2</u>. <u>Second order sufficient condition</u>.

Suppose Assumption 4.1 holds. Then, if for every $\xi \in P$ there exist $\lambda \geq 0$, $\lambda_k \geq 0$, $k \in K_s$, μ_j, such that (2.8) holds and, if $\xi \neq 0$, $q(\lambda, \lambda_k, \mu_j, \xi) < 0$, $\bar{x}$ is a strict local maximum.

We remark, that these conditions imply those in [2], [3] for the special case of Chebyshev approximation problems.

5. On the convergence of some numerical methods.

Sufficient conditions of optimality are important in proving convergence of numerial methods as will be illustrated now.

Condition C. Let $\bar{x} \in X$. Condition C holds at $\bar{x}$ if Assumption 4.1 is satisfied, the gradients $g_x(\bar{x},\bar{y}^k)$, $k \in K$, $e_x^j(\bar{x})$, $j = 1,\ldots,p$, are linearly independent, and the second order sufficient conditions of Theorem 4.2 hold with $\lambda = 1$, $\lambda_k = \bar{\lambda}_k > 0$, $k \in K$, and $\mu_k = \bar{\mu}_k$.

Analogously to [12] for finite problems, [1] for linearly constrained semi-infinite problems and [4] for nonlinear Chebyshev approximation, we consider the Newton method (assuming Condition C) applied to the following system of nonlinear equations in order to compute $\bar{x}$:

$$\left.\begin{aligned} f_x(x) - \sum_{k\in K} \lambda_k g_k(x,y^k) - \sum_{j=1}^{p} \mu_j e_x^j(x) &= 0 \\ g(x,y^k) &= 0, \quad k \in K \\ e^j(x) &= 0, \quad j = 1, \ldots, p \\ g_y(x,y^k) - \sum_{\ell\in L(\bar{y}^k)} \chi_{k\ell} h_y^\ell(y^k) &= 0, \quad k \in K \\ h^\ell(y^k) &= 0, \quad \ell \in L(\bar{y}^k), \ k \in K \end{aligned}\right\}. \quad (5.1)$$

These are $n + |K| + p + m|K| + \sum_{k\in K} |L(\bar{y}^k)|$ nonlinear equations for an equal number of unknowns x, λ_k, μ_j, y^k, $\chi_{k\ell}$. In the same way as in [4] it can be shown that the Jacobian of (5.1) is regular at $\bar{x}$, $\bar{\lambda}_k$, $\bar{\mu}_j$, $\bar{y}^k$, $\bar{\chi}_{k\ell}$ if Condition C holds at $\bar{x}$.

Theorem 5.1. If Condition C holds at $\bar{x} \in X$, the Newton method applied to the system (5.1) converges for sufficiently good starting values. The convergence is at least superlinear and quadratic if additional conditions hold (for instance that f, g, e^j, h^k have continuous derivatives up to order three).

In [10] Robinson for finite optimization studies the dependence of the solution on parameters under an assumption which is for finite problems the same as Condition C. With the results obtained he is able to prove convergence for a whole class of methods. In fact, making use of the corresponding results of this paper or [13], the results of Robinson can be generalized to the semi-finite case.

References.

[1] Gustafson, S.A., Kortanek, K.O.: Numerical treatment of a class of semi-infinite programming problems, Nav. Res. Log. Quart., 20(1973), pp. 477-504.

[2] Hettich, R.: Kriterien zweiter Ordnung für lokal beste Approximationen, Numer. Math., 22 (1974), pp. 409-417.

[3] Hettich, R.: Kriterien erster und zweiter Ordnung für lokal beste Approximationen bei Problemen mit Nebenbedingungen, Numer. Math., 25 (1975), pp. 109-122.

[4] Hettich, R.: A Newton-method for nonlinear Chebyshev approximation, In: Approximation Theory, Lect. Notes in Math. 556 (1976), Schaback, R., Scherer, K., eds., Springer, Berlin-Heidelberg-New York, pp. 222-236.

[5] Hettich, R., Jongen, H.Th.: On first and second order conditions for local optima for optimization problems in finite dimensions, to appear in Proceedings of a conference on Operations Research at Oberwolfach, August 1976.

[6] John, F.: Extremum problems with inequalities as side-conditions, Studies and Essays, Courant Anniversary Volume, Friederichs, K.O., Neugebauer, O.E., Stocker, J.J., eds., Wiley, New York, 1948, pp. 187-204.

[7] Kuhn,H.W., Tucker, A.W.: Nomlinear Programming, Proc. Second Berkeley Symposium on Math. Statistics and Probability, Univ. of California Press, Berkeley, California, 1951, pp. 481-492.

[8] Mangasarian, O.L., Fromovitz, S.: The Fritz John necessary optimality conditions in the presence of equality and inequality constraints, J. Math. Anal. Appl., 17 (1967), pp. 37-47.

[9] McCormick, G.P.: Second order conditions for constrained minima, SIAM J. Appl. Math., 15 (1967), pp. 641-652.

[10] Robinson, S.M.: Perturbed Kuhn-Tucker points and rates of convergence for a class of nonlinear programming algorithms, Math. Programming 7 (1974), pp. 1-16.

[11] Taylor, G.D.: Uniform approximation with side conditions. In: Approximation theory, G.G. Lorentz (ed.), Academic Press, New-York -London, 1973, pp. 495-503.

[12] Wetterling, W.: Über Minimalbedingungen und Newton-Iteration bei nichtlinearen Optimierungsaufgaben. In: Iterationsverfahren, Numerische Mathematik, Approximationstheorie, ISNM,15 (1970), Birkhäuser Verlag, Basel und Stuttgart, pp. 93-99.

[13] Wetterling, W.: Definitheitsbedingungen für relative Extrema bei Optimierungs- und Approximationsaufgaben, Numer. Math., 15 (1970), pp. 122-136.

[14] Zoutendijk, G.: Methods of feasible directions, Elsevier Publishing Company, Amsterdam 1960.

ON EQUIWELLSET MINIMUM PROBLEMS

T. Zolezzi
Laboratorio per la Matematica Applicata del C.N.R.
University of Genova - Italy.

The following problem arises very often in both theoretical and applied optimization. A sequence of real-valued functionals I_n is given on some metric space X, together with a fixed functional I_o. Here I_n are to be considered as (variational) perturbations of the "limit" functional I_o. We are interested in approximating the optimal objects of I_o by solving the (more tractable) minimization problems for I_n. When any approximate minimization on I_n automatically converges to the optimal objects of I_o?

Such a problem shows many interesting connections with recently developed theories about variational convergence of functionals by Mosco, De Giorgi-Franzoni-Spagnolo.

The sequence I_n along with I_o is called here <u>equiwellset</u> when the following is true:

1) every I_n is minimized on X in an unique point x_n;

2) $\min I_n \longrightarrow \min I_o$;

3) if y_n is an asymptotically minimizing sequence for I_n, i.e.
$I_n(y_n) - \min I_n \longrightarrow 0$, then $y_n \longrightarrow x_o$.

In the particular case $I_n = I_o$ for all n, the above definition specializes to minimum problems for a single functional which are well set in the sense of Tyhonov.

Some characterizations are obtained of equiwellposedness, of a metric and differential nature. As a byproduct, characterizations of Tyhonov's well set minimum problems are obtained extending criteria of Vajnberg and Poracká-Divis. Some applications are given to an abstract epsilon problem, and to the perturbations of the plant in an abstract linear-quadratic problem.

SECOND-ORDER NECESSARY AND SUFFICIENT OPTIMALITY CONDITIONS FOR INFINITE-DIMENSIONAL PROGRAMMING PROBLEMS

H. Maurer and J. Zowe

Institut für Numerische und Instrumentelle Mathematik, Westfälische Wilhelms-Universität, Roxeler Straße 64, 44 Münster, F.R.G.,

Institut für Angewandte Mathematik und Statistik, Universität Würzburg, Am Hubland, 87 Würzburg, F.R.G.

ABSTRACT

Second-order necessary and sufficient optimality conditions are given for infinite-dimensional programming problems with constraints defined by closed convex cones. The necessary conditions are an immediate generalization of those known for the finite-dimensional case. However, this does not hold for the sufficient conditions as shown by a counterexample. Modified sufficient conditions are developed for the infinite-dimensional case.

1. INTRODUCTION

Throughout this paper let X and Y be Banach spaces, let $f: X \to \mathbb{R}$ and $g: X \to Y$ be mappings and let $K \subset Y$ be a closed convex cone with vertex at the origin. We consider the nonlinear programming problem

(P) $\qquad$ minimize $f(x)$ subject to $g(x) \in K$.

A point $x_0 \in X$ is called optimal for (P) if $g(x_0) \in K$ and if f restricted to the feasible set $M = g^{-1}(K)$ assumes a local minimum at x_0 . We will derive second-order necessary and sufficient conditions for the optimality of x_0 . Such conditions are well-known for finite-dimensional spaces; see Guignard [2] for finite-dimensional X and see e.g. Fiacco/McCormick [1] and Hestenes [3] for finite-dimensional X and Y and K the standard cone defining equalities and inequalities.

The second-order necessary conditions immediately carry over to infinite-dimensional spaces X , Y and arbitrary K . However, this is not true for the finite-dimensional second-order sufficient conditions as will be shown by an example in section 2 . Instead, we give modified second-order sufficient conditions for (P) in section 4 which are based on a strengthening of the usual assumption on the second derivative of the Lagrangian associated with (P). The proof techniques exhibit a fundamental difference between the finite-dimensional and the infinite-dimensional case. In the finite-dimensional case the proofs proceed indirectly using in a decisive way the compactness of the unit sphere of X . Since for infinite-dimensional X the unit sphere is not compact in the norm-topology, one has to devise a direct proof. The main technical tool for this is provided by section 3 .

2. PRELIMINARIES

The topological duals of X and Y are denoted by X^* and Y^* . The dual cone of K is denoted by K^+ , i.e., $K^+ = \{ l \in Y^* \mid ly \geq 0 \text{ for all } y \in K \}$. We always assume that the first and second Fréchet-derivatives $f'(x_o)$, $g'(x_o)$ and $f''(x_o)$, $g''(x_o)$ of f and g exist at the points $x_o \in M = g^{-1}(K)$ under consideration. The maps $f''(x_o)$ and $g''(x_o)$ are interpreted as bilinear forms on $X \times X$.

An element $l \in K^+$ is called a normal Lagrange-multiplier for (P) at $x_o \in M$ if

(2.1) $$f'(x_o) - lg'(x_o) = 0 \quad , \quad lg(x_o) = 0 \; .$$

Conditions which assure the existence of a normal Lagrange-multiplier for (P) at an optimal point x_o are given by various authors; cf. Kurcyusz [5] , Lempio [6] , Robinson [7] . With a normal Lagrange-multiplier l one associates the Lagrangian function $F(x) = f(x) - lg(x)$. In terms of F the condition (2.1) becomes

(2.2) $$F'(x_o) = 0 \quad , \quad lg(x_o) = 0 \; .$$

We now define two cones which approximate the feasible set $M = g^{-1}(K)$ at a given point $x_o \in M$; cf. Kurcyusz [5] :

(2.3) $$T(M,x_o) = \{ h \in X \mid h = \lim_{n\to\infty} (x_n - x_o)/t_n \; , \; x_n \in M \; , \; t_n > 0 \; , \; t_n \to 0 \} \; ,$$

(2.4) $$L(M,x_o) = \{ h \in X \mid g'(x_o)h \in K_{g(x_o)} \} = g'(x_o)^{-1}(K_{g(x_o)}) \; .$$

Here $K_{g(x_o)}$ denotes the conical hull of $K - g(x_o)$. $T(M,x_o)$ is called the sequential tangent cone and $L(M,x_o)$ the linearizing cone of M at x_o .

Sometimes we will require that the cone $K_{g(x_o)}$ is closed; see Theorem 3.2(ii). Note that this holds if $Y = \mathbb{R}^k \times \mathbb{R}^n$ and $K = \{0\} \times \mathbb{R}^n_+$ where $\mathbb{R}^n_+ = \{ y \in \mathbb{R}^n \mid y_i \geq 0,\ i=1,..,n\}$. It is easy to verify that one always has $T(M,x_o) \subset L(M,x_o)$ if $K_{g(x_o)}$ is closed. The reverse inclusion $L(M,x_o) \subset T(M,x_o)$ holds if x_o is a <u>regular point</u> of M, i.e. if

$$(2.5) \qquad 0 \in \operatorname{int} \{ g(x_o)+g'(x_o)h-k \mid h \in X\ ,\ k \in K \}$$

where int denotes the topological interior; cf. Robinson [8,Thm.1] .

Now suppose (2.1) holds with l and put

$$(2.6) \qquad K_1 = K \cap \{ y \mid ly = 0 \}\ , \quad S = g^{-1}(K_1) \quad .$$

Obviously, K_1 is again a closed convex cone and we have $x_o \in S$ since $lg(x_o)=0$. We can define approximating cones $T(S,x_o)$ and $L(S,x_o)$ of S at x_o by replacing in (2.3),(2.4) M by S and K by K_1. Using (2.6) one easily verifies that

$$(2.7) \qquad L(S,x_o) = \{ h \in L(M,x_o) \mid lg'(x_o)h = 0 \}\ .$$

The inclusion $L(S,x_o) \subset T(S,x_o)$ holds if x_o is a regular point of S which means according to (2.5):

$$(2.8) \qquad 0 \in \operatorname{int} \{ g(x_o)+g'(x_o)h-k \mid h \in X\ ,\ k \in K_1 \}\ .$$

For <u>finite</u>-dimensional spaces X and Y with $Y = \mathbb{R}^k \times \mathbb{R}^n$ and $K = \{0\} \times \mathbb{R}^n_+$ Hestenes [3] shows that

$$(2.9) \qquad F''(x_o)(h,h) \geq 0 \quad \text{for all } h \in T(S,x_o)$$

is a <u>necessary</u> condition while

$$(2.10) \qquad F''(x_o)(h,h) > 0 \quad \text{for all } h \in L(S,x_o) \setminus \{0\}$$

is a <u>sufficient</u> condition (provided (2.8) holds) for x_o to be optimal for (P); cf. also Fiacco/McCormick [1] .

For arbitrary spaces X , Y the condition (2.9) remains necessary (see Theorem 4.1) while condition (2.10) is <u>not</u> sufficient as shown by the following

Counterexample:

Let $X = Y = \{ x = (x_n) \mid x^T x = \Sigma x_n^2 < \infty \}$, let $K = \{x \in X \mid x_n \geq 0$ for all $n\}$ and define $g(x) = x$, $f(x) = 1^T x - x^T x$ for some $1 \in K$ with $1_n > 0$ for all n. With this 1 (2.1) holds at $x_o = 0$ and we have $M = L(M,x_o) = K$ and $L(S,x_o) = K \cap \{ h \mid 1^T h = 0 \} = \{0\}$. Thus (2.10) is satisfied. Nevertheless, x_o is not optimal for (P). To see this, consider the sequence $x^k = (x_n^k)$ with $x_n^k = 2 1_k \delta_{kn}$ (δ_{kn} : Kronecker-symbol). Then $x^k \in M$ for all k and $x^k \to 0$, but $f(x^k) = 2 1_k^2 - 4 1_k^2 < 0 = f(0)$ for all k.

This counterexample motivates the fact that in order to derive second-order sufficient conditions one needs a larger cone than the cone $L(S,x_o)$. For every constant $\beta \geq 0$ we define the following cone which will be important for the main result in Theorem 4.3 :

(2.11) $$L_\beta(S,x_o) = \{ h \in L(M,x_o) \mid 1 g'(x_o)h \leq \beta \, ||h|| \} .$$

3. APPROXIMATION PROPERTY

For the proof of the second-order sufficient condition in the next section we have to guarantee that the linearizing cone $L(M,x_o)$ is a 'good approximation' of the feasible set M at $x_o \in M$.

Definition 3.1: The feasible set M is said to be approximated at x_o by $L(M,x_o)$, if there is a map $h : M \to L(M,x_o)$ such that

(3.1) $$||h(x) - (x-x_o)|| = o(||x-x_o||) \quad \text{for } x \in M .$$

Theorem 3.2: Each of the following conditions implies that M is approximated at x_o by $L(M,x_o)$.

(i) x_o is a regular point of M, i.e., (2.5) holds.

(ii) X is finite-dimensional and $K_{g(x_o)}$ is closed.

(iii) X, Y are finite-dimensional, $Y = \mathbb{R}^k \times \mathbb{R}^n$ and $K = \{0\} \times \mathbb{R}^n_+$.

Proof: Condition (i) follows from Robinson [8,Corollary 2] . We sketch the proof of (ii). Without restriction assume $x_o = 0$. For every $x \in M$ select $h(x)$ in $L(M,0)$ such that

$$\|h(x) - x\| = \min\{ \|h - x\| \mid h \in L(M,0) , \|h\| \le 2\|x\| \} .$$

$h(x)$ exists since, by assumption, $K_{g(0)}$ is closed and thus for fixed $x \in M$ the set $\{ h \in L(M,0) \mid \|h\| \le 2\|x\| \}$ is a compact subset of the finite-dimensional space X . One then shows via an indirect proof that the above map $x \to h(x)$ satisfies (3.1). The condition (iii) follows from (ii).

The following technical result, which will be needed in the proof of Theorem 4.3, can be easily deduced from (3.1).

Lemma 3.3: *Let* $h : M \to L(M,x_o)$ *be a map for which* (3.1) *holds. Then for every* $\gamma > 0$ *there is* $\rho > 0$ *such that* $\|h(x) - (x-x_o)\| \le \gamma \|h(x)\|$ *for all* $x \in M$ *with* $\|x-x_o\| \le \rho$.

4. SECOND-ORDER NECESSARY AND SUFFICIENT CONDITIONS

The proof of the following necessary conditions proceeds along the lines of the one given by Hestenes [3,Ch.4.7] in the finite-dimensional case and will not be repeated.

Theorem 4.1: *Let* x_o *be optimal for* (P) *and let* $F(x) = f(x) - lg(x)$ *be a Lagrangian for* (P) *at* x_o . *Then*

$$F''(x_o)(h,h) \ge 0 \quad \text{for all} \quad h \in T(S,x_o) .$$

If futhermore (2.8) *holds then*

$$F''(x_o)(h,h) \ge 0 \quad \text{for all} \quad h \in L(S,x_o) .$$

We now show that a suitable strengthening of this necessary condition is also sufficient for the optimality of x_o . This requires the following technical result whose proof is not difficult and will be omitted.

Lemma 4.2: *Let* B *be a continuous symmetric bilinear form on* $X \times X$, H *a subset of* X *and* $\delta > 0$ *with*

$$B(h,h) \ge \delta \|h\|^2 \quad \text{for all} \quad h \in H .$$

Then there are $\delta_o > 0$ *and* $\gamma > 0$ *such that*

$$B(h+z,h+z) \geq \delta_o \|h+z\|^2 \quad \underline{\text{for all}} \quad h \in H \ , \ z \in X \quad \underline{\text{and}} \quad \|z\| \leq \gamma \|h\| \ .$$

For the next result, recall that in Theorem 3.2 conditions are given which guarantee that M is approximated at x_o by $L(M,x_o)$. Recall also the definition of the cone $L_\beta(S,x_o)$ in (2.11).

Theorem 4.3: *Let $x_o \in M$ and assume that M is approximated at x_o by $L(M,x_o)$. Let $F(x) = f(x) - lg(x)$ be a Lagrangian for (P) at x_o and suppose that there exist $\delta > 0$ and $\beta > 0$ such that*

$$F''(x_o)(h,h) \geq \delta \|h\|^2 \quad \text{for all} \quad h \in L_\beta(S,x_o) \ . \tag{4.1}$$

Then there exist $\alpha > 0$ and $\rho > 0$ such that

$$f(x) \geq f(x_o) + \alpha \|x-x_o\|^2 \quad \text{for all} \quad x \in M \quad \text{and} \quad \|x-x_o\| \leq \rho \ .$$

In particular f has a strict local minimum on M at x_o.

Proof: For simplicity we assume $x_o = 0$. By assumption on M and $L(M,0)$ there is a map $h : M \to L(M,0)$ such that each $x \in M$ can be written as

$$x = h(x) + z(x) \ , \tag{4.2}$$

where the remainder term $z(x)$ satisfies $\|z(x)\| = o(\|x\|)$. In what follows we will show that one has with suitable real positive $\alpha_1, \alpha_2, \rho_1, \rho_2$:

$$f(x) \geq f(0) + \alpha_1 \|x\|^2 \quad \text{for all} \quad x = h(x)+z(x) \in M, \ \|x\| \leq \rho_1 \ \text{and} \ lg'(0)h(x) \leq \beta \|h(x)\| \ , \tag{4.3}$$

and

$$f(x) \geq f(0) + \alpha_2 \|x\| \quad \text{for all} \quad x = h(x)+z(x) \in M, \ \|x\| \leq \rho_2 \ \text{and} \ lg'(0)h(x) > \beta \|h(x)\| \ . \tag{4.4}$$

The assertion of the theorem then follows directly from this. In order to prove (4.3) note that by definition of the second Fréchet-derivative one has for all $x \in M$:

$$\begin{aligned} f(x) &\geq f(x) - lg(x) = F(x) = F(0) + F'(0)x + \tfrac{1}{2} F''(0)(x,x) + r(x) \\ &= f(0) + \tfrac{1}{2} F''(0)(x,x) + r(x) \ , \end{aligned} \tag{4.5}$$

where $|r(x)| = o(\|x\|^2)$. Here we have used that $lg(x) \geq 0$ for $x \in M$ and that $F'(0) = 0$, $lg(0) = 0$ by (2.2). Now put $B = F''(0)$ and $H = L_\beta(S,0)$ in Lemma 4.2 so that we have with suitable $\delta_o > 0$ and $\gamma > 0$:

$$F''(0)(x,x) \geq \delta_o \|x\|^2 \quad \text{for all } x = h(x)+z(x) \in M \text{ with}$$
$$lg'(0)h(x) \leq \beta \|h(x)\| \quad \text{and} \quad \|z(x)\| \leq \gamma \|h(x)\| .$$

Making use of Lemma 3.3 we see that with a sufficiently small $\rho_1 > 0$

(4.6) $\|z(x)\| \leq \gamma \|h(x)\|$ whenever $x = h(x)+z(x) \in M$ and $\|x\| \leq \rho_1$

and thus

(4.7) $F''(0)(x,x) \geq \delta_o \|x\|^2$ for all $x = h(x)+z(x) \in M$ with $\|x\| \leq \rho_1$ and $lg'(0)h(x) \leq \beta \|h(x)\|$.

Moreover, choose ρ_1 so small such that the remainder term in (4.5) satisfies

$$|r(x)| \leq \frac{\delta_o}{4} \|x\|^2 \quad \text{for} \quad \|x\| \leq \rho_1 .$$

Then (4.3) follows from (4.5) and (4.7) with $\alpha_1 = \delta_o/4$.

Now consider the case $x = h(x)+z(x) \in M$, $\|x\| \leq \rho_1$, but $lg'(0)h(x) > \beta \|h(x)\|$. Using $f'(0) = lg'(0)$ we get

$$\text{(4.8)} \qquad \begin{aligned} f(x) - f(0) &= f'(0)x + r(x) = lg'(0)h(x) + lg'(0)z(x) + r(x) \\ &\geq \beta \|h(x)\| + r_1(x) , \end{aligned}$$

where $|r(x)| = o(\|x\|)$ and $r_1(x) = lg'(0)z(x) + r(x)$. Since $\|z(x)\| = o(\|x\|)$ one has $|r_1(x)| = o(\|x\|)$. By (4.6)

$$\|x\| \leq \|h(x)\| + \|z(x)\| \leq (1+\gamma) \|h(x)\| \quad \text{for} \quad \|x\| \leq \rho_1$$

and thus (4.8) becomes

$$f(x) \geq f(0) + (\beta/(1+\gamma)) \|x\| + r_1(x) .$$

Obviously, (4.4) follows from this for $\rho_2 \in (0,\rho_1]$ sufficiently small.

We note that Theorem 4.3 contains the sufficient condition of Ioffe/Tikhomirov [4,p.307] for equality constraints since $L_\beta(S,x_o) = L(M,x_o)$ for $K = \{0\}$. Moreover, the next lemma shows that in the finite-dimensional case Theorem 4.3 reduces to the sufficient condition (2.10).

<u>Lemma 4.4</u>: <u>If</u> $\dim X < \infty$ <u>and</u> $K_{g(x_o)}$ <u>is closed then</u> (4.1) <u>holds with suitable</u> $\delta > 0$ <u>and</u> $\beta > 0$ <u>if</u> $F''(x_o)(h,h) > 0$ <u>for all</u> $h \in L(S,x_o) \setminus \{0\}$.

<u>Proof</u>: Suppose that the assertion is false. Then for every $n \in \mathbb{N}$ there is some $h_n \in L(M,x_o)$ such that

$$(4.9) \qquad lg'(x_o)h_n \leq \frac{1}{n}\|h_n\| \quad \text{and} \quad F''(x_o)(h_n,h_n) < \frac{1}{n}\|h_n\|^2 .$$

Obviously $h_n \neq 0$ for all n . As X is finite-dimensional, a subsequence of $\{h_n / \|h_n\|\}_{n \in \mathbb{N}}$ converges to some $h \in L(M,x_o)$ with $\|h\| = 1$. Note that $L(M,x_o)$ is closed since $K_{g(x_o)}$ is assumed to be closed. From (4.9) we obtain for this h :

$$lg'(x_o)h = 0 \quad \text{and} \quad F''(x_o)(h,h) \leq 0 .$$

The first equality implies $h \in L(S,x_o) \setminus \{0\}$, but then $F''(x_o)(h,h) \leq 0$ contradicts the assumption $F''(x_o)(h,h) > 0$.

REFERENCES

1. FIACCO,A.V. and McCORMICK,G.P.: Nonlinear Programming: Sequential unconstrained minimization techniques, John Wiley, New York, 1968.

2. GUIGNARD,M.: Generalized Kuhn-Tucker conditions for mathematical programming problems in a Banach space, SIAM Journal on Control 7 (1969), 232-241.

3. HESTENES,M.R.: Optimization Theory. The finite-dimensional case, John Wiley, New York, 1975.

4. IOFFE,A.D. and TIKHOMIROV,W.M.: Theory of extremals (in Russian), Nauka, Moscow, 1974.

5. KURCYUSZ,S.: On the existence and nonexistence of Lagrange multipliers in Banach spaces, J. of Optimization Theory and Applications 20 (1976), 81-110.

6. LEMPIO,F.: Bemerkungen zur Lagrangeschen Funktionaldifferentialgleichung, International Series of Numerical Analysis 19 (1974), 141-146.

7. ROBINSON,S.M.: First order conditions for general nonlinear optimization, SIAM J. Appl. Math. 30 (1976), 597-607.

8. ROBINSON,S.M.: Stability theory for systems of inequalities, Part II: Differentiable nonlinear systems, SIAM J. Numer. Anal. 13 (1976), 497-513.

AN UNDERRELAXED GAUSS-NEWTON METHOD FOR EQUALITY CONSTRAINED NONLINEAR LEAST SQUARES PROBLEMS

P. Deuflhard
Technische Universität München
Institut für Mathematik
Postfach 202420
D-8 München 2

V. Apostolescu
Leibniz-Rechenzentrum der
Bayerischen Akademie der Wissenschaften
Barerstraße 21
D-8 München 2

0. INTRODUCTION

This paper is concerned with the numerical solution of the following special class of constrained optimization problems:

(0.1) For $x \in \mathbb{R}^n$, minimize

$$T(x) = \sum_{i=1}^{m} g_i^2(x)$$

subject to

$$h_j(x) = 0 \qquad j = 1,\dots,p$$

where usually

$$p \le n \le m+p$$

A convenient derivation and analysis of the algorithm to be proposed can be given in terms of an embedding of the above *constrained* problem (0.1) into the following ω-family of *unconstrained* problems:

(0.2) Minimize

$$T_\omega(x) := \sum_{i=1}^{m} g_i^2(x) + \omega^2 \sum_{j=1}^{p} h_j^2(x) .$$

Then, for each fixed weight parameter ω, there exists an associated Gauss-Newton iteration, say GN_ω:

(0.3) $x_\omega^0 := x^0 , \quad x_\omega^{k+1} := x_\omega^k + \Delta x_\omega^k \quad$ for $k = 0,1,\dots$

where Δx_ω^k denotes the ordinary *Gauss-Newton correction*. The paper deals with the algorithm GN_∞, i.e. with the Gauss-Newton type iteration obtained from (0.3) in the limiting case $\omega \longrightarrow \infty$.

In chapter 1, the realization of GN_∞ is shown to require the solution of a *sequence of linear least squares problems with linear equality constraints* – in contrast to penalty methods (see Fiacco/Mc Cormick [8]) where a sequence of unconstrained nonlinear subproblems for increasing values of ω needs to be solved. In chapter 2, the performance of the algorithm is illustrated by a realistic numerical example.

1. REALIZATION AND PROPERTIES OF THE ALGORITHM GN_∞

Let $D_i \in \mathbb{R}^n$, $i = 1,2$, $D_0 := D_1 \cap D_2 \neq \emptyset$ open, $H : D_1 \longrightarrow \mathbb{R}^p$, $H \in C^1(D_1)$, and $G : D_2 \longrightarrow \mathbb{R}^m$, $G \in C^1(D_2)$ with $p \leq n \leq m+p$.

In this notation, problem (0.1) is written as

(1.1) Minimize $\|G(x)\|_2^2$ subject to $H(x) = 0$,

which is embedded into the ω-family of the unconstrained problems

(1.2) Minimize $\left[\|G(x)\|_2^2 + \omega^2\|H(x)\|_2^2\right]$

At some point $x^0 \in \mathbb{R}^n$, the *ordinary Gauss-Newton* (GN_ω) *correction vector* for (1.2) is

(1.3) $\Delta x_\omega^0 := -J_\omega(x^0)^\dagger F_\omega(x^0)$

where

$$F_\omega(x) := \begin{pmatrix} \omega H(x) \\ G(x) \end{pmatrix} , \quad J_\omega(x) := \begin{pmatrix} \omega H'(x) \\ G(x) \end{pmatrix}$$

and the superscript $\dagger$ denotes the Penrose pseudo-inverse (cf. [12]).

Linear special case. As a preparation, let

(1.4) $H(x) := Ax-c$, $G(x) := Bx-d$

with A,B matrices and c,d vectors of appropriate size.

LEMMA 1.1. Let A have maximum rank p. Then the limiting vector

$$x_{\infty}^{*} := \lim_{\omega \to \infty} \begin{pmatrix} \omega A \\ B \end{pmatrix}^{\dagger} \begin{pmatrix} \omega c \\ d \end{pmatrix}$$

exists and is the best least squares solution of the constrained linear problem (1.1)/(1.4).

For a proof of this lemma, see chapter 22 in the book of Lawson/Hanson [11] (with $\epsilon \Rightarrow 1/\omega$). Note, however, that - in contrast to the method described in that chapter - numerical *ill-conditioning is not necessarily caused* by the limiting process $\omega \longrightarrow \infty$! A rather elegant, numerically stable, and compact algorithm was brought to the authors' knowledge by Chr. Zenger: this algorithm may be obtained from the algorithm due to Businger/Golub [3] by performing the limiting process *analytically*. Of course, *any* numerically stable algorithm for the solution of the linearly constrained linear least squares problem can be regarded as a numerical realization of the limiting process. For example, if *linear inequality constraints* are included, one may use the algorithm due to Stoer [13] which might be preferable in connection with a numerically stable realization of exchange steps.

Ordinary constrained Gauss-Newton method GN_{∞}. This algorithm requires the computation of

$$(1.3') \quad \Delta x_{\infty}^{k} := -\lim_{\omega \to \infty} J_{\omega}(x^{k})^{\dagger} F_{\omega}(x^{k})$$

at each iterate x^k. Now, from the preceding study of the linear special case, it is clear that the realization of GN_{∞} means to solve a *sequence of linear least squares problems with linear equality constraints* - using *any* of the above mentioned algorithms at each iterative step. To simplify the subsequent presentation, the following preparation will be helpful.

LEMMA 1.2. Assume that, for all $x \in D_0$, $H'(x)$ has maximum rank p. Then the following forms exist for all $x,y \in D_0$:

$$(1.5) \quad \Delta(x,y) := -\lim_{\omega \to \infty} J_{\omega}(x)^{\dagger} F_{\omega}(y) ,$$

$$P(x,y) := \lim_{\omega \to \infty} J_{\omega}(x)^{\dagger} J_{\omega}(y) .$$

Proof. Lemma 1.1 clearly indicates that $\Delta(x,y)$ exists for $H'(x)$ being of rank p. As a consequence, the existence of $P(x,y)$ can be shown columnwise. ■

In the above introduced notation, the Penrose axioms for $J_\omega(x)$ with ω *finite* can be analyzed in view of the limiting process $\omega \longrightarrow \infty$. One readily verifies that

(1.6) a) $P(x,x)^T = P(x,x)$, $P(x,x)^2 = P(x,x)$,
b) $P(x,x)\Delta(x,x) = \Delta(x,x)$.

Moreover, for x^* to be a solution of the constrained nonlinear least squares problem (1.1), the following is easily shown to be a necessary condition

(1.7) $\Delta(x^*,x^*) = 0$.

Hence, x^* is a fixed point of the GN_∞ iteration. As for the convergence of the iterates $\{x_\infty^k\}$, Theorem 4 due to [7] can be applied, since the crucial condition (1.6.b) holds.

THEOREM 1.3. Notation as just introduced. Let $H'(x)$ have maximum rank p for all $x \in D_0$. Moreover, assume that

(1.7) a) $\|\Delta(x^0,x^0)\| \le \alpha_0$ for $x^0 \in D_0$,

b) $\|P(z,y)-P(z,x)\| \le \gamma\|y-x\|$ for all $x,y,z \in D_0$

c) $\|\Delta(y,x)-P(y,x)\Delta(x,x)\| \le \kappa(x)\|y-x\|$

for all $x,y \in D_0$ (yielding (1.6.b) for $y = x$)

d) $\kappa(x) \le \bar{\kappa} < 1$ for all $x \in D_0$

$h_0 := \frac{1}{2}\alpha_0\gamma < 1-\bar{\kappa}$

e) $S_\rho(x^0) := \{x \in \mathbb{R}^n \mid \|x-x^0\| \le \rho\}$

for

$$\rho := \frac{\alpha_0}{1-\bar{\kappa}-h_0},$$

$\overline{S_\rho}(x^0) \subset D_0$

Then the GN_∞ iterates defined by

(1.8) $x^{k+1} := x^k + \Delta(x^k, x^k) \quad k = 0,1,\ldots$

remain in $\overline{S_\rho}(x^o)$ and converge to a point x^* with

$$\Delta(x^*,x^*) = 0$$

Moreover, for $k = 0,1,\ldots$:

$$\|\Delta(x^{k+1},x^{k+1})\| \le \frac{\gamma}{2}\|\Delta(x^k,x^k)\|^2 + \kappa(x^k)\|\Delta(x^k,x^k)\|$$

Proof. One first writes down the proof of Theorem 4 due to [7] when applied to GN_ω *for finite* ω. Upon carefully analyzing each step of the proof for $\omega \longrightarrow \infty$, the above convergence theorem for GN_∞ is verified. ■

As in the unconstrained case, the statements of the above theorem can be expressed in the following way: for GN_∞, *local convergence* is guaranteed, if the constrained nonlinear least squares problem (1.1) is *adequate* in some sense defined similarly as in [5] for unconstrained problems (essentially requiring the existence of some domain $D_0 \supset x^*$ with $\bar{\kappa} < 1$). For inadequate problems, divergence may occur.

Underrelaxed constrained Gauss-Newton method. With the aim of expanding the domain of convergence, the ordinary GN_∞ iteration (1.8) is replaced by

(1.9) $x^{k+1} := x^k + \lambda_k \Delta(x^k,x^k) \quad \text{for } k = 0,1,\ldots ,$

where $\lambda_k \in \left]0,1\right]$ denotes the (under-) relaxation parameter. By virtue of the embedding idea, the associated concepts and techniques for the unconstrained case can be studied in the limit $\omega \longrightarrow \infty$. The main result is that the usual monotonicity test for finite ω

(1.10) $\|F_\omega(x^{k+1})\|_2 \le \|F_\omega(x^k)\|_2$

is not useful for $\omega \longrightarrow \infty$, since both sides are unbounded. However, the *naturally scaled monotonicity test*

(1.11) $\|\Delta(x^k,x^{k+1})\|_2 \le \|\Delta(x^k,x^k)\|_2$

remains a valid condition to determine the parameters λ_k. Moreover, the following techniques known from the unconstrained case (cf. [5]) can be transferred to the constrained case:

(I) *relaxation strategy* as presented in chapter 2 of [5]: for *adequate* problems, this strategy will produce $\lambda_k = 1$, if x^k, x^{k-1} are "sufficiently close" to x^*,

(II) *rank-1 approximations* of the Jacobian matrices (cf. Broyden [2]): computing time may be saved, but at the expense of a less reliable solution (see also the numerical results in §2),

(III) *rank-strategy* for ill-conditioned problems (cf. [4]),

(IV) *termination criteria:* in order to avoid waste of computing time for incompatible problems, one may terminate the iteration with less than prescribed relative accuracy (in view of the rounding error analysis due to Bjoerck [1]).

For more details, the reader may refer to the more extensive presentation in [6].

2. NUMERICAL RESULTS

For lack of space, only one realistic test problem is presented to illustrate the performance of our algorithm. Two different versions have been programmed in ALGOL 60 (38 bit mantissa):

MGN: modified (= underrelaxed) GN_∞ method, Jacobian matrices by numerical differences (counted for $\geq n$ function evaluations),

MGN1: as above, but rank-1 approximations of the Jacobian matrix at selected iterates.

As usual, the total number of function evaluations was taken as a measure of the computing time needed to solve each problem.

The test problem arises in chemistry (enzyme reaction problem, $p = 0$ due to Kowalik/Osborne [10], p. 127). For details and notation see [5], example 2. The experimental measurements (t_i, y_i), $i = 1,\ldots,11$ are to be fitted by means of the 4-parameter *ansatz function*

$$\varphi(x;t) := \frac{x_1(t^2+x_2t)}{t^2+x_3t+x_4} .$$

In addition to the unconstrained nonlinear least squares problem ($p = 0$), two constrained versions of the problem ($p = 1,2$) have been constructed. Notation:

$$p = 0: \quad g_i(x) = \varphi(x;t_i) - y_i , \qquad i = 1,\dots,11 ,$$

$$p = 1: \quad h_1(x) = \varphi(x;t_1) - y_1 ,$$

$$g_i(x) = \varphi(x;t_{i+1}) - y_{i+1} , \qquad i = 1,\dots,10 ,$$

$$p = 2: \quad h_1(x) = \varphi(x;t_1)-y_1, \quad h_2(x) = \varphi(x;t_{11})-y_{11} ,$$

$$g_i(x) = \varphi(x;t_{i+1})-y_{i+1} , \qquad i = 1,\dots,9 .$$

The results of our computations are listed in Table A. Two features seem to be indicated:

(I) MGN1 may save computing time when compared with MGN .

(II) The final levels of the objective function obtained from MGN1 may be slightly *higher* than those obtained from MGN. Moreover, MGN1 may produce different final levels for different starting points (an explanation of this undesirable feature was given in [5]).

TABLE A. Amount of computation needed to solve the test problem in the unconstrained case $(p = 0)$ and in two constrained cases $(p = 1,2)$

	$x^0 = (0,0,0,0)$ [a]			$x^0 = (0.25,0.39,0.415,0.39)$		
	$p = 0$	$p = 1$	$p = 2$	$p = 0$	$p = 1$	$p = 2$
MGN	35	70	55	69	58	78
MGN1	27	53	33	42	51	52
obtained levels of objective function (and constraints):						
MGN	$3.104_{10}-4$	$3.128_{10}-4$	$4.130_{10}-4$	$3.104_{10}-4$	$3.128_{10}-4$	$4.130_{10}-4$
		$1_{10}-14$	$2_{10}-14$		$7_{10}-14$	$3_{10}-14$
MGN1	$3.104_{10}-4$	$3.131_{10}-4$	$4.130_{10}-4$	$3.105_{10}-4$	$3.128_{10}-4$	$4.131_{10}-4$
		$3_{10}-16$	$3_{10}-15$		$7_{10}-15$	$2_{10}-13$

[a] Jacobian matrix singular at x^0

In order to give some insight into the performance of the algorithms MGN and MGN1, the iterative values of the objective function $\|G(x^k)\|$ together with a measure $\|H(x^k)\|$ for the constraints are represented in Fig. 1/2 (starting point $x^0 = (0.25, 0.39, 0.415, 0.39)$). The figures illustrate nicely that, globally, the modified GN_∞ algorithms *simultaneously* reduce the objective function and aim at a point on the manifold given by the nonlinear equality constraints. One must not expect, however, that *both* $\|H(x)\|$ *and* $\|G(x)\|$ are reduced si-

Algorithm MGN :

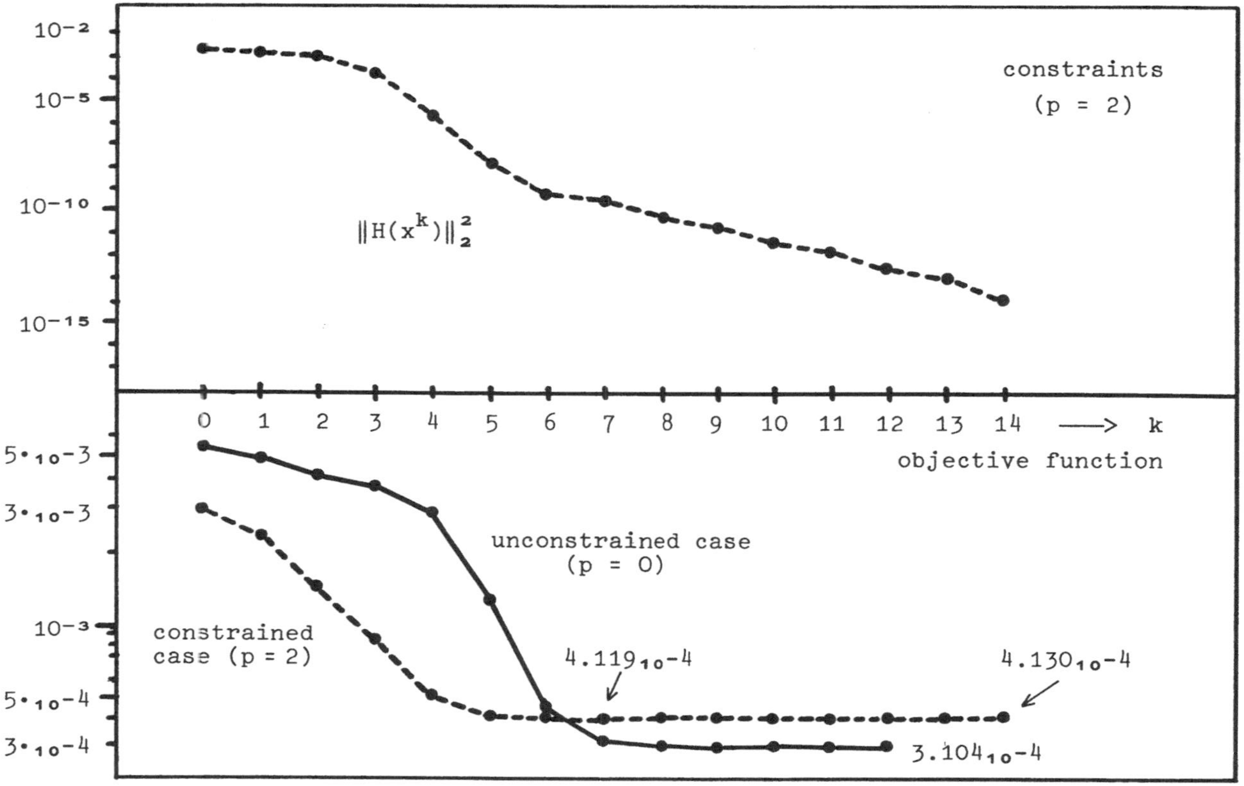

- Fig. 1 -

Algorithm MGN1 :

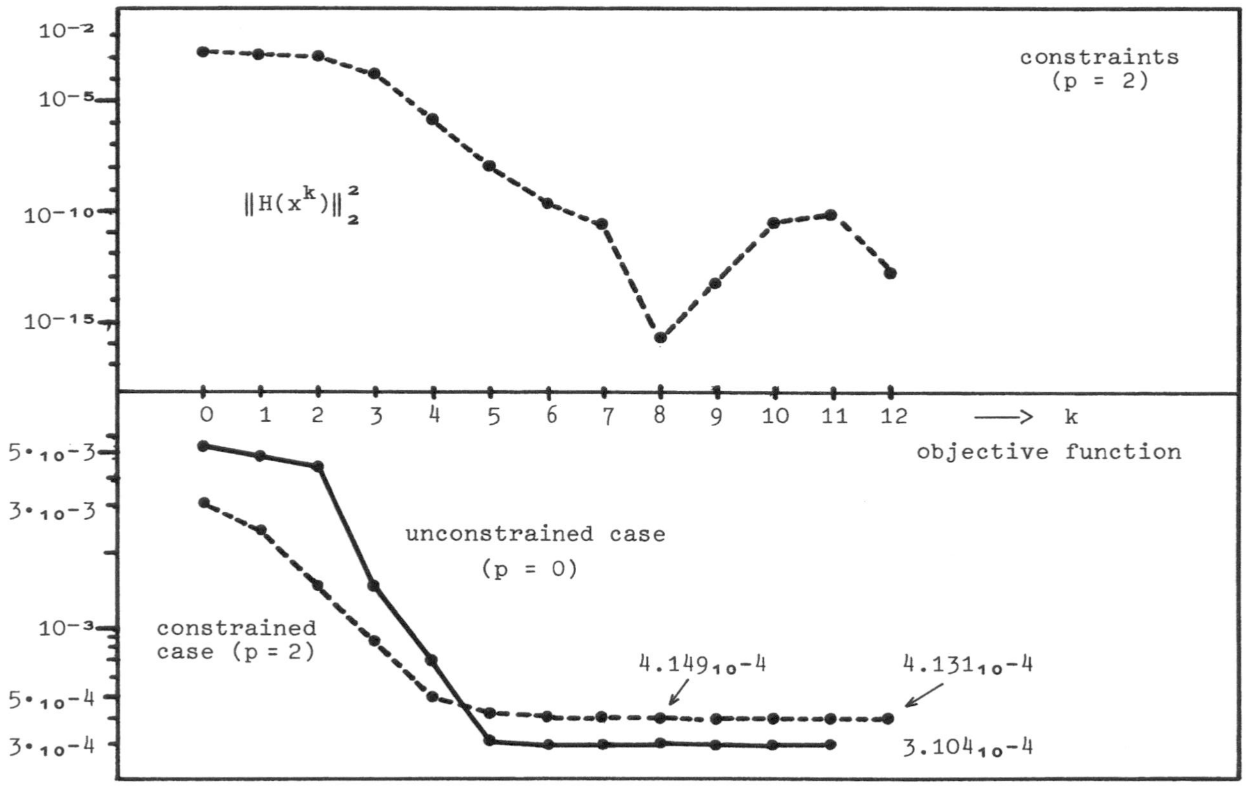

- Fig. 2 -

multaneously at *each* iterative step, since neither of these level functions is *appropriate* in the sense introduced in [4] - see also [6]. In Fig. 1 (for MGN), the iterative minimum of the objective function is obtained at $k = 7$. The finally accepted value of the objective function is somewhat larger, but the constraints are more accurately satisfied. In Fig. 2 (for MGN1), the complementary situation occurs: at $k = 8$, the solution for $p = 2$ is not yet obtained, even though the constraints are most accurately satisfied. In fact, the accociated norms of the correction vectors are

$$\|\Delta(x_\infty^8, x_\infty^8)\|_2 = 2\cdot 10^{-1}, \quad \|\Delta(x_\infty^{12}, x_\infty^{12})\|_2 = 1\cdot 10^{-4} .$$

The control of the iteration is totally done by applying the *naturally scaled* monotonicity test (1.11).

Finally, for comparison purposes, the solution points from MGN for $p = 0,1,2$ are presented below together with the finally achieved relative precision eps :

$$\begin{aligned} p = 0 :\quad & x^* = (0.19, 0.19, 0.12, 0.14), \quad eps = 2\cdot 10^{-2} , \\ p = 1 :\quad & x_\infty^* = (0.19, 0.18, 0.13, 0.13), \quad eps = 2\cdot 10^{-2} , \\ p = 2 :\quad & x_\infty^* = (0.19, 0.40, 0.27, 0.21), \quad eps = 1\cdot 10^{-3} . \end{aligned}$$

The computation of more accurate results seemed to be not reasonable with this method (in this example) in view of the rounding error analysis due to Golub/Wilkinson [9] and Bjoerck [1].

An interesting feature of the method GN_∞ proposed here is that the amount of computation needed to solve the *constrained* problem approximately equals the amount needed to solve the *unconstrained* problem. This feature seems to stand out in comparison with standard constrained optimization routines.

Acknowledgments. The authors wish to thank Dr. Chr. Zenger for many helpful discussions. Also they are grateful to Dr. K. Horn for her careful reading of the manuscript. The numerical experiments were run on the TR440 of the *Leibniz-Rechenzentrum der Bayerischen Akademie der Wissenschaften*. Last not least, the authors want to thank Mrs. A. Bußmann for her careful and patient typing of the present manuscript.

REFERENCES

[1] Bjoerck, A.: Iterative refinement of linear least squares solutions II., BIT 8, 8-30 (1968)

[2] Broyden, C.G.: A class of methods for solving nonlinear simultaneous equations, Math. Comp. 19, 577-583 (1965)

[3] Businger, P., Golub, G.H.: Linear least squares solutions by Householder Transformations, Num. Math. 7, 269-276 (1965)

[4] Deuflhard, P.: A Modified Newton Method for the Solution of Ill-Conditioned Systems of Nonlinear Equations with Application to Multiple Shooting, Num. Math. 22, 289-314 (1974)

[5] Deuflhard, P., Apostolescu, V.: A Study of the Gauss-Newton Algorithm for the Solution of Nonlinear Least Squares Problems, submitted for publication

[6] Deuflhard, P., Apostolescu, V.: A Gauss-Newton Method for Nonlinear Least Squares Problems with Nonlinear Equality Constraints, TUM-Report 7607

[7] Deuflhard, P., Heindl, G.: Affine Invariant Convergence Theorems for Newton's Method with Extensions to Related Methods, TUM-Report 7723

[8] Fiacco, A.V., Mc Cormick, G.P.: Nonlinear Programming: Sequential Unconstrained Minimization Techniques, New York-London-Sydney-Toronto: J. Wiley 1968

[9] Golub, G.H., Wilkinson, J.H.: Note on Iterative Refinement of Least Squares Solution, Num. Math. 9, 139-148 (1966)

[10] Kowalik, J., Osborne, M.R.: Methods for Unconstrained Optimization Problems, New York: American Elsevier Publ. Comp. Inc. (1968)

[11] Lawson, C.L., Hanson, R.J.: Solving Least Squares Problems, Englewood Cliffs: Prentice Hall (1974)

[12] Penrose, R.: A generalized inverse for matrices, Proc. Cambridge Philos. Soc. 51, 406-413 (1955)

[13] Stoer, J.: On the Numerical Solution of Constrained Least Squares Problems, SIAM J. Num. Anal. 4, 27-36 (1967)

A MODIFICATION OF ROBINSON'S ALGORITHM FOR GENERAL NONLINEAR PROGRAMMING PROBLEMS REQUIRING ONLY APPROXIMATE SOLUTIONS OF SUBPROBLEMS WITH LINEAR EQUALITY CONSTRAINTS

Jürgen Bräuninger

Mathematisches Institut A
Universität Stuttgart

Pfaffenwaldring 57
D-7000 Stuttgart 80

1. Introduction

One of the best methods to solve nonlinearly constrained optimization problems is Robinson's quadratically-convergent algorithm for general nonlinear programming problems. This algorithm constructs a sequence of subproblems with linear equality and inequality constraints. Assuming that the exact solutions of these subproblems can be obtained and that the process is started sufficiently close to a strict second-order Kuhn-Tucker point, Robinson proved R-quadratic convergence of the subproblem-solutions to that point.

Here a modification of Robinson's algorithm is presented which is more suitable for implementation. This modification uses only subproblems with linear equality constraints. For such problems very efficient algorithms are available.

Furthermore not the exact solutions of these subproblems have to be obtained but only approximations of them which have a certain accuracy. If the process starts sufficiently close to a strict second-order Kuhn-Tucker point, it can explicitly be tested if an approximation is good enough or if it has to be improved.

For the modification the same local convergence properties are proved as for Robinson's original algorithm.

2. Formulation of the problem, definitions and notations

Let $f(x)$, $h_i(x)$ $(i=1,\dots,m+p)$ be real valued functions $\mathbb{R}^n \to \mathbb{R}$.

If f is differentiable at a point x_j, we denote its gradient at x_j by $\nabla f(x_j)$.

For $x \in \mathbb{R}^n$ let $\| x \|$ denote the Euclidean norm and x' the transpose of the column vector x.

We consider the problem

(1) $$\min \{f(x) \mid h_i(x) \leq 0 \ (i=1,\dots,m),\ h_i(x) = 0 \ (i=m+1,\dots,m+p)\} .$$

First we introduce some additional notations:

A Kuhn-Tucker point of problem (1) is a point $x \in \mathbb{R}^n$ such that there is $u \in \mathbb{R}^{m+p}$ with:

$$\nabla f(x) = \sum_{i=1}^{m+p} u_i \nabla h_i(x) , \quad u_i h_i(x) = 0 \ (i=1,\dots,m) , \quad h_i(x) = 0 \ (i=m+1,\dots,m+p) ,$$

$$h_i(x) \leq 0 \ (i=1,\dots,m) , \quad u_i \leq 0 \ (i=1,\dots,m) .$$

The vector $\binom{x}{u} \in \mathbb{R}^{n+m+p}$ is then called a Kuhn-Tucker solution of (1), the u_i $(i=1,\dots,m+p)$ are called Lagrangian multipliers. We write z for $\binom{x}{u}$, z_r for $\binom{x_r}{u_r}$ etc. $(u_r)_i$ is the i-th component of the vector u_r.

$I(z)$ is the set of indices $\{i \mid h_i(x) = 0\}$.

We say that a Kuhn-Tucker solution $\bar{z}$ of (1) satisfies assumption A, if

(i) the second-order sufficiency conditions for (1) are satisfied (see [1]) ;

(ii) strict complementary slackness holds (i.e. either $\bar{u}_i < 0$ or $h_i(\bar{x}) < 0$ for $i=1,\dots,m$) ;

(iii) the gradients to the active constraints are linearly independent ;

(iv) f, h_i $(i=1,\dots,m+p)$ are twice continuously differentiable in an open neighborhood about $\bar{x}$.

For the linearization of a function $h_i(x)$ at x_r we write

$$Lh_i(x_r, x) := h_i(x_r) + \nabla h_i(x_r)'(x - x_r) .$$

As objective function in subproblems we need

$$H(x, z_r) := f(x) - \sum_{i=1}^{m+p} (u_r)_i \ [h_i(x) - Lh_i(x_r, x)] .$$

$F(z)$ and $d(z,\tilde{z})$ are defined by

$$F(z) := \begin{pmatrix} \nabla f(z) - \sum_{i=1}^{m+p} u_i \nabla h_i(x) \\ u_1 \ h_1(x) \\ \vdots \\ u_m \ h_m(x) \\ h_{m+1}(x) \\ \vdots \\ h_{m+p}(x) \end{pmatrix} \qquad d(z,\tilde{z}) = F(\tilde{z}) - \begin{pmatrix} \sum_{i=1}^{m+p} (u_i - \tilde{u}_i)(\nabla h_i(\tilde{x}) - \nabla h_i(x)) \\ \tilde{u}_1(h_1(\tilde{x}) - Lh_1(x,\tilde{x})) \\ \vdots \\ \tilde{u}_m(h_m(\tilde{x}) - Lh_m(x,\tilde{x})) \\ h_{m+1}(\tilde{x}) - Lh_{m+1}(x,\tilde{x}) \\ \vdots \\ h_{m+p}(\tilde{x}) - Lh_{m+p}(x,\tilde{x}) \end{pmatrix}$$

$S(z_r)$ is the Kuhn-Tucker solution z of the problem

$$\min \{H(x,z_r) \mid Lh_i(x_r,x) \leq 0 \ (i=1,\dots,m),\ Lh_i(x_r,x) = 0 \ (i=m+1,\dots,m+p)\} \ ^{*)}$$

$S(z_r,I)$ with $I \subset \{1,\dots,m+p\}$ is the Kuhn-Tucker solution $z \in \mathbb{R}^{n+m+p}$ (consisting of $x \in \mathbb{R}^n$ and $u \in \mathbb{R}^{m+p}$ where u_i are the Lagrangian multipliers for $i \in I$ and $u_i = 0$ for $i \notin I$) of the problem

$$\min \{H(x,z_r) \mid Lh_i(x_r,x) = 0 \ (i \in I)\} \ ^{*)}$$

From the Kuhn-Tucker conditions it is easy to see, that $d(z_r,S(z_r)) = 0$ and $d(z_r,S(z_r,I)) = 0$ for any $I \supset \{m+1,\dots,m+p\}$.

The end of a proof is denoted by Δ .

3. Robinson's algorithm

With the notations from section 2 Robinson's algorithm is as follows:

Algorithm 1

Step 1: Start with a given $z_0 \in \mathbb{R}^{n+m+p}$, set $r := 0$.
Step 2: $z_{r+1} := S(z_r)$.
Step 3: Set $r := r+1$ and go to Step 2 .

If $S(z_r)$ should be undefined, the algorithm itself is regarded as undefined. Of course, in practice we will test for convergence in Step 3 and eventually stop the algorithm.

In [5] Robinson showed, that, if this algorithm is started with z_0 close enough to a Kuhn-Tucker solution $\bar{z}$ of (1) satisfying assumption A, the algorithm is defined and converges R-quadratically to $\bar{z}$. The exact formulation of this result is the same as we will get for the modification of the algorithm and show in Theorem 2.

Note that in Step 2 Robinson's algorithm requires the exact solution of a subproblem with linear equalities and inequalities as constraints. Algorithms which solve such subproblems (e.g. [2, 3, 4]) will in general only give a sequence of points converging to the solution, but not the exact solution itself.

---

*) If there is more than one such z, choose any Kuhn-Tucker solution which is next to z_r. We assume here that a Kuhn-Tucker solution exists. Otherwise $S(z_r)$ or $S(z_r,I)$ is undefined.

4. The modification of Robinson's algorithm

In the following modification of Robinson's algorithm we don't need the exact solutions of the subproblems.

First in Step 1 the set I_0 is determined, which contains the indices of the constraints which are regarded as active at the solution $\bar{z}$. We will see later, that, if $\| z_0 - \bar{z} \|$ is small enough, then $I_0 = I(\bar{z})$. Assuming that this is true, z_1 is nearer to $\bar{z}$ than z_0 and it is reasonable to start with z_1.

In Step 2 we look for an approximation to $S(z_r, I_0)$. This means we have to solve a subproblem with only linear equalities as constraints and we are contended, if we get an approximation to the solution, which is good enough.

The modification of Robinson's algorithm is ($c > 0$ is a fixed parameter):

Algorithm 2

Step 1: Start with $z_0 \in \mathbb{R}^{n+m+p}$.

Determine $I_0 := \{i \mid -(u_0)_i > | h_i(x_0) | / \| \nabla h_i(x_0) \| \} \cup \{m+1, \dots, m+p\}$.

Set $x_1 := x_0$, $(u_1)_i = \begin{cases} (u_0)_i, & \text{if } i \in I_0 \\ 0, & \text{if } i \notin I_0 . \end{cases}$

Set $r := 1$.

Step 2: Determine any z_{r+1} such that

$$\| z_{r+1} - S(z_r, I_0) \| \leq c \, \| F(z_r) \|^2 \tag{2}$$

Step 3: Set $r := r+1$ and got to Step 2.

Again, if $S(z_r, I_0)$ should be undefined, the algorithm itself is regarded as undefined. In Step 3 one will in practice test for convergence and eventually stop the algorithm.

In practice we cannot test, if (2) is fulfilled, because $S(z_r, I_0)$ is unknown. But it can be shown that there is $c_1 > 0$ such that $\| z - S(z_r, I_0) \| \leq c_1 \| P_r \nabla_x H(x, z_r) \|$ if z and z_r are close enough to $\bar{z}$. ($P_r y$ denotes the projection of y on the space spanned by the constraints $Lh_i(x_r, x)$, $i \in I_0$.) Assuming that we are close enough to $\bar{z}$ we will test, if $\| P_r \nabla_x H(x_{r+1}, z_r) \| \leq c_2 \| F(z_r) \|^2$ for some $c_2 > 0$, and thus guarantee, that (2) holds with $c = c_1 c_2$.

5. Theoretical results

As a first result about the sequence $\{z_r\}$ produced by algorithm 2 we get:

Theorem 1: Let f, h_i $(i=1,\dots,m+p)$ be continuously differentiable. If $z_r \to \bar{z}$ and $S(z_r, I_0) - z_{r+1} \to 0$, then $\bar{z}$ is a Kuhn-Tucker solution of the problem $\min\{f(x) \mid h_i(x) = 0 \ (i \in I_0)\}$.

Proof: With $\tilde{z}_r = S(z_r, I_0)$ we have from the Kuhn-Tucker conditions for $S(z_r, I_0)$:

$$\nabla f(\tilde{x}_r) - \sum_{i \in I_0} (u_r)_i \left[\nabla h_i(\tilde{x}_r) - \nabla h_i(x_r) \right] = \sum_{i \in I_0} (\tilde{u}_r)_i \nabla h_i(x_r)$$

and $h_i(x_r) + \nabla h_i(x_r)'(\tilde{x}_r - x_r) = 0$ for $i \in I_0$.

From $\tilde{z}_r = S(z_r, I_0) - z_{r+1} + z_{r+1} \to \bar{z}$ and $z_r \to \bar{z}$ it follows, that

$$\nabla f(\bar{x}) = \sum_{i \in I_0} \bar{u}_i \nabla h_i(\bar{x}) \quad \text{and} \quad h_i(\bar{x}) = 0 \ (i \in I_0) \qquad \Delta$$

Next we state some results, which have been shown by Robinson [5], in the following lemma.

Lemma 1: Let $\bar{z}$ be a Kuhn-Tucker solution of (1) satisfying assumption A. Then:

a) $\beta := \| F'(\bar{z})^{-1} \| > 0$;

b) there exist $\mu > 0$, $M > 0$ such that for $z_1, z_2 \in B(\bar{z}, \mu)$:

(i) $\| F(z_2) - d(z_1, z_2) \| \leq M \| z_1 - z_2 \|^2$

(ii) $h_i(\bar{x}) < 0$ implies $Lh_i(x_1, x_2) < 0$

(iii) $\bar{u}_i < 0$ implies $(u_1)_i < 0$

(iv) $\| F(z_2) - F(z_1) \| \leq M \| z_2 - z_1 \|$

(v) If $\hat{z} \in B(\bar{z}, \frac{\mu}{2})$, such that $4\beta \| F(\hat{z}) \| \leq \mu$, then there is $\tilde{z} \in B(\hat{z}, \frac{\mu}{2})$ such that $\tilde{z}$ is the unique zero of $d(\hat{z}, \cdot)$ in $B(\hat{z}, \frac{\mu}{2})$ and such that $\| \tilde{z} - \hat{z} \| \leq 2\beta \| F(\hat{z}) \|$.

Part (v) of Lemma 1 b) can be extended in the following way.

Lemma 2: Let $\bar{z}$ be a Kuhn-Tucker solution of (1) satisfying assumption A. If $\hat{z} \in B(\bar{z}, \frac{\mu}{2})$ such that $4\beta \| F(\hat{z}) \| \leq \mu$, then there is a unique $\tilde{z} \in B(\hat{z}, \frac{\mu}{2})$ with $\tilde{z} = S(\hat{z}) = S(\hat{z}, I(\bar{z}))$ and $\| \tilde{z} - \hat{z} \| \leq 2\beta \| F(\hat{z}) \|$.

Proof: From Lemma 1 b) (iv) we have the existence of $\tilde{z} \in B(\hat{z}, \frac{\mu}{2})$ such that $\| \tilde{z} - \hat{z} \| \leq 2\beta \| F(\hat{z}) \|$ and $d(\hat{z}, \tilde{z}) = 0$. Hence $\tilde{z}$ satisfies the equalities of the Kuhn-Tucker conditions for $S(\hat{z})$. Then $\tilde{u}_i Lh_i(\hat{x},\tilde{z}) = 0$ for $i=1,\dots,m+p$. From the strict complementary slackness and (ii), (iii) of Lemma 1 b) it follows that

(3) $$\text{for } i \in I(\bar{z}),\ i \le m:\quad \bar{u}_i < 0,\ \tilde{u}_i < 0,\ Lh_i(\hat{x},\tilde{x}) = 0;$$
$$\text{for } i \notin I(\bar{z}),\ i \le m:\quad h_i(\bar{x}) < 0,\ Lh_i(\hat{x},\tilde{x}) < 0,\ \tilde{u}_i = 0.$$

Since there is no other zero of $d(\hat{z},\cdot)$ in $B(\hat{z},\frac{\mu}{2})$, we have $\tilde{z} = S(\hat{z})$. Furthermore we see from (3) that z satisfies the Kuhn-Tucker conditions for $S(\hat{z}, I(\bar{z}))$ as well with $\tilde{u}_i = 0$ for $i \notin I(\bar{z})$. △

Next we show, that for an appropriate z_0 we get $I_0 = I(\bar{z})$.

Lemma 3: Let $\bar{z}$ be a Kuhn-Tucker solution of (1) satisfying assumption A. If $z_0 \in B(\bar{z},\frac{\mu}{2})$ and $4\beta \| F(z_0) \| \le \mu$, then $I_0 = I(\bar{z})$.

Proof: From Lemma 2 we have $\| S(z_0) - z_0 \| \le \frac{\mu}{2}$. For i=1,...,m let z(i) be any optimal solution of $\min_z \{ \| z - z_0 \| \mid u_i Lh_i(x_0,x) = 0 \}$.

We have $\| z(i) - \bar{z} \| \le \| z(i) - z_0 \| + \| z_0 - \bar{z} \|$

(4) $$\le \| S(z_0) - z_0 \| + \| z_0 - \bar{z} \| \le \mu .$$

There are two possibilities for z(i):

1) $z(i) = \begin{pmatrix} x_0 \\ u \end{pmatrix}$ with $u_k = \begin{cases} (u_0)_k, & \text{for } k \ne i \\ 0, & \text{for } k = i . \end{cases}$

Then $\| z(i) - z_0 \| = | (u_0)_i |$.

2) $$z(i) = \begin{pmatrix} x_0 - \dfrac{h_i(x_0)}{\| \nabla h_i(x_0) \|^2} \nabla h_i(x_0) \\ u_0 \end{pmatrix} .$$

Then $$\| z(i) - z_0 \| = \frac{| h_i(x_0) |}{\| \nabla h_i(x_0) \|} .$$

From Step 1 of algorithm 2, Lemma 1 b) (ii) and (4) it follows:

If $i \in I_0$ $(i \le m)$, then $(u_0)_i = (u(i))_i < 0$, $Lh_i(x_0,x(i)) = 0$ and $h_i(\bar{x}) = 0$;

If $h_i(\bar{x}) = 0$ $(i \le m)$, then from the complementary slackness $\bar{u}_i < 0$ and thus from Lemma 1 b) (iii) $(u(i))_i < 0$, which means z(i) is the second possibility, and so $i \in I_0$.

Altogether $i \in I_0$, if and only if $h_i(\bar{x}) = 0$. △

Now we can show the same theorem for the modification as is shown by Robinson [5] for his original algorithm.

Theorem 2: Let $\bar{z}$ be a Kuhn-Tucker solution of (1) satisfying assumption A. Then there is $\delta > 0$ such that for $z_0 \in B(\bar{z},\delta)$ algorithm 2 is defined and the sequence $\{z_r\}$ converges R-quadratically to $\bar{z}$. In particular there is $Q > 0$ such that

$$\| z_r - \bar{z} \| \leq Q \left(\tfrac{1}{2}\right)^{2^r} \quad \text{for } r=1,2,\ldots .$$

Proof: Choose $\delta \in (0,\frac{\mu}{4})$ such that for all $z \in B(\bar{z},\delta)$:

$$\| F(z) \| \leq \frac{\eta^2}{M(c+4\beta^2)}$$

where $\eta = \min\left(\frac{1}{2}, \frac{\mu M(c+4\beta^2)}{3\cdot 4\beta}, \sqrt{\frac{\beta M(c+4\beta^2)}{c}}\right)$.

Then for all $z \in B(\bar{z},\delta)$: $\| F(z) \| \leq \frac{\beta}{c}$ and $4\beta \| F(z) \| \leq \frac{\eta\mu}{3} \leq \frac{\mu}{3}$.

Let z_0 be in $B(\bar{z},\delta)$. Then from Lemma 3 we have $I_0 = I(\bar{z})$ and from $\| z_1 - \bar{z} \| \leq \| z_0 - \bar{z} \|$ that $z_1 \in B(\bar{z},\delta)$.

From Lemma 2 we have $\| S(z_1,I_0) - z_1 \| \leq 2\beta \| F(z_1) \|$.

Then z_2 exists and

$$\begin{aligned} \| z_2 - z_1 \| &\leq \| z_2 - S(z_1,I_0) \| + \| S(z_1,I_0) - z_1 \| \\ &\leq c \, \| F(z_1) \|^2 + 2\beta \, \| F(z_1) \| \\ &\leq c \, \frac{\beta}{c} \, \| F(z_1) \| + 2\beta \, \| F(z_1) \| \\ &= 3\beta \, \| F(z_1) \| \leq \frac{\mu}{4} . \end{aligned}$$

Hence $z_2 \in B(\bar{z},\mu)$ and with Lemma 1 and $\tilde{z} = S(z_1,I_0)$:

$$\begin{aligned} \| F(z_2) \| &\leq \| F(z_2) - F(\tilde{z}) \| + \| F(\tilde{z}) - d(z_1,\tilde{z}) \| \\ &\leq M \, \| z_2 - \tilde{z} \| + M \, \| z_1 - \tilde{z} \|^2 \\ &\leq Mc \, \| F(z_1) \|^2 + M4\beta^2 \, \| F(z_1) \|^2 \\ &\leq \frac{\eta^{(2^2)}}{M(c+4\beta^2)} . \end{aligned}$$

Now suppose that for $r = 2,3,\ldots,k$ we have :

(i) $$\| z_r - z_{r-1} \| \leq 3\beta \, \| F(z_{r-1}) \|$$

(ii) $$\| F(z_r) \| \leq \frac{\eta^{(2^r)}}{M(c+4\beta^2)} .$$

Then $\| z_r - z_{r-1} \| \leq \frac{3\beta\, \eta^r}{M(c+4\beta^2)} \leq \frac{\mu}{4}\, \eta^{r-1} \leq \frac{\mu}{4} (\frac{1}{2})^{r-1}$.

Hence $\| z_k - \bar{z} \| \leq \| z_1 - \bar{z} \| + \sum_{r=2}^{k} \| z_r - z_{r-1} \| \leq \frac{\mu}{4} + \frac{\mu}{4} \sum_{r=2}^{k} (\frac{1}{2})^{r-1} < \frac{\mu}{2}$.

From (ii) we have $4\beta \| F(z_k) \| \leq \frac{4\beta\,\eta}{M(c+4\beta^2)}\, \eta^{(2^k-1)} \leq \frac{\mu}{3}$ and from

Lemma 2 it follows that $\| S(z_k,I_o) - z_k \| \leq 2\beta \| F(z_k) \|$.

Then, since from (ii) $\| F(z_k) \| \leq \frac{\beta}{c}$, it follows as above, that z_{k+1} exists with

$$\| z_{k+1} - z_k \| \leq 3\beta \| F(z_k) \| \leq \frac{\mu}{4}$$

and

$$\begin{aligned} \| F(z_{k+1}) \| &\leq M(c+4\beta^2)\ \|F(z_k)\|^2 \\ &\leq \frac{\eta^{(2^{k+1})}}{M(c+4\beta^2)} . \end{aligned}$$

Hence we have shown by induction that the algorithm 2 is defined and (i), (ii) hold for all $r \geq 2$.

Thus $\{z_r\}$ is a Cauchy sequence and therefore converges to a point $z' \in B(\bar{z}, \frac{\mu}{2})$. Since $z_{r+1} - S(z_r,I_o) \to 0$, it follows from Theorem 1, that z' is a Kuhn-Tucker solution of

$$\min \{f(x) \mid h_i(x) = 0,\ \ i \in I(\bar{z})\} .$$

Since $z' \in B(\bar{z}, \frac{\mu}{2})$ and $u_i' = 0$ for $i \notin I(\bar{z})$ it follows from Lemma 1 that $z' = \bar{z}$.

Furthermore

$$\begin{aligned} \| z_r - \bar{z} \| &\leq 3\beta \sum_{k=r}^{\infty} \| F(z_k) \| \leq \frac{3\,\beta}{M(c+4\beta^2)} \sum_{k=r}^{\infty} (\tfrac{1}{2})^{(2^k)} \\ &\leq \frac{3\,\beta}{M(c+4\beta^2)} \ \Big(\sum_{k=0}^{\infty} (\tfrac{1}{2})^{(2^k)} \Big) \ (\tfrac{1}{2})^{(2^r)} \end{aligned}$$

Δ

References:

[1] A.V.Fiacco, G.P.McCormick: *Nonlinear programming: sequential unconstrained minimization techniques* (Wiley, New York, 1968).

[2] G.P.McCormick: *A second-order method for the linearly constrained nonlinear programming problem,* in: Nonlinear Programming, Eds. J.B.Rosen, O.L.Mangasarian, K.Ritter (Academic Press, New York, 1970), pp. 207-243.

[3] K.Ritter: *A superlinearly convergent method for minimization problems with linear inequality constraints,* Mathematical Programming 4(1973), 44-71.

[4] K.Ritter: *A method of conjugate directions for linearly constrained nonlinear programming problems,* SIAM J.Numer.Anal. 12(1975), 273-303.

[5] S.M.Robinson: *A quadratically-convergent algorithm for general nonlinear programming problems,* Mathematical Programming 3(1972), 145-156.

On a Minimization Problem in Structural Mechanics

Ulrich Eckhardt
KFA Jülich GmbH - ZAM
D-5170 Jülich

0. Introduction The computation of the static and dynamic behaviour of geometrically nonlinear networks is a very essential problem in structural mechanics, especially since these networks are simple models for more complicated geometrically nonlinear structures. On the other hand, from the more mathematical point of view these networks are also of great interest. They provide simple mechanical models for catastrophe theory [6,8], they can be used as application of algebraic geometry [7] and they show all effects which are typical for nonlinear problems (bifurcation points, multiple solutions, hysteresis etc.) without being too complicated for a thorough theoretical investigation. Finally, the numcerical treatment of large nonlinear networks is a challenging topic for numerical analysts. The careful analysis of the numerical methods invented and used by structural engineers can stimulate new ideas and give new insights for practical work.

1. Illustrative Examples A. The most prominent example of a geometrically nonlinear network is the olympic tent in Munich [1,2]. It consists of a large number of elastic trusses which are linked together at the ends to form a network. The mathematical problem arising with this network is to find the absolute minimum of the potential energy. If the structure is at a relative minimum, an exceptional load (wind, snow etc.) can cause it to jump catastrophically to the absolute minimum.

B. Consider the structure in Fig. 1 with fixed nodes a and b and free node c which is loaded by a force P. This structure can be interpreted as the tip of the arm of a crane. In $\mathbb{R}^2$ we can easily find two minima of the potential energy (for formulas see example D. below). The dashed line indicates an alternative minimum at c' which is, however, not realizable in most technical applications. Mathematically one wants to find a minimum in a certain neighbourhood of a certain configuration, regardless whether it is a relative or an absolute one.

Fig. 1: Tip of a crane

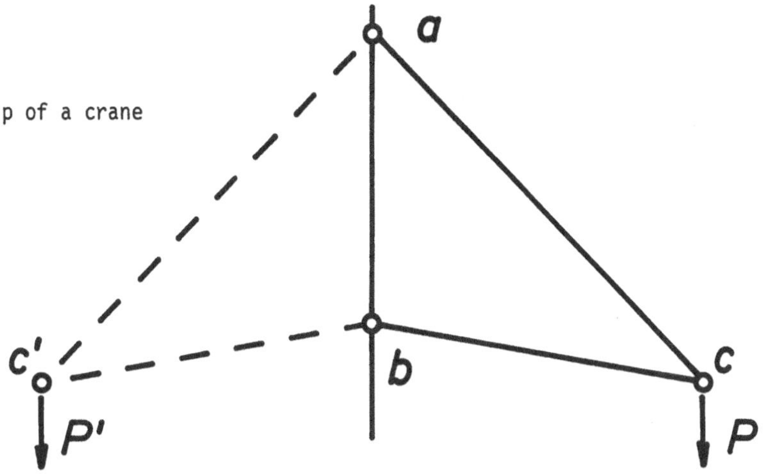

C. [2] When the load of the braced frame in Fig. 2. exceeds a certain critical load the structure will undergo deformations perpendicular to the direction of P, in analogy to Euler's buckling load. In this case one wants to find the critical load belonging to a bifurcation point, and for a greater load all solutions, or a least all stable solutions are desired.

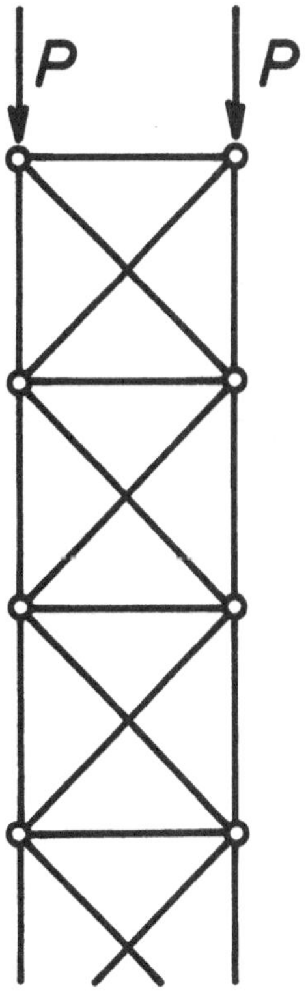

Fig. 2: Braced frame

D. The system of Thomas [7]. Let $a_1,\dots,a_n$ be given points in $\mathbb{R}^d$ (d = 2 or 3). At each point a_j an elastic spring with spring constant k_j and (unloaded) length l_j is attached with one end. The other ends of the springs are fixed together in a point x. Let P be a force acting at x, then the potential energy of the system is

$$U(x) = \sum_{j=1}^{n} \frac{k_j}{2} \cdot \left[\|x - a_j\| - l_j \right]^2 - P^T x \qquad (1)$$

where $\|\cdot\|$ is the Euclidean vector norm. In order to find a stationary point of U one has to solve $\nabla U = \Theta$ which is an algebraic system of equations. The system of Thomas can be considered as a simple prototype of a geometrically nonlinear network. When it is possible to find a complete theory for this case then it is easily possible to extend it to general networks.

E. A special case of Thomas' system is the system of v. Mises consisting of two (three) springs in $\mathbb{R}^2$ ($\mathbb{R}^3$) such that the a_j are in general position. This system is a simple illustrative example showing all relevant features of more general systems. Let in $\mathbb{R}^2$ $a_1 = \binom{0}{0}$, $a_2 = \binom{1}{0}$, $l_1 = l_2 = 1$, $k_1 = k_2 = 1$. For $P = \binom{0}{0}$ there are five stationary points of U: $(\frac{3}{2}, 0)$, $(\pm \frac{1}{2}, 0)$ and $(\frac{1}{2}, \pm \frac{\sqrt{3}}{4})$. For forces of the form $P = \binom{\mu}{0}$ there are bifurcation points at $\binom{1 + \sqrt{0.5}}{0}$ and $\binom{-\sqrt{0.5}}{0}$. For $P = \binom{0}{\mu}$ the force-displacement diagram showing the solutions of the form $\binom{0}{y}$ is sketched in Fig. 3. At the critical points $(0, \pm \frac{1}{2} \cdot \sqrt{\sqrt[3]{4} - 1})$ a snap-through occurs. If μ varies from large positive values to large negative values and then back, hysteresis occurs. Near the origin the system behaves like a spring with negative constant and it is sometimes used in engineering for simulating such a spring.

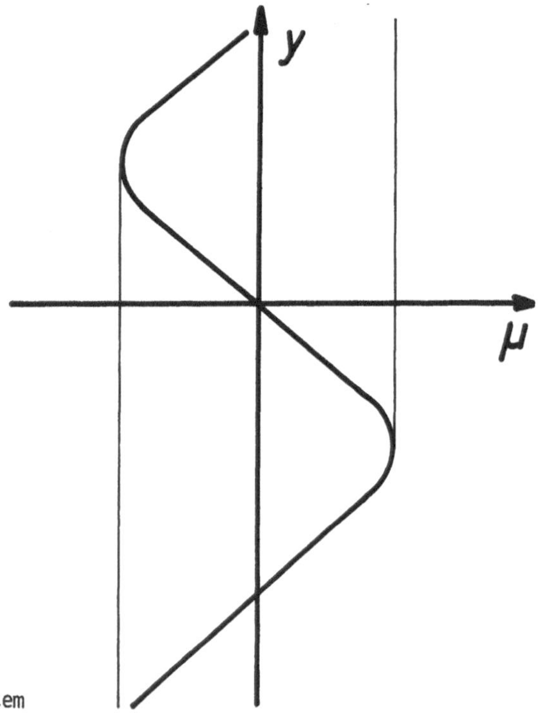

Fig. 3: v. Mises' system

F. The general geometrically nonlinear network. Let $x_1,\dots,x_n$ be the nodes of a network in $\mathbb{R}^3$. $i \sim j$ means that x_i and x_j are joined by a spring. Let k_{ij} be the spring constant and l_{ij} the length of this spring and let P_i be an exterior force acting at x_i. Then the potential energy of this system is

$$U(x_1,\dots,x_n) = \sum_{i \sim j} \frac{k_{ij}}{2} \cdot \left[\|x_i - x_j\| - l_{ij}\right]^2 - \sum_{i=1}^{n} P_i\, x_i \; . \qquad (2)$$

Some of the x_i are subjected to equality (or inequality) constraints.

G. A very interesting application was pointed out to the author by C. Witzgall. In modern geodesy it is possible to measure distances very accurately by means of optical methods. Let l_{ij} be the measured distance from x_i to x_j. Then the x_i have to be arranged in such a way that the mean square error is minimal which leads to the minimization of a functional of the form (2) with $P_i = \Theta$.

2. Linearization For the numerical treatment the function U in (1) (or (2)) is approximated by a quadratic function $\psi(x)$ which coincides with $U(x)$ in a certain point x_0 and whose gradient at x_0 is equal to the gradient of U at x_0. Such a linearization $\psi(x)$ leads in a natural way to an iteration method:

0. Given an approximation x_0 to the solution, put $r := 0$.
1. Calculate the linearization ψ to U in x_r and compute the minimum x_{r+1} of ψ (if it exists).
2. Replace r by r + 1 and go to 1.

This method may converge or not, depending on the linearization ψ.

In order to introduce some linearizations we concentrate ourselves on a unloaded single spring with length 1 and unit spring constant whose one end is attached to the origin. If the other end is at point x then its potential energy is

$$\phi(x) = \frac{1}{2} \cdot \left[\|x\| - 1\right]^2 .$$

The linearization of a general network then can be found by superposition.

The most common linearization in numerical analysis is Newton's

$$\phi_N = \frac{1}{2} \cdot \frac{\|x_0\| - 1}{\|x_0\|} \cdot \|x\|^2 + \frac{1}{2 \cdot \|x_0\|^3} \langle x_0, x\rangle^2 - \frac{1}{\|x_0\|} \cdot \langle x_0, x\rangle + \frac{1}{2} \cdot 1^2$$

($\langle \cdot,\cdot \rangle$ = scalar product in $\mathbb{R}^d$). Note that $x_0 \neq \Theta$ is required since ϕ is not differentiable at $x = \Theta$.

The linearization which was used to calculate the olympic tent in Munich is

$$\phi_u = \frac{1}{2}\cdot \|x\|^2 - \frac{1}{\|x_0\|}\cdot \langle x_0,x\rangle + \frac{1}{2}\cdot 1^2 .$$

Another linearization is frequently used in practice:

$$\phi_1 = \frac{1}{2}\cdot \frac{\|x_0\| - 1}{\|x_0\|}\cdot \|x\|^2 - \frac{1}{2}\cdot 1\cdot \|x_0\| + \frac{1}{2}\cdot 1^2 .$$

ϕ_u and ϕ_1 have interesting properties:

Theorem 1: $\phi_1(x) \leq \phi(x) \leq \phi_u(x)$ for all x.

Let $\psi(x)$ be a quadratic function.

$\phi_1 \leq \psi \leq \phi \qquad \psi = \phi_1,$

$\phi \leq \psi \leq \phi_u \qquad \psi = \phi_u.$

Proof: 1. $\phi - \phi_1 = \frac{1}{2\cdot\|x_0\|}(\|x_0\| - \|x\|)^2 \geq 0.$

For the quadratic function $q = \Psi - \Phi_1$ is

$$0 \leq q(x) \leq \frac{1}{2\cdot\|x_0\|}\cdot(\|x_0\| - \|x\|)^2$$

and $q(x) = 0$ for $\|x\| = \|x_0\|$. For $x_0 \neq \Theta$ this is only possible if $q(x) = 0$ for all x.

2. $\phi_u - \phi = \frac{1}{\|x_0\|}\cdot(\|x_0\|\cdot\|x\| - \langle x_0,x\rangle) \geq 0$ by Schwarz' inequality. The quadratic function $q = \phi_u - \psi$ has the property

$$0 \leq q(x) \leq \frac{1}{\|x_0\|}\cdot(\|x_0\|\cdot\|x\| - \langle x_0,x\rangle).$$

$q(\mu\cdot x)$ is a quadratic function of μ which is bounded from above and from below by a linear function, hence it is itself linear. Since $q(\Theta) = 0$ and $q(x) \geq 0$ for all x we conclude $q(x) = 0$ for all x.

3. Convergence The convergence properties of Newton's method are well known and will not be discussed here (see e.g. [5]). We only mention that this method does not always converge when applied to our problem, it even does not work in all cases because of the nondifferentiability of U at certain points. If it converges, however, the convergence is usually very fast.

The iteration method which is defined by the linearization ϕ_l also does not converge in general. If it converges, the convergence is also very fast. It may happen, that it converges to a local maximum of U. No convergence proof is known to the author.

The iteration with ϕ_u has very favourable properties exept that the convergence is in general extremely slow as compared with the other methods. First we note that ϕ_u is always defined if for $x \to 0$ the direction of approach $x_o / \|x_o\|$ is known.

Let $U_u(x)$ be the approximation for $U(x)$ in x_o which is obtained from (1) by replacing the potential energies of the springs of the system by their linearizations ϕ_u in x_o. Let $x_o, x_1, \ldots$ be the iterates of an iteration process as defined above. We have

$$U(x_1) \leq U_u(x_1) = U(x_o) + (\nabla U(x_o))^T (x_1 - x_o) + \\ + \frac{1}{2} \cdot (x_1 - x_o)^T H (x_1 - x_o)$$

where H is the Hessian of U_u. H is a matrix which is positive definite and does not depend on x_r. By minimality of $U_u(x_1)$ this implies

$$U(x_o) - U(x_1) \geq \frac{1}{2} \cdot (x_1 - x_o)^T H (x_1 - x_o) \geq \frac{\lambda}{2} \cdot \|x_1 - x_o\|^2 \qquad (3)$$

$\lambda > 0$ being the smallest eigenvalue of H. This proves the

<u>Theorem 2:</u> $U(x_{r+1}) \leq U(x_r)$ for all r.

Moreover, we can state

<u>Theorem 3:</u> $\lim x_r = x^*$ and x^* is a stationary point of U but not a maximum.

<u>Proof:</u> The set $\{x \mid U(x) \leq C\}$ is bounded (or empty) for each C. Theorem 2 and (3) imply because of $\nabla U(x_r) = \nabla U_u(x_r) = -H \cdot (x_{r+1} - x_r)$ that $\nabla U(x^*) = 0$ for each accumulation point x^* of $\{x_r\}$. (3) further implies that there is only one accumulation point which is no maximum by Theorem 2.

In the general case of a geometrically nonlinear network (2) the Theorem remains true. The method even converges if H is only positive semidefinite i.e. if the network is a statically undetermined mechanism.

4. Numerical Properties If the linearization ϕ_u is used for iteration, then the so called stiffness matrix (i.e. the Hessian) does not depend on the point of linearization x_0 so that it has to be inverted (or triangularized) only once at the beginning of the iterations. Only the linear term in ϕ_u depends on x_0 that means in the terminology of mechanics that the iteration is performed by introducing auxiliary forces. For ϕ_1 the stiffness matrix changes at each iteration but no auxiliary forces are introduced. In Newton's method everything changes.

A very essential point in practical calculations is the mechanical interpretation of the different linearization functionals. If it is possible to interpret a specific linearization by a simple mechanical device then this linearization can be easily implemented into existing finite element computer codes thus leading to considerable savings in setup time. The simplest mechanical device is the linear spring which is a spring with length zero. The potential ϕ_u can be interpreted as a linear spring with fixed positive constant. ϕ_1 is a linear spring with effective constant $(\|x_0\| - 1)/ \|x_0\|$ which can vanish or even become negative. The quadratic term in ϕ_N is the superposition of the latter linear spring with an "exotic" element with potential energy $\langle x_0, x\rangle^2$ being highly anisotropic which causes troubles if implemented into a finite element package.

Although the convergence with ϕ_u is very slow, this method is preferred in engineering practice because of its simplicity in implementation and because it is always convergent in contrast to the other methods. Since engineers usually know very clever approximations to the desired solution and since in engineering an accuracy of some percent is sufficient, the disadvantage of slow convergence is not so serious.

A very similar situation arises with the so called Weber problem in economical location theory [3]. This problem also is usually solved by a very slowly convergent method having the advantage of great simplicity.

5. Conclusions The scope of this article was limited to numerical considerations. A very exciting topic, however, is the investigation of theoretical properties of the given problem. Since geometrically nonlinear networks are very simple prototypes of more general geometrically nonlinear structures (shells and plates undergoing buckling) they deserve the attention of mathematicians. It would be highly desirable to find characterizations of those configurations which represent absolute minima, relative minima, unstable points, and bifurcation points. It is also of interest to have topological properties of the sets of these points, e.g. whether they are connected or not.

There are many related problems in different fields of application. In solid state physics the numerical simulation of the behaviour of crystal lattices leads to similar minimization problems since the potential of the interatomic forces (e.g. Born-Mayer potential) has similar features as the spring potential $\phi(x)$. Another application is the description of elastic-plastic behaviour, phase transitions, and hysteresis effects by systems of springs [4].

References

1. J.H. Argyris, P.C. Dunne and T. Angelopoulos: Die Lösung nicht-linearer Probleme nach der Methode der finiten Elemente. In: J. Albrecht und L. Collatz, eds.: Finite Elemente und Differenzenverfahren. ISNM 28, 15 - 52. Basel und Stuttgart: Birkhäuser Verlag 1975.

2. J.H. Argyris, M. König, D.A. Nagy, M. Haase and G. Malejnnakis: Geometric nonlinearity and the finite element displacement method. In: Finite Element Linear and Nonlinear Analysis: Methods and General Purpose Programs. pp. 451 - 535. Politecnico di Milano, University Extension Program. Milano 1975.

3. U. Eckhardt: On an optimization problem related to minimal surfaces with obstacles. In: R. Bulirsch, W. Oettli and J. Stoer, eds.: Optimization and Optimal Control. Lecture Notes in Mathematics, Vol. 477, pp. 95 - 101. Berlin, Heidelberg, New York: Springer-Verlag 1975.

4. I. Müller and P. Villagio: A model for an elastic-plastic body. Arch. Rational Mech. Anal. 65, 25 - 46 (1977).

5. J.M. Ortega and W.C. Rheinboldt: Iterative Solution of Nonlinear Equations in Several Variables. New York and London: Academic Press 1970.

6. M.J. Sewell: On Legendre transformations and elementary catastrophes. Math. Proc. Cambridge Philos. Soc. 82, 147 - 163 (1977).

7. J. Thomas: Zur Statik eines gewissen Federsystems im E_n. Math. Nachr. 23, 185 - 195 (1961).

8. J.M.T. Thompson: Catastrophe theory and its role in applied mechanics. In: W.T. Koiter, ed.: Theoretical and Applied Mechanics. Proceedings of the 14th IUTAM Congress, pp. 451 - 458. Amsterdam, New York, Oxford: North-Holland Publishing Company 1977.

NON-LINEAR LEAST SQUARES INVERSION OF AN INTEGRAL EQUATION USING FREE-KNOT CUBIC SPLINES.

J N Holt
Department of Mathematics
University of Queensland
St Lucia, 4067
Queensland
AUSTRALIA

1. Introduction

A stable iterative inversion technique has recently been developed by Holt and Jupp (1978) for the equation describing the limb-darkening effect on radiation from the solar disc, which is a Fredholm integral equation of the first kind.

The purpose of the present paper is to demonstrate the usefulness of the inversion method by describing its application to another equation, one which represents the class of problems arising from the smearing of a signal by convolution with the instrumental profile of a measuring device of finite resolution. Such problems arise in optical and x-ray spectroscopy, instrumental optics etc. (see Turchin et al (1971)).

In the inversion, the unknown function is approximated by a cubic spline function whose knots are variables of the problem. A basis of B-splines is used to represent the spline, and the algorithm seeks the optimal set of parameters defining the spline in the sense of minimizing the sum of squares of deviations between the actual and model computed data.

The minimization is carried out by the method of Jupp and Vozoff (1975) which provides stable convergence and also information on the number and nature of the important parameters of the spline model.

It is important to consider inversions of noisy data, as most practical applications will use data subject to some kind of errors. Calculations with synthetic noise are therefore described to show the stability of the method.

2. The Inversion Problem

Consider the first kind Fredholm integral equation

$$g(s) = \int_{t_{min}}^{t_{max}} k(s,t)\, f(t)\, dt \qquad -(1)$$

for $s \in [s_{min}, s_{max}]$.

Suppose the only data given consists of values of $g(s)$ at a discrete set of points $s_i (i=1,\cdots,m)$ in $[s_{min}, s_{max}]$. The given values may contain a random noise component as stated earlier, and it will be assumed that such noise is normally distributed with zero mean. The kernel function $k(s,t)$ is assumed known and continuous, and it is desired to recover an approximation to the unknown function $f(t)$.

In particular, consider the following specific problem. Let

$$k(s,t) = \begin{cases} 1 + \cos((s-t)\pi/3) & \text{for } |s-t| \le 3 \\ 0 & |s-t| > 3 , \end{cases} \qquad -(2)$$

$$[t_{min}, t_{max}] = [s_{min}, s_{max}] = [-6, 6] , \qquad -(3)$$

and

$$f(t) = k(0,t) . \qquad -(4)$$

Then by integrating,

$$g(s) = (6-|s|)(1+\tfrac{1}{2}\cos(\pi s/3)) + \text{sign}(s)\sin(\pi s/3)\frac{9}{2\pi} \qquad -(5)$$

This problem has been treated by Phillips (1962), and by Hanson and Phillips (1975). $k(s,t)$ may be thought of as an unnormalized cosine-bell instrumental profile of finite width, and $f(t)$ as a signal pulse being measured. Synthetic data may be generated from (5), and pseudo-random noise added if desired. The result of applying the inversion method to this data can then be compared with the known $f(t)$.

3. Review of the Inversion Method.

(i) Spline representation of $f(t)$.

The method of Holt and Jupp (1978) approximates $f(t)$ over the region of integration by a polynomial spline of order 4 (of degree 3) which shall be written $\sigma(t,\underset{\sim}{\tau})$, where $\underset{\sim}{\tau} = (\tau_1,\cdots,\tau_N)^T$ is an N-component vector of knots, with $\tau_i \in (t_{min}, t_{max})$. Knots may be coalescent, but the maximum knot multiplicity for a cubic spline is 4, corresponding to a discontinuity in the spline. $\sigma(t,\underset{\sim}{\tau})$ is a piecewise polynomial of degree ≤ 3 .

In the interval $[t_{min}, t_{max}]$, any spline of order 4 on the set of knots $\underset{\sim}{\tau}$ has a unique representation in terms of a set of basis splines. A convenient basis results from appending 4 arbitrary runout knots at each end of the interval, i.e.

$$\tau_{-3} \le \tau_{-2} \le \tau_{-1} \le \tau_0 \le t_{min} ,$$

and

$$t_{max} \le \tau_{N+1} \le \tau_{N+2} \le \tau_{N+3} \le \tau_{N+4} . \qquad -(6)$$

Then

$$\sigma(t,\tau) = \sum_{i=1}^{N+4} \alpha_i M_i(t,\underset{\sim}{\tau}) \qquad (7)$$

for $t \in [t_{min}, t_{max}]$, with $\underset{\sim}{\tau}$ extended to include the extra knots. The $M_i(t,\underset{\sim}{\tau})$'s are basic or B-splines. A full discussion of their properties, and of the material leading up to their introduction here, may be found in Curry and Schoenberg (1966).

From (7) it can be seen that the parameters which are required to completely describe $\sigma(t,\underset{\sim}{\tau})$ consist of the constants α_i, and the variable knots τ_1 to τ_N, assuming the runout knots to remain fixed. There are then $2N + 4$ variables in the problem.

Holt and Jupp have shown how certain advantages result from working, in the minimization, with new variables z_i which are logarithmic transformations of the free knots. They have also shown that if the desired function $f(t)$ is known a-priori to be non-negative, the iterations are enhanced by using $\log \alpha_i$ rather than α_i, thus ensuring non-negativity since the B-splines are non-negative. Both of these devices are used here, but it should be remarked that the latter is certainly not an essential ingredient of the method.

(ii) Determination of Optimal Parameters.

The inversion procedure of Holt and Jupp minimizes the sum of squares of deviations between the given discrete data values $g(s_i)$ and the model data computed by using the spline approximation to $f(t)$ in the right-hand side of equation (1). This model data is

$$h(s_i) = \int_{t_{min}}^{t_{max}} k(s_i,t) \sum_{j=1}^{N+4} \alpha_j M_j(t,\underset{\sim}{\tau})dt$$

$$= \sum_{j=1}^{N+4} \alpha_j \Psi_j(s_i,\underset{\sim}{\tau}) \qquad -(8)$$

where

$$\Psi_j(s_i,\underset{\sim}{\tau}) = \int_{t_{min}}^{t_{max}} k(s_i,t)M_i(t,\underset{\sim}{\tau})dt . \qquad -(9)$$

A useful property of the integration of cubic B-splines is

$$\int_{-\infty}^{\infty} c^{(4)}(t)M_i(t,\underset{\sim}{\tau})dt = 3!\, c[\tau_{i-4},\ldots,\tau_i] . \qquad -(10)$$

(see de Boor and Lynch (1966)). Here, the function on the right hand side is the 4th order divided difference of the function $c(t) \in C^{(4)}$ on the 5 knots τ_{i-4} to τ_i over which M_i is non-zero.

If the runout knots are defined by

$$\tau_{-3} = \cdots = \tau_0 = t_{min} \text{ and } \tau_{N+1} = \cdots = \tau_{N+4} = t_{max} , \qquad -(11)$$

then using

$$k(s,t) = c^{(4)}(s,t)$$

where

$$c(s,t) = \begin{cases} \frac{y^4}{4!} + \left(\frac{3}{\pi}\right)^4 \cos\left(\frac{y\pi}{3}\right) & |y| \le 3 \\ \frac{1}{2}\mathrm{sign}(y)y^3 + \frac{1}{2}\left[\left(\frac{3}{\pi}\right)^2 - \frac{9}{2}\right]y^2 + \mathrm{sign}(y)\left[3\left(\frac{3}{\pi}\right)^2 - \frac{9}{2}\right]y + \left[\left(\frac{3}{\pi}\right)^2\left(\frac{9}{2} - \left(\frac{3}{\pi}\right)^2\right) - \frac{81}{24}\right] & |y| > 3 \end{cases}$$

with $y = s-t$, -(12)

$\Psi_j(s_i,\tau)$ can be evaluated by a confluent divided difference algorithm. (See Holt and Jupp for details and for the evaluation of derivatives of the $\Psi_j(s_i,\tau)$'s with respect to the knots.)

$$F = \sum_{i=1}^{m} w_i(g(s_i) - h(s_i))^2$$

is carried out by a modified Gauss-Newton-Marquardt method due to Jupp and Vozoff (1975) (JV). For a discussion of the method and the attributes which make it particularly suitable to inversion problems such as this, Jupp and Vozoff (1975) and Holt and Jupp (1978), should be consulted. Only a brief, informal description is given here.

One of the problems with inversions is that no theory provides the correct number of parameters to use for optimal modelling of the unknown function. In this case, that means how many variable knots to take. As a consequence, it is possible to specify too many parameters in the model and obtain no unique solution. In non-linear iterative methods of Gauss-Newton type which use local linear approximations to the model data, this overspecification is reflected in a Jacobian matrix J which is rank deficient, or nearly so.

The Gauss-Newton method seeks to find, at each iteration, an increment in the vector of unknown parameters which minimizes a local quadratic approximation to F with Hessian matrix J^TJ . The directions of the principal axes of the ellipsoidal contours of constant F are the eigenvectors of J^TJ. The lengths of the axes are proportional to the reciprocals of the square roots of the eigenvalues. Very small eigenvalues thus correspond to long flat valleys. The JV method effects a rotation of axes in parameter space to these orthonormal eigenvectors, and the components of the increment in parameters in directions corresponding to small, unimportant eigenvalues are damped in comparison to the values from the normal Gauss-Newton method. Hence, a bound on the size of the allowable step is imposed.

As the process nears a solution, the damping is decreased to some threshold value determined by the estimated signal-to-noise ratio of the given data. This is specified in terms of a relative damping threshold μ .

At the solution, the gradient of F is orthogonal to the directions in which the important parameters can change the model data.

By rotating back to the original axes using the Jacobian at the solution, the important parameters may be expressed as linear combinations of the parameters originally used to specify the model.

The damping makes the method stable in the presence of unimportant parameters.

The computational procedure used in the JV method employs a singular value decomposition of the Jacobian matrix to obtain the required eigenvectors and eigenvalues. J^TJ is, of course, never explicitly formed. A damped generalized inverse of J for the linear least squares problem at each iteration is computed once the singular value decomposition is known.

4. Numerical Results and Discussion.

Figure 1 shows the final spline approximations to f(t) using 2, 3 and 5 knots with no noise on the data. A damping threshold of 10^{-4} was used here. For all inversions, 21 data values were used at equally spaced points in [-6,6] .

The values of F in the three cases show remarkable improvement - 6.72, 2.23 and 0.705×10^{-2} for the 2, 3 and 5 knot models respectively. The 5 knot approximation to f(t) is effectively exact throughout [-6,6].

Of interest is the final placement of knots in the three cases. For 2 knots, $\tau_1 = .569$ and $\tau_2 = .750$, an unexpectedly unsymmetric result, although the resulting spline is quite symmetric. For 3 knots, there appears to be a discontinuity in slope at the origin indicating a knot of multiplicity 3. In fact, $\tau_1 = -.10 \times 10^{-1}$, $\tau_2 = .7 \times 10^{-5}$, $\tau_3 = .92 \times 10^{-2}$. The 5 knot result is of greatest importance since this fits f(t) so well. Here the knots are $(-3.04, -2.90, 4.3 \times 10^{-5}, 2.90, 3.04)$. The clustering of knots around ± 3 allows considerable flexibility in the second derivative in those regions, a feature required if f(t) is to be fitted well, as it has a discontinuous second derivative at ± 3 .

To test the effects of noise on the inversion, pseudo-random noise of r.m.s. amplitude approximately 1% of the maximum data value was added to the data, and 5 knot solutions sought. Figure 2 shows two such spline fits for μ equal to 10^{-2} and 10^{-3}, revealing the stability of the inversion against noise of this amplitude. The choice of threshold does not drastically impair the fit but does, as will be discussed, effect the convergence.

One difficulty which is yet to be totally overcome is the establishing of satisfactory convergence criteria. The 5 knot splines in Figures 1 and 2 were obtained after 35 to 40 iterations. This is considered excessive as there are only 14 variables in the 5 knot model. Each iteration requires approximately 10 seconds of CPU time on a time-sharing PDP-10 system.

An examination of the progress of the iteration reveals that there is not a great deal gained after the first 15 to 20 iterations. F appears to display linear convergence thereafter. The parameter values (i.e. knots, and α_i's) may change considerably, but the shape of the resulting spline $\sigma(t,\underset{\sim}{\tau})$ and the value of F show only minor changes. Significantly, the important combinations of parameters display almost no change. The iterations should not then have been allowed to continue for up to 40 iterations. The quantities used to define convergence are the tolerances specified for the absolute and relative changes in F. Their satisfactory specification, together with a consistent μ value may be the keys to defining convergence successfully. In the calculations shown here, they were arbitrarily set at reasonable values.

Jupp has developed a linear statistical analysis permitting a consistent determination of μ and the tolerances. The method works well on problems of least squares data fitting, but has not yet been adequately tested on this problem, and further work is needed to determine whether the present difficulty can be overcome in this way. It may be that another aspect of the method is producing the slow convergence, such as the use of the logarithmic knot transformations. These have the one undesirable property of not allowing a multiple knot situation to be exactly achieved, thereby possibly retarding convergence in cases where such an arrangement is optimal.

A noteworthy feature of the method is its ability to determine the important combinations of parameters at the solution. For the 5 knot, 1% noise case for example, there are 8 important parameters at the 1% threshold level. The most important is found to be

$$p_1 \approx \sum_{i=1}^{5} \tau_i ,$$

the mean knot position. The next most important is

$$p_2 \approx (\tau_4 + \tau_5) - (\tau_1 + \tau_2) ,$$

the separation of the means of the positive and negative knots, followed by

$$p_3 \sim .39\alpha_4 + .48\alpha_5 + .39\alpha_6 ,$$

a symmetric, weighted mean of the three linear coefficients expressing the amplitudes of the B-splines covering the region containing the non-zero part of $f(t)$. The remaining important parameters do not have the immediate interpretation of these three.

5. Conclusions

A 5 knot cubic spline model has been found to reproduce the unknown function $f(t)$. The non-linear least squares algorithm used for the optimization has provided stable convergence in the presence of noise, and also indicates the important combination of model parameters in so far as fitting the given data is concerned. Further work needs to be done on the rate of convergence. It is planned to remove the logarithmic knot transformations by introducing constraints explicitly into the non-linear least squares formulation, and carrying out comparative studies.

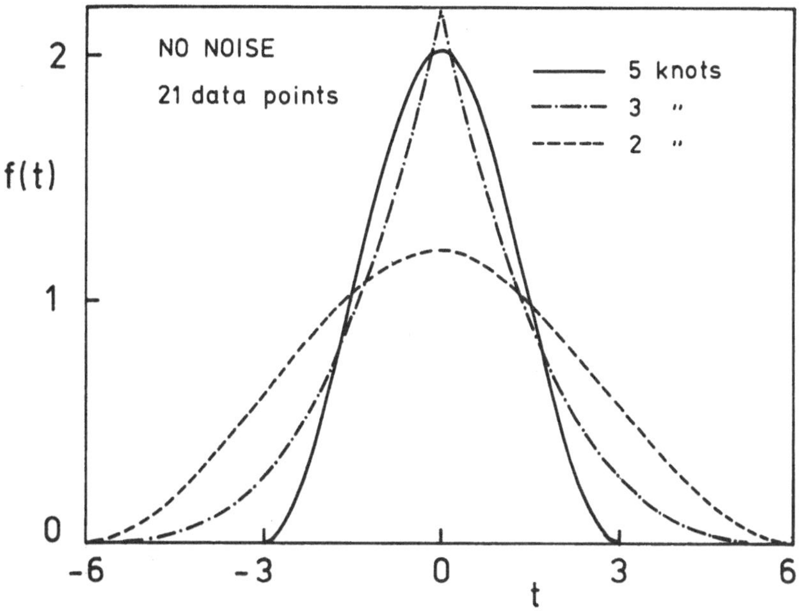

Figure 1 : 2, 3 and 5 knot fits, no noise.

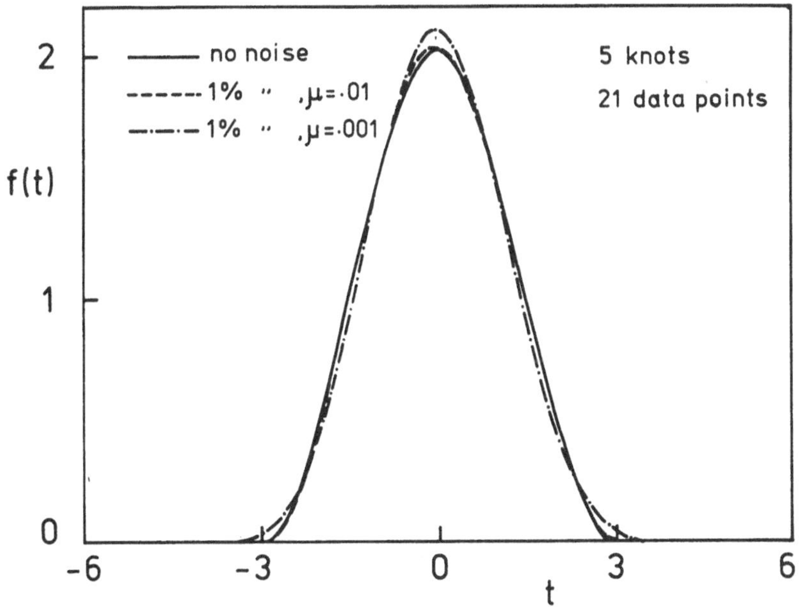

Figure 2 : 5 knot fits with noise.

Acknowledgement: Much of the computing for this paper was performed by Robyn Priddle.

References

1. Curry, H.B. and Schoenberg, I.J. 1966 J. Anal. Math. 17, 71.

2. de Boor, C. and Lynch, R.E. 1966 J. Math. Mech. 15, 953.

3. Hanson, R.J. and Phillips, J.L. 1975 Numer. Math. 24, 291.

4. Holt, J.N. and Jupp, D.L.B. 1978 (accepted for publication in J.I.M.A.)

5. Jupp, D.L.B. and Vozoff, K. 1975 Geophys. J.R. Astr. Soc. 42, 957.

6. Phillips, D.L. 1962 J. Assoc. Comput. Mach. 9, 84.

7. Turchin, V.F., Kozlov, V.P. and Malkevich, M.S. 1971 Soviet Phys. Usp. 13. 681.

ON THE CONVERGENCE OF THE MODIFIED TUI ALGORITHM FOR MINIMIZING A CONCAVE FUNCTION ON A BOUNDED CONVEX POLYHEDRON

by

S. Bali
Bell Laboratories
Holmdel, NJ 07733
USA

S. E. Jacobsen
Department of System Science
University of California
Los Angeles, CA 90024 / USA

1. INTRODUCTION

The Tui "cone-splitting" algorithm [2] for minimizing a concave function on a convex polyhedron has been shown by Zwart [3] to be nonconvergent. Subsequently, Zwart [4] produced a modification of Tui's important idea and demonstrated, by a clever geometric argument, that his modification would produce, in a finite number of steps, a point which is at least as good as some point in an ε-neighborhood of an optimal point. The parameter $\varepsilon > 0$ is crucial to that demonstration and the parameter essentially represents a lower bound for how far solution points of certain subproblems are from the various hyperplanes that are generated during the course of the procedure. Zwart has not proven convergence of his modification when the "geometric tolerance" parameter $\varepsilon = 0$ although he does point out [4] that his algorithm never failed to work in the case $\varepsilon = 0$.

In this paper we present a modification of Tui's algorithm which is slightly different from that of Zwart but which, of course, incorporates his basic observation. In particular, it is indicated, for the modification, why (i) the algorithm is convergent for $\varepsilon = 0$, and (ii) degeneracy of the polyhedron presents virtually no difficulty.

In Section 2 we state the basic ideas and present the algorithm. Section 3 discusses the algorithm and its relationships and differences with the modification of Zwart. Section 4 deals with convergence.

2. ALGORITHM

This paper is concerned with finding a globally optimal solution for

$$\min_{x \in X} f(x) \qquad (P)$$

where $f : R^n \to R^1$ is a concave function, and $X \subset R^n$ is a nonempty compact convex polyhedron. Of course, the latter imply that (P) possesses an optimal vector which is a vertex of X.

Before stating the algorithm, we present some basic ideas. Let $z^1,\ldots,z^n$ be n linearly independent vectors in R^n and let $C = \text{cone}\{z^1,\ldots,z^n\}$ denote the convex cone generated by $z^1,\ldots,z^n$. That is, C is the set of all nonnegative linear combinations of the vectors $z^1,\dots,z^n$. Let the defining matrix of C be the nxn matrix D; that is, $D = [z^1,\dots,z^n]$. Then $C = \{x|D^{-1}x \geq 0\}$. Also, if we let 1 denote an n-dimensional row vector of ones, we see that $\{x|1D^{-1}x = 1\}$ is the hyperplane which passes through the vectors $z^1,\dots,z^n$.

The modification of Tui's cone-splitting algorithm depends upon the following simple geometric fact which is nothing more than an interpretation of the pivot operation of linear programming.

Proposition 2.1. Let $z^1,\dots,z^n$ be n linearly independent vectors in R^n. Assume the nonzero vector $z \in C = \text{cone}\{z^1,\dots,z^n\}$ and define $P(z) = \{j|\mu_j > 0\}$ where $\mu \geq 0$ is the vector of weights by which z can be expressed in terms of $z^1,\dots,z^n$. Define $C_j = \text{cone}\{z^1,\dots,z^{j-1}z,z^{j+1},\dots,z^n\}$. Then $C = \bigcup_{j \in P(z)} C_j$.

Proof: Since it is immediate that $C_j \subset C$ we need to show that $\bigcup_{j\in P(z)} C_j \supset C$. Let $x \in C$ and let $\lambda \geq 0$ be the vector so that $x = \lambda_1 z^1+\dots+\lambda_n z^n$. Let $k \in P(z)$ be an index such that $\lambda_k/\mu_k = \min\{\lambda_i/\mu_i | i \in P(z)\}$. Since $\mu_k > 0$ we have, by the usual substitution arguments (i.e., the pivot operation), that $x = \sum_{i\neq k}^{n}(\lambda_i - \frac{\lambda_k}{\mu_k}\mu_i)z^i + \frac{\lambda_k}{\mu_k} z^k$. But the definition of k (i.e., the min ratio test) implies that each of the above weights is nonegative and therefore $x \in C_k$ which, in turn, implies that $x \in \bigcup_{j\in P(z)} C_j$.

Definition 2.1. We say the vector $\bar{x} \in X$ is an *absolute local minimum vertex* (almv) if $\bar{x}$ is a vertex of X and for any neighboring vertex y we have $f(\bar{x}) \leq f(y)$.

In what follows we will assume that (i) X is contained in the nonnegative orthant R^n_+, (ii) $\bar{x} = 0$ is a vertex of X, (iii) $\bar{x} = 0$ has n neighboring vertices, each of which is proportional to a unit vector in R^n (i.e., the neighboring vertices are on the n coordinate axes), and (iv) $\bar{x} = 0$ is an almv.

Assumptions (i) and (ii) are innocuous. In particular, let D be an mxk matrix (k > m) of full rank and consider the feasible region $Y = \{y \in R^k|Dy = d, y \geq 0\}$ which we assume to be bounded. Let $\bar{y}$ be any nondegenerate basic solution. Then, by translation to the space of the nonbasic variables, the feasible region Y is equivalent to a feasible region X with properties (i) and (ii). In particular, let B be the basis matrix associated with $\bar{y}$ and let N denote the matrix composed of the remaining columns of D. Let $b = B^{-1}d$, $A = B^{-1}N$, and let the n-vector x correspond to the nonbasic variables (n = k - m). Then $X = \{x \in R^n|Ax \leq b, x \geq 0\}$ has properties (i) and (ii) and is equivalent to the feasible region Y [5].

Assumptions (iii) and (iv) imply that we can find a nondegenerate vertex of Y which is also an almv. Often it is the case that we can easily find an almv. For example, assume F is a concave differentiable function on R^k and the optimization problem is $\min\{F(y)|y \in Y\}$. An almv can be found by utilizing the Frank-Wolfe algorithm [6]. In particular, let y^ν be a vertex of Y and solve $\min\{\nabla F(y^\nu)y|y \in Y\}$. If y^ν is an optimizer of this linear program, then find all neighboring vertices of y^ν. If $F(y^\nu) \leq F(z)$ for all neighbors z then y^ν is an almv. If there is a neighbor z such that $F(z) < F(y^\nu)$, set $y^{\nu+1} = z$, $\nu = \nu + 1$, and return to the linear program. On the other hand, if y^ν is not an optimizer, let $y^{\nu+1}$ be one and set $\nu = \nu + 1$ and return to the linear program. We therefore see that degeneracy of Y can be a computational problem. In the first place there may be no nondegenerate almv and, in the second place, degeneracy of a y^ν implies a computational burden for finding all of its neighbors.

We will have more to say about degeneracy later in the paper.

In what follows we will denote the convex hull of a set $A \subset R^n$ by convA. Also, we will utilize the following definition.

Definition 2.2. The vector z is said to be the α *extension* of the vector y if (i) $f(y) \geq \alpha$, and (ii) $z = \theta y$ where $\theta = \max\{\gamma \geq 1 | f(\gamma y) \geq \alpha\}$.

As mentioned above, we assume X satisfies assumptions (i)-(iv). Also, we assume all required α extensions exist. We can now state the algorithm.

Algorithm

Step 0: Set $q = 1$, $\bar{x}^q = 0$, an almv, $\alpha_q = f(\bar{x}^q)$. Let $y_1{}^{q,1},\ldots,y_1{}^{q,n}$ be, respectively, the α_q extensions of the n neighboring vertices of $\bar{x}^q$. Set $P_q = \{1\}$. Go to Step 1.

Step 1: For each $k \in P_q$ let $C_{q,k} = \text{cone}\{y_k^{q,1},\ldots,y_k^{q,n}\}$ and let $D = [y_k^{q,1},\ldots,y_k^{q,n}]$. For each $k \in P_q$ solve the linear program $\max\{1D_{q,k}^{-1}x | x \in X \cap C_{q,k}\}$ to obtain optimal vectors $\bar{x}^{q,k}$, $k \in P_q$. Let $0_q = \{j \in P_q | 1D_{q,j}^{-1}\bar{x}^{q,j} > 1\}$. If $0_q = \phi$, stop; $\bar{x}^q$ is optimal. If $0_q \neq \phi$, go to Step 2.

Step 2: Set $\alpha_{q+1} = \min\{\alpha_q; f(\bar{x}^{q,j}), j \in 0_q\}$ and let $\bar{x}^{q+1} = \arg\alpha_{q+1}$. For each $j \in 0_q$, define $P(\bar{x}^{q,j}) = \{\nu | (D_{q,j}^{-1}\bar{x}^{q,j})_\nu > 0\}$ and, for each $\nu \in P(\bar{x}^{q,j})$, define the cones $C_{q,j,\nu} = \text{cone}\{y_j^{q,1},\ldots,y_j^{q,\nu-1}, \bar{x}^{q,j}, y_j^{q,\nu+1},\ldots,y_j^{q,n}\}$. Now, α_{q+1} extend these vectors (if necessary) to define the vectors

$$y_j^{q+1,1},\ldots,y_j^{q+1,\nu-1},x^{q,j},y_j^{q+1,\nu+1},\ldots,y_j^{q+1,n},$$

respectively, and therefore

$$C_{q,j,\nu} = C_{q+1,j,\nu} = \text{cone}\{y_j^{q+1}, \ldots,x^{q,j},\ldots,y_j^{q+1,n}\}.$$

For each index (j,ν) define an index k and set

$$(y_k^{q+1}, \ldots,y_k^{q+1,\nu-1},y_k^{q+1,\nu},\ldots,y_k^{q+1,n}) = (y_j^{q+1,1},\ldots,y_j^{q+1,\nu-1},x^{q,j},\ldots,y_j^{q+1,n}).$$

Therefore, $C_{q+1,k} = \text{cone}\{y_k^{q+1,1},\ldots,y_k^{q+1,n}\}$. Set $P_{q+1} = \{k\}$, $q = q + 1$, and return to Step 1.

3. DISCUSSION

Step 0 initiates the algorithm by setting $\alpha_1 = f(0)$ as an initial upper bound for $\min\{f(x) | x \in X\}$. Also, each of the n neighboring vertices (which are on the coordinate axes because of the nondegeneracy assumption) is α_1 extended (recall, by assumption $\bar{x}^1$ is at least as good as all of its neighbors).

We first enter Step 1 with the task of solving one linear program

$\max\{1D_{1,1}^{-1}x | x \in X \cap C_{11}\}$. But $C_{11} = R_+^n$ and, therefore, this linear program decides whether or not there is a vector in X which is on the opposite side, from the origin $\bar{x}^1 = 0$, of the hyperplane $\{x | 1D_{1,1}^{-1}x - 1 = 0\}$. That is, if $1D_{1,1}^{-1}\bar{x}^{1,1} \le 1$ then X is contained in $C_{11} \cap \{x | 1D_{1,1}^{-1}x \le 1\} = \text{conv}\{0, y_1^{1,1}, \ldots, y_1^{1,n}\} = S_1$. But, by construction of $y_1^{1,j}$, $j = 1, \ldots, n$, and quasi-concavity of f, we have that $\bar{x}^1 = 0$ solves $\min\{f(x) | x \in S_1\}$ and therefore $X \subset S_1$, implies that $\bar{x}^1 = 0$ also solves (P). On the other hand, if $1D_{1,1}^{-1}\bar{x}^{1,1} > 1$ then there are points of X on the other side, from the origin, of the hyperplane determined by $y_1^{1,1}, \ldots, y_1^{1,n}$ and therefore $X \not\subset S_1$. This means that we cannot conclude that $\bar{x}^1$ solves (P) even though it may. At this point we enter Step 2.

At Step 2 we set $\alpha_2 = \min\{\alpha_1, f(\bar{x}^{1,1})\}$ and we α_2 extend, if necessary, the vectors $y_1^{1,1}, y_1^{1,2}, \ldots, y_1^{1,n}, \bar{x}^{1,1}$ to get what we have called $y_1^{2,1}, \ldots, y_1^{2,n}, x^{1,1}$. We form $P(\bar{x}^{1,1})$, the set of indices of positive weights by which $\bar{x}^{1,1}$ can be expressed in terms of the linearly independent vectors $y_1^{1,1}, \ldots, y_1^{1,n}$. For convenience, assume $P(\bar{x}^{1,1}) = \{1, 2, \ldots, \ell\}$. We then form the new cones, for $\nu \in P(\bar{x}^{1,1})$,

$$C_{1,1,\nu} = \text{cone}\{y_1^{2,1}, \ldots, y_1^{2,\nu-1}, x^{1,1}, y_1^{2,\nu+1}, \ldots, y_1^{2,n}\}.$$

Of course, by Proposition 2.1, we have that $\bigcup_{\nu \in P(\bar{x}^{1,1})} C_{1,1,\nu} = C_{1,1} = R_+^n$. Notationally, we redefine each of these cones to be $C_{2,k} = \text{cone}\{y_k^{2,1}, \ldots, y_k^{2,n}\}$, $k \in P_2 = \{1, \ldots, \ell\}$ and therefore $\bigcup_{k \in P_2} C_{2,k} = C_{1,1}$. We now return to Step 2.

As we re-enter Step 2, we have ℓ linear programs $\max\{1D_{2,k}^{-1}x | x \in X \cap C_{2,k}\}$, $k \in P_2$, to solve in order to obtain the vectors $\bar{x}^{2,k}$, $k \in P_2$. If $1D_{2,k}^{-1}\bar{x}^{2,k} \le 1$ for all $k \in P_2$ (i.e., $O_2 = \phi$) then $\bar{x}^2 = \arg \alpha_2$ is optimal for (P). To see this we proceed as follows. Let S_1^e be the α_2 extension of $S_1 = \text{conv}\{0, y_1^{1,1}, \ldots, y_1^{1,n}\}$. That is, at this iteration, $S_1^e = \text{conv}\{0, y_1^{2,1}, \ldots, y_1^{2,n}\}$. Now, let $S_2 = \text{conv}\{S_1^e \cup \{x^{1,1}\}\}$. Then we see that $S_2 \supset S_1$ and, by construction and quasi-concavity of f, $\bar{x}^2$ solves $\min\{f(x) | x \in S_2\}$. Therefore, we show that $O_2 = \phi$ implies that $X \subset S_2$ and then $\bar{x}^2$ must solve (P). Let $\bar{x} \in X$; then there exists $k \in P_2$ so that $\bar{x} \in C_{2,k}$ and therefore $1D_{2,k}^{-1}\bar{x} \le 1$. This implies that $\bar{x} \in C_{2,k} \cap \{x | 1D_{2,k}^{-1} \le 1\} = \text{conv}\{0, y_k^{2,1}, \ldots, y_k^{2,n}\} \subset S_2$. On the other hand, if $O_2 \ne \phi$, then *each* $C_{2,j}$, $j \in O_2$, is split into a number of cones and that number is equal to the number of indices in the set $P(\bar{x}^{2,j})$. If we let $|A|$ denote the cardinality of the set A, the number of linear programs to be solved, at iteration $q = 3$, will then be $\sum_{j \in O_2} |P(\bar{x}^{2,j})|$. The process continues in the above fashion.

With the above as motivation, we are now in a position to justify the termination condition of Step 2. Let S_q^e denote the α_{q+1} extension of S_q; that is, α_{q+1} extend each of the nonzero vectors for which S_q is the convex hull. Define $S_{q+1} = \text{conv}\{S_q^e \cup \{x^{q,j}, j \in O_q\}\}$ for $q \ge 1$ and $S_1 = \text{conv}\{0, y_1^{1,1}, \ldots, y_1^{1,n}\}$. Then $S_{q+1} \supset S_q$ and, by construction and quasi-concavity of f, $\bar{x}^q = \arg \alpha_q$ solves $\min\{f(x) | x \in S_q\}$. We show that $O_q = \phi$ implies $X \subset S_q$ and, therefore, $\bar{x}^q$ is optimal for (P).

Proposition 3.1. If $O_q = \phi$ then $X \subset S_q$.

Proof: Let $\bar{x} \in X$ and assume there is a $k \in P_q$ so that $\bar{x} \in C_{q,k}$. Then $\bar{x} \in C_{q,k} \cap \{x \mid 1D_{q,k}^{-1}x \le 1\} = \text{conv}\{0, y_k^{q,1}, \ldots, y_k^{q,n}\} \subset S_q$. On the other hand, suppose $\bar{x} \notin \bigcup_{k \in P_q} C_{q,k}$. Let ν denote the latest iteration (before q) for which $\bar{x} \in \bigcup_{k \in P_\nu} C_{\nu,k}$. Then, for some $j \in P_\nu$, $\bar{x} \in C_{\nu,j}$ and, by the definition of ν, $1D_{\nu,j}^{-1}\bar{x}^{\nu,j} \le 1$ and hence $1D_{\nu,j}^{-1}\bar{x} \le 1$. Then, $\bar{x} \in C_{\nu,j} \cap \{x \mid 1D_{\nu,j}^{-1}x \le 1\} = \text{conv}\{0, y_j^{\nu,1}, \ldots, y_j^{\nu,n}\} \subset S_\nu \subset S_q$.

The algorithm we have described essentially follows that of Tui with the exception, of course, that we have incorporated Zwart's observation that the constraints which define the cone to be searched need to be explicitly incorporated at Step 1.

The algorithm is slightly different from that of Zwart in that we seek an almv only to initiate the algorithm, while Zwart seeks such points often. Therefore, the probable degeneracy problem which arises in seeking such points (i.e., neighboring vertices do not necessarily correspond to neighboring bases) is not an integral part of our procedure.

4. CONVERGENCE

Let $||\cdot||$ denote a norm in R^n and let T be a compact subset of R^n. Let

$$\mathscr{A}(T) = \{A \mid A \subset T,\ A \neq \phi \text{ and closed}\}.$$

Hausdorff [1] has defined a metric for $\mathscr{A}(T)$ as follows. Let $A,B \in \mathscr{A}(T)$ and define

$$\rho(A,B) = \max_{x \in A} \min_{y \in B} ||x-y|| \ ;$$

the Hausdorff metric is

$$d(A,B) = \max\{\rho(A,B),\ \rho(B,A)\}$$

and hence $(\mathscr{A}(T),d)$ is a metric space. Note that $A \subset B$ if, and only if, $\rho(A,B) = 0$ and, therefore, $d(A,B) = \rho(B,A)$.

The following proposition will help to clarify what needs to be shown in order to establish convergence.

Proposition 4.1.

(i) Let $F : R^n \to R^1$ be continuous and define $g : \mathscr{A}(T) \to R^1$ by $g(A) = \min\{F(x) \mid x \in A\}$. Then g is continuous.

(ii) Let (A_n) be a sequence of elements of $\mathscr{A}(T)$ so that $A_n \subset A_{n+1}$ all n. Then

$A_n \to \overline{\bigcup_r A_r} = A_\infty$, where $\bar{A}$ denotes the closure of the set A.

Proof: (i) Let $A_0 \in \mathscr{A}(T)$ and let (A_n) be a sequence in $\mathscr{A}(T)$ so that $A_n \to A_0$. Then, for each n, there is an $x^n \in A_n$ so that $g(A_n) = F(x^n)$. Since the sequence (x^n) is in a compact set we have, on a subsequence if necessary, $x^n \to x_0 \in T$. Now $\rho(A_n, A_0)$ must converge to zero and

$$\rho(A_n,A_\theta) = \max_{x\in A_n} \min_{y\in A_\theta} ||x-y|| \geq \min_{y\in A_\theta} ||x^n-y||.$$

Let $y^n \in A_\theta$ be such that $||x^n-y^n|| = \min_{y\in A_\theta} ||x^n-y||$; then $||x^n-y^n|| \to 0$ and therefore $y^n \to x^\theta$. But the sequence (y^n) is contained in $A_\theta \in \mathscr{A}(T)$ and hence $x^\theta \in A_\theta$. Also, $\rho(A_\theta,A_n)$ must converge to zero; let $\bar{x} \in A_\theta$ and then

$$\rho(A_\theta,A_n) = \max_{x\in A_\theta} \min_{y\in A_n} ||x-y|| \geq \min_{y\in A_n} ||\bar{x}-y||.$$

Let $y^n \in A_n$ be such that $||\bar{x}-y^n|| = \min_{y\in A_n} ||\bar{x}-y||$. Then $y^n \to \bar{x}$. But $y^n \in A_n$ and therefore $F(y^n) \geq F(x^n)$; continuity of F then implies $F(\bar{x}) \geq F(x^\theta)$ and therefore $g(A_\theta) = F(x^\theta)$ and this establishes the continuity of g.

(ii) This part follows directly from a result of Hausdorff [1] which states that $\rho(L,A_n) \to 0$ where L denotes the "lower closed limit" of the sequence A_n (e.g., $x \in L$ if every neighborhood of x intersects all but a finite number of the A_n). It is immediate, in this case, that $L = A_\infty$.

In order to utilize Proposition 4.1 we compactify as follows. Let $\bar{x}^{1,1}$ solve $\max\{1D_{1,1}^{-1}x | x \in X\}$ and let the number K be chosen so that $K > 1D_{1,1}^{-1}\bar{x}^{1,1}$. Define

$$T = \{x \in R_+^n \mid 1D_{1,1}^{-1}x \leq K\}.$$

It is immediate that X is contained in the simplex T. Also, we modify Definition 2.2 slightly. The vector z is said to be the (α,T) extension of the vector $y \in T$ if (i) $f(y) \geq \alpha$, and (ii) $z = \theta y$ where $\theta = \max\{\alpha \geq 1 | f(\gamma y) \geq \alpha,\ \gamma y \in T\}$. It is then to be understood that all extensions at Step 2 are (α,T) extensions.

The situation is now as follows. For each q, $\bar{x}^q \in X \cap S_q$ solves $\min\{f(x) | x \in S_q\}$ and $S_q \subset S_{q+1}$. If $O_q \neq \phi$ for all q, let x^* be any limit point of the sequence $(\bar{x}^q)$. By Proposition 4.1 we have that $S_q \to \overline{\bigcup_r S_r} = S_\infty$, $x^* \in X \cap S_\infty$, and x^* solves $\min\{f(x) | x \in S_\infty\}$. Therefore, if we can show that $X \subset S_\infty$ it will immediately follow that x^* solves (P).

In what follows, all limits are subsequential, if necessary, and from now on we assume that $||\cdot||$ denotes the Euclidean norm on R^n. Assume

$$0 < \rho(X,S_q) = \max_{x\in X} \min_{y\in S_q} ||x-y||$$

for all iterations q. Then, since

$$\phi_q(x) = \min_{y\in S_q} ||x-y||$$

is convex in x, there must exist a vertex of X which repeats infinitely often (which we abbreviate i.o.). That is

$$\rho(X,S_q) \overset{i.o.}{=} \min_{y\in S_q} ||\bar{x}-y||$$

Now, $\rho(X,S_q) > 0$ implies $\bar{x} \notin S_q$ and therefore, by Proposition 2.1, there exists a sequence of cones C_{q,k_q}, each derived from the previous cone via Step 2, so that $\bar{x} \in C_{q,k_q}$, all q. For convenience, we will abbreviate the sequence as $C_{q,k_q} = C_q$. Now, abbreviate also $CB_{q,k_q} = CB_q$ and we have

$$0 < \rho(X,S_q) \overset{i.o.}{=} \min_{y\in S_q}||\bar{x}-y|| \le \min_{y\in CB_q}||\bar{x}-y||,$$

where the latter inequality holds because $S_q \supset CB_q$. Also observe that $\rho(X,S_\infty) \le \rho(S,S_{q+1}) \le \rho(X,S_q)$ for all q.

Definition 4.1. Let A_q be a sequence of sets in R^n. Let $\overline{\lim}_{q\to\infty} A_q$ be the set of points x such that every neighborhood of x intersects infinitely many of the sets A_q. That is, for all $\varepsilon > 0$, $N_\varepsilon(x) \cap A_q \ne \phi$ i.o.

It is easy to check that

$$\overline{\lim_{q\to\infty}} A_q = \bigcap_{q=1}^{\infty} \overline{\bigcup_{m=q}^{\infty} A_m}$$

Now let $\{A_q\}$ be a sequence of nonempty compact subsets of the compact set $T \subset R^n$, and let f be a continuous function on T. Let $x^q \in A_q$ solve $\min_{x\in A_q} f(x)$. Then, we must have

$$\underline{\lim}_{q\to\infty} f(x^q) \le \min_{x\in\overline{\lim}_{q\to\infty} A_q} f(x)$$

Now this latter fact gets used as follows. We have

$$\rho(X,S_q) \le \min_{y\in CB_q}||\bar{x}-y||$$

and $\bar{x} \in \bigcap_q C_q \Rightarrow \bar{x} \in \overline{\lim}_{q\to\infty} C_q$. Therefore, if $\bar{x} \in \overline{\lim}_{q\to\infty} CB_q$, we would have

$$\lim_{q\to\infty} \rho(X,S_q) \le \underline{\lim}_{q\to\infty} \min_{y\in CB_q}||\bar{x}-y|| \le \min_{y\in\overline{\lim}_{q\to\infty} CB_q}||\bar{x}-y|| = 0$$

This then would imply that $\rho(X,S_\infty) = 0$ which, in turn, implies $X \subset S_\infty$.

Compactness of T can be used to establish the following result. Recall, $\bar{x} \in C_q$ all q. It can be shown by an application of the Schwarz inequality that

$$\frac{1D_q^{-1}\bar{x} - 1}{||1D_q^{-1}||} \to 0$$

subsequentially, if necessary. That is, the Euclidean distance of the point $\bar{x}$ from the planes

$$P_q = \{x \mid 1D_q^{-1}x - 1 = 0\},$$

which bound the cones C_q (to form CB_q), converges to zero. However, this is not enough to establish convergence since the point $y^q \in P_q$, to which $\bar{x}$ is closest, may not be in C_q. However, if it turned out that infinitely many of these y^q are in the corresponding C_q (and hence in $P_q \cap CB_q$), then we would have that $\bar{x} \in \overline{\lim}_{q\to\infty} CB_q$ (since $y^q \to \bar{x}$) and the proof would be complete. However, there appears to be no way that one can be assured of this.

Another way to proceed is as follows. Observe that $CB_{q+1} \supset CB_\nu \cap C_{q+1}$ for all $\nu \le q$. Let $R(\bar{x})$ denote the ray through $\bar{x}$. Also, let $\lambda_q = (1D_q^{-1}\bar{x})^{-1}$. Note, $\lambda_q\bar{x} \in P_q \cap CB_q$ and therefore $\lambda_q < 1$. Using $CB_{q+1} \supset CB_\nu \cap C_{q+1}$ it can be shown that λ_q is nondecreasing. Of course, if $\lambda_q \to 1$, then, again, $\bar{x} \in \overline{\lim}_{q\to\infty} CB_q$ and the proof would be complete. However, in the case $\lambda_q \to \lambda^* < 1$, then we can use the fact

$$CB_{q+1} \supset CB_\nu \cap C_{q+1}\ , \quad \nu \le q,$$

to establish that inifinitely many of the y^q are in CB_q to again establish that $\bar{x} \in \overline{\lim_{q\to\infty}} CB_q$.

The above sketch serves merely to indicate that $S_\infty \supset X$ does hold, which in turn validates Zwart's remark that $\varepsilon = 0$ presented no problem for his computations.

REFERENCES

[1] Hausdorff, F., *Set Theory,* 2nd edition, Chealsea Publishing Co., New York, 1962.

[2] Tui, H., "Concave Programmming Under Linear Constraints," *Soviet Mathematics,* July-December, 1964.

[3] Zwart, P., "Nonlinear Programming: Counterexamples to Global Optimization Algorithms by Ritter and Tui," *Operations Research,* Vol. 21, 1973.

[4] Zwart, P., "Global Maximization of a Convex Function with Linear Inequality Constraints," *Operations Research,* Vol. 22, May-June, 1974.

[5] Dantzig, G.B., *Linear Programming and Extensions,* Princeton University Press, 1963.

[6] Frank, M. and P. Wolfe, "An Algorithm for Quadratic Programming," *Naval Research Logistics Quarterly,* vol. 3, 1956.

Acknowledgement: This work partially supported by NSF Grant ENG 76-12250.

A CLASS OF ALGORITHMS FOR THE DETERMINATION OF A SOLUTION OF A SYSTEM OF NONLINEAR EQUATIONS

B.D. Kiekebusch-Müller

Institut für Angewandte Mathematik
und Statistik der Universität Würzburg

Am Hubland, D-8700 Würzburg
Federal Republic of Germany

Consider the problem of finding a solution x^* of the in general non-linear system of equations $g(x)=0$, where $g:\mathbb{R}^n \to \mathbb{R}^k$ $(n \geq k)$ is a continuously differentiable function with Jacobian G. Assume furthermore that G has the following properties:

(1) There exists $p > 0$ such that
$\|G(x)^T w\|^2 \geq p \|w\|^2$ for all $x \in \mathbb{R}^n$, $w \in \mathbb{R}^k$,

(2) G is Lipschitz-continuous, i.e. there exists $L > 0$ such that
$\|G(x) - G(\bar{x})\| \leq L \|x - \bar{x}\|$ for all $x, \bar{x} \in \mathbb{R}^n$.

($\|.\|$ denotes the Euclidean vector norm and the corresponding lub-norm for matrices, respectively).

Problems of this type arise for example at the determination of saddle points or minimization of functions $f:\mathbb{R}^n \to \mathbb{R}$, if these problems are reduced to finding a zero of the gradient g of f. But in general g need not be a gradient. Also we do not require conditions such as the positive-definiteness of the Jacobian.

Nevertheless it is of course always possible to reformulate the problem in terms of minimization: Find $\min \Phi(x)$, where $\Phi:\mathbb{R}^n \to \mathbb{R}$ is any function with the properties:

(3) $\Phi(x) \geq 0$ for all x
$\Phi(x) = 0 \iff g(x) = 0$.

In case of n=k Deuflhard [5] proposed to use an iterative procedure, in which at each iterate a so-called "level-function" $T(x,\tilde{A})$ is minimized. These level-functions, which are of type (3), have the form

(4) $T(x,A) := \frac{1}{2} \|\tilde{A}\, g(x)\|^2$,

where $\tilde{A}$ is an appropriate non-singular matrix. Deuflhard proposed to use for $\tilde{A}$ the inverse Jacobian, if it exists, because this leads to the Newton-direction, if the method of steepest descent is applied to the corresponding level-function:

(5) $D_xT(x,\tilde{A})^T = G(x)^T\tilde{A}^T\tilde{A}g(x)$

$$= G(x)^{-1}g(x) \qquad \text{if } \tilde{A} = \tilde{A}(x) := G(x)^{-1}$$

(D_xT denotes the gradient of T with respect to x).

Now, using the iterative procedure $x_{i+1}:=x_i-\lambda_i G(x_i)^{-1}g(x_i)$, Deuflhard's method consists in determing a step-length λ_i such that the value of the actual level-function $T(x, \tilde{A}(x_i))$ decreases:

$$T(x_{i+1},\tilde{A}(x_i)) < T(x_i,\tilde{A}(x_i)) .$$

As a consequence, at each iterate a different level function is minimized. Moreover, in real-life problems it is usually too complicated to compute the inverse Jacobian at each iterate. Therefore approximation techniques have been introduced. These techniques have turned out to be quite successful in practice (see Deuflhard [5]), but their convergence has only been proved for the special case of Newton's method, $\tilde{A}=\tilde{A}(x)=G(x)^{-1}$. These facts were our motivation to consider iterative algorithms having variable level-functions, i.e. at each iterate x_i a different level function is minimized. We try to establish conditions on the sequence of level functions and on the search directions selected at each step such that the resulting algorithm converges.

For convenience, let us assume the sequence of level functions to be given in the form

(6) $\Phi_i(x) := \frac{1}{2}\, g(x)^T A_i\, g(x)$

where the sequence of matrices $\{A_i\}$ satisfies two conditions:

i) The matrices A_i are uniformly positive definite:

(7) There exists $q>0$ such that
$w^T A_i w \geq q \|w\|^2$ for all $w\in\mathbb{R}^k$ and for all i.

ii) They are uniformly bounded from above:

(8) There exists $Q\geq q>0$ such that
$\|A_i\|\leq Q$ for all i.

Note that Deuflhards level functions are covered by this approach, setting: $\Phi_i(x) = T(x,\sqrt{A_i})$

Now considering the iterative procedure

(9) $x_{i+1} = x_i - \sigma_i s_i\,, \qquad i=0,1,2,\ldots$

we have to find at each iterate x_i a search direction s_i and a steplength σ_i, which are related to the actual level function Φ_i. For that purpose we define a set of feasible search directions by

(10) $D(m,x) := \{s \in \mathbb{R}^n \,/\, \|s\|=1 \text{ and } D\Phi(x)s \geq m\sqrt{\Phi(x)}\}, \quad m>0$.

Here the index i is omitted for simplicity. Then -s, $s \in D(m,x)$, is some sort of approximation to the direction of steepest descent with respect to the level function Φ.

This way of defining sets of search directions is not the usual one - usually we have the condition: $D\Phi(x)s \geq m\|D\Phi(x)\|$, where $0<m\leq 1$. But beside some technical advantages, it has also interesting geometric properties. Some of its properties are listed in (11) to (15):

(11) a) $D(m,x) \neq \emptyset \quad \Longleftrightarrow \quad 0 \leq m \leq \dfrac{\|D\Phi(x)\|}{\sqrt{\Phi(x)}}$

b) $m = \dfrac{\|D\Phi(x)\|}{\sqrt{\Phi(x)}} \quad \Rightarrow \quad D(m,x) = \left\{\dfrac{D\Phi(x)^T}{\|D\Phi(x)\|}\right\}$

(12) For all $x \in \mathbb{R}^n$ there exists $m>0$ such that $D(m,x) \neq \emptyset$, for example $m = \sqrt{2pq}$, because if $g(x) \neq 0$ then

$$\frac{\|D\Phi(x)\|}{\sqrt{\Phi(x)}} \geq \sqrt{2pq} .$$

(13) $D(m,x)$ is a compact, connected set.

(14) $m \leq m' \quad \Rightarrow \quad D(m',x) \subseteq D(m,x)$.

(15) Let α denote the maximal angle between two directions of $D(m,x)$. Then

$$\alpha = 2 \arccos \left(m \frac{\sqrt{\Phi(x)}}{\|D\Phi(x)\|}\right)$$

The first four properties are more of technical interest. (15) has a nice geometrical consequence. It shows that the definition of D(m,x) contains some sort of scaling of the set of search directions. To demonstrate this let us assume that N is a level set of Φ, i.e. $\Phi(x) = c$ for all $x \in N$, where c is some constant. Suppose furthermore m to be small enough that D(m,x) is non-empty for all $x \in N$. Then the maximal angle α between two directions of D(m,x) is a monotone increasing function of the length of the gradient of Φ on N. And this has geometrically the consequence that the directions of D(m,x) approximate the gradient direction the better, the flatter the function Φ is at the point x. From the geometrical aspect such a scaling seems to be very natural (see figure 1).

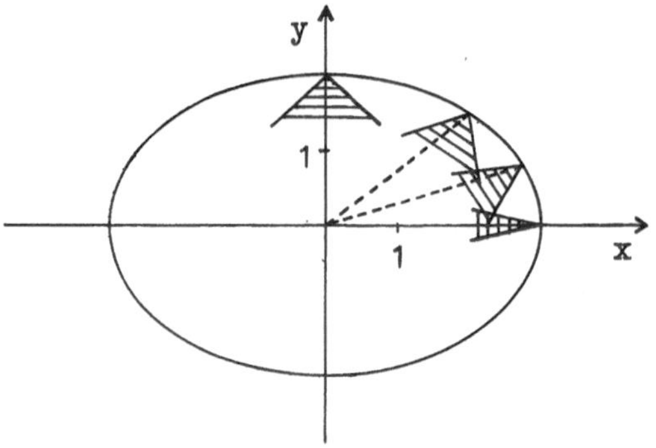

Figure 1: The sets D(m,x) at different points for the function $\Phi(x):=4x^2+9y^2$ at $N=\{(x,y)/\Phi(x)=36\}$ and for m=3.9 .

Now using these search directions we can get a step-length estimate of the following kind:

Lemma: Let $g:\mathbb{R}^n \to \mathbb{R}^k$ $(n \geq k)$ be a continuously differentiable function satisfying (1) and (2) and let A be a positive definite matrix satisfying (7) and (8). Furthermore let $m \geq \underline{m} > 0$ and assume that $g(\bar{x}) \neq 0$, $\|g(\bar{x})\| \leq \rho$ and $\|G(\bar{x})\| \leq P$.
Then there exists a number $\Lambda = \Lambda(P,q,Q,L,\underline{m},\rho) > 0$ such that

(16) $$\Phi(x-\sigma s) \leq \Phi(x) - \frac{1}{4}\sigma m \sqrt{\Phi(\bar{x})}$$

for all $\|x-\bar{x}\| \leq \Lambda\|g(\bar{x})\|$, $0 \leq \sigma \leq \Lambda\|g(\bar{x})\|$, $s \in D(m,x)$.

The assumptions about the constants $\underline{m}$, ρ and P seem to be trivial. They are formulated in this way, because the lemma will be used for iterative procedures, where $\bar{x}$ is replaced by the iterate x_i and m by m_i, whereas $\underline{m}$, ρ, P remain independent from i.

The situation of this lemma is shown in figure 2.

Inequalities of type (16) will be important during the following considerations. For given $m_i > 0$ and $s_i \in D_i(m_i,x_i)$ let us denote by $\mu_i(x_i)$ the largest value of the parameter σ such that (16) is satisfied for all $0 \leq \sigma \leq \mu_i(x_i)$:

(17) $$\mu_i(x_i) := \max\left\{\mu>0 \;\Big/\; \begin{array}{r}\Phi_i(x_i-\sigma s_i) \leq \Phi_i(x_i) - \frac{1}{4}\sigma m_i\sqrt{\Phi_i(x_i)} \\ \text{for all } 0 \leq \sigma \leq \mu\end{array}\right\}$$

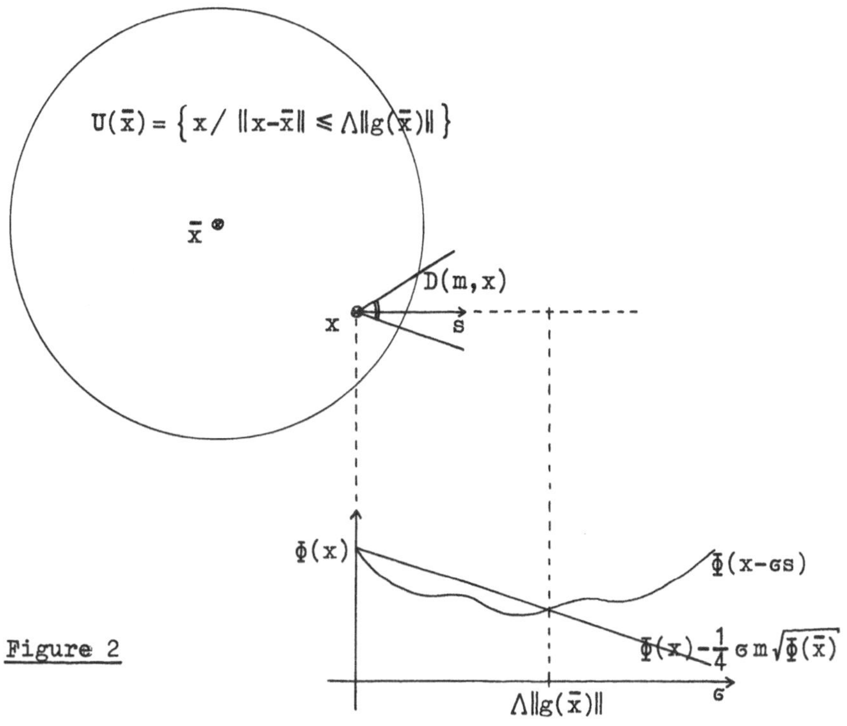

Figure 2

Now, $\mu_i(x_i)$ can be estimated from below and from above:

(18) $\quad \Lambda\|g(x_i)\| \leq \mu_i(x_i) \leq \frac{4}{m_i}\sqrt{\Phi_i(x_i)} \leq \frac{4}{\underline{m}}\sqrt{\Phi_i(x_i)}$

(The estimate from above holds because $\Phi_i(x) \geq 0$ for all x).

After these preliminaries we can state the class of algorithms considered:

Algorithm:

a) Let be given x_0 such that $g(x_0) \neq 0$, $\gamma, \tau \in (0,1]$, $0 < q \leq Q$ and let A_0 be a symmetric matrix having the property $q\|w\|^2 \leq w^T A_0 w \leq Q\|w\|^2$ for all $w \in \mathbb{R}^k$.

b) For i=0,1,2,... do

(i) Take $m_i \geq \underline{m} := \gamma\sqrt{2pq}$ such that $D_i(m_i, x_i) \neq \emptyset$.

(ii) Choose $s_i \in D_i(m_i, x_i)$.

(iii) Take $\sigma_i \in [\tau\mu_i(x_i), \mu_i(x_i)]$.

(iv) $x_{i+1} := x_i - \sigma_i s_i$.

(v) If $\Phi_i(x_{i+1})=0$ then STOP (x_{i+1} is a solution of $g(x)=0$).

(vi) Choose a symmetric matrix A_{i+1} such that

$$\alpha)\ q\|w\|^2 \leq w^T A_{i+1} w \leq Q\|w\|^2 \quad \text{for all } w\in\mathbb{R}^k ,$$

$$\beta)\ w^T(A_{i+1}-A_i)w \leq r_i \|w\|^2 \quad \text{for all } w\in\mathbb{R}^k ,$$

$$\text{where} \quad r_i := \frac{q\, m_i\, \sigma_i}{8\sqrt{\Phi_i(x_i)}} .$$

Steps (i) - (v) look the same as in usual minimization algorithms. Of course the step-length condition (iii) is of somewhat different kind. In step (vi) the variable level functions are introduced. The conditions are such that the sequence $\{\Phi_i(x_i)\}$ is strictly monotone decreasing, which is used for the proof of the following convergence theorem.

Theorem: Let $g:\mathbb{R}^n \to \mathbb{R}^k$ $(n \geq k)$ be a continuously differentiable function satisfying (1) and (2). Furthermore assume that

(i) $g(x_o) \neq 0$, $x_o\in\mathbb{R}^n$,

(ii) with respect to the constant $q>0$ and the matrix A_o as given in the algorithm the following holds:
The connected component of the set

$$\{x\in\mathbb{R}^n \,/\, \|g(x)\|^2 \leq \tfrac{2}{q}\,\Phi_o(x_o)\} ,$$

that contains x_o, is bounded.

Then the algorithm is defined for all i and the sequence $\{x_i\}$ generated has the following property:
Either there exists an index i_o such that $g(x_{i_o})=0$ or the sequence $\{x_i\}$ has at least one cluster point and all cluster points of the sequence $\{x_i\}$ are solutions of the system of equalities $g(x)=0$.

Remark: Assume that all level sets $\{x\in\mathbb{R}^n \,/\, \|g(x)\| \leq R\}$, $R\geq 0$ of the functional $\|g\|$ are compact. Then assumption (ii) of the theorem is automatically satisfied for all points $x_o\in\mathbb{R}^n$.

The idea of the proof is the following: At first it is shown that for all i holds

$$(19) \qquad \Phi_{i+1}(x_{i+1}) \leq \Phi_i(x_i) - \tfrac{1}{8}\,\sigma m_i \sqrt{\Phi_i(x_i)} \qquad \text{for all } 0 \leq \sigma \leq \sigma_i .$$

For this the conditions on the sequence of matrices $\{A_i\}$ respectively the sequence of level functions $\{\Phi_i\}$ are needed. It follows that the sequence $\{\Phi_i(x_i)\}$ is monotone decreasing. Hence there exists a limit $\bar{\Phi} \geq 0$. From (19) and because of the assumptions it follows that the iterates x_i are contained in a compact set and therefore there exists at least one cluster point. Now, assume that one cluster point $\bar{x}$ is not a solution of $g(x)=0$. Then $\bar{\Phi} > 0$ and the step-length condition in connection with the lemma gives the contradiction $\Phi_i(x_i) \rightarrow -\infty$.

One difficulty of the algorithm still consists in the step-length condition (iii), which firstly does not give an explicite formula for the step-length and secondly even requires that inequality (16) holds for all $0 \leq \sigma \leq \sigma_i$ (see (17)). In the case that the functional $\|g\|$ has compact level sets, it is possible to replace step (iii) of the algorithm by the more practicable rule:

(iii*) Choose σ_i such that

a) $\sigma_i \geq \tau \mu_i(x_i)$,

b) $\Phi_i(x_i - \sigma_i s_i) \leq \Phi_i(x_i) - \frac{1}{4}\sigma_i m_i \sqrt{\Phi_i(x_i)}$.

In this case it is not necessary to require that
$\Phi_i(x_i - \sigma s_i) \leq \Phi_i(x_i) - \frac{1}{4}\sigma m_i \sqrt{\Phi_i(x_i)}$ for all $0 \leq \sigma \leq \sigma_i$.
This is possible because condition (iii) is needed only to ensure that the sequence $\{x_i\}$ remains in a compact set, which holds under the above assumption anyway.

(iii*) can be realized by a simple procedure of Armijo-type:

(20) a) Set $\tilde{\sigma}_i := \frac{4}{m_i} \sqrt{\Phi_i(x_i)}$.

b) Take the least non-negative integer j such that

$$\Phi_i(x_i - 2^{-j}\tilde{\sigma}_i s_i) \leq \Phi_i(x_i) - 2^{-j}\tilde{\sigma}_i \frac{m_i}{4} \sqrt{\Phi_i(x_i)}$$
$$= \Phi_i(x_i)\ (1 - 2^{-j})$$

c) Set $\sigma_i := 2^{-j}\tilde{\sigma}_i$.

Here $\frac{4}{m_i} \sqrt{\Phi_i(x_i)}$ is the upper bound for σ_i , which is known from (18).

There are two main reasons to consider algorithms as described above. Firstly there are a lot of different methods, which fit in this scheme: For Newton's method, which is given by $A_i := (G(x_i)^{-1})^2$, we could reestablish the results of Deuflhard [5]. Now, in practice this method usually is replaced by Quasi-Newton methods. This technique can be justified by the following new result:

> Let the level functions, i.e. the matrices A_i, be chosen according to the updating rules of Quasi-Newton methods. For some of these methods, including those observed by Broyden, Dennis, Moré [1], it could be shown that they are members of our class of algorithms and therefore have the convergence property of our theorem.

Moreover there are also other methods contained in our class of algorithms such as for example the method of Dem'yanov [2] for finding saddle points or the method of Levenberg-Marquardt [4] for nonlinear least squares. From this point of view minimizing variable level functions seems to be a quite important principle behind many well known methods.

Another interesting aspect lies in the step-length estimate. If the above defined set of search directions (see (10)) is used, then estimate (18) gives:

(21) There exist positive numbers Λ_1 and Λ_2 such that

$$\Lambda_1 \|g(x_i)\| \leqslant \sigma_i \leqslant \Lambda_2 \|g(x_i)\| \quad \text{for all } i,$$

whenever σ_i is a feasible step-length for the algorithm.

Now, for many methods including the above mentioned it is possible to get enough information about these constants Λ_1 and Λ_2, such that at least after a finite number of "first iterations" a step-length of the form $\lambda \|g(x_i)\|$ with fixed λ can be used. And this is a chance to avoid the crucial line search at each iterate.

References:

1. C.G.BROYDEN, J.E.DENNIS Jr., J.J.MORÉ, On the Local and Superlinear Convergence of Quasi-Newton Methods; J.Inst.Maths Applics 12 (1973) 223-245

2. V.F.DEM'YANOV, K Razyskaniju Sedlovych Toček (Russian); Vestnik Leningradskogo Universiteta 22 (1967) no.19, 25-34

3. V.F.DEM'YANOV, A.B.PEVNYI, Numerical Methods for Finding Saddle Points; USSR Comp.Math.math.Phys. 12,5 (1972) 11-52

4. J.E.DENNIS Jr., Some Computational Techniques for the Nonlinear Least Squares Problem; published in: Numerical Solution of Systems of Nonlinear Algebraic Equations (edts. G.D.Byrne, C.A.Hall), Academic Press, New York- London (1973) 157-183

5. P.DEUFLHARD, A Modified Newton Method for the Solution of Ill-conditioned Systems of Nonlinear Equations with Application to Multiple Shooting; Num.Math. 22 (1974) 289-315

6. B.D.KIEKEBUSCH-MÜLLER, Eine Klasse von Verfahren zur Bestimmung von stationären Punkten, insbesondere Sattelpunkten; Würzburg (1976) (Thesis)

7. J.M.ORTEGA, W.C.RHEINBOLDT, Iterative Solutions of Nonlinear Equations; Academic Press, New York- London (1970)

STOCHASTIC LINEAR PROGRAMS WITH RANDOM DATA HAVING STABLE DISTRIBUTIONS

Kurt Marti

Institut für Operations Research
der Universität Zürich
Weinbergstrasse 59, CH-8006 Zürich

1. Introduction. In a linear program

$$\min c'x \text{ s.t. } Ax=b,\ x\in D \ (=\text{convex polyhedron}) \tag{1}$$

describing a concrete situation from economics or engineering some or all of the elements c_k, a_{ik}, b_i, $i=1,\dots,m$, $k=1,\dots,n$ of the n-vector c, the (m,n)-matrix A and the m-vector b may be random variables, hence $A=A(\omega)$, $b=b(\omega)$, $c=c(\omega)$, where ω is an element of a probability space (Ω,A,P). Selecting a convex function $q: R^m\to R$ such that the number $q(A(\omega)x-b(\omega))$ measures the costs for violating the equality constraint $A(\omega)x=b(\omega)$ related to the decision situation that the random variable ω is revealed only after the selection of $x\in D$, the original problem (1) is converted in many cases, see e.g. [2],[4],[9] into the convex minimization problem

$$\text{minimize } F(x) \text{ s.t. } x\in D, \tag{2}$$

where the objective F of (2) is given by the expression

$$F(x)=E(c(\omega)'x+q(A(\omega)x-b(\omega))= \bar{c}'x + Eq(A(\omega)x-b(\omega)),$$

where E denotes the expectation operator and $\bar{c}:=Ec(\omega)$ is assumed to be finite. Obviously by (2) we want to minimize the expected total costs arising from the choice of a $x\in D$ and the subsequent realization of the random variable ω.

In this paper we consider now the case that all elements $\xi=\xi_{ij}$ of the random (m,n+1)-matrix $\Xi=(A(\omega),b(\omega))$ have a stable distribution P_ξ, i.e. P_ξ has a log characteristic function given by

$$\log\phi(t) = i\bar{\xi}t - \gamma|t|^\alpha(1+\beta\frac{t}{|t|}tg(\tfrac{1}{2}\pi\alpha)),\ t\in R, \tag{3}$$

where $\alpha,\beta,\gamma,\bar{\xi}$ are the parameters of P_ξ satisfying the relations $0<\alpha\leq 2$, $-1\leq\beta\leq 1$, $\gamma\geq 0$ and $\bar{\xi}\in R$. The parameter α is called the characteristic exponent of P_ξ, if $\beta=0,\alpha=1$ resp. $\alpha=2$, then P_ξ is the Cauchy resp. normal distribution; The parameter β is related to the skewness of the distribution P_ξ, symmetric distributions have $\beta=0$; The parameter γ is a scale factor and refers to the dispersion of P_ξ, if P_ξ is nondegenerate, then $\gamma>0$; Finally $\bar{\xi}$ is a location parameter. It is known [1] that a) $E|\xi|^\tau<+\infty$ for all $0\leq\tau<\alpha$ and $\bar{\xi}=E\xi(\omega)$ if $1<\alpha\leq 2$, b) P_ξ is infinitely divisible, c) the subclass of nonsymmetric stable laws with agreeing parameters α, β is closed with respect to linear combinations $a_1\xi_1+a_2\xi_2$, $a_i\geq 0, i=1,2,$

d) the subclass of symmetric stable laws with agreeing characteristic exponents α is closed with respect to arbitrary linear combinations $a_1\xi_1+a_2\xi_2$, $a_i\in R, i=1,2$, e) the stable laws arise as limit distributions in central limit theorems: A distribution possesses a domain of attraction if and only if it is stable, f) $E|\xi|^{\tau}=+\infty$ for $\tau\geq\alpha, \alpha<2$, hence also cases with infinite variances become tractable by selecting some of the following α-values within the open interval (0,2), g) stable laws are absolutely continuous with respect to the Lebesgue measure, their density functions f may be differentiated any number of times and if $1<\alpha\leq 2$, then f is strictly positive, see [3].

In the following we assume $1<\alpha\leq 2$ for all characteristic exponents.

The distribution of Ξ. Up to now we described the distribution of an individual element ξ_{ik}, in the following the joint stable distribution of $\Xi=(\xi_{ik})$, $1\leq i\leq m, 1\leq k\leq n+1$ will be defined: For each row index i=1, 2,...,m let $\{s_{ij}: j\in J_i\}$ be a partition of the index set {1,2,...,n+1} labelling the columns of $\Xi=(A(\omega),b(\omega))$ and assume that the $|s_{ij}|$-subvectors $\eta_{ij}:=(\xi_{ik})_{k\in s_{ij}}, j\in J_i$ of each row i=1,2,...,m are given by

$$\eta_{ij}:=(\xi_{ik})_{k\in s_{ij}} = (\bar{\xi}_{ik})_{k\in s_{ij}} + \Gamma_{ij}(\xi^0_{ik})_{k\in s_{ij}},\ j\in J_i, 1\leq i\leq m \tag{4}$$

where $(\bar{\xi}_{ik})_{k\in s_{ij}}$ is the mean of η_{ij}, the $(|s_{ij}|,|s_{ij}|)$-matrix Γ_{ij} is related to the dispersion of η_{ij} and $\{\xi^0_{ik}: 1\leq i\leq m, 1\leq k\leq n+1\}$ are mutually independent random variables having stable distributions P_{ik} with parameters $\alpha_{ik}=\alpha_{ij}\in(1,2]$, $\beta_{ik}=\beta_{ij}\in[-1,+1]$ for each $k\in s_{ij}$, $j\in J_i, 1\leq i\leq m$ and $\gamma_{ik}=1$, $\xi^0_{ik}=0$, $1\leq i\leq m, 1\leq k\leq n+1$.

Note. a) The original idea, see [5], leading finally to this definition was the collection of stable random variables $\xi_{ik}, 1\leq k\leq n+1$ with agreeing characteristic exponents to the same class (s_{ij}). b) Sometimes it is reasonable to assume from the beginning that $A(\omega), b(\omega)$ are stochastically independent; Introducing then the convex function Q on R^m

$$Q(z) = Eq(z-b(\omega)),\ z\in R^m,$$

where $-\infty<Q(z)<+\infty, z\in R^m$ - together with a strict convexity condition, which we need later on - is the only condition restricting the distribution of $b(\cdot)$, the objective function F satisfies the equation

$$F(x) = \bar{c}'x + EQ(A(\omega)x),\ x\in R^n \tag{5}$$

In this case of independence between $A(\omega), b(\omega)$ the above definition concerns therefore only $\Xi=A(\omega)$, i.e. $\{s_{ij}: j\in J_i\}$ is then a partition of {1,2,...,n} only.

2. The basic theorem. In the following we set $x_{n+1}\equiv -1$, except that F is given by (5), and for each $x\in R^n$ the (n+1)-vector $(x_1,\dots,x_n,-1)'$

is again denoted by x, furthermore we use the notations $\eta_{ij}'x = \sum_{k\in s_{ij}}\xi_{ik}x_k$, and $\Gamma_{ij}'x=\Gamma_{ij}'(x_k)_{k\in s_{ij}}$, $j\in J_i, 1\leq i\leq m$, hence we have that $\Xi_i x=\sum_{j\in J_i}\eta_{ij}'x$, where Ξ_i is the ith row of Ξ. Because of (4) we find

$$\eta_{ij}'x = \bar{\eta}_{ij}'x + (\Gamma_{ij}'x)'\eta_{ij}^o,$$

where $\bar{\eta}_{ij}=(\bar{\xi}_{ik})_{k\in s_{ij}}$ and $\eta_{ij}^o=(\xi_{ik}^o)_{k\in s_{ij}}$, hence from the above assumptions and the properties of stable laws stated in §1 it follows that each $\eta_{ij}'x, j\in J_i, 1\leq i\leq m$ has a stable distribution with the parameters $\alpha_{ij},\beta_{ij},\gamma_{ij}=\|\Gamma_{ij}'x\|_{\alpha_{ij}}^{\alpha_{ij}}$ and $\bar{\eta}_{ij}'x$, where $\|\Gamma_{ij}'x\|_{\alpha_{ij}}$ is the α_{ij}-norm

$$\|\Gamma_{ij}'x\|_{\alpha_{ij}} := \left(\sum_{k\in s_{ij}}|y_k|^{\alpha_{ij}}\right)^{1/\alpha_{ij}} \text{ with } y=\Gamma_{ij}'x,$$

provided that $\Gamma_{ij}'x\geq 0$ for each pair (i,j) such that $\beta_{ij}\neq 0$. From the independence of ξ_{ik}^o, $1\leq i\leq m, 1\leq k\leq n+1$(n, if (5) holds) follows that $\eta_{ij}'x$, $j\in J_i, 1\leq i\leq m$ as also the components $\Xi_1 x,\dots,\Xi_m x$ of

$$\Xi x=A(\omega)x-b(\omega) \text{ (resp. } A(\omega)x \text{ only if F is given by (5))}$$

are mutually independent random variables. Furthermore F(x) is given by

$$F(x)=\bar{c}'x+Eu(\Xi_1 x,\dots,\Xi_m x)=\bar{c}'x+Eu\left(\sum_{j\in J_1}\eta_{1j}'x,\sum_{j\in J_2}\eta_{2j}'x,\dots,\sum_{j\in J_m}\eta_{mj}'x\right),$$

where u=q resp. u=Q if (5) holds. In order to formulate manageable conditions concerning $x,y\in R^n$ which imply $F(y)\leq(<)F(x)$ (again is $y_{n+1}\equiv -1$, except (5) is true) we need some more definitions. Given a particular $(r,s), s\in J_r, 1\leq r\leq m$ and an arbitrary partition K_1,K_2 of $\{(i,j): j\in J_i, 1\leq i\leq m, (i,j)\neq(r,s)\}$ we define for given $x,y\in R^n$ the convex function $F_{K_{1,2}}^{r,s}$ by

$$F_{K_{1,2}}^{r,s}(t)=\bar{c}'x+Eu\Big(\sum_{\substack{j\in J_1\\(1,j)\in K_1}}\eta_{1j}'x+\sum_{\substack{j\in J_1\\(1,j)\in K_2}}\eta_{1j}'y,\dots,\sum_{\substack{j\in J_r\\(r,j)\in K_1}}\eta_{rj}'x+t+$$

$$+\sum_{\substack{j\in J_r\\(r,j)\in K_2}}\eta_{rj}'y,\dots,\sum_{\substack{j\in J_m\\(m,j)\in K_1}}\eta_{mj}'x+\sum_{\substack{j\in J_m\\(m,j)\in K_2}}\eta_{mj}'y\Big),\ t\in R.$$

Now we can state the following theorem which is basic for our present considerations; Let $\bar{A}=EA(\omega)$. For simplification let $\beta_{ij}=0, j\in J_i, 1\leq i\leq m$.

<u>Theorem 2.1.</u> a) If the n-vectors x,y are related such that

$$\bar{c}'y\leq\bar{c}'x,\ \bar{A}y=\bar{A}x,\ \|\Gamma_{ij}'y\|_{\alpha_{ij}}\leq\|\Gamma_{ij}'x\|_{\alpha_{ij}},\ j\in J_i,\ i=1,2,\dots,m, \tag{6}$$

then $F(y)\leq F(x)$; b) If (6) holds with $\bar{c}'y<\bar{c}'x$ and $F(y)<+\infty$, then $F(y) < F(x)$; c) If in addition to (6) there exists at least one pair (r,s), $s\in J_r, 1\leq r\leq m$ such that $\|\Gamma_{rs}'y\|_{\alpha_{rs}}<\|\Gamma_{rs}'x\|_{\alpha_{rs}}$, then $F(y)<F(x)$, provided that $F(y)<+\infty$ and u=q resp. Q is strictly convex or there is a partition K_1,K_2 of $\{(i,j): j\in J_i, 1\leq i\leq m, (i,j)\neq(r,s)\}$ such that $F_{K_{1,2}}^{r,s}$ is strictly convex.

Proof: See [8].

Provided that F, $F_{K_{1,2}}^{r,s}$ are real-valued convex functions and the conditions of strict convexity concerning q,Q or $F_{K_{1,2}}^{r,s}$ are fulfilled, then the assumptions in Theorem 2.1 are reduced to a simple comparison of $p(x)$ and $p(y)$, where $p(x)=(\|\Gamma_{ij}'x\|_{\alpha_{ij}}, j\in J_i, 1\leq i\leq m, \bar{c}'x, \bar{A}x)$, $x\in R^n$ encloses the relevant parameters of the distribution of Ξx.

An important class of models, including stochastic programs with complete fixed recourse [9], for the substitution of a linear program (1) with random datas, which shows these integrability and strict convexity conditions consists in the problem (2) having a sublinear costfunction q, i.e. $q(\lambda z)=\lambda q(z)$, $q(z_1+z_2)\leq q(z_1)+q(z_2)$ for all $\lambda\geq 0$ and $z,z_1,z_2\in R^m$. Note that a sublinear q is not strictly convex.

Lemma 2.1. a) Let q be sublinear. Then F and each $F^{r,s}_{K_{1,2}}$ are real valued, convex functions; b) Let $u=q$ and the sublinear q be defined by $q(z)=\max_{1\leq k\leq\rho} f_k'z$, $z\in R^m$, where $\rho>2$ and $f_1,f_2,\ldots,f_\rho$ are given m-vectors such that $f_k\neq f_\kappa, k\neq\kappa$. If J_r contains more than one element and there is $(r,j)\in K_1$ or $(r,j)\in K_2$ such that $\|\Gamma_{rj}'x\|_{\alpha_{rj}}\neq 0$ resp. $\|\Gamma_{rj}'y\|_{\alpha_{rj}}\neq 0$, then $F^{r,s}_{K_{1,2}}$ is strictly convex.

Proof. a) Since $|q(z)|\leq\|q\|\cdot\|z\|$, $z\in R^m$, where $\|q\|=\sup\{|q(z)|: \|z\|=1\}$ is the norm of q and therefore $|Q(z)|\leq\|q\|(\|z\|+E\|b(\omega)\|)$ assuming that $E\|b(\omega)\|<+\infty$, for $u=q$ resp. Q we have that

$$|u(\Xi x)|\leq\|q\|(\sum_{i=1}^{m}\|e_i\|\sum_{j\in J_i}|\eta_{ij}'x| \text{ (resp. } +E\|b(\omega)\|)),\ x\in R^n$$

where e_i is the ith unit vector of R^m. Since the distribution of $\eta_{ij}'x$ is stable with characteristic exponent $\alpha_{ij}>1, j\in J_i, 1\leq i\leq m$, all absolute moments $E|\eta_{ij}'x|$ exist, hence F is finite for each $x\in R^n$; A slight modification of this argument shows that also each $F^{r,s}_{K_{1,2}}$ is a real-valued function. b) From the assumptions in part (b) follows first that $F^{r,s}_{K_{1,2}}(t)=\bar{c}'x + Eq(a(\omega)+te_r)$, where $a(\omega)$ is a random m-vector whose distribution P_a has a probability density which is positive everywhere. Secondly it holds $q(z)=f_k'z$, $z\in C_k$, where $C_k=\{z\in R^m: f_j'z\leq f_k'z, 1\leq j\leq\rho\}$, $k=1, 2,\ldots,\rho$ and $\bigcup_{k=1}^{\rho}C_k=R^m$. Assuming that $f(t):=Eq(a(\omega)+te_r)$ is not strictly convex, we find numbers $t_1\neq t_2$ and $0<\mu<1$ such that $f(\mu t_1+(1-\mu)t_2)=\mu f(t_1)+(1-\mu)f(t_2)$. Because of $q(\mu z_1(\omega)+(1-\mu)z_2(\omega))\leq\mu q(z_1(\omega))+(1-\mu)q(z_2(\omega))$, where $z_i(\omega)=a(\omega)+t_ie_r$, $i=1,2$ we get then that $q(\mu z_1(\omega)+(1-\mu)z_2(\omega)) = \mu q(z_1(\omega))+(1-\mu)q(z_2(\omega))$ with probability 1, hence $\bigcup_{k=1}^{\rho}\{\omega: z_1(\omega)\in C_k, z_2(\omega)\in C_k\}=\bigcup_{k=1}^{\rho}\{\omega: a(\omega)\in(-t_1e_r+C_k)\cap(-t_2e_r+C_k)\}$ must have probability one. But this contradicts to the properties of P_a stated above.

A first consequence of Theorem 2.1: The calculation of descent directions of F without using the gradient ∇F of F. If the n-vectors x,y are related according to the assumptions in Theorem 2.1.b or 2.1.c, then $h=y-x$ is a direction of decrease of F at x obtained without use of ∇F.

In [6],[7] we used this first consequence of Theorem 2.1 in order to construct algorithms of the feasible direction type.

3. Necessary conditions for a solution of (2) not involving the gradient ∇F of the objective function F of (2). According to our basic Theorem 2.1 we suppose that the following implication hold:

If $x,y\in R^n$ are related according to (6) and if "<" holds at least once in the $1+\sum_{i=1}^{m}|J_i|$ inequalities in (6), then $F(y)<F(x)$. (7)

In order to derive now necessary conditions for a solution x^0 to our optimization problem (2) we consider the following class of auxiliary convex programs related to (7) and depending on arbitrary but fixed numbers $\rho\geq 0,\sigma_{ij}\geq 0,j\in J_i,1\leq i\leq m$ such that $\rho+\sum_{j\in J_i,1\leq i\leq m}\sigma_{ij}>0$.

$$\begin{aligned}&\text{minimize } \rho\bar{c}'x+\sum_{(i,j)}\sigma_{ij}\|\Gamma_{ij}'x\|_{\alpha_{ij}} && (8)_p^{\rho,\sigma_{ij}}\\ &\text{s.t. } \|\Gamma_{ij}'x\|_{\alpha_{ij}}\leq c_{ij},\ j\in J_i,1\leq i\leq m,\\ &\qquad \bar{c}'x\leq\gamma,\\ &\qquad \bar{A}x = z,\\ &\qquad x\in D,\end{aligned}$$

where the "right hand side" $p=((c_{ij}),\gamma,z)$ is a given vector of parameters $c_{ij}\geq 0$ for all (i,j), $\gamma\in R$, $z\in R^m$; In the important special case $\rho=1$ and $\sigma_{ij}=0,j\in J_i,1\leq i\leq m$ we don't need the second inequality constraint, hence program (8) is then reduced to this auxiliary convex program:

$$\begin{aligned}&\text{minimize } \bar{c}'x && (9)_p\\ &\text{s.t.} \|\Gamma_{ij}'x\|_{\alpha_{ij}}\leq c_{ij},\ j\in J_i,i=1,2,\dots,m\\ &\qquad \bar{A}x = z\\ &\qquad x\in D,\end{aligned}$$

where in this case p is given by $p=((c_{ij}),z)$.

Because of the validity of the implication (7) - justified by Theorem 2.1 - this class of convex programs $(8)_p,(9)_p$ enables now the following necessary condition for a solution to the original program (2).

Theorem 3.1. Let x^0 be a solution to (2). Then x^0 solves also the convex programs $(8)_p$ resp. $(9)_p$, provided that $c_{ij}=\|\Gamma_{ij}'x^0\|_{\alpha_{ij}}$, $j\in J_i$, $1\leq i\leq m$, $\gamma=\bar{c}'x^0$, $z=\bar{A}x^0$ resp. $c_{ij}=\|\Gamma_{ij}'x^0\|_{\alpha_{ij}},j\in J_i,1\leq i\leq m$, $z=\bar{A}x^0$.

Proof. Assuming that x^0 does not solve $(8)_p$ resp. $(9)_p$ with p given above, we find $\tilde{x}\in D$ such that $\tilde{x}$ satisfies (i) the constraints of $(8)_p$ resp. $(9)_p$ and (ii) $\rho\bar{c}'\tilde{x}+\sum_{(i,j)}\sigma_{ij}\|\Gamma_{ij}'\tilde{x}\|_{\alpha_{ij}}<\rho\bar{c}'x^0+\sum_{(i,j)}\sigma_{ij}\|\Gamma_{ij}'x^0\|_{\alpha_{ij}}$ resp. $\bar{c}'\tilde{x}<\bar{c}'x^0$. In the second case concerning $(9)_p$ we have then that $\bar{c}'\tilde{x}<\bar{c}'x^0$, $\bar{A}x=z=\bar{A}x^0$ and $\|\Gamma_{ij}'x\|_{\alpha_{ij}}\leq c_{ij}=\|\Gamma_{ij}'x^0\|_{\alpha_{ij}}$, $j\in J_i,1\leq i\leq m$, hence the left hand side of (7) is fulfilled and it follows that $F(\tilde{x})<F(x^0)$ indi-

cating that x^0 is not a solution of (2), which is a contradiction to the optimality of x^0. In the first case related to $(8)_p$ from the strict inequality above and the holding constraints $\|\Gamma_{ij}'\tilde{x}\|_{\alpha_{ij}} \leq c_{ij} = \|\Gamma_{ij}'x^0\|_{\alpha_{ij}}$ for all $j \in J_i, 1 \leq i \leq m$, $\bar{c}'\tilde{x} \leq \gamma = \bar{c}'x^0$, $\bar{A}\tilde{x} = z = \bar{A}x^0$ we obtain next to that $\bar{c}'\tilde{x} < \bar{c}'x^0$ (possible if $\rho > 0$) or $\|\Gamma_{ij}'\tilde{x}\|_{\alpha_{ij}} < \|\Gamma_{ij}'x^0\|_{\alpha_{ij}}$ (possible if $\sigma_{ij} > 0$) for at least one pair $(i,j)=(r,s)$. Hence the left hand side of (7) is fulfilled again implying that $F(\tilde{x}) < F(x^0)$, which is again a contradiction to the optimality of x^0.

Note. Observe that the two classes of programs $(8)_p, (9)_p$ do <u>not</u> depend on the costfunction q, hence Theorem 3.1 is valid for each problem (2) such that the implication (7) holds.

An immediate consequence of Theorem 3.1 is this result.

<u>Corollary 3.1.</u> By solving $(8)_p$ resp. $(9)_p$ parametrically for all values of the parameter $p=((c_{ij}),\gamma,z)$ resp. $p=((c_{ij}),z)$, where $c_{ij} \geq 0$, $\gamma \in R, z \in R^m$ the solutions x^0 to (2) may be determined up to a certain parameter p.

Proof. By Theorem 3.1 any solution x^0 to (2) must solve a program $(8)_p$ resp. $(9)_p$ for a certain parameter value p^0 of p.

Also by Theorem 3.1 is suggested the next definition.

<u>Definition 3.1.</u> Let $\Phi=\Phi(x)$ denote the set of solutions to $(8)_p$ resp. $(9)_p$ for the parameter $p=p(x):=((\|\Gamma_{ij}'x\|_{\alpha_{ij}}),\bar{c}'x,\bar{A}x)$ resp. $p(x)=((\|\Gamma_{ij}'x\|_{\alpha_{ij}}),\bar{A}x)$. An element $x \in D$ such that $x \in \Phi(x)$ is called a <u>critical</u> decision related to $(8)_p$ resp. $(9)_p$.

Obviously Theorem 3.1 reads now as follows.

<u>Theorem 3.1'.</u> If x^0 solves (2), then x^0 must be a critical decision.

Note. The above version of Theorem 3.1 indicates that it is sufficient to limit consideration to only those decisions that are critical, i.e. <u>fixed points</u> of the set-valued map $x \to \Phi(x)$ on D.

Further necessary optimality conditions not involving ∇F may be obtained by means of the following definition.

<u>Definition 3.2.</u> A decision $x \in D$ is called <u>efficient</u> (or <u>minimal</u>) if there is no $y \in D$ such that x,y satisfies the left hand side of (7).

The relevance of this notion is seen from the next theorem.

<u>Theorem 3.2.</u> Each solution x^0 of (2) is also efficient.

Proof. Assuming that x^0 is not efficient we find $y \in D$ such that x^0,y satisfies the left hand side of (7). By the validity of this implication we obtain that $F(y)<F(x)$ contradicting to the optimality of x^0.

Efficient decisions may be determined as follows.

<u>Theorem 3.3.</u> a) Each solution $\tilde{x}$ of a convex program of the type

minimize $\rho\bar{c}'x+\sum_{j\in J_i,1\leq i\leq m}\sigma_{ij}\|\Gamma_{ij}'x\|_{\alpha_{ij}}$ subject to $x\in D$, (10)

where $\rho>0,\sigma_{ij}>0$ for all $j\in J_i,1\leq i\leq m$ is efficient; b) Each solution $\tilde{x}$ to $(8)_p^{\rho,\sigma_{ij}}$ is efficient if $\rho>0,\sigma_{ij}>0$ for all $j\in J_i,1\leq i\leq m$.

Proof. a) Assuming that $\tilde{x}$ is not efficient, there is $y\in D$ such that $\tilde{x},y$ satisfy the left hand side of the implication (7). Since $\rho>0,\sigma_{ij}>0$ for all (i,j) this implies that $\rho\bar{c}'y+\sum_{(i,j)}\sigma_{ij}\|\Gamma_{ij}'y\|_{\alpha_{ij}}<\rho\bar{c}'\tilde{x}+\sum_{(i,j)}\sigma_{ij}\cdot\|\Gamma_{ij}'\tilde{x}\|_{\alpha_{ij}}$ contradicting to the optimality of $\tilde{x}$. b) Assuming again that $\tilde{x}$ is not efficient we have $y\in D$ such that $\tilde{x},y$ satisfy the left hand side of (7). From this follows that $\bar{c}'y\leq\bar{c}'\tilde{x}\leq\gamma,\bar{A}y=\bar{A}\tilde{x}=z,\|\Gamma_{ij}'y\|_{\alpha_{ij}}\leq\|\Gamma_{ij}'\tilde{x}\|_{\alpha_{ij}}\leq c_{ij}$ for all (i,j) and $\rho\bar{c}'y+\sum_{(i,j)}\sigma_{ij}\|\Gamma_{ij}'y\|_{\alpha_{ij}}<\rho\bar{c}'\tilde{x}+\sum_{(i,j)}\sigma_{ij}\|\Gamma_{ij}'\tilde{x}\|_{\alpha_{ij}}$ which contradicts to the optimality of the vector $\tilde{x}$ since y is admissible in the program $(8)_p^{\rho,\sigma_{ij}}$.

Note to program (10). Since the constraint $\bar{A}y=z$ corresponds to the two opposite inequalities $\bar{A}y\leq z,-\bar{A}y\leq -z$ the equality constraint $\bar{A}y=z$ (or $\bar{A}y=\bar{A}x$) has no impact to the objective function of (10).

By solving $(8)_p^{\rho,\sigma_{ij}}$ resp. (10) parametrically for all $\rho>0,\sigma_{ij}>0$, $p=((c_{ij}),\gamma,z)$ resp. for all $\rho>0,\sigma_{ij}>0$ we may calculate efficient decisions. Comparing "critical" and "efficient" decisions we find this next result.

<u>Theorem 3.4.</u> a) If $\tilde{x}$ is a critical decision related to $(8)_{p_{\tilde{x}}}^{\rho,\sigma_{ij}}$ where $\rho>0,\sigma_{ij}>0$ for all (i,j), then $\tilde{x}$ is also efficient; b) If $\tilde{x}$ is efficient, then $\tilde{x}$ is also critical related to both programs $(8)_p^{\rho,\sigma_{ij}}$, $(9)_p$, where $\rho\geq0,\sigma_{ij}\geq0,\rho+\sum_{(i,j)}\sigma_{ij}>0$.

Proof. a) According to our presumptions $\tilde{x}$ solves $(8)_p^{\rho,\sigma_{ij}}$ for $p=p(\tilde{x})$. Since $\rho>0,\sigma_{ij}>0$ for all (i,j) the assertion follows now from Theorem 3.3.b. b) Assuming that $\tilde{x}$ is not critical, i.e. $\tilde{x}\notin\Phi(\tilde{x})$ there is $y\in D$ such that $\|\Gamma_{ij}'y\|_{\alpha_{ij}}\leq\|\Gamma_{ij}'\tilde{x}\|_{\alpha_{ij}}$, $j\in J_i,1\leq i\leq m$, $\bar{c}'y\leq\bar{c}'\tilde{x}$, $\bar{A}y=\bar{A}\tilde{x}$ and $\rho\bar{c}'y+\sum_{(i,j)}\sigma_{ij}\|\Gamma_{ij}'y\|_{\alpha_{ij}}<\rho\bar{c}'\tilde{x}+\sum_{(i,j)}\sigma_{ij}\|\Gamma_{ij}'\tilde{x}\|_{\alpha_{ij}}$ resp. $\bar{c}'y<\bar{c}'\tilde{x}$ corresponding to program $(8)_{p(\tilde{x})}^{\rho,\sigma_{ij}}$ resp. $(9)_{p(\tilde{x})}$. Therefore the vectors $\tilde{x},y$ must satisfy the left hand side of (7) and $\tilde{x}$ is then not efficient what contradicts to our assumptions in Theorem 3.4.b.

<u>4. Extension to arbitrary β_{ij}, $-1\leq\beta_{ij}\leq+1$, $j\in J_i,1\leq i\leq m$.</u> Up to now we discussed the symmetric case $\beta_{ij}=0,j\in J_i,1\leq i\leq m$. In order to generalize our theory to the more general case $-1\leq\beta_{ij}\leq+1,j\in J_i,1\leq i\leq m$ we remember first that if ξ has a stable distribution with parameters $\alpha,\beta\neq0,\gamma,\bar{\xi}$, then for each nonnegative $u\geq0$ the random variable $u\xi$ is stable distributed with parameters $\alpha,\beta,u^{\alpha},u\bar{\xi}$, hence if $\xi\rightsquigarrow u\xi,u\geq0$, then the skewness parameter β remains unchanged.

Since (see §2) $\eta_{ij}'x=\bar{\eta}_{ij}'x + (\Gamma_{ij}'x)'\eta_{ij}^0$, where the elements $\xi_{ik}^0, k\in s_{ij}$ of ξ_{ij}^0 have stable distributions with parameters $\alpha_{ij},\beta_{ij}\in[-1,1]$, $\gamma_{ij}=1$, $\bar{\xi}_{ik}=0$ we obtain that $\eta_{ij}'x$ has a stable distribution with the parameters $\alpha_{ij},\beta_{ij},\|\Gamma_{ij}'x\|_{\alpha_{ij}}^{\alpha_{ij}},\bar{\eta}_{ij}'x$ provided that the coordinates of $\Gamma_{ij}'x$ are nonnegative, i.e. $\Gamma_{ij}'x\geq 0$. Therefore we have this generalization of Theorem 2.1.

Theorem 4.1. Let $-1\leq\beta_{ij}\leq 1, j\in J_i, 1\leq i\leq m$. The statements in Theorem 2.1 remains valid if to (6) there are added the constraints $\Gamma_{ij}'y\geq 0,\Gamma_{ij}'x\geq 0$ for each pair (i,j) such that $\beta_{ij}\neq 0$.

Denote by W the set of all pairs (i,j) such that $\beta_{ij}\neq 0$. The next modification for $W\neq\emptyset$ concerns the implication (7) and the related programs (8),(9),(10): A) According to Theorem 4.1 to the left hand side of (7) we have just to add the two conditions $\Gamma_{ij}'y\geq 0,\Gamma_{ij}'x\geq 0$ for each $(i,j)\in W$; B) To the constraints of the programs (8),(9),(10) we have to add the constraints $\Gamma_{ij}'x\geq 0$ for each $(i,j)\in W$; C) Finally the concept of an "efficient" decision must be modified as follows.

Definition 3.2'. A decision $\tilde{x}$ is called efficient if $\Gamma_{ij}'\tilde{x}\geq 0,(i,j)\in W$ and if there is no $y\in D$ such that $\tilde{x},y$ satisfy the left hand side of the (modified) implication (7).

Theorem 4.2. Let $-1\leq\beta_{ij}\leq +1, j\in J_i, 1\leq i\leq m$ and consider here only solutions x^0 to (2) with $\Gamma_{ij}'x^0\geq 0,(i,j)\in W$. Then all Theorems and Corollaries in §3 remains valid if we follow the above modifications.

5. The determination of components of solutions to the original problem (2) by solving auxiliary programs of the family $(8)_p^{\rho,\sigma ij},(9)_p$.

We have already mentioned the basic fact for this consideration here that by solving one of the convex programs $(8)_p^{\rho,\sigma ij},(9)_p$ parametrically for all values of the parameter p, the solutions x^0 to (2) (such that $\Gamma_{ij}'x^0\geq 0,(i,j)\in W$) may be determined up to a parameter $\tilde{p}$; Hence (i) some components or "parts" of x^0 are then known and (ii) the still missing vector $\tilde{p}$ may be determined in a second optimization-step, where the corresponding second-stage-optimization-problem shows often a reduced complexity compared with the given problem (2). In order to discuss this in more detail we have to determine first the solutions to a program of the type $(8)_p^{\rho,\sigma ij}$. Since these programs are convex we establish for this purpose the corresponding local Kuhn-Tucker-conditions, where we suppose in this § that the set D representing the deterministic constraints of (1) is given by

$$D = \{x\in R^n: Hx\leq g\},$$

where H resp. g is a known (r,n)-matrix resp. r-vector. Furthermore we suppose in this § that some of the columns of $\Xi=(A(\omega),b(\omega))$ are non-stochastic. By Ξ_I resp. Ξ_{II} we denote then the deterministic resp. stochastic part of Ξ and in the same way we decompose also the decision x into two parts $x_I=(x_k)_{k\in K_I}$ resp. $x_{II}=(x_k)_{k\in K_{II}}$ refering to the deterministic resp. stochastic columns of Ξ. We use the notations $\Gamma_{ij}=(\gamma_{k\ell}^{ij})$, $k,\ell\in s_{ij}$, $\Gamma_{ij}'x=\Gamma_{ij}'x_{II}=\Gamma_{ij}'(x_k)_{k\in s_{ij}}$ and γ_ℓ^{ij} denotes the ℓth column of Γ_{ij}. The Lagrangian related to program $(8)_p^{\rho,\sigma_{ij}}$, $p=((c_{ij}),\gamma,z)$ is then given by the equation

$$L=L_p^{\rho,\sigma_{ij}}(x,(u_{ij}),v^1,v^2,w,(z_{ij})_{(i,j)\in W})=\rho\bar{c}'x+\sum_{(i,j)}\sigma_{ij}\|\Gamma_{ij}'x\|_{\alpha_{ij}}+$$
$$+\sum_{(i,j)}u_{ij}(\|\Gamma_{ij}'x\|_{\alpha_{ij}}-c_{ij})+v^{1\prime}(\bar{c}'x-\gamma)+v^{2\prime}(\bar{A}x-z)+w'(Hx-g)+$$
$$+\sum_{(i,j)\in W}z_{ij}'(-\Gamma_{ij}'x),$$

where $u_{ij}\geq 0, j\in J_i, 1\leq i\leq m, v^1\geq 0, v^2\in R^m, w\in R_+^r, z_{ij}\in R^{|s_{ij}|}, (i,j)\in W$. Note that according to the definition of $(9)_p$ in §3 the second and fourth term of L can be omitted in the case $\rho>0, \sigma_{ij}=0, j\in J_i, 1\leq i\leq m$. Since $\|\Gamma_{ij}'x\|_{\alpha_{ij}}$ depends only on x_{II}, the local Kuhn-Tucker conditions of $(8)_p^{\rho,\sigma_{ij}}$ are given now as follows.

$$0=\frac{\partial L}{\partial x_k}=\rho\bar{c}_k+v^1\bar{c}_k+(\bar{A}'v^2)_k+(H'w)_k,\ k\in K_I, \tag{11.1}$$

$$0=\frac{\partial L}{\partial x_k}=\rho\bar{c}_k+\sum_{\substack{(i,j)\\ k\in s_{ij}}}\sigma_{ij}\delta_{ij}+\sum_{\substack{(i,j)\\ k\in s_{ij}}}u_{ij}\delta_{ij}+v^1\bar{c}_k+(\bar{A}'v^2)_k+(H'w)_k+\sum_{(i,j)\in W}(-\Gamma_{ij})z_{ij},\ k\in K_{II} \tag{11.2}$$

$$0\geq\frac{\partial L}{\partial u_{ij}}=\|\Gamma_{ij}'x_{II}\|_{\alpha_{ij}}-c_{ij},\ 0=u_{ij}\frac{\partial L}{\partial u_{ij}}=u_{ij}(\|\Gamma_{ij}'x_{II}\|_{\alpha_{ij}}-c_{ij}),\ u_{ij}\geq 0,\ j\in J_i, 1\leq i\leq m \tag{11.3}$$

$$0\geq\frac{\partial L}{\partial v^1}=\bar{c}'x-\gamma,\ 0=v^1\frac{\partial L}{\partial v^1}=v^1(\bar{c}'x-\gamma),\ v^1\geq 0, \tag{11.4}$$

$$0=\frac{\partial L}{\partial v^2}=\bar{A}x-z, \tag{11.5}$$

$$0\geq\frac{\partial L}{\partial w}=Hx-g,\ 0=w'\frac{\partial L}{\partial w}=w'(Hx-g),\ w\geq 0 \tag{11.6}$$

$$0\geq\frac{\partial L}{\partial z_{ij}}=-\Gamma_{ij}'x_{II},\ 0=z_{ij}'\frac{\partial L}{\partial z_{ij}}=z_{ij}'(-\Gamma_{ij}'x_{II}),\ z_{ij}\geq 0,\ (i,j)\in W \tag{11.7}$$

where each δ_{ij} such that k is contained in s_{ij} is given by

$$\delta_{ij}=\frac{\partial}{\partial x_k}\|\Gamma_{ij}'x_{II}\|_{\alpha_{ij}}=\|\Gamma_{ij}'x_{II}\|_{\alpha_{ij}}^{1-\alpha_{ij}}\sum_{\ell\in s_{ij}}\gamma_{k\ell}^{ij}|\gamma_\ell^{ij\prime}x_{II}|^{\alpha_{ij}-1}\cdot\operatorname{sgn}(\gamma_\ell^{ij\prime}x_{II}) \tag{12}$$

for $\Gamma_{ij}'x_{II}\neq 0$: If $\Gamma_{ij}'x_{II}=0$, then δ_{ij} denotes the subgradient of $\|\cdot\|_{\alpha_{ij}}$ at $z=0$. Now we have the following criterion.

Lemma 5.1. Let $(\tilde{x},(\tilde{u}_{ij}),\tilde{v}^1,\tilde{v}^2,\tilde{w},(\tilde{z}_{ij})_{(i,j)\in W})$ be a variable/covariable system fulfilling the conditions (11), then $\tilde{x}$ is a solution to $(8)_p^{\rho,\sigma_{ij}}$. Conversely if $\tilde{x}$ is a solution to $(8)_p^{\rho,\sigma_{ij}}$ such that $c_{ij}>\|\Gamma_{ij}'x\|_{\alpha_{ij}}$, $j\in J_i, 1\leq i\leq m$, then there are covariables $(\tilde{u}_{ij}),\tilde{v}^1,\tilde{v}^2,\tilde{w},\tilde{z}_{ij}$, $(i,j)\in W$ such that $(\tilde{x},(\tilde{u}_{ij}),\tilde{v}^1,\tilde{v}^2,\tilde{w},(\tilde{z}_{ij})_{(i,j)\in W})$ satisfies (11).

For a given vector $\tilde{p}$ of "right hand sides" $\tilde{p}=((\tilde{c}_{ij}),\tilde{\gamma},\tilde{z})$, $\tilde{c}_{ij}\geq 0$, $\tilde{\gamma}\in R$, $\tilde{z}\in R^m$ let now $(\tilde{x},(\tilde{u}_{ij}),\tilde{v}^1,\tilde{v}^2,\tilde{w},(\tilde{z}_{ij})_{(i,j)\in W})$ be a variable/covariable-system which satisfies (11). Refering to the partition of x into x_I and x_{II} we consider now the vector-family $\tilde{x}[x_I,\tau], x_I \in R^{|K_I|}, \tau>0$, where

$\tilde{x}[x_I,\tau]_I = x_I$ and $\tilde{x}[x_I,\tau]_{II} = \tau\tilde{x}_{II}$,

while fixing the Lagrange multipliers $\tilde{u}_{ij}, \tilde{v}^1, \tilde{v}^2, \tilde{w}, \tilde{z}_{ij}$. According to (12) δ_{ij} is positively homogeneous of degree zero in x_{II}, furthermore $\Gamma_{ij}'x$ depends not on x_I. Therefore we have this lemma.

Lemma 5.2. The variable/covariable-system $\tilde{S}(x_I,\tau) = (\tilde{x}[x_I,\tau], (\tilde{u}_{ij}), \tilde{v}^1, \tilde{v}^2, \tilde{w}, (\tilde{z}_{ij})_{(i,j)\in W})$ fulfills conditions (11.1), (11.2) and (11.7) for all $x_I \in R^{|K_I|}$ and $\tau>0$.

This lemma has then the following consequence.

Corollary 5.1. The variable/covariable-system $\tilde{S}(x_I,\tau)$ satisfies (11) having "right hande side" $p=((c_{ij}),\gamma,z)$ if $\hat{x}_I \in R^{|K_I|}, \hat{\tau}>0$ solves the following system of relations:

$$\begin{aligned} &\text{i)}\ c_{ij} = \tau\tilde{c}_{ij}, j\in J_i, 1\leq i\leq m \\ &\text{ii)}\ \gamma \geq \bar{c}_I'x_I + \tau\bar{c}_{II}'\tilde{x}_{II},\ 0 = \tilde{v}^1(\bar{c}_I'x_I + \tau\bar{c}_{II}'\tilde{x}_{II}) \\ &\text{iii)}\ z = \bar{A}_I x_I + \tau\bar{A}_{II}\tilde{x}_{II} \\ &\text{iv)}\ g \geq H_I x_I + \tau H_{II}\tilde{x}_{II},\ 0 = w'(H_I x_I + \tau H_{II}\tilde{x}_{II} - g). \end{aligned} \qquad (13)_p$$

According to Lemma 5.1 $\tilde{x}[\hat{x}_I,\hat{\tau}]$ is then a solution to $(8)_p^{\rho,\sigma ij}$ with p mentioned in Corollary (5.1).

Assume now that x^0, $\Gamma_{ij}'x^0 \geq 0, (i,j)\in W$, solves (2). According to Theorem 3.1 and its generalization in Theorem 4.2 we know that x^0 solves also program $(8)_p^{\rho,\sigma ij}$ for $p=p(x^0)$ (see Definition 3.1).

Theorem 5.1. Let x^0 be determined as above and assume that $(13)_p$, $p=p(x^0)$ has a solution $\hat{x}_I, \hat{\tau}>0$. Then $x^0_{II} = \hat{\tau}\tilde{x}_{II}$ if the II-part of a solution to $(8)_{p(x^0)}^{\rho,\sigma ij}$ is uniquely determined.

Proof. By the above considerations x^0 and $\tilde{x}[\hat{x}_I,\hat{\tau}]$ are solutions to $(8)_p^{\rho,\sigma ij}, p=p(x^0)$, what yields the assertion.

Concerning the uniqueniss of the II-part of a solution to $(8)_p^{\rho,\sigma ij}$, $p=p(x^0)$, x^0 solves (2), we have still this result.

Lemma 5.3. Assume that $x_{II} \to \sum_{(i,j)} \sigma_{ij}\|\Gamma_{ij}'x_{II}\|_{\alpha_{ij}}$ is strictly convex. Then the II-part $\bar{x}_{II}$ of any admissible $\bar{x}\in D$ of $(8)_{p(x^0)}^{\rho,\sigma ij}$, where x^0 is a solution to (2) satisfying $\Gamma_{ij}'x^0 \geq 0, (i,j)\in W$ is uniquely determined.

Proof. Let $\bar{x},\bar{y}$ be admissible vectors of $(8)_p^{\rho,\sigma ij}, p=p(x^0)$ such that $\bar{x}_{II} \neq \bar{y}_{II}$. Consider then $\hat{x} = \frac{1}{2}\bar{x} + \frac{1}{2}\bar{y}$. We find $\hat{x}\in D, \|\Gamma_{ij}'\hat{x}\|_{\alpha_{ij}} \leq \|\Gamma_{ij}'x^0\|_{\alpha_{ij}}, j\in J_i, 1\leq i\leq m$, $\bar{c}'\hat{x} \leq \bar{c}'x^0$, $\bar{A}\hat{x} = \bar{A}x^0$, $\Gamma_{ij}'\hat{x} \geq 0, (i,j)\in W$ and $\sum_{(i,j)} \sigma_{ij}\|\Gamma_{ij}'\hat{x}_{II}\|_{\alpha_{ij}} < \sum_{(i,j)} \sigma_{ij}\|\Gamma_{ij}'x^0_{II}\|_{\alpha_{ij}}$. This implies $\|\Gamma_{rs}'\hat{x}_{II}\|_{\alpha_{rs}} < \|\Gamma_{rs}'x^0\|_{\alpha_{rs}}$ for at least one pair (r,s). Now implication (7) (in its modified form) yields $F(\hat{x}) < F(x^0)$ contradicting to the optimality of x^0.

Note. Obviously, under the assumptions of Theorem 5.1 the direction of the II-part x_{II}^{o} of a solution x^{o} to (2) such that $\Gamma_{ij}'x^{o} \geq 0$, $(i,j) \in W$ is determined already by $(8)_{\tilde{p}}^{\rho,\sigma ij}$; Hence the direction of x_{II}^{o} is in this case <u>independent on the costfunction q</u>. The detailed discussion of the assumptions in the important Theorem 5.1, allowing the optimization of (2) in two different steps:

A) Find $\tilde{x}_{II}$ by solving $(8)_{\tilde{p}}^{\rho,\sigma ij}$ with given $\tilde{p}$;

B) Find the (remaining) optimal values $\hat{x}_I, \hat{\tau}$ in $\tilde{x}[x_I, \tau]$

is given in a subsequent paper.

Extensions to the non-stable distributed case: The proof of our basic Theorem 2.1 (see [8]) is based on the concept of "stochastic dominance" between probability measures. Since this concept is defined for any two probability distributions, Theorems similar to Theorem 2.1 and its consequences may be obtained for various non-stable distributed random data $(A(\omega), b(\omega))$.

References

[1] Feller, W.: An Introduction to Probability Theory and its Applications, Volume II. New York-London-Sydney: Wiley 1966

[2] Kall, P.: Stochastic linear programming. Reihe Oekonometrie und Unternehmensforschung XXI. Berlin-Heidelberg-New York: Springer 1976

[3] Lévy, P.: Calcul des Probabilités. Paris: Gauthier-Villars 1925, Chapitre VI, 252-277

[4] Marti, K.: Approximationen der Entscheidungsprobleme mit linearer Ergebnisfunktion und positiv homogener, subadditiver Verlustfunktion. Z.Wahrscheinlichkeitstheorie verw.Geb. 31, 203-233 (1975)

[5] Marti, K., Riepl, R.-J.: Optimale Portefeuilles mit stabil verteilten Renditen. ZAMM 57, T337-T339 (1977)

[6] Marti, K.: Stochastische Dominanz und Stochastische Lineare Programme. Operations Research Verfahren 1977

[7] Marti, K.: On approximative solutions of stochastic programming problems by means of stochastic dominance and stochastic penalty methods. Proceedings IXth Math.Progr.Symp. Budapest 1976. Amsterdam: North Holland Publ.Comp. 1978

[8] Marti, K.: Approximationen stochastischer Optimierungsprobleme mit Sensitivitätsuntersuchungen bezüglich der Verteilung des zufälligen Parameters; Manuskript. Zürich: Institut für Operations Research der Universität Zürich 1977

[9] Walkup, D.W., Wets, R.: Stochastic programs with recourse. SIAM J. Appl. Math. 15, 1299-1314 (1967)

METHODS OF FEASIBLE DIRECTIONS WITH INCREASED GRADIENT MEMORY

Gerard G. L. Meyer
Electrical Engineering Department
The Johns Hopkins University
Baltimore, Maryland 21218

INTRODUCTION

The class of feasible directions methods is a powerful tool for solving constrained minimization problems, min-max problems, and unconstrained minimization problems in the absence of continuity of the gradient. The different versions proposed either involve all the gradients and do not require "antizigzagging precautions" or involve only the constraints in a neighborhood of the current feasible point and do require "antizigzagging precautions" [1-10].

This paper considers the class of feasible directions methods which do require "antizigzagging precautions". Such methods require that at each iteration the gradient of the cost function, the gradients of the active constraints, and the gradients of the "almost" active constraints be evaluated. The method we propose only requires that at each iteration, the gradient of the cost function and the gradient of the active constraints, be recomputed, and we use the previously computed gradients of the almost active constraints, provided that these gradients are not too "old". This results in a considerable saving of computational time when the constraint functions are complicated.

PRELIMINARIES

1. Problem: Given m+1 maps $f^0(.)$, $f^1(.)$,..., $f^m(.)$ from E^n into E, let T be the subset of E^n defined by $T = \{z \in E^n \mid f^j(z) \leq 0,$ $j = 1,\ldots, m\}$. Find a point z in T so that $f^0(z) \leq f^0(z')$ for all z' in T.

2. Hypothesis:
   - (i) The map $f^0(.)$ is continuously differentiable and strictly convex.
   - (ii) The maps $f^j(.)$, j = 1,2,..., m are continuously differentiable and convex.

(iii) The set T is non-empty and compact.

(iv) For every z in T, the set $\{\nabla f^j(z) \mid f^j(z) = 0,\ j = 1,2,\dots,\ m\}$ is linearly independent.

Necessary and sufficient conditions of optimality for Problem 1 are well known.

3. Lemma: A point z in T is a solution of Problem 1 if and only if

$$\min_{h \varepsilon S} \max_{j \varepsilon J(z,\infty)} \langle \nabla f^j(z),\ h \rangle = 0,$$

where S is a compact neighborhood of the origin in E^n and $J(z,\infty)$ is the index set defined by

$$J(z,\infty) = \{j \varepsilon \{1,2,\dots,\ m\} \mid f^j(z) = 0\} \cup \{0\}.$$

The notations used in this paper are standard with the following possible exceptions:

(i) Given a sequence $\{u_i\}$ and an index set K, $\{u_i\}_K$ denotes the subsequence of $\{u_i\}$ consisting of all points in $\{u_i\}$ with index i in K;

(ii) Given $\alpha>0$, we define the set $J(z,\alpha)$ to be the set containing the indices of the cost function, the active constraints and the almost active constraints, i.e.,

$$J(z,\alpha) = \{j \varepsilon \{1,2,\dots,\ m\} \mid f^j(z)+1/\alpha \geq 0\} \cup \{0\};$$

(iii) $[a;b]$ denotes the set $\{x \mid a \leq x \leq b\}$;

(iv) E is the real line, and E^n is the Euclidean space of dimension n .

A METHOD OF FEASIBLE DIRECTIONS WITH INCREASED GRADIENT MEMORY

The algorithm proposed in this paper to solve Problem 1 uses an initial point z_1 in T, a compact neighborhood S of the origin in E^n, two maps p(.) and q(.) from the positive integers into the reals, a real ρ , and an integer s.

4. Algorithm:

Step 0: Set $r_1 = 1$, set $c_1^j = +\infty$ for all $j = 0,1,\dots,\ m$, and set $i = 1$.

Step 1: Set $b_i^j = \nabla f^j(z_i)$ and $c_i^j = 0$ for all j in $J(z_i,\infty) \cup \{j \varepsilon J(z_i,p(r_i)) \mid c_i^j > s\}$

Step 2: Compute h_i in S so that for all h in S

$$\max_{j\varepsilon J(z_i,p(r_i))} \langle b_i^j,h_i\rangle \le \max_{j\varepsilon J(z_i,p(r_i))} \langle b_i^j,h\rangle.$$

<u>Step 3</u>: If $\max_{j\varepsilon J(z_i,p(r_i))} \langle b_i^j,h_i\rangle \le -1/q(r_i)$, go to Step 4; otherwise, set

$\lambda_i = 0$, $\mu_i = 0$, $z_{i+1} = z_i$, $r_{i+1} = r_i + 1$, $c_{i+1}^j = c_i^j$ and $b_{i+1}^j = b_i^j$ for all $j = 0,1,\ldots, m$, $i = i+1$, and go to Step 1.

<u>Step 4</u>: Compute $\lambda_i = \max \{\lambda\varepsilon[0;\rho] \mid f^j(z_i+\lambda h_i) \le 0,\ j = 1,2,\ldots, m\}$

<u>Step 5</u>: Compute μ_i in $[0;\lambda_i]$ so that for all μ in $[0;\lambda_i]$

$f^0(z_i+\mu_i h_i) \le f^0(z_i+\mu h_i)$.

<u>Step 6</u>: Set $z_{i+1} = z_i+\mu_i h_i$, $r_{i+1} = r_i$, $c_{i+1}^j = c_i^j + 1$ and $b_{i+1}^j = b_i^j$ for all $j = 0,1,\ldots, m$, $i = i+1$, and go to Step 1.

We note that if s is picked equal to 0, then Algorithm 4 reduces to the drivable method of feasible directions [6]. By choosing s larger than 0, we may therefore reduce the number of gradients of almost active constraints we have to compute in Step 1. The integers c_i^j are used to indicate the "age" of the quantities b_i^j. For example, if $c_i^j = 0$, then we know that $b_i^j = \nabla f^j(z_i)$, if $c_i^j = 1$, then we know that $b_i^j = \nabla f^j(z_{i-1})$, etc..., and if $b_i^j = +\infty$, then we know that $\nabla f^j(.)$ has not been evaluated at iteration $1,2,\ldots, i$. The quantity ρ may be chosed as large as wished but must be finite.

5. <u>Hypothesis</u>:
   (i) $\rho>0$.
   (ii) $s\geq 0$.
   (iii) Given any m in E, there exists k in E, depending on m, so that $p(r)\geq m$, and $q(r)\geq m$ for all $r\geq k$.

6. <u>Lemma</u>: Suppose that Hypotheses 2 and 5 are satisfied, and let $\{z_i\}$ be a sequence generated by Algorithm 4, then the sequence $\{\| z_{i+1} - z_i \|\}$ converges to 0.

<u>Proof</u>: Suppose that the sequence $\{\| z_{i+1} - z_i\|\}$ does not converge to 0, then there exist an infinite subset K of the integers and $\alpha>0$ so that

$\{z_i\}_K$ converges to some z_* (1)

$\{z_{i+1}\}_K$ converges to some z_{**}, and (2)

$\| z_{**} - z_*\| \geq \alpha$. (3)

By construction

$$\langle b_i^0, h_i \rangle \leq 0, \text{ and} \tag{4}$$

$$b_i^0 = \nabla f^0(z_i) \text{ for all } i. \tag{5}$$

It follows that

$$\langle \nabla f^0(z_i),\ z_{i+1} - z_i \rangle \leq 0 \text{ for all } i. \tag{6}$$

Clearly, if

$$\langle \nabla f^0(z_*),\ z_{**} - z_* \rangle < 0\ , \tag{7}$$

then the sequence $\{f^0(z_i)\}$ will be unbounded from below, and this cannot happen because $f^0(.)$ is continuous and T is compact. We conclude that

$$\langle \nabla f^0(z_*),\ z_{**} - z_* \rangle = 0 \tag{8}$$

The sequence $\{f^0(z_i)\}$ is monotonocally decreasing, and therefore

$$f^0(z_{**}) = f^0(z_*). \tag{9}$$

Equations (3), (8), and (9) imply that $f^0(.)$ is not strictly convex. But the map $f^0(.)$ is assumed to be strictly convex, and we conclude that the sequence $\{z_i\}$ is asymptotically regular, i.e., the sequence $\{\| z_{i+1} - z_i \|\}$ converges to 0.

7. <u>Lemma</u>: Suppose that Hypotheses 2 and 5 are satisfied, and let $\{z_i\}$ be a sequence generated by Algorithm 4, then

(i) If z_* is a cluster point of $\{z_i\}$, there exist an infinite subset L of the integers, and a subset $\hat{J}$ of $\{0,1,\ldots, m\}$ so that

$$\{z_i\}_L \text{ converges to } z_*,$$
$$J(z_i, p(r_i)) = \hat{J} \text{ for all } i \text{ in } L,$$

(ii) the sequence $\{r_i\}$ "converges" to $+\infty$, and

(iii) there exists an infinite subset M of the integers so that the subsequence

$$\{ \max_{j \varepsilon J(z_i, p(r_i))} \langle b_i^j, h_i \rangle \}_M$$

converges to 0.

<u>Proof</u>: (i) Let z_* be a cluster point of a sequence $\{z_i\}$ generated by Algorithm 4, then there exists an infinite subset K of the integers so that the subsequence $\{z_i\}_K$ converges to z_*. The set $J(z_i, p(r_i))$

is a subset of the finite set $\{0,1,\ldots, m\}$ and therefore there exist an infinite subset L of K, and a subset $\hat{J}$ of $\{0,1,\ldots, m\}$ so that

$$\{z_i\}_L \text{ converges to } z_*, \text{ and} \tag{10}$$

$$J(z_i,p(r_i)) = \hat{J} \text{ for all } i \text{ in } L \tag{11}$$

Let j be in $\hat{J}$, then $b_i^j = \nabla f^j(z_{i-t})$ for some t in $[0;s]$. The sequence $\{z_i\}$ is asymptotically regular, and this implies immediately that

$$\{b_i^j\}_L \text{ converges to } \nabla f^j(z_*). \tag{12}$$

(ii) Suppose that the sequence $\{r_i\}$ is bounded from above. Then, there exist k_1 and r so that for all $i \geq k_1$,

$$r_i = r \tag{13}$$

$$\langle b_i^j,h_i\rangle \leq -1/q(r) \text{ for all } j \text{ in } J(z_i,p(r)), \text{ and} \tag{14}$$

$$f^j(z_i) \leq -1/p(r) \text{ for all } j \text{ not in } J(z_i,p(r)). \tag{15}$$

Equations (11) and (12) imply that there exists k_2 so that for all i in L, $i \geq k_2$,

$$\langle\nabla f^j(z_i),h_i \leq -1/2q(r) \text{ for all } j \text{ in } \hat{J}, \text{ and} \tag{16}$$

$$f^j(z_i) \leq -1/p(r) \text{ for all } j \text{ not in } \hat{J}. \tag{17}$$

Using (16) and (17), it is not difficult to show that the sequence $\{f^0(z_i)\}$ is unbounded from below. The map $f^0(.)$ is continuous and T is compact, and therefore, the assumption that $\{r_i\}$ is bounded from above leads to a contradiction. We conclude that $\{r_i\}$ "converges" to $+\infty$.

(iii) The fact that the sequence $\{r_i\}$ converges to $+\infty$ implies immediately that there exists an infinite subset M of the integers so that the subsequence

$$\{ \max_{j\varepsilon J(z_i,p(r_i))} \langle b_i^j,h_i\rangle \}_M \tag{18}$$

converges to 0.

8. <u>Lemma</u>: Suppose that Hypotheses 2 and 5 are satisfied, and let $\{z_i\}$ be a sequence generated by Algorithm 4, then at least one cluster point of $\{z_i\}$ is solution to Problem 1.

<u>Proof</u>: Let M be an infinite subset of the integers so that

$$\{ \max_{j\varepsilon J(z_i,p(r_i))} \langle b_i^j,h_i\rangle \}_M \text{ converges to } 0. \tag{19}$$

The set T is compact, and therefore there exists an infinite subset K of M so that

$$\{z_i\}_K \quad \text{converges to some } z_*. \tag{20}$$

Part (i) of Lemma 7 implies that there exist an infinite subset L of K, and a subset $\hat{J}$ of $\{0,1,\ldots, m\}$ so that

$$J(z_i,p(r_i)) = \hat{J} \text{ for all } i \text{ in } L. \tag{21}$$

It follows from (19) and (21) that

$$\{ \max_{j\varepsilon\hat{J}} \langle b_i^j,h_i\rangle \}_L \quad \text{converges to } 0. \tag{22}$$

But we know from Equation (12) that if j is in $\hat{J}$, then $\{b_i^j\}$ converges to $\nabla f^j(z_*)$, and we conclude that

$$\{ \max_{j\varepsilon\hat{J}} \langle\nabla f^j(z_*),h_i \}_L \quad \text{converges to } 0. \tag{23}$$

The sequence $\{r_i\}$ converges to $+\infty$, and it is not difficult to see that this implies that

$$\hat{J} \subseteq J(z_*,\infty). \tag{24}$$

It follows that

$$\{ \min_{h\varepsilon S} \max_{j\varepsilon J(z_*,\infty)} \langle\nabla f^j(z_*),h\rangle \}_L \quad \text{converges to } 0, \tag{25}$$

i.e.,

$$\min_{h\varepsilon S} \max_{j\varepsilon J(z_*,\infty)} \langle\nabla f^j(z_*),h\rangle = 0 \tag{26}$$

and z_* is solution to Problem 1.

9. Theorem: Suppose that Hypotheses 2 and 5 are satisfied, and let $\{z_i\}$ be a sequence generated by Algorithm 4, then $\{z_i\}$ converges to the solution of Problem 1.

Proof: The result is a direct consequence of Lemma 8 and the fact that, when Hypothesis 2 is satisfied, Problem 1 has one and only one solution.

REFERENCES

[1] J. Cullum, W. E. Donath and P. Wolfe, An Algorithm for Minimizing Certain Nondifferentiable Convex Functions, I.B.M. Research Report RC 4611, November 6, 1973.

[2] A. M. Geoffrion, Primal Resource-Directive Approaches for Optimizing Nonlinear Decomposable Systems, Operations Research, 18 (1970), pp. 375-403.

[3] P. Huard, Tour d'Horizon: Programmation Non Lineaire, Revue Française de Recherche Operationnelle, 5 R-1 (1971), pp. 3-48.

[4] D. G. Luenberger, Introduction to Linear and Nonlinear Programming, Addison-Wesley Publishing Company Inc., 1973.

[5] G. G. L. Meyer, An Open Loop Method of Feasible Directions for the Sixth Annual Princeton Conference on Information Sciences and Systems (1972), pp. 679-680.

[6] G. G. L. Meyer, "A Drivable Method of Feasible Directions," SIAM J. Control, 11 (1973), pp. 113-118.

[7] E. Polak, Computational Methods in Optimization: A Unified Approach, Academic Press, New York, 1971.

[8] D. M. Topkis and A. Veinott, "On the Convergence of Some Feasible Directions Algorithms for Nonlinear Programming," SIAM J. Control, 5 (1967), pp. 268-279.

[9] W. I. Zangwill, Nonlinear Programming: A Unified Approach, Prentice Hall, Inc., Englewood Cliffs, New Jersey, 1969.

[10] G. Zoutendijk, Methods of Feasible Directions, Elsevier Publishing Co., Amsterdam, 1960.

THE CONTINUOUS METHOD OF STEEPEST DESCENT AND ITS DISCRETIZATIONS

PETER MITTER, CHRISTOPH W. UEBERHUBER

1. INTRODUCTION

Consider the unconstrained minimization problem

$$f(x) \to \min, \quad f : \mathbb{R}^n \to \mathbb{R}; \quad f \in C^2$$

$g(x)$ being the gradient of g, $g'(x)$ its Hessian. The efficiency of a method to solve this problem depends very much on individual properties of f as well as on the initial point x_0. Concerning starting points near the minimum x^* one hardly has difficulties. But very frequently one has no initial estimate at all or just a very bad one. It may be so bad that high speed iteration methods like Newton's method do not converge. In this situation the method of steepest descent seems to be appropriate, convergence is guaranteed at least theoretically. But indeed convergence can be extremely slow, thus it possibly is not of practical value. The method of steepest descent corresponds to the application of Euler's method to the ordinary differential equation $\dot{x} = -g(x)$. As long as this differential equation is not stiff, there will be no difficulties. But as soon as it becomes stiff (which might be checked by "stiffness detectors" e.g., see Shampine [11]), Euler's method is of little value because of its small region of stability. Stiff equations $\dot{x} = -g(x)$ correspond to functions f with a minimum x^* lying in a narrow valley with steep sides (Kowalik-Osborne [5]), and in this case the method of steepest descent is known to be very inefficient. Therefore we are investigating other techniques, especially the implicit version of Euler's method will be shown to be an efficient scheme for this type of problems.

After a short presentation of imbedding techniques Chapter 2 deals with the continuous method of steepest descent, given by the differential equation $\dot{x} = -g(x)$. Main attention is focused on the domain of attraction of stationary points and on the speed and direction of convergence of solutions approaching this point. Chapter 3 deduces stability properties which are to be satisfied by suitable discrete methods and states convergence theorems for these methods; a numerical example is presented.

2. THE CONTINUOUS METHOD OF STEEPEST DESCENT:

A very elegant general approach in numerical analysis is the technique of imbedding. Given a problem P with unknown solution (e.g. a system of equations $P(x) = 0$), a one parameter family of problems P_t, $0 \leq t \leq T$

(sometimes the range of the parameter is infinite), is constructed in such a way that P_o is a problem the solution of which is known, and that P_T is the original problem P. Then one tries to find a representation of the one-parameter family of the related solutions. One expects, of course, that neighbouring parameter values correspond to neighbouring solutions, which is true under certain circumstances (see e.g. Ortega-Rheinboldt [10], Dennis [4]). In our case of a minimization problem one possibility would be the evaluation of stationary points of f by means of an imbedding technique yielding a function x(t), $x(t) \to x^*$ if $t \to \infty$ (Boggs [1]). Our approach, however, is different from that. We are defining a function x(t) with this property without making this detour via a continuum of problems. Consider the method of steepest descent

$$x_{k+1} = x_k - hg(x_k).$$

For $h \to 0$ the points x_k tend to the trajectory of steepest descent, which is everywhere orthogonal to the level-surfaces. This trajectory is the solution of the initial value problem

$$\dot{x} = -g(x) \qquad x(o) = x_o.$$

In the continuous method of steepest descent stationary points of the above differential equation are calculated (without discretizations e.g. on an analog computer). In an analogous way continuous Quasi-Newton methods can be defined and investigated (Mitter and Ueberhuber [7], for a special case of such a method see Boggs [2]).

Theorem 1. If the function $f \in C^1(\mathbb{R}^n)$ satisfies the assumptions

(i) f is bounded below

(ii) every level set of f is bounded

(iii) f has exactly one stationary point

then the convergence of the continuous method $\dot{x} = -g(x)$, $x(0) = x_o$ is guaranteed from any starting point $x_o \in \mathbb{R}^n$.

Proof: See Ueberhuber [12].

Note, that f is *not* assumed to be convex. The next two theorems concern with the manner, in which the solution trajectories $x(t;x_o)$ tend to x^*. Assume that the differential equation can be linearized in the manner given below, and, for simplicity, assume the minimum x^* to be shifted into the origin.

Theorem 2. For the continuous method

$$\dot{x} = -g(x) = Bx + r(x), \qquad r(x) = o(||x||) \text{ for } ||x|| \to 0$$

with the eigenvalues of B $\lambda_1 \leq \ldots \leq \lambda_n < 0$ the subset E_a of the domain of attraction of $x^* = 0$ whose points z are characterized by

$$\| x(t;z) \| \leq c \cdot \exp((a + e(t))t)$$
$$\text{(with } e(t) \to 0 \text{ as } t \to \infty \text{ and } c \in \mathbb{R})$$

is an m-parametric manifold of $\mathbb{R}^n$, where m is defined by $\lambda_1 \leq \ldots \leq \lambda_m \leq a < \lambda_{m+1} \leq \ldots \leq \lambda_n$.

Proof: This theorem is an immediate consequence of Theorem IV 1.45 in Nemytskii, Stepanov [8].

The set E_a (a negative) is the set of all starting values such that the trajectories starting from those points approach the minimum asymptotically faster than exp(at). As E_a has maximum dimension only if $\lambda_n \leq a < 0$, this condition is practically necessary for arbitrarily chosen starting points to lie in E_a. Or, intuitively spoken, the rate of convergence for $x(t;x_o) \to x^*$ is mainly determined by the smallest eigenvalue of the Hessian $g'(x^*)$. This eigenvalue λ_n not only determines the rate of convergence, but also the direction along which the trajectory approaches the minimum.

Theorem 3. For the continuous method

$$\dot{x} = -g(x) = Bx + r(x), \quad r(x) = o(\|x\|) \quad \text{for} \quad \|x\| \to 0$$

with the eigenvalues of B $\lambda_1 \leq \ldots \leq \lambda_n < 0$ it holds that

(i) $\exists i: \frac{1}{t} \ln(\|x(t;x_o)\| \to \lambda_i$ as $t \to \infty$

(ii) $\|P_\pm x(t;x_o)\| = o(\|Px(t;x_o)\|)$ as $t \to \infty$

where P_-, P, P_+ denote the projections onto the eigenspaces with respect to $\{\lambda_1, \ldots, \lambda_{i-1}\}$, $\{\lambda_i\}$, $\{\lambda_{i+1}, \ldots, \lambda_n\}$.

Proof: Coppel [3] Theorem IV.5.

Combined with Theorem 2, Theorem 3 says that for "almost all" initial values x_o the solution $x(t;x_o)$ approaches x^* in a subspace of $\mathbb{R}^n$ which is the eigenspace belonging to λ_n. This result can be used for an acceleration (or extrapolation) technique. Take e.g. Rosenbrock's function

$$f(x_1,x_2) = 100(x_1^2 - x_2)^2 + (1 - x_1)^2$$

and the continuous method of steepest descent

$$\dot{x}_1 = -400(x_1^2 - x_2)x_1 + 2(1 - x_1)$$
$$\dot{x}_2 = 200(x_1^2 - x_2)$$

The matrix of the linearized problem (at $(x_1^*, x_2^*) = (1,1)$) has eigenvalues

$\lambda_1 \approx -1001.6$
$\lambda_2 \approx -\ \ 0.4$

If the initial point lies on the curve $E_{-1001.6}$, the speed of convergence is considerably high, but for all other initial points, i.e. for "almost all", it is asymptotically low ($\sim \exp(-0.4t)$), moreover, the direction of the path lies (asymptotically) in the eigenspace belonging to $\lambda_2 = -0.4$.

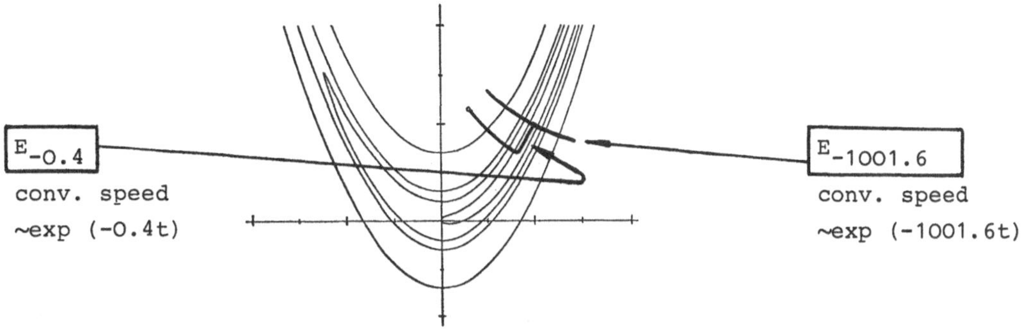

3. DISCRETIZATIONS

The application of an integration method to $\dot{x} = -g(x)$ now yields an optimization method. As long as the function f to be minimized is well-behaved, there will be now problem. Problems arise, if f is not well-behaved in the sense, that the minimum x* lies in a narrow valley with steep sides, and this is the situation we are considering in this section. If one defines a condition measure

$$\text{cond } f = \frac{\text{largest eigenvalue of } g'(x^*)}{\text{smallest eigenvalue of } g'(x^*)},$$

badly-behaved functions ("ill-conditioned" problems) correspond to large condition numbers. In this case the steepest descent differential equation $\dot{x} = -g(x)$ is said to be *stiff*. Stiff differential equations have the property, that many common integration methods fail, because they must be applied with extremely small step sizes to avoid exploding error terms. Euler's method (classical steepest descent) is an example of a method which is *not* suitable in this situation. To obtain a downhill sequence, step sizes $h \approx 1/\lambda_1$ have to be used, λ_1 being the *largest* eigenvalue of $g'(x)$. But as we have seen in section 2, the corresponding continuous solution is converging very slow due to the smallest eigenvalue λ_n of $g'(x)$. We are following a slowly converging curve using small time steps, therefore we cannot be successful. Which criteria have thus to be fulfilled by a suitable method?

(i) The method should be applicable with arbitrary step size. The class of A(O)-stable methods guarantees this fact at least in the case of quadratic functions (concerning the stability concepts used in this section, the reader is refered to the Appendix of Mitter, Ueberhuber [7]).

(ii) The choice of a larger step size should yield a better result than the choice of a shorter one. Again in the case of quadratic functions, this criterion is fulfilled by strongly A(O)-stable methods, at least for sufficiently large h. The well-known trapezoidal rule, for example, is A(O)-stable, but not strong. It results in the formula

$$x_{k+1} = x_k - h(g(x_k) + g(x_{k+1}))$$

which easily leads to a zig-zagging sequence x_k, as can be seen by simple examples.

(iii) We are not interested in high order (=high accuracy) methods. High accuracy of strongly A(O)-stable methods generally implies expensive calculations, and moreover it may impede fast convergence to x^*, because the numerical solution is kept near the slowly convergent solution of the differential equation.

The simplest method due to these considerations is implicit Euler's method

$$x_{k+1} = x_k + h\dot{x}_{k+1}.$$

Applied to $\dot{x} = -g(x)$ it yields the implicit method of steepest descent

$$x_{k+1} = x_k - hg(x_{k+1}).$$

Note, that the subsequent iterative is not computed by linear search or something like that, but by the solution of a system of equations.

Theorem 4. Assume that

(i) $f: \mathbb{R}^n \to \mathbb{R}$ has a bounded, nonempty level set L

(ii) $f \in C^2$ in a neighbourhood of L

(iii) $g'(x)$ is positive semidefinite in L

(iv) $x_o \in L$ arbitrary, $g(x_o \neq 0$

Then it holds that

(i) the equation $y = x_o - h.g(y)$ has a solution y_h for all $h \geq 0$

(ii) $0 \leq h_1 < h_2 \Rightarrow f(y_{h_2}) < f(y_{h_1})$

(iii) $0 \leq h_1 < h_2 \Rightarrow \|g(y_{h_2})\| \leq \|g(y_{h_1})\|$

(iv) $\lim_{h\to\infty} \|g(y_h)\| = 0$

Proof: Mitter [6]

To estimate the minimum x^* with a given tolerance, one single step with a sufficiently large step size h would suffice. Unfortunately, the condition number of $y = x_o - hg(y)$ when solved numerically is increasing with the step size. Thus we would replace an ill-conditioned problem (the original optimization problem) by another one. By the choice of appropriate step sizes, however, we can replace the ill-conditioned optimization problem by a sequence of well-conditioned systems of nonlinear equations

$$x_{k+1} = x_k - h_{k+1}\, g(x_{k+1}).$$

From Theorem 4 it follows that the sequences $f(x_k)$ and $\|g(x_k)\|$ are monotonely decreasing.

Theorem 5. Assume the conditions of Theorem 4 to be valid and let the sequence x_k, $k \geq 1$ be defined by $x_{k+1} = x_k - h_{k+1}\, g(x_{k+1})$, using a convergent sequence of step sizes h_k with limit $h_o > 0$. Then x_k converges to a stationary point of f.

Proof: immediate consequence of proposition IV.4 in Mitter [6].

As a numerical example, consider again Rosenbrock's function with the usual starting values $x_1 = -1.2$, $x_2 = 1.0$. The stiffness (condition number) of this differential equation at the point $x^* = (1,1)$ is ~ 2500. The method of steepest descent requires step sizes $h \approx 0.001$. The implicit method of steepest descent on the other hand needed 5 implicit steps to reach the approximate result (0.99998, 0.99996). Computation needed 36 gradient evaluations (including 24 gradient evaluations to approximate the Hessian numerically).

REFERENCES

[1] P.T. Boggs: The solution of Nonlinear Systems of Equations by A-stable Integration Techniques; SIAM Journal on Numerical Analysis Vol. 8 (1971), pp. 767-785.

[2] P.T. Boggs: The Convergence of the Ben-Israel Iteration for Nonlinear Least Squares Problems; Math. Comp. Vol. 30 (1976), pp. 512-522.

[3] W.A. Coppel: Stability and Asymptotic Behaviour of Differential Equations; Heath, Boston, 1965.

[4] J.E. Dennis, Jnr.: Non-linear Least Squares and Equations; in D. Jacobs (Ed.): The State of the Art in Numerical Analysis, Academic Press, London, 1977.

[5] J. Kowalik, M.R. Osborne: Methods for Unconstrained Optimization Problems; Elsevier, New York, 1968.

[6] P. Mitter: Minimization Methods based on A(O)-stable Runge-Kutta Processes; In W. Gröbner et. al.:
The Method of Lie Series for Differential Equations and its Extensions; Techn. Report of the European Research Office, London, 1973.

[7] P. Mitter, C.W. Ueberhuber: Analysis of Continuous Methods for Unconstrained Optimization and their Discretizations, Forschungsbericht Nr. 121, Institute for Advanced Studies, Vienna, 1977.

[8] V.V. Nemytskii, V.V. Stepanov: Qualitative Theory of Differential Equations; Princeton University Press, 1960.

[9] J.M. Ortega: Stability of Difference Equations and Convergence of Iterative Processes; SIAM Journal on Numerical Analysis Vol. 10 (1973), pp. 268-282.

[10] J.M. Ortega, W.C. Rheinboldt: Iterative Solution of Nonlinear Equations in Several Variables; Academic Press, New York, 1970.

[11] L.F. Shampine: Stiffness and Nonstiff Differential Equation Solvers, II: Detecting Stiffness with Runge-Kutta Methods; ACM Trans. Math. Software Vol. 3 (1977), pp. 44-53.

[12] C.W. Ueberhuber: Optimierung mittels kontinuierlicher Quasi-Newton-Verfahren und deren Diskretisierungen; Institut für Numerische Mathematik, TU-Wien, Bericht Nr. 19/76, Vienna, 1976.

CONVERGENCE RATE RESULTS
FOR A PENALTY FUNCTION METHOD

V. Hien Nguyen and J.-J. Strodiot
Department of Mathematics
Facultés Universitaires de Namur
B-5000 Namur / BELGIUM

ABSTRACT

This paper considers some convergence aspects of an optimization algorithm, whose basic idea is closely related to *penalty* and *augmented Lagrangian* methods, proposed by Kort and Bertsekas in 1972. We prove, *without convexity* assumptions, that the algorithm has a *parametrically superlinear root convergence rate*. We also give a *partial global convergence* result for the algorithm considered.

1. DESCRIPTION OF THE KORT-BERTSEKAS ALGORITHM

Let f and $g_i : \mathbb{R}^n \to \mathbb{R}$, $1 \leq i \leq m$, $m+1$ continuously differentiable functions.

We are concerned with the *nonconvex* minimization problem :

$$(1.1) \qquad \begin{cases} \text{minimize} \quad f(x) \\ \text{subject to} \quad g_i(x) \leq 0 \ , \quad 1 \leq i \leq m \ . \end{cases}$$

In order to describe the algorithm, we consider a function $\Phi : \mathbb{R} \to \mathbb{R}$ which is supposed to be two times continuously differentiable and to satisfy the following assumptions (where $' \equiv d/dt$) :

$$(1.2) \qquad \begin{cases} \text{(i)} \quad \Phi(0) = 0 \ , \\ \text{(ii)} \quad \lim_{t \to -\infty} \Phi(t) = -1 \ , \\ \text{(iii)} \quad \Phi'(0) = 1 \ , \\ \text{(iv)} \quad \lim_{t \to \infty} \Phi'(t) = \infty \ , \\ \text{(v)} \quad \Phi''(t) > 0 \quad \text{for all} \quad t \in \mathbb{R} \ . \end{cases}$$

Remarks 1.1. 1° The hypotheses (1.2) imply that the derivative of Φ is everywhere (strictly) positive.

2° For example, all the hypotheses of (1.2) are satisfied by the following functions (due to Kort and Bertsekas) :

(1.3)
$$\begin{cases} \text{(a)} \quad \Phi^1(t) = \exp(t) - 1 & \text{for all } t \in \mathbb{R} \; ; \\ \text{(b)} \quad \Phi^2(t) = \begin{cases} t + t^2 & \text{if } t \geq 0 \; , \\ t/(1-t) & \text{if } t < 0 \; ; \end{cases} \\ \text{(c)} \quad \Phi^3(t) = \begin{cases} t + 3\,t^2/4 & \text{if } t \geq 0 \; , \\ -1 + 1/(1-(t/2)^2 & \text{if } t < 0 \; . \end{cases} \end{cases}$$

Observe that they are all of exponential type.□

We set :

$$\Phi_k(t) = \frac{1}{k}\,\Phi(k\,t)$$

($k > 0$, penalty parameter that must be choosen *sufficiently large, but finite*).

We then introduce the extended Lagrangian :

(1.4)
$$F(x, u, k) = f(x) + \sum_{i=1}^{m} u_i\,\Phi_k[g_i(x)]$$

($u_i \geq 0$, approximate Lagrange multiplier for (1.1)).

Remark 1.2. If we take $\Phi = \Phi^1$ (defined by (1.3a) in (1.4), we obtain the extended Lagrangian due to Gould and Howe (Ref. 1).□

Under these conditions, the Kort-Bertsekas algorithm (Ref. 2) is based on the construction of two sequences of elements $x^j \in \mathbb{R}^n$, $u^j \in \mathbb{R}^m_+$, defined in the following way : we select $k > 0$ and we start with any :

(1.5)
$$u_i^o \in \mathbb{R}^*_+ \; , \qquad 1 \leq i \leq m \; ;$$

we calculate x^o , then u^1 , x^1 , etc. The general rule is :

(1.6)
u^j being known, we determine x^j as the element of $\mathbb{R}^n$ which satisfies $\nabla_1 F(x^j, u^j, k) = 0$, where ∇_1 is the gradient whith respect to the first argument x .

Then we define :

(1.7)
$$u_i^{j+1} = u_i^j\,\Phi_k'[g_i(x^j)] \; , \qquad 1 \leq i \leq m \; .$$

Remarks 1.3. 1° If x^j is not unique in (1.6), we take a closest x^j to x^{j-1} in some norm.

2° The algorithm defined by (1.5)-(1.7) is an algorithm of penalty-duality type (see, for example, Refs. 3, 4 and 5).

3° If we take $\Phi = \Phi^1$ in the algorithm, we obtain the algorithm proposed independently by Hartman (Ref. 6).□

2. PARAMATRIC SUPERLINEAR CONVERGENCE OF THE ALGORITHM

Let $x^* \in \mathbb{R}^n$ be a local optimal solution for (1.1). We make the following classical hypotheses (see, for example, Ref. 3) :

(2.1) $\begin{cases} f \text{ and } g_i , \ 1 \leq i \leq m \text{, are two times continuously} \\ \text{differentiable in an open ball with center } x^* \text{ and} \\ \text{radius } \varepsilon > 0 \ ; \end{cases}$

(2.2) $\begin{cases} \text{there is a unique optimal Lagrange multiplier } u^* \in \mathbb{R}^m \\ \text{such that } (x^*, u^*) \text{ satisfy the Kuhn-Tucker conditions} \\ \text{and also the second order sufficiency conditions} \\ \text{discussed by Fiacco and McCormick in Ref. 3;} \end{cases}$

(2.3) $\begin{cases} \text{the strict complementarity condition does hold at} \\ (x^*, u^*) \ . \end{cases}$

We are now ready for the main result of this paper :

Theorem 2.1. *Under hypotheses (2.1)-(2.3), if $k > 0$ is sufficiently large but finite, then the algorithm defined by (1.5)-(1.7) is locally and superlinearly convergent in the following sense :*

(a) Local Convergence : There exists an open neighborhood $V(u^*)$ of u^* such that, for all $u^o \in V(u^*)$ with $u_i^o > 0$, $1 \leq i \leq m$, there exists $x^o \in \mathbb{R}^n$ satisfying $\nabla_1 F(x^o, u^o, k) = 0$, and the iterates (x^j, u^j) generated by the algorithm are well-defined and converge to (x^*, u^*) as $j \to \infty$;

(b) Parametric R-Superlinear Convergence : Moreover, we have :

(2.4) $$\|(x^j, u^j) - (x^*, u^*)\| \leq (c/k)^j \qquad \text{for } j \geq j_o ,$$

for some integer j_o and some constant $c > 0$, where $\|\cdot\|$ is the l_2-norm on $\mathbb{R}^{n+m}$, provided that in addition to hypotheses (1.2) the function Φ satisfies the following additional assumption :

(2.5) $\begin{cases} \text{there exists a constant } \alpha > 0 \text{ such that :} \\ \qquad \Phi'(t) \leq \alpha / t^2 \qquad \text{for all } t < 0 \end{cases}$.

Remarks 2.1. 1° We call the convergence rate (2.4) parametrically R-superlinear, because by increasing k it can be made better than any linear root rate convergence : $\|(x^j, u^j) - (x^*, u^*)\| \leq (\gamma)^j$, where γ is some fixed number < 1 (see, for instance, Refs. 7 and 8).

2° Observe that the functions Φ^1 , Φ^2 and Φ^3 defined by (1.3) satisfy (2.5).□

3. THE IDEA OF THE PROOF OF THEOREM 2.1

We begin by modifying slightly the original algorithm : we replace the extended Lagrangian F (defined by (1.4)) by the function $G : \mathbb{R}^n \times \mathbb{R}^m_+ \times \mathbb{R}_+ \to \mathbb{R}$ given by :

$$G(y, v, k) = f(y) + \sum_{i=1}^{m} (v_i)^2 \, \Phi_k [g_i(y)] \quad . \tag{3.1}$$

The points x^o , x^j , u^o and u^j are replaced by y^o , y^j , v^o and v^j , respectively, whereas the updating formula (1.7) becomes (as a consequence of the replacement of F by G) :

$$v_i^{j+1} = v_i^j \, \{\Phi_k' [g_i(y^j)]\}^{1/2} \quad , \qquad 1 \leq i \leq m \quad .$$

We then apply techniques due to Mangasarian (Ref. 7) to establish the following theorem concerning the *modified* algorithm (compare with Theorem 2,1) :

Theorem 3.1. *Under the stated conditions, if* $k > 0$ *is sufficiently large but finite, then there exists an open neighborhood* $W(v^*)$ *of* $v^* \equiv (u_1^{*1/2}, \dots, u_m^{*1/2})$ *such that, for all* $v^o \in W(v^*)$ *with* $v_i^o > 0$, $1 \leq i \leq m$, *there exists* $y^o \in \mathbb{R}^n$ *satisfying* $\nabla_1 G(y^o, v^o, k) = 0$, *and the sequence* $\{(y^j, v^j)\}_{j \geq 0}$ *of iterates obtained by the* modified *algorithm is well-defined and converges to* (x^*, v^*) *at a parametrically superlinear root convergence rate, provided that the function* Φ *satisfies hypotheses (1.2) and (2.5).*

The proof of this theorem is based on the fact that the spectral radius of the gradient of the functional concerning the v-iteration is (strictly) less than 1 and on the

Lemme 3.1. *(a) For all* $k > 0$, *we have :*

$$\nabla_1 G(x^*, v^*, k) = 0 \quad ;$$

(b) There exists a constant $k_o > 0$ *such that, for all* $k \geq k_o$, *the Hessian with respect to the first argument* x , $\nabla_{11} G(x^*, v^*, k)$, *is positive definite.*

Remark 3.1. (Motivation of the modification.) It is the introduction of the square on v in the expression (3.1) of the new extended Lagrangian G which allows one to easily compute the spectral radius (see above). This justifies the modification of the original algorithm.□

Finally, we go back to the original algorithm with the help of

Proposition 3.1. *If the starting points* (x^o, u^o) *and* (y^o, v^o) *of the modified and original algorithms are chosen such that* $x^o = y^o$ *and* $u_i^o = (v_i^o)^2$, $1 \leq i \leq m$, *then, for all* j, *we have* $x^j = y^j$ *and* $u_i^j = (v_i^j)^2$, $1 \leq i \leq m$.

Theorem 2.1 can now be seen to hold by Proposition 3.1, Theorem 3.1 and the inequality :

$$\| u^j - u^* \| \leq d \, \| v^j - v^* \| \qquad , \qquad d > 1 \quad .$$

4. A PARTIAL RESULT

Note that Theorem 2.1 is a convergence theorem of local type. The problem of the *global* convergence of the algorithm, in the *nonconvex* case, is therefore an *open question*, at least by the method of proof proposed here. However, we can show the following

Proposition 4.1. *Under conditions described in Section 1, we have :*

(a) If the sequence of iterates (x^j, u^j) *is bounded and if* $M \equiv \{i \mid \liminf_{j\to\infty} u_i^j = 0 , \; 1 \leq i \leq m\}$, *then* $\liminf_{j\to\infty} g_i(x^j) \leq 0$ *for all* $i \in M$.

(b) If, in addition to (a), the whole sequence $\{u^j\}_j$ *converges to* $\bar{u}$, *then any limit point* $(\bar{x}, \bar{u})$ *of* $\{(x^j, u^j)\}_j$ *satisfies the first Kuhn-Tucker equation* [$\nabla_1 L(\bar{x}, \bar{u}) = 0$, *where* $L(x, u)$ *denotes the classical Lagrangian associated to (1.1)*] *and is such that* $\bar{u}_i \, g_i(\bar{x}) = 0$ *for all* $1 \leq i \leq m$.

(c) If, in addition to (a) and (b), the whole sequence $\{x^j\}_j$ *converges to* $\bar{x}$, *then any limit point* $(\bar{x}, \bar{u})$ *of* $\{(x^j, u^j)\}_j$ *satisfies all the Kuhn-Tucker conditions.*

A complete presentation of the results described in this note and additional numerical experience will appear elsewhere.

REFERENCES

1. F.J. Gould and S. Howe, "A New Result on Interpreting Lagrange Multipliers as Dual Variables", Institute of Statistics Mimeo Series No. 738, Dept. of Statistics, U. of North Carolina at Chapel Hill, Jan. 1971.

2. B.W. Kort and D.P. Bertsekas, "A New Penalty Function Method for Constrained Minimization", *Proc. 1972 I.E.E.E. Conf. on Decision and Control*, pp. 162-166, New Orleans, LA., Dec. 1972.

3. A.V. Fiacco and G.P. McCormick, *Nonlinear Programming : Sequential Unconstrained Minimization Techniques*, Wiley, New York, NY., 1968.

4. D.P. Bertsekas, "Multiplier Methods : A Survey", *Automatica* 12 (1976), pp. 133-145.

5. R.T. Rockafellar, "Lagrange Multipliers in Optimization", *Proc. Symp. Appl. Math. A.M.S. and S.I.A.M.*, pp. 145-168, New York, NY., March 1975.

6. J.K. Hartman, "Iterative Determination of Parameters for an Exact Penalty Function", *J. Optimization Theor. Appls.* 16 (1975), pp. 49-66.

7. O.L. Mangasarian, "Unconstrained Methods in Nonlinear Programming", *Proc. Symp. Appl. Math. A.M.S. and S.I.A.M.*, pp. 169-184, New York, NY., March 1975.

8. J.M. Ortega and W.C. Rheinboldt, *Iterative Solution of Nonlinear Equations in Several Variables*, Academic Press, New York, NY., 1970.

A COMBINED VARIABLE METRIC - CONJUGATE GRADIENT ALGORITHM FOR A CLASS OF LARGE SCALE UNCONSTRAINED MINIMIZATION PROBLEMS

Shmuel S. Oren
Xerox Palo Alto Research Center
3333 Coyote Hill Road
Palo Alto, California

Abstract

An algorithm is being presented for a special class of unconstrained minimization problems. The algorithm exploits the special structure of the Hessian in the problems under consideration. It is based on applying Bertsekas' [1] Scaled Partial Conjugate Gradient method with respect to a metric that is updated by the Rank One update, using gradients obtained in the preceeding steps. Two classes of problems are presented having the structure assumed in designing the proposed algorithm. In both cases the algorithm uses only first derivative information. Furthermore, it possesses quadratic termination in considerably fewer steps than the number of variables.

1. Introduction

Variable Metric algorithms are considered to be the most advanced methods for solving unconstrained minimization problems of the form: min $f(x)$ where $x \in R^n$ and $f \in C^2$. The basic recursion in these algorithms is analog to the one used in Newton Raphson method having the form:

$$x_{k+1} = x_k - \alpha_k D_k g_k \tag{1}$$

In this recursion x_k denotes the kth approximation to the minimum, g_k is the gradient at x_k, α_k is a stepsize parameter selected to ensure some convergence criteria, while D_k is an nxn matrix approximating the inverse Hessian $[\nabla^2 f(x)]^{-1}$. The approximations D_k are inferred from the gradients at previous iterations and updated as new gradients become available so as to satisfy the "quasi Newton condition"

$$D_k(g_k - g_{k-1}) = x_k - x_{k-1} \quad . \tag{2}$$

The main motivation underlying such procedures is to capture the second order convergence properties of Newton's method while avoiding the expensive calculation of second derivatives.

The first Variable Metric algorithm was invented by Davidon [4] and further developed and simplified by Fletcher and Powell [5]. Since then a vast literature has been published on this subject. Many of these contributions propose alternative updating procedures for D_k and contain computational results comparing the various computational schemes. However, practically all the theoretical and computational work in this area has been directed toward solving small problems in which the number of variables rarely exceeds fifty.

It is evident even from the above brief description of Variable Metric methods, that the use of such algorithms for large scale problems is limited by the computational and storage requirement involved in maintaining D_k. In such cases it becomes advantageous to use Conjugate Gradient algorithms such as Fletcher Reeves [6] method. These algorithms are usually slower than variable metric methods as they lack the memory features of the later techniques. On the other hand, conjugate gradient methods have the advantage of generating the search directions directly, avoiding the need to store and update an nxn matrix which becomes prohibitive for large n.

The above considerations are relevant as long as no structural information about $f(x)$ is being utilized. Fortunately, in many of the large scale problem the objective function has some special structure. The expense involved in solving such problems, and computational feasibility considerations, justify the development of special purpose algorithms that exploit the special structure of the objective function. One of the central themes of large scale mathematical programming has been to develop such special purpose algorithms. This approach, however, has not influenced yet the development of Variable Metric type algorithms for large scale problems.

This paper attempts to follow the aforementioned theme of large scale mathematical programming and proposes an algorithm for a special class of unconstrained minimization problems. More specifically, we focus on problems where the Hessian matrix $\nabla^2 f(x) = M+R$ where M is a block diagonal matrix with blocks of dimension m or less and R is a matrix of rank r, with m and r significantly lower than the dimension of x. Such functions arise for instance from a special class of control problems or in solving certain resource allocation problems by penalty or multiplier methods. Bertsekas [1] who addressed the aforementioned class of optimal control problems, proposed an algorithm in which the directions of search are generated using Fletcher Reeves [6] algorithm with respect to the metric M^{-1} restarted every $r+1$ steps. The matrix M^{-1}, is evaluated in this method at the beginning of each cycle from second derivative information. Bertsekas has shown that this algorithm converges superlinearly and for a quadratic function it terminates in one cycle (i.e. $r+1$ steps).

The algorithm proposed in this paper relates to Bertsekas [1] method in the sense that Variable Metric algorithms relate to Newton's method. The search direction at each step are generated using Fletcher Reeves' [6] algorithm with respect to a metric D, restarted every $r+1$ steps. The $n \times n$ matrix D is an approximation to the matrix M^{-1} updated by the Broyden's [2] Rank-One updating formula using the gradients computed at each step. Since M^{-1} is block diagonal we force D to have the same structure which enables us to update and store each block individually. Consequently, for a quadratic function, $D=M^{-1}$ after at most m steps implying "quadratic termination" in significantly fewer steps than n.

Following is an outline for the remainder of this paper. In section 2 we present the theoretical foundation and a conceptual outline of the proposed algorithm. In Sections 3 and 4 we specialize the algorithm to a class of resource allocation problems and to the optimal control problems considered by Bertsekas. Section 5 contains the conclusions and some remarks on the proposed method.

2. Theoretical Foundation and the Conceptual Algorithm

The Fletcher Reeves [6] conjugate gradient algorithm can be described as follows: Starting with an initial point x_0 and $d_0 = -g_0$,

$$x_{k+1} = x_k + \alpha_k d_k \tag{3}$$

where

$$\alpha_k = \arg\min f(x_k + \alpha d_k) \tag{4}$$

$$d_k = -g_k + \beta_{k-1} d_{k-1} \tag{5}$$

and

$$\beta_{k-1} = \| g_k \|^2 / \| g_{k-1} \|^2 \tag{6}$$

It is shown in Luenberger [7] that if $f(x)$ is a positive definite quadratic function and $\nabla^2 f(x)$ has s distinct eigenvalues then the above procedure converges to the minimum of $f(x)$ in at most s steps. When $\nabla^2 f(x) = M+R$ where M is positive definite and R has rank r we can define $y = M^{1/2}x$. Then,

$$\nabla_y f(M^{-1/2}y) = M^{-1/2} g_k \tag{7}$$

and

$$\nabla_y^2 f(M^{-1/2}y) = M^{-1/2} \nabla_x^2 f(x) M^{-1/2} = I + M^{-1/2} R M^{-1/2} \tag{8}$$

Clearly $\nabla_y^2 f(M^{-1/2}y)$ has only $r+1$ distinct eigenvalues. Thus, applying Fletcher Reeves algorithm after changing the variables from x to y will yield the minimum in at most $r+1$ steps. The above change of variables can be implementd implicitly by writing eq. (3) to (6) for y_k and then substituting $y_k = M^{1/2}x_k$. The resulting algorithm is similar to the original one, but now $d_0=-M^{-1}g_0$ and for $k>0$:

$$d_k = -M^{-1}g_k + \beta_{k-1}d_{k-1} \tag{9}$$

with

$$\beta_{k-1} = (g_k' M^{-1}g_k) / (g_{k-1}'M^{-1}g_{k-1}) \tag{10}.$$

The above algorithm can be generalized to non quadratic functions by restarting it every $r+1$ steps with M^{-1} evaluated at the beginning of each cycle and kept fixed during the entire cycle.

The above implementation which has been proposed by Bertsekas [1] results in superlinear convergence but requires the evaluation and inversion of the second derivative matrices composing M at the beginning of each cycle. The alternative approach proposed in this paper avoids the need for second derivative information as well as matrix inversions. In our implementation the matrix M^{-1} is substituted by an approximation D inferred from gradients generated in preceeding iterations and updated successively as new gradients become available.

The following theorem states the properties of Broyden's [2] Rank-One updating formula that forms the basis for the proposed algorithm.

Theorem 1

Let H *be a positive definite symmetric* nxn *matrix and* $\{r_0, ..., r_{n-1}\}$ *and* $\{v_0, ..., v_{n-1}\}$ *sequences of linearly independent vectors such that* $v_k = Hr_k$ *for* $k = 0 ... n-1$. *Let* D_k *be* nxn *matrices such that*

$$D_{k+1} = D_k + (r_k - D_kv_k)(r_k - D_kv_k)' / (r_k - D_kv_k)'v_k \tag{11}$$

and D_0 *is an arbitrary* nxn *positive semi-definite symmetric matrix. Then,* $D_{k+1}v_j = r_j$ *for* $j\leq k$.

The above theorem is well known (see for example Luenberger [7]) and its proof will hence be ommited. In particular the theorem implies that $D_n = H^{-1}$; i.e., the nth approximation will be identical to H^{-1} regardless of the initial approximation D_0.

In the specific problem under consideration, assuming the function is quadratic, we have $(M+R)p_k=q_k$ where $p_k = x_k-x_{k-1}$ and $q_k = g_k-g_{k-1}$. Consequently,

$$Mp_k = \hat{q}_k = q_k - Rp_k \tag{12}$$

We shall assume that either Rp_k is available or $\hat{q}_k$ can be obtained directly. In view of (12) and Theorem 1 we can then obtain M^{-1} by repeated application of (11) with p_k and $\hat{q}_k$ taking the role of r_k and v_k respectively. This would require, however, n updates and considerable storage. The computational and storage requirement can be radically reduced by exploiting the fact that M is block diagonal. If we partition the vectors p_k and $\hat{q}_k$ into n segments corresponding to the blocks in M then eq. (12) can be written as

$$M^i p^i_k = \hat{q}^i_k \qquad i = 1, ..., h \tag{13}$$

where M^i is the ith block and p^i_k, $\hat{q}^i_k$ are the corresponding segments of p_k and $\hat{q}_k$. Consequently, we can use (11) to obtain each block $(M^i)^{-1}$ individually. Following this procedure enables us to obtain M^{-1} using only m pairs of vectors (p_k,q_k) where m is the dimension of the largest block in M.

In the remainder of this section we outline a conceptual algorithm based on the above observations. The algorithm is designed to minimize an unconstrained function $f(x)$ whose Hessian $\nabla^2 f(x) = M+R$ where M is a symmetric block diagonal matrix consisting of h blocks having dimensions m^i $(i = 1, ..., h)$, while R has rank no greater than r. We use the notation x^i

to denote the segment of the vector x corresponding to M^i. Thus, $x = (x^1, x^2, \dots x^h)$ and $x^i \in R^{m_i}$.

Algorithm 1.

Start with an initial point $x_0 = (x^1_0, x^2_0, \dots x^h_0)$ and h positive definite symmetric $m_i x m_i$ matrices D^i_0 for $i = 1, \dots, h$.

Step 1: Obtain $g_0 = \nabla f(x_0)$ and set $d^i_0 = -D^i_0 g^i_0$ for $i=1, \dots, h$

Step 2: Compute

$$x_{k+1} = x_k + \alpha_k d_k \tag{14}$$

where α_k minimize $f(x_k + \alpha d_k)$ with respect to α

Step 3: Obtain $g_{k+1} = \nabla f(x_{k+1})$

$$p_k = x_{k+1} - x_k \tag{15}$$

$$q_k = g_{k+1} - g_k - Rp_k \tag{16}$$

Step 4: For $i = 1, \dots, h$ compute

$$v^i_k = p^i_k - D^i_k q^i_k \tag{17}$$

$$\left.\begin{array}{ll} D^i_{k+1} = D^i_k & \text{if } v^i_k{}' q^i_k \leq 0 \\ D^i_{k+1} = D^i_k + v^i_k v^i_k{}' / v^i_k{}' q^i_k & \text{otherwise} \end{array}\right\} \tag{18}$$

Step 5: If $k<r$, compute for $i=1, \dots, h$

$$d^i_{k+1} = -D^i_0 g^i_{k+1} + \beta_k d^i_k \ , \tag{19}$$

where

$$\beta_k = \left(\sum_{i=1}^{h} g^i_{k+1}{}' D_0{}^i g^i_{k+1}\right) / \sum_{i=1}^{h} g^i_k{}' D_0{}^i g_k{}^i \ . \tag{20}$$

Then increment k by 1 and go to Step 2.

Step 6: If $k=r$, reset k to 0, set $x_0 = x_{r+1}$, and $D^i_0 = D^i_{r+1}$ for $i=1, \dots, h$, then go to Step 1.

We note that though the matrices D^i_k are updated on each iteration, the matrices used in the calculation of β_k are kept fixed during a cycle. This is required in order not to destroy the conjugency of the search direction that is needed to assure quadratic termination. It should be also noted that in Step 4 we do not update D^i_k unless the denominator in the rank one correction term is positive. This rule is a crude stabilization device included just to indicate the need for some device that will assure positive definitness of the D^i_k matrices. In implementing Algorithm 1, one can use any of the stabilization approaches proposed in the literature for the Rank-One update. Such approaches have been suggested for instance by Murtagh and Sargent [8] and more recently by Cullum and Brayton [3].

3. A Resource Allocation Problem

We now consider the class of resource allocation problem having the form

$$\left.\begin{array}{l} \min f(x) = \sum_{i=1}^{h} f_i(x^i) \\ \text{subject to} \\ Ax = b \end{array}\right\} \tag{21}$$

where $x = (x^1, x^2, \dots x^h)$, $x \in R^n$, $x^i \in R^{m_i}$, $b \in R^r$ and A is an rxn matrix.

In this problem the objective function is partially separable in the sense that the decision vector x can be partitioned into segments each of which affects only one term in the objective function. All the decision variables, however, are related through a relatively small number of linear constraints. This would be a typical formulation in a multiperiod resource allocation problem in which the variables have to satisfy some linear resource or budget constraints.

One possible approach for solving this type of problem is using a multiplier or penalty function method. Such an approach involves the repeated unconstrained minimization of a penalty or penalized Lagrangian function having the form

$$L(x) = \sum_{i=1}^{h} f_i(x^i) + \mu \| Ax - b \|^2 \tag{22}$$

or

$$L(x) = \sum_{i=1}^{h} f_i(x^i) + \lambda' Ax + \mu \| Ax - b \|^2 \tag{23}$$

For either of the functions (22) or (23), the Hessian is:

$$\nabla^2 L(x) = \nabla^2 f(x) + 2\mu A'A$$

The matrix $\nabla^2 f(x)$, however, is block diagonal with the ith block being $\nabla^2 f_i(x^i)$. We thus have the structure assumed in the design of Algorithm 1 with $M_i = \nabla^2 f_i(x^i)$ and $R = 2\mu A'A$. Consequently, the penalty or penalized Lagrangian function given by (22) or (23) can be minimized using Algorithm 1 restarted every $r+1$ steps (i.e., number of constraints plus one).

4. A Class of Optimal Control Problems

We address now the class of optimal control problems considered by Bertsekas [1]. These problems have the form

$$\left.\begin{aligned} &J(u^0, \dots, u^N) = G(x^N) + \sum_{k=0}^{N-1} l_i(u^i) \ . \\ &\text{Subject to} \\ &x^{i+1} = A_i x^i + f_i(u^i) \qquad i = 0, 1, \dots, N-1; \qquad x^0 \text{ given.} \end{aligned}\right\} \tag{24}$$

Here $x^i \in R^n$ denotes the state, $u^i \in R^m$ is the control, A_i is an $n x n$ matrix, f_i: $R^m \to R^n$, $G{:}R^n \to R$ and $l_i{:}R^m \to R$.

The Hamiltonian for this problem is

$$H_i(x^i, u^i, \lambda^{i+1}) = l_i(u^i) + \lambda^{i+1\,'} [A_i x^i + f_i(u^i)], \qquad i = 0, \dots, N-1. \tag{25}$$

λ^i denotes here the costate and is defined by the adjoint equations

$$\left.\begin{aligned} &\lambda^i(u) = A_i' \lambda^{i+1}(u) \ , \qquad i = 1, \dots, N-1 \\ &\lambda^N(u) = \partial G/\partial x^N (x^N(u)) \end{aligned}\right\} \tag{26}$$

The gradient of the cost functional J with respect to the mN dimensional control vector $\mathbf{u}$ is given by

$$\nabla J(\mathbf{u}) = \left[\frac{\partial H_0}{\partial u^0}(\mathbf{u}), \ \dots, \ \frac{\partial H_{N-1}}{\partial u^{N-1}}(\mathbf{u}) \right] \tag{27}$$

where

$$\frac{\partial H_i}{\partial u^i}(\mathbf{u}) = \frac{\partial l_i}{\partial u^i}(u^i) + \lambda^{i+1}(\mathbf{u})' \frac{\partial f_i}{\partial u^i}(u^i) \tag{28}$$

In eq. (28) $\partial l_i/\partial u^i$ is a row vector denoting the gradient of l_i with respect to u^i and $\partial f_i/\partial u^i$ is the Jacobian of f_i with respect to u^i. The Hessian of J has the form

$$\nabla^2 J = \frac{\partial^2 H}{\partial u^2}(u) + M(u)\frac{\partial^2 G}{\partial (x^N)^2}(u)\, M(u)' \tag{29}$$

where $\partial^2 H/\partial u^2(u)$ is the block diagonal matrix

$$\frac{\partial^2 H}{\partial u^2}(u) = \begin{bmatrix} \frac{\partial^2 H_0}{\partial (u^0)^2}(u) & 0 & & \\ 0 & \frac{\partial^2 H_1}{\partial (u^1)^2}(u) & & 0 \\ & & \ddots & \\ & 0 & & \frac{\partial^2 H_{N-1}}{\partial (u^{N-1})^2}(u) \end{bmatrix} \tag{30}$$

and $M(u)'$ is the nxNm matrix

$$M(u)' = \left[A_{N-1} \cdots A_1 \frac{\partial f_0}{\partial u^0}, \ldots, A_{N-1}\frac{\partial f_{N-2}}{\partial u^{N-2}}, \frac{\partial f_{N-1}}{\partial u^{N-1}}\right]. \tag{31}$$

In view of this structure of the Hessian, the problem described in (24) can be conceptually solved by Algorithm 1 with a cycle length equal to the rank of $\partial^2 G/\partial (x^N)^2$ plus one. Such a procedure, however, would be impractical as it requires the evaluation of $M(u)[\partial^2 G/\partial (x^N)^2]M(u)'$ at each iteration. Fortunately, this can be avoided by replacing (16) with a scheme that evaluates q directly as a first order approximation to $(\partial^2 H/\partial u^2)\Delta u$.

The first order approximation to the change in $\nabla J(u)$ due to a change Δu in the control is given by

$$\nabla J(u+\Delta u) - \nabla J(u) \simeq \Delta u' \frac{\partial^2 H}{\partial u^2}(u) + \left[[\lambda^1(u+\Delta u) - \lambda^1(u)]' \frac{\partial f_0}{\partial u^0}(u^0), \ldots, [\lambda^N(u+\Delta u) - \lambda^N(u)]' \frac{\partial f_{N-1}}{\partial u^{N-1}}(u^N)\right] \tag{32}$$

Consequently q can be obtained by subtracting the second term in the right hand side of (32) from the gradient difference. Due to the special structure of $\partial^2 H/\partial u^2$ and in view of (27) and (28), eq. (32) can be decomposed into N equations of the form

$$\Delta u_i' \frac{\partial^2 H_i}{\partial (u^i)^2}(u) \simeq \frac{\partial l_i}{\partial u^i}(u^i+\Delta u^i) - \frac{\partial l_i}{\partial u^i}(u^i) + \lambda^{i+1}(u+\Delta u)'\left[\frac{\partial f_i}{\partial u^i}(u^i+\Delta u^i) - \frac{\partial f_i}{\partial u^i}(u^i)\right], \qquad i=0, 1, \ldots, N-1 \tag{33}$$

The right hand side in (33) is then an expression for the segment of q corresponding to the ith block in $\partial^2H/\partial u^2$.

Based on the above considerations we describe now a more specific version of Algorithm 1 designed for the optimal control problems addressed in this section. In this algorithm we use subscripts to denote iteration number and superscripts to denote time period. The rank of the matrix $\partial^2G/\partial x^N$ is denoted by r. For convenience, we have also changed the order of steps.

Algorithm 2

Start with a positive definite simetric matrix D^i_0 and any u^i_0 for $i=0, ..., N-1$

Step 1: Calculate x^i_k, $i = 0, ..., N-1$, using

$$x^{i+1}_k = A_i x^i_k + f_i(u^i_k), \text{ with } x^0_k = x^0 \tag{34}$$

and λ^i_k, $i = 1, ..., N-1$, using

$$\lambda^i_k = A_i'\lambda^{i+1}_k, \text{ with } \lambda^N_k = \partial G/\partial x^N(x^N_i) \tag{35}$$

Step 2: For $i = 0, ..., N-1$, calculate

$$L^i_k = \frac{\partial l_i}{\partial u^k}(u^i_k)', \quad \text{and} \quad F^i_k = \frac{\partial f_i}{\partial u^i}(u^i_k)' \quad . \tag{36}$$

then obtain

$$\nabla J^i_k = L^i_k + F^i_k\,\lambda^{i+1}_k \tag{37}$$

Step 3: If $k=0$ then go to Step 6, otherwise for $i = 0, ..., N-1$, calculate

$$\Delta u^i_k = u^i_k - u^i_{k-1} \tag{38}$$

$$q^i_k = L^i_k - L^i_{k-1} + (F^i_k - F^i_{k-1})\lambda^{i+1}_k \tag{39}$$

and

$$v^i_k = \Delta u^i_k - D^i_k\, q^i_k \tag{40}$$

Step 4: For $i = 0, ..., N-1$;

$$\left.\begin{array}{ll} \text{if} & v^i_k{}'\, q^i_k \leq 0 \text{ set } D^i_k = D^i_{k-1} \\ \text{otherwise} & D^i_k = D^i_{k-1} + ({}^1/_{v^i_k{}' q^i_k})\, v^i_k\, v^i_k{}' \end{array}\right\} \tag{41}$$

Step 5: If $k < r+1$, go to Step 6. Otherwise, reset k to 0, then set $D^i_0 = D^i_{r+1}$, $u^i_0 = u^i_{r+1}$ and $\nabla J^i_0 = \nabla J^i_{r+1}$ for $i = 0, ..., N-1$.

Step 6: For $i = 0, ..., N-1$; if $k = 0$ set $d^i_0 = -D^i_0\nabla J^i_0$, otherwise

$$d^i_k = -D^i_0\nabla J^i_k + \beta_{k-1}\, d^i_{k-1} \tag{42}$$

where

$$\beta_{k-1} = \Big(\sum_{i=0}^{N-1} \nabla J^i_k{}'\, D^i_0 \nabla J^i_k\Big) \Big/ \sum_{i=0}^{N-1} \nabla J^i_{k-1}{}'\, D^i_0\, \nabla J^i_{k-1} \tag{43}$$

Step 7: For $i = 0, ..., N-1$ calculate

$$u^i_{k+1} = u^i_k + \alpha_k\, d^i_k \tag{44}$$

where α_k minimizes $J(u^0_k+\alpha d^0_k, ..., u^{N-1}_k+\alpha d^{N-1}_k)$.

Then, increment k by 1 and return to Step 1.

It should be noted that since D^i_0 is replaced only at the end of r+1 steps the updating performed in Step 3 and 4 could be done after the cycle is completed. Such an approach, however, would require more storage since then we have to store Δu^i_k and q^i_k for all $i=0,...,N-1$ and $k=1,...,r+1$. For a linear quadratic problem where $f_i(u^i) = B_i u^i$ and $l_i(u^i) = u^{i'} R_i u^i$ we have $q^i_k = R_i \Delta u^i_k$. Thus, by Theorem 1 it will take m updatings to obtain R_i^{-1}. Assuming that $r+1 \geq m$, the second cycle will be a properly scaled partial conjugate gradient cycle and will thus converge to the exact minimum by the end of that cycle.

5. Conclusions

General purpose Variable Metric algorithms are impractical for large scale optimization problems due to their high computational and storage requirements. These costs, however, can be reduced by specializing such algorithms to specific classes of problems and exploiting the special structure of such problem to reduce computational and storage requirements. The paper implements this philosophy for a class of problems in which the Hessian matrix consist of a sum of a block diagonal matrix and a low rank matrix. We use a rank-one updating formula to approximate the inverse of the first part of the Hessian. We then use that approximation in a Scaled Partial Conjugate Gradient algorithm restarted every r+1 steps where r is the rank of the second term in the Hessian. Due to the block diagonal form of the first part of the Hessian being approximated, the block's can be updated and stored individually. This procedure considerably reduces the storage requirements by maintaining sparcity. Furthermore for a quadratic problem the approximation becomes exact after as many steps as the dimension of the largest blocks. The resulting algorithm possesses "quadratic termination" in a number of step significantly lower than the number of variables.

The use of the Rank One update was motivated by the fact that it permits the use of arbitrary independent updating vectors. This property is crucial since the search directions in our approach are different from the vectors obtained by multiplying the current matrix approximation times the gradient. On the other hand, the use of the Rank One formula raises stabilization problems as it does not guarantee the positive definitness of the approximations even when applied to a positive definite quadratic function. Various stabilization schemes have been proposed in the literature and can be used in implementing our approach. Much of the stabilization problems, however, can be avoided by using positive definite initial approximations whose eigenvalues are all below the eigenvalues of the matrix being approximated. It can be shown (see Luenberger [7]) that in the quadratic case such an initial approximation guarantees that all the rank one correction terms will have positive denominators, and hence, the approximations will be positive definite. A simple initial approximation satisfying the above requirement would be, for instance, the identity scaled through division by an upper bound on the norm of the matrix being approximated.

We presented two classes of problems leading to the structure under consideration and discussed the potential implementation of the proposed approach to these problems. In both cases the algorithm uses only first derivatives and doesn't require any matrix inversions. Numerical experiments, however, with this method are limited so far and it still remains to demonstrate practical value of the proposed method.

References:

1. Bertsekas, D. P., "Partial Conjugate Gradient Methods for a Class of Optimal Control Problems", *IEEE Transaction on Automatic Control*, Vol. AC-19, pp. 209-217, 1974.

2. Broyden, C. G., "Quasi-Newton Methods and Their Application to Function Minimization", *Math. Comp.*, Vol. 21, pp 368-381, 1967.

3. Cullum, Jane and Brayton, R. K., Some Remarks on the Symmetric Rank One Update, IBM Research Center Report 6157, Yorktown Heights, New York, 1976.

4. Davidon, W. C., "Variable Metric Method for Minimization," A.E.C. Research and Development Rep. ANL-5990 (Rev.), 1959.

5. Fletcher, R. and Powell, M. J. D., "A Rapidly Convergent Descent Method for Minimization", *Comp. Jnl.*, Vol. 6, pp. 163-168, 1963.

6. Fletcher, R. and Reeves, C. M., "Function Minimization by Conjugate Gradients", *Comp. Jnl.*, Vol. 7, pp. 149-154, 1964.

7. Luenberger, D. G., *"Introduction to Linear and Nonlinear Programming"*, Reading, Mass: Addison-Wesley 1973.

8. Murtagh, B. A. and R. W. H. Sargent, Computational Experience with Quadratically Convergent Minimization Methods, *Comp. Jnl.* 13, 185-194, 1970.

SIMPLEX METHOD FOR DYNAMIC LINEAR PROGRAM SOLUTION

V.E. Krivonozhko and A.I. Propoi
Institute for Systems Studies
Moscow, USSR
and
International Institute for Applied Systems Analysis
Laxenburg, Austria 2361

1. INTRODUCTION

In this paper, the extension of the simplex-method, one of the most effective linear programming methods [1,2] to dynamic linear programming [3] is described. The main concept of the static simplex-method -- the "global" basis -- is replaced by the set of local (for each time period t) bases. It allows us to develop a whole group of finite-step DLP methods: primal, dual and primal-dual methods, each yielding the same solution path as the corresponding static version of the simplex-method. The methods are closely related to the basis factorization approach to DLP problems. We consider a DLP problem in the form:

Problem 1: Find a control $u = \{u(0),\ldots,u(T-1)\}$ *and a trajectory* $x = \{x(0),\ldots,x(T)\}$, *satisfying the state equation*

$$x(t+1) = A(t)x(t) + B(t)u(t) \tag{1}$$

with initial condition

$$x(0) = x^0 \tag{2}$$

and constraints

$$G(t)x(t) + D(t)u(t) = f(t) \tag{3}$$

$$u(t) \geq 0 \tag{4}$$

which maximize the objective function

$$J_1(u) = a(T)x(T) \quad . \tag{5}$$

Here $x(t)$ is the n-vector of state variables; $u(t)$ is the r-vector of control variables; $f(t)$ is the given m-vector $(t=0,1,\dots,T-1)$.

This model is flexible enough and allows various extensions and modifications. The results stated below for Problem 1 can be used with minor changes for these extensions and modifications (see Section 3 and [4]).

Along with the primary Problem 1, statement of the dual problem will be necessary [4].

Problem 2: Find a dual control $\lambda = \{\lambda(T-1),\dots,\lambda(0)\}$ *and a dual trajectory* $p = \{p(T),\dots,p(0)\}$, *satisfying the costate equations*

$$p(t) = p(t+1)A(t) - \lambda(t)G(t) \qquad (t = T-1,\dots,1,0) \tag{6}$$

with boundary condition

$$p(T) = a(T) \tag{7}$$

and constraints

$$p(t+1)B(t) - \lambda(t)D(t) \leq 0 \tag{8}$$

which minimize the performance index

$$J_2(\lambda) = p(0)x^0 + \sum_{t=0}^{T-1} \lambda(t)f(t) \quad . \tag{9}$$

For this pair of dual problems the conventional duality realizations hold [4].

2. DYNAMIC SIMPLEX METHOD

Problem 1 can be considered as a "large" static LP problem with a constraint matrix of dimension $(m+n)T \times (n+r)T$ and solved by standard LP methods. This straightforward approach is, however, uneffective due to the dimension of the "global" basic matrix. More natural for DLP problems is to introduce in some way a local basis (of dimension $m \times m$) for each time period $t = 0,\dots,T-1$ and operate only with this set of T local bases [5]. Here we interpret this approach as a block factorization method.

Let $\bar{B}$ represent the "global" basic matrix of the Problem 1. It has the same stair-case structure as the constraint matrix of Problem 1 and dimension $(m+n)T \times (m+n)T$. The basic matrix $\bar{B}$ can be represented as follows:

$$\bar{B} = B^*V_{T-2}U_{T-2} \cdots V_tU_t \cdots V_0U_0 = B^*U \qquad (10)$$

The matrix B* has a structure

$$B = \begin{bmatrix} \hat{D}_{0B}(0) & & & & & & \\ \hat{B}_{0B}(0) & -I & & & & & \\ & G(1) & \hat{D}_{0B}(1) & & & & \\ & A(1) & \hat{B}_{0B}(1) & -I & & & \\ & & & \ddots & & & \\ & & & & G(T-1) & \hat{D}_{0B}(T-1) & \\ & & & & A(T-1) & \hat{B}_{0B}(T-1) & -I \end{bmatrix} \qquad (11)$$

where $\hat{D}_{0B}(t)$ $(t=0,\dots,T-1)$ is a square nonsingular matrix of dimension $m \times m$ and formed either by columns of the matrix $D(t)$ or by some columns of matrices $D(\tau)$ $(\tau=0,\dots,t-1)$, which are recomputed to step t during factorization process [5]. We shall call matrices $\hat{D}_{0B}(t)$ $(t=0,\dots,T-1)$ the *local bases*.

The matrices U_t and V_t are

$$U_t = \begin{bmatrix} 1 & & & \\ & \ddots & & \Phi(t) \\ 0 & & \ddots & \\ & & & 1 \end{bmatrix} \qquad V_t = \begin{bmatrix} 1 & & & \\ & \ddots & & \Psi(t) \\ 0 & & \ddots & \\ & & & 1 \end{bmatrix}$$

where

$$\Phi(t) = [\Phi_0^{t+1}(t) \ \dots \ 0 \ \dots \ \Phi_i^j(t) \ \dots \ 0 \ \dots \ \Phi_t^{T-1}(t)]$$

$$\Psi(t) = [-B_0^{t+1}(t) \ \dots \ 0 \ \dots \ -B_i^j(t) \ \dots \ 0 \ \dots \ -B_t^{T-1}(t)]$$

and $\Phi_i^j(t)$ and $-B_i^j(t)$ correspond to those basic control variables $u_B(i)$ which enter local basis $\hat{D}_{0B}(j)$. Location of raws of submatrices $\Phi(t)$ and $\Psi(t)$ in U_t and V_t corresponds to the location of raws of submatrices $\hat{D}_{0B}(t)$ and $\hat{B}_{0B}(t)$ in B*.

Taking into account the permutations of basis columns in the factorization process, we can write the basic variables as

$$\{u_B, x\} = \{\hat{u}_{0B}(0), x(1), \hat{u}_{0B}(1), \dots, \hat{u}_{0B}(T-1), x(T-1)\}$$

where vectors $\hat{u}_{0B}(t)$ correspond to matrices $\hat{D}_{0B}(t)$, free variables $x(t)$ are always in the basis.

Using the representation (10), a basic feasible control u_B (primal solution) can be computed in an effective way by the recurrent formulas: first, at forward run $(t=0,\dots,T-1)$ we obtain

$$\hat{u}^*_{0B}(t) = \hat{D}^{-1}_{0B}(t)[f(t) - G(t)x^*(t)]$$

$$x^*(t+1) = A(t)x^*(t) + \hat{B}_{0B}(t)u^*_{0B}(t)$$

$$x^*(t) = x(0)$$

and second, at backward run $(t = T-1,\dots,0)$ we obtain the values of $\hat{u}_{0B}(t)$, which are a basic feasible control of Problem 1:

$$x(T) = x^*(T); \quad u(T-1) = u^*(T-1)$$

$$x(t) = x^*(t) + \sum_{i=0}^{t-1} \sum_{j=t}^{T-1} [B_i^j(t) \vdots 0]\hat{u}_{0B}(j)$$

$$\hat{u}_{0B}(t) = \hat{u}^*_{0B}(t) - \sum_{i=0}^{t} \sum_{j=t+1}^{T-1} [\Phi_i^j(t) \vdots 0]\hat{u}_{0B}(j)$$

where notation $[B_i^j(t) \vdots 0]$ denotes that submatrix $B_i^j(t)$ is augmented by zeros if necessary for correctness of multiplication.

The coefficients which represent a column to be introduced in the basis are computed similarly to primal solution procedure.

The *dual solution* $\{\lambda(t), p(t+1)\}$ can be also obtained using only the set of local bases:

$$\lambda(t) = p(t+1)\hat{B}_{0B}(t)\hat{D}_{0B}^{-1}(t)$$
$$p(t) = p(t+1)A(t) - \lambda(t)G(t) \quad ; \quad p(T) = a(T) \quad .$$

It is shown [5] that $\{\lambda, p\}$ are simplex-multipliers for the given set of local bases $\{\hat{D}_{0B}(t)\}$. Hence, we can use them for *pricing* out the columns.

The *updating* procedure is one of the most crucial. In the algorithm, an effective procedure of the updating of local bases is suggested. We describe the idea briefly. Let a variable to be removed from the basis belong to local basis $\hat{D}_{0B}(t_1)$ and let submatrix $\Phi(t_1)$ contain a nonzero pivot element corresponding to a variable from local basis $\hat{D}_{0B}(t_1+1)$. At the interchange of these variables, the inverse $\hat{D}_{0B}^{-1}(t_1)$ is updated by multplying from the left on an elementary column matrix, and the inverse $\hat{D}_{0B}^{-1}(t_1+1)$ -- by multiplying from the left on an elementary raw matrix. The updating of $\Phi(t_1)$, $\Psi(t_1)$ and $\Phi(t_1+1)$ is carried out in a similar way. The other local bases are not changed in this case.

Now let a nonzero pivot element of $\Phi(t_1)$ correspond to a variable from local basis $\hat{D}_{0B}(t_2)$ $(t_2>t_1+1)$, and let all elements of the pivot raw of $\Phi(t_1)$, which correspond to variable from local bases $\hat{D}_{0B}(\tau)$, $(t_1<\tau<t_2)$ be zero. Then at variable interchange, the inverses of local bases $\hat{D}_{0B}^{-1}(t_1)$ and $\hat{D}_{0B}^{-1}(t_2)$ are updated as was described above, the others are not changed.

By these successive interchanges, the variable to be removed from the basis is transformed to such a local basis $\hat{D}_{0B}(t_3)$ $(t_3\leq T-1)$, for which the pivot raw of $\Phi(t_3)$ is zero.

A column to be introduced into the basis is recomputed to the step t_3, then the updated column is introduced to the local basis $\hat{D}_{0B}(t_3)$.

We described briefly the basic procedures of the dynamic simplex-method. One can see that for realization of the algorithm, it is sufficient to operate only with matrices $\hat{D}_{0B}^{-1}(t)$, $\Phi(t)$, $\hat{B}_{0B}(t)$, $\Psi(t)$, $G(t)$ and $A(t)$ $(t=0,\dots,T-1)$. It is shown in [5] that the number of nonzero columns of matrices $\Phi(t)$ or $\Psi(t)$ does not exceed n. Hence, the algorithm operates only with the set of T matrices, each containing no more than m or n columns.

Thus, instead of dealing with the 'global' $(m+n)T \times (m+n)T$ basis, we can effectively write all the simplex operations for the set of T local $(m \times m)$ bases. It simplifies operations to a great extent. By updating the inverse of a local basis, multiplication is carried out on an elementary matrix, therefore the local bases can be kept in multiplicative form and standard invert-procedures can be used here. In addition, it should be noted that not all local bases are updated at each iteration.

3. EXTENSIONS

The approach considered above is flexible and allows different extensions and generalisations. Below we describe briefly two of them. First, in Problem 1 the state variables $x(t)$ are considered to be free. The case when $x(t) \geq 0$ or $0 \leq x(t) \leq \alpha(t)$ can be treated by the approach very easily. In fact, from the point of view of the computer implementation of the algorithm, it is better to handle the multiplicative form of the inverse of

$$\tilde{D}_{0B}(t) = \begin{pmatrix} \hat{D}_{0B}(t) & 0 \\ B_{0B}(t) & -I \end{pmatrix}$$

rather than $D_{0B}^{-1}(t)$ because the addition of the unit matrix $-I$ does not generate additional zeros in the eta-file. If $x(t)$ are not constrained, then by handling with the inverse of $\tilde{D}_{0B}(t)$ we can consider the raws corresponding to low blocks of $\tilde{D}_{0B}(t)$ as free.

In this case, all $x(t)$ are in the basis. If $x(t) \geq 0$, then the state variables $x(t)$ should be handled in the same way as control variables $u(t) \geq 0$. In this case, not all $x(t)$ will be in the basis. Evidently, this includes the case, when both state and control variables have upper bound constraints. (The inclusion of generalized upper bound constraints is also possible).

The second case, which is very important for practice, is DLP with time delays. Instead of (1) and (3) we have

$$x(t+1) = \sum_{\nu} A(t,\tau_\nu)x(t-\tau_\nu) + \sum_{\mu} B(t,\tau_\mu)u(t-\tau_\mu) \qquad (12)$$

$$\sum_{\nu} G(t,\tau_\nu)x(t-\tau_\nu) + \sum_{\mu} D(t,\tau_\mu)u(t-\tau_\mu) = f(t) \qquad (13)$$

with given values for $x(t)$, $u(t)$, $t \leq 0$. Here $\{\tau_\nu\}$, $\{\tau_\mu\}$ are given sets of integers.

The new submatrices will appear to the left from the main staircase of the diagonal of B^* in (11). As this main staircase structure does not change in this case, we can use the same procedures as in the case without time delays. But now local basis $\hat{D}_{0B}(t)$ will contain recomputed columns both from previous steps $\tau < t$ and columns from time delayed matrices $D(t,\tau)$ $\tau \leq t$, which enter the constraints (13) at step t.

Thus, both of these extensions of Problem 1 can be handled by the algorithm almost without any modifications.

4. DUAL ALGORITHMS

The introduction of local bases and technique of their handling allows us to develop dual and primal-dual versions of the dynamic simplex-method. The main advantage of using the dual methods is that the dual statements of many problems have explicit solutions. The other is connected with the choice of different selection strategies to the vector pair which enters and leaves the basis.

In the primal version of the dynamic simplex-method, there are some options for choice of a column with the most negative price from all non-basic columns or from some set of these columns, etc. But a column to be removed from the basis is unique in the nondegenerative case.

Contrarily, in dual methods there are options in the choice of a column to be removed from the basis. It can be effectively used in dual versions of the method. In practical problems, local bases $\{\hat{D}_{0B}(t)\}$ can be rather large, therefore part of the local bases should be stored at the external storage capacities. Input-Output operations are comparitively time-consuming. Hence, to reduce the total solution time, it is desirable to have more pivoting operations with a given local basis.

Thus, the usage of different dual and primal-dual strategies allows us to adjust the algorithm to the specifics of the computer to be used and to the problem to be solved.

REFERENCES

[1] Dantzig, G.B., *Linear Programming and Extensions*, Princeton Univ. Press, Princeton, N.J., 1963.

[2] Kantorovich, L.V., *The Best Use of Economic Resources*, Harvard University Press, Cambridge, Mass., 1965.

[3] Propoi, A.I., *Problems of DLP*, RM-76-78, International Institute for Applied Systems Analysis, Laxenburg, Austria, 1976.

[4] Propoi, A.I., *Dual Systems of DLP*, RR-77-9, International Institute for Applied Systems Analysis, Laxenburg, Austria, 1977.

[5] Propoi, A.I. and V.E. Krivonozhko, *The Dynamic Simplex Method*, RM-77-24, International Institute for Applied Systems Analysis, Laxenburg, Austria, 1977.

AN ADAPTIVE PRECISION METHOD FOR THE NUMERICAL SOLUTION OF CONSTRAINED OPTIMIZATION PROBLEMS APPLIED TO A TIME-OPTIMAL HEATING PROCESS

K. Schittkowski
Institut für Angewandte Mathematik und Statistik
Universität Würzburg
D-87 Würzburg, W.Germany

1. Introduction

Consider the general minimization problem

$$\inf_{x \in C} \varphi(x) \tag{1}$$

where φ is a real-valued function defined on a normed linear space E and C is a subset of E. It is presumed that

$$\inf \{\varphi(x) : x \in C\} > -\infty .$$

In many applications it is not possible to compute φ exactly or a precise evaluation is extremely expensive, for example if we try to solve min-max, variational, or control problems. The main advantage of the subsequent algorithm will be that it is possible to vary the accuracy of the evaluation of φ or its derivative. Specially it is allowed to use low accuracy while far from a solution and to improve it step by step as a solution is approached.

The minimization procedure is derived from an unconstrained version defined in arbitrary normed linear spaces [3] which allows very general gradient type search directions only restricted by the condition that they are bounded away from orthogonality with the gradient. Furthermore the steplength procedure is a simple bisection method leading to short programming codes.

If restrictions appear, i.e. if $C \neq E$, then they are handled by defining penalty functions φ^n. The degree of penalizing the function φ, the index n, is controlled by the algorithm and raised infinitely when reaching a solution of (1). So we get an algorithm which controls both the accuracy of the approximations and the degree of penalizing φ. It is possible to develop global convergence theorems in the sense that a critical point is approximated.

To show the efficiency of the proposed method, we apply it to the numerical solution of a time-optimal parabolic boundary value control problem resulting from a one-dimensional heat diffusion process. In this case the variables are switching times of bang-bang solutions, the cost function φ is known exactly, and the set C is described by one highly nonlinear inequality constraint. The corresponding restriction function is defined by an infinite series and has to be approximated by finite series. A quite different approach to solve this problem numerically with respect to the L_∞-norm is outlined in [2].

Since this heat diffusion process leads to a finite dimensional problem, we restrict ourselves to the case $E = \mathbb{R}^n$, a generalization to arbitrary normed linear spaces is found in [4]. In the next section all assumptions are gathered to define the penalty functions, the accuracy of the approximations, and to present the algorithm. Section 3 contains some global convergence results. It is shown in section 4, how this method can be applied to the solution of the time-optimal heating process mentioned above. In the last section some numerical results are listed.

2. The algorithm

Let $E := \mathbb{R}^n$ be supplied with the Euclidean norm $\|.\|$ and $d(x,U)$ denote the distance of a point x of E from a subset $U \subset E$. The subsequent algorithm is based on the idea to penalize the function φ, if an iterate leaves the set C.

Definition 2.1. A sequence of real-valued functions $\{\varphi^n\}$ is called a sequence of penalty functions for problem (1), if the following conditions are valid:

a) Each function φ^n is continuously differentiable on E.

b) Let L be a compact subset of E, $\epsilon > 0$, and C_ϵ be any neighbourhood of C with $d(z,C) \leq \epsilon$ for all $z \in C_\epsilon$. Then there is a $\delta_\epsilon > 0$ and a $n_\epsilon \in N$ with

$$\|D\varphi^n(x)\| \geq \delta_\epsilon \qquad (2)$$

for all $n \geq n_\epsilon$ and $x \in L \setminus C_\epsilon$.

Let $\{\varphi^n\}$ be a sequence of penalty functions. Now it is possible to state the assumptions about the accuracy of the approximations.

Assumption 2.2. For each $n,l \in N$ and $x \in E$ there are subsets $\Phi_l^n(x) \subset \mathbb{R}$, $\Phi_l^{n'}(x) \subset E$ such that the convergence of the approximations of φ^n and $D\varphi^n$ is uniform on every compact subset U of E, that means for all $\varepsilon > 0$ there exists a $l_\varepsilon \in N$ with

$$|\varphi_l^n(x) - \varphi^n(x)| < \varepsilon$$
$$\|\varphi_l^{n'}(x) - D\varphi^n(x)\| < \varepsilon \qquad (3)$$

for all $l \geq l_\varepsilon$, $n \in N$, $\varphi_l^n(x) \in \Phi_l^n(x)$, $\varphi_l^{n'}(x) \in \Phi_l^{n'}(x)$, $x \in U$.

The approximations of φ^n and $D\varphi^n$ are defined by certain subsets and not by functions. This makes sense for example when solving min-max problems, i.e. $\varphi(x) := \max\{f(x,y) : y \in D\}$, $C := E$. For the l-th approximation of $\varphi^n = \varphi$ we would use the results of the l-th iteration step of an optimization procedure to maximize $f(x,y)$, $y \in D$. But the iterates are in general not uniquely determined, they depend on a lot of parameters, for example the choice of initial values.

Now we are able to develop an algorithm for solving the minimization problem (1). A version for the unconstrained case is found in [3], an extension to the general model is presented in [4]. For the implementation of the procedure select positive real numbers with

$$1 \geq \gamma_j \geq \gamma > 0 \qquad j \in N,$$
$$\sigma^* \geq \sigma_j \geq \sigma > 0 \qquad (4)$$

and

$$\lim_{j\to\infty} \mu_j = 0 .$$

The algorithm proceeds as follows:

Algorithm 2.3. Choose $q(0), n(0) \in N$, $x_0 \in E$. For $k = 0,1,2,\ldots$ compute $q(k+1), n(k+1)$, and $x_{k+1} \in E$ as follows:

1) Denote $l := q(k)$, $n := n(k)$.
2) If $\|\varphi_l^{n'}(x_k)\| \leq \mu_n$ for a $\varphi_l^{n'}(x_k) \in \Phi_l^{n'}(x_k)$, let $n := n+1$.
3) Let $l := l+1$.
4) Determine a $\varphi_l^n(x_k) \in \Phi_l^n(x_k)$ and a $\varphi_l^{n'}(x_k) \in \Phi_l^{n'}(x_k)$,

further $\rho_k^l := \sigma_k \|\varphi_l^{n'}(x_k)\|$.

5) Compute a search direction $s_k \in E$ with $\|s_k\| \leq 1$ and
$$\varphi_l^{n'}(x_k)^T s_k \geq \gamma_k \|\varphi_l^{n'}(x_k)\| \ . \qquad (5)$$

6) Evaluate the smallest nonnegative integer $j \leq l$ with
$$\varphi_l^n(x_k - 2^{-j}\rho_k^l s_k) \leq \varphi_l^n(x_k) - \frac{1}{2} 2^{-j}\rho_k^l \gamma_k \|\varphi_l^{n'}(x_k)\| \ , \qquad (6)$$
where $\varphi_l^n(x_k - 2^{-j}\rho_k^l s_k) \in \Phi_l^n(x_k - 2^{-j}\rho_k^l s_k)$ arbitrarily chosen.

7) If such a j does not exist, goto step 3.

8) Define the new iterate
$$x_{k+1} := x_k - 2^{-j}\rho_k^l s_k$$
and let $q(k+1) := l$, $n(k+1) := n$.

In other words, the algorithm works as follows: Starting from an initial point, 2.3. yields iterates $x_k \in E$ and two monotone increasing sequences of positive integers $\{q(k)\}$, $\{n(k)\}$ with $q(k) \to \infty$ for $k \to \infty$ and, this has to be shown, $n(k) \to \infty$ for $k \to \infty$. In each iteration step the accuracy of the approximations will be raised at least by 1. This might be replaced by a more general condition which only requires that the increase of the accuracy, defined by the integer l, is unbounded.

The computation of the search direction s_k can be characterized as a gradient-type method. It is a "downhill" direction only restricted by condition (5) which requires that s_k is bounded away from orthogonality with the gradient. The steplength parameters 2^{-j} can be replaced by some sequence $\{\alpha_j\}$ with $\lim_{j\to\infty} \alpha_j = 0$ and $\alpha_{j+1} > \alpha_j > 0$ for all j.

Since the approximations φ_l^n are in general not defined by functions, it is not possible to guarantee the existence of a j implementing inequality (6). Therefore the restriction "$j \leq l$" is mandatory and theorem 3.4. of [4] shows the finiteness of the loop between step 7 and step 3 of algorithm 2.3. for each k with $\|D\varphi^n(x_k)\| > 0$.

3. Convergence results

In this section we gather some global convergence results developed in [4] where the corresponding proofs are presented. The fundamental convergence statement is the following one:

Theorem 3.1. Let $\{x_k\}$ be an iteration sequence constructed by algorithm 2.3. subject to a sequence of penalty functions $\{\varphi^n\}$ with $\|D\varphi^{n(k+1)}(x_k)\| > 0$ for all k. Assume that there is a compact subset L of E with $x_k \in L$ for all k. Then we have

$$\lim_{k\to\infty} n(k) = \infty$$

and there is a subsequence of $\{x_k\}$, i.e. an infinite subset S of N with

$$\begin{aligned} &\lim_{k\in S} \|D\varphi^{n(k)}(x_k)\| = 0 , \\ &\lim_{k\in S} d(x_k,C) = 0 . \end{aligned} \tag{7}$$

To get some further results concerning the convergence of $\{\varphi^{n(k)}(x_k)\}$, we need more restrictive assumptions about the penalty functions:

Theorem 3.2. Let C be bounded and $\{\varphi^n\}$ be a sequence of convex penalty functions with $\varphi^n(x) \geq \varphi(x)$ for all $x \notin C$ and $\varphi^n(x) \to \varphi(x)$ uniformly on C for $n \to \infty$. Furthermore let $\{x_k\}$ be an iteration sequence of algorithm 2.3. with $x_k \in L$ for all k and a compact subset L of E, and with

$$\|D\varphi^{n(k+1)}(x_k)\| > 0$$

for all k. Then there is an infinite subset S of N with

$$\lim_{k\in S} \varphi^{n(k)}(x_k) = \inf_{x\in C} \varphi(x) =: m . \tag{8}$$

If, in addition, $\lim_{\|x\|\to\infty} \varphi^n(x) = \infty$ for all n, then

$$\lim_{n\to\infty} \inf_{x\in E} \varphi^n(x) = \underline{\lim}_{k\to\infty} \varphi^{n(k)}(x_k) = m . \tag{9}$$

4. Application to the solution of a time-optimal heating process

The adaptive precision method outlined in the last two sections is applied to the numerical solution of a one-dimensional heat diffusion process. A thin rod shall be heated at one end point such that a given temperature distribution $k_o \in L_2[0,1]$ will be approximated subject to the L_2-norm with a given accuracy ϵ as soon as possible. This leads to the following time-optimal parabolic boundary value control problem, where $y(s,t)$ denotes the temperature at a point $s \in [0,1]$ and a time $t \in [0,T]$:

Minimize the control time T under the restrictions that there is a $u \in L_2[0,T]$ with

$$\begin{aligned} & y_t(s,t) - y_{ss}(s,t) = 0 \\ & y_s(0,t) = 0, \; y(s,0) = 0 \\ & y(1,t) + \alpha y_s(1,t) = u(t) \\ & \|y(.,T) - k_o(.)\|_2 \leq \epsilon \\ & -1 \leq u(t) \leq 1 \; . \end{aligned} \qquad \begin{aligned} & t \in [0,T] \\ & s \in [0,1] \end{aligned} \qquad (10)$$

α is a constant heat transfer coefficient. For every $u \in L_2[o,T]$ the solution $y(s,t,u)$ of the above boundary value problem is given by

$$y(s,t,u) = \sum_{j=1}^{\infty} A_j \mu_j^2 \cos(\mu_j s) \int_0^t u(\tau) \exp(-\mu_j^2(t-\tau)) \, d\tau, \quad (11)$$

where $A_j := 2\sin\mu_j/(\mu_j + \sin\mu_j \cos\mu_j)$, $j = 1,2,\ldots$, and $\{\mu_j\}$ is the sequence of all positive solutions of the equation $\mu \tan\mu = 1/\alpha$, confer Yegorov [5].

We know that the above control problem is solvable and we proceed from the assumption that the minimal control time T_o is positive. Furthermore the optimal control is a uniquely determined bang-bang function whose switching times accumulate at most in T_o, confer Yegorov [5]. If we assume now that this optimal solution has at most k switching times, we get a finite dimensional optimization problem of the kind

$$\begin{gathered} \min \; T \\ (T,t) := (T,t_1,\ldots,t_k): \; \|y(.,T,u(T,t)) - k_o(.)\|_2 \leq \epsilon \quad (12) \\ 0 \leq t_1 \leq \ldots \leq t_k \leq T \end{gathered}$$

The control $u(T,t)$ is bang-bang with jumps at $t_1,\dots,t_k$, i.e. the variables are the switching times of bang-bang functions.

In order to eliminate the restriction $0 \le t_1 \le \dots \le t_k \le T$, we use the transformation $x_1^2 := t_1$, $x_i^2 := t_i - t_{i-1}$, $i=2,\dots,k$, and $x_{k+1}^2 := T - t_k$, to get the problem

$$\begin{gathered}\min \|x\|^2 \\ x \in \mathbb{R}^{k+1}:\ g(x) \le \varepsilon^2\end{gathered} \tag{13}$$

with $g(x) := \int_0^1 \left(\sum_{j=1}^{\infty} \alpha_j(s)\rho_j(x) - k_o(s)\right)^2 ds$,

$$\alpha_j(s) := \mu_j^2 A_j \cos(\mu_j s) ,$$

$$\rho_j(x) := -\mu_j^{-2}\left(\exp\left(-\mu_j^2 \sum_{i=1}^{k+1} x_i^2\right) - 2\exp\left(-\mu_j^2 \sum_{i=2}^{k+1} x_i^2\right) + \dots + (-1)^k 2\exp\left(-\mu_j^2 x_{k+1}^2\right) + (-1)^{k+1}\right)$$

for all $s \in [0,1]$, $x := (x_1,\dots,x_{k+1})^T \in \mathbb{R}^{k+1}$.

Defining $E := \mathbb{R}^{k+1}$, $\varphi(x) := \|x\|^2$, and $C := \{x \in E:\ g(x) \le \varepsilon^2\}$, we get an optimization problem of the kind (1). Furthermore let

$$\varphi^n(x) := \|x\|^2 + r_n \chi(x) \tag{14}$$

with $r_n \in \mathbb{R}$, $r_n \to \infty$ for $n \to \infty$, and $\chi(x) := \max(0, g(x) - \varepsilon^2)^2$. Since we consider only problems with a positive control time T_o, we can expect without loss of generality that the minimization procedure 2.3. does not approximate the origin. It is easy to see that each φ^n is continuously differentiable on $E \setminus \{0\}$ and that 2.1.b) is valid for each closed and bounded subset of $E \setminus \{0\}$. So we are able to regard $\{\varphi^n\}$ as a sequence of penalty functions.

For a numerical evaluation of φ^n, we have to replace the infinite series defining φ^n by a finite one. Let

$$\varphi_l^n(x) := \|x\|^2 + r_n \chi_l^n(x),\quad x \in E, \tag{15}$$

with $\chi_l^n(x) := \max(0, g_l^n(x) - \varepsilon^2)^2$

and $g_l^n(x) := \int_0^1 \left(\sum_{j=1}^{j(l,n)} \alpha_j(s)\rho_j(x) - k_o(s)\right)^2 ds$,

where $j(l,n)$ satisfies $l r_n \le \beta(j(l,n) - 1)$ for all $l,n \in \mathbb{N}$ and a constant real number $\beta > 0$.

Since $g_l^n(x)$ is differentiable on E, we define $\varphi_l^{n'}(x) := D\varphi_l^n(x)$ to approximate $D\varphi^n$. Furthermore let $\Phi_l^n(x) := \{\varphi_l^n(x)\}$ and $\Phi_l^{n'}(x) := \{\varphi_l^{n'}(x)\}$ for all $x \in E$. This leads to the result:

Theorem 4.1. The approximations defined by $\varphi_l^n(x)$ and $\varphi_l^{n'}(x)$ satisfy assumption 2.2. for each compact subset $U \subset E \setminus \{0\}$.

Proof. Let U be a compact subset of $E \setminus \{0\}$ and $x \in U$. The orthogonality of $\{\cos(\mu_j s)\}$ leads to

$$|g_l^n(x) - g(x)|$$
$$= \Big|\sum_{j=j(l,n)+1}^{\infty} \Big({}_0\int^1 \alpha_j{}^2(s)ds\rho_j(x) - 2\,{}_0\int^1 \alpha_j(s)k_o(s)ds\Big)\rho_j(x)\Big)\Big|$$
$$\le p_o \sum_{j=j^*}^{\infty} |\rho_j(x)| \le p_1 \sum_{j=j^*}^{\infty} \mu_j{}^{-2} \le p_2 \sum_{j=j^*}^{\infty} j^{-2} \le p_3/(lr_n)$$
$$\le p_4/l \tag{16}$$

with suitable constants p_i, $i=0,..,4$, and $j^* := j(l,n) + 1$.
The derivatives are estimated using

$$\|Dg(x) - Dg_l^n(x)\|^2 = 4\sum_{r=1}^{k+1} |(f_{j(l,n)}^r, f_{j(l,n)}) - (f_\infty^r, f_\infty)|^2$$

with $f_i(s) := \sum_{j=1}^{i} \alpha_j(s)\rho_j(x) - k_o(s)$, $f_i^r(s) := \sum_{j=1}^{i} \alpha_j(s)(\rho_j)_{x_r}(x)$

for $i \in \{j(l,n),\infty\}$, $s \in [0,1]$. $(.,.)$ denotes the usual inner product in $L_2[0,1]$. For the following estimates we use the notation $j' := j(l,n)$: From

$$|(f_{j'}^r, f_{j'}) - (f_\infty^r, f_\infty)| = |(f_{j'}^r - f_\infty^r, f_{j'}) + (f_\infty^r, f_{j'} - f_\infty)|$$
$$\le m_o\|f_{j'}^r - f_\infty^r\| + m_1\|f_{j'} - f_\infty\|$$

and $\|f_{j'}^r - f_\infty^r\|^2 = \sum_{j=j'+1}^{\infty} {}_0\int^1 \alpha_j(s)^2 ds(\rho_j)_{x_r}(x)^2$

$$\le m_2 \sum_{j=j'+1}^{\infty} \exp(-\mu_j{}^2 m_3) \le m_4 \exp(-\mu_{j'}{}^2 m_3) ,$$

$$\|f_{j'} - f_\infty\|^2 = \sum_{j=j'+1}^{\infty} {}_0\int^1 \alpha_j(s)^2 ds\, \rho_j(x)^2 \le m_5 \sum_{j=j'+1}^{\infty} \mu_j{}^{-4} ,$$

we get $\|Dg(x) - Dg_l^n(x)\|^2 \le m_6\, \mu_{j'}{}^{-2}$

or $$\|Dg(x) - Dg_l^n(x)\| \le m_7/(j' - 1) \le m_8/(lr_n) \tag{17}$$

with suitable constants $m_0,\dots,m_8$.

Now we have to distinguish between three possible cases:

a) $g(x) > \varepsilon^2$: From (16) we have $g_l^n(x) > \varepsilon^2$ for all $l \geq l_o$ and for all n, leading to

$$|\varphi_l^n(x) - \varphi^n(x)| = r_n|(g_l^n(x) - \varepsilon^2)^2 - (g(x) - \varepsilon^2)^2|$$

$$\leq M_o r_n|g_l^n(x) - g(x)| \leq M_1/l \quad \text{for all n, confer (16),}$$

$$\|\varphi_l^{n'}(x) - D\varphi^n(x)\| = 2r_n\|(g_l^n(x) - \varepsilon^2)Dg_l^n(x) - (g(x) - \varepsilon^2)Dg(x)\|$$

$$\leq M_2 r_n|g_l^n(x) - g(x)| + M_3 r_n\|Dg_l^n(x) - Dg(x)\| \leq M_4/l \text{ for all}$$

$n \in N$, confer (16) and (17), $M_o,\dots,M_4$ suitable constants.

b) $g(x) = \varepsilon^2$: We get

$$|\varphi_l^n(x) - \varphi^n(x)| = r_n|\chi_l^n(x)| \leq r_n(g_l^n(x) - \varepsilon^2)^2 \leq M_5/l,$$

$$\|\varphi_l^{n'}(x) - D\varphi^n(x)\| = r_n\|D\chi_l^n(x)\| \leq r_n(g_l^n(x) - \varepsilon^2)\|Dg_l^n(x)\|$$

$\leq M_6/l$ for all n, M_5, M_6 suitable constants.

c) $g(x) < \varepsilon^2$: Since $g_l^n(x) < \varepsilon^2$ for all n and $l \geq l_o$, we have

$$|\varphi_l^n(x) - \varphi^n(x)| = \|\varphi_l^{n'}(x) - D\varphi^n(x)\| = 0.$$

This completes the proof.

5. Numerical results

In the last section we showed that algorithm 2.3. is applicable to the solution of the time-optimal heat diffusion problem (13). The test data are determined in the following way: For the temperature distribution we choose $k_o(s) := 0.5 - 0.5s^2$, $s \in [0,1]$, the desired accuracy ε is given by $\varepsilon := 0.01$, and the heat transfer coefficient α is set to $\alpha := 1$. The computations have been performed on a TR440 of the computing center of the Würzburg university.

Step 6 of algorithm 2.3. requires the determination of a search direction $s_k \in E$. These directions are computed using the conjugate gradient method of Fletcher and Reeves [1] combined with a restart as soon as condition (5) is violated.

Algorithm 2.3. was initialized with $\gamma_j = \gamma = 0.01$, $\sigma_j = \sigma = 1$, $x_o = (0.5,-0.5,\dots,(-1)^k 0.5)$, $l(0) = n(0) = 1$, $\mu_n = 4/n$,

$r_n = 1000*10^n$, and $j(l,n) = l$, $l=1,\ldots,30$, $n \in N$, $j(l,n) = 30$ for $l > 30$, $n \in N$. The iteration terminates as soon as the condition

$$|g_1^n(x)^{1/2} - \varepsilon| < 5*10^{-10} \tag{18}$$

is satisfied for a current iterate x.

The subsequent table shows some numerical results. k denotes the number of the jumps, T_o the computed minimal control times, and the k-vector t_o the switching times of the computed optimal bang-bang controls. In all cases, the penalty parameter r_n raised until 10^{13} or 10^{14}, respectively. Obviously the control with one jump seems to be optimal, for k > 1 we get nearly coincident jumps.

k	t_o						T_o
1					1.154		1.363
2					1.154	1.363	1.363
3		.490	.490		1.156		1.364
4		.494	.494		1.156	1.365	1.365
5	.281	.282	.657	.657	1.158		1.366

Table 1: Results for the time-optimal heat diffusion process

To show the efficiency of the adaptive precision approach we solved the problem with a fixed degree of approximation, i.e. we used $j(l,n) = 30$ as a fixed length of the series defining $g_1^n(x)$. Table 2 gives the numerical results of algorithm 2.2. after 30 iterations. Q denotes the quotient of the CPU-times of the adaptive and the fixed precision method, PHI the value of the penalty function, and ERR the violation of the constraint as defined by (18).

		Adaptive precision		Fixed precision	
k	Q	PHI	ERR	PHI	ERR
1	.60	1.362	3.2E-5	1.419	1.5E-3
2	.57	1.370	2.7E-4	1.664	7.5E-3
3	.56	1.366	2.6E-4	1.380	4.9E-4
4	.57	1.381	5.8E-4	1.382	3.3E-4
5	.59	1.383	2.2E-5	1.360	1.8E-3

Table 2: Results for the adaptive and fixed precision method

To test the advantage of an adaptive evaluation of the penalty parameters r_n, we tried to solve the problem with a fixed parameter $r = 10^{13}$. Table 3 presents the computed minimal values of the penalty functions.

k	1	2	3	4	5
PHI	13.85	1.42	1.40	1.44	1.67

Table 3: Results for a fixed penalty parameter

References.

[1] R. Fletcher, C. Reeves, Function minimization by conjugate gradients, British Comp. J., Vol.7 (1964), 145-154

[2] K. Schittkowski, The numerical solution of a time-optimal parabolic boundary value control problem, to appear: Journal of Optimization Theory and Applications

[3] - , A global minimization algorithm in a normed linear space using function approximations with adaptive precision, Preprint No.16, Institut für Angewandte Mathematik und Statistik, Universität Würzburg, 1976

[4] - , An adaptive precision method for nonlinear optimization problems, submitted for publication

[5] Y.V. Yegorov, Some problems in the theory of optimal control, USSR Computational Mathematics and Mathematical Physics, Vol.3 (1963), 1209-1232

GENERALIZED INVERSES AND A NEW STABLE SECANT TYPE MINIMIZATION ALGORITHM

A. Friedlander, J.M. Martínez and H.D. Scolnik
Fundación Bariloche, Argentina, and
Universidade Cândido Mendes, Rio de Janeiro, Brazil

1.- Introduction

Let $f: R^n \longrightarrow R$, $f \in C^1(R^n)$ and $g(x)$ be the gradient of f. Most minimization algorithms generate a sequence $x^k \in R^n$ by means of

$$x^{k+1} = x^k - \lambda_k d_k, \tag{1.1}$$

where d_k is a descent direction and λ_k is a suitable relaxation parameter. Among the methods requiring only gradient information, the Quasi-Newton algorithms play a key role (see [3], [4], [6], [7], [14], etc.). Briefly, these methods consist of iterating according to:

$$x^{k+1} = x^k - \lambda_k H_k g_k, \tag{1.2}$$

where g_k will denote from hereafter the gradient $g(x^k)$, and

$$H_{k+1} = H_k \Delta g_k \Delta g_k^t H_k / \Delta g_k^t H_k \Delta g_k + \Delta x_k \Delta x_k^t / \Delta x_k^t \Delta g_k + \beta_k w_k w_k^t, \tag{1.3}$$

$$w_k = H_k \Delta g_k / \Delta g_k^t H_k \Delta g_k - \Delta x_k / \Delta x_k^t \Delta g_k, \tag{1.4}$$

$$\Delta x_k = x^{k+1} - x^k, \tag{1.5}$$

$$\Delta g_k = g_{k+1} - g_k, \tag{1.6}$$

and β_k is a free parameter. By respectively taking $\beta_k = 0$ and

$$\beta_k = \Delta g_k^t H_k \Delta g_k \Delta x_k^t \Delta g_k / (\Delta g_k^t H_k \Delta g_k - \Delta x_k^t \Delta g_k), \tag{1.7}$$

the Davidon-Fletcher-Powell ([7]) and Broyden's rank one formula are obtained (see [12]). These methods have the quadratic termination property when λ_k is chosen as the value which minimizes

$$F(\lambda) = f(x^k - \lambda H_k g_k), \ \lambda > 0. \tag{1.8}$$

Broyden's rank one updating formula does not require line searches for obtaining the quadratic termination property. Unfortunately, this formula is neither always well defined nor stable.

Given $H_o > 0$, most updating schemes derived from (1.3)-(1.7) generate in the absence of rounding errors a sequence of $H_k > 0$, and thus the corresponding search directions are downhill. This property can be guaranteed in real computations using matrix factorizations ([9]).

Dixon ([5]) proved that if f is of class C^2 and in each iteration λ_k is taken as a local minimum of (1.8) according to a consistent rule, then given x^o, the sequence x^k is independent of the parameters β_k in

(1.3). However, it is generally accepted that it is best to use the value of β_k which gives

(1.9) $$H_{k+1} = (I - \Delta x_k \Delta g_k^t)/\Delta x_k^t \Delta g_k) H_k (I - \Delta g_k \Delta x_k^t/\Delta x_k^t \Delta g_k) + \Delta x_k \Delta x_k^t/\Delta x_k^t \Delta g_k$$

(see [10] and [16]), named the Broyden-Fletcher-Goldfarb-Shanno (BFGS) formula (see [3], [6], [11], [20]). Moreover, one-dimensional minimization is expensive and unpractical and so, modern algorithms use other relaxation criteria ([6], [14]).

Let us now suppose we have an unstable problem in a neighborhood of a solution (a singular or ill-conditioned hessian matrix). Unfortunately, the formulae of the family (1.3) do not provide numerically stable algorithms for computing the minimum, as shown in the numerical experiences of Section 3. In this paper a new algorithm is presented, which is able to deal efficiently with rank defficient situations by using generalized inverses and a secant scheme. Its performance is compared with that of VA13A ([15]), a Harwell's routine which uses BFGS formula, Fletcher-Powell's factorizations ([8]) and Powell's ([14]) one-dimensional bracketing scheme for finding λ_k. According to Schuller's results ([17]) we conjecture that the R-order of convergence of VA13A is the positive root of $t^{n+1} - t^n - 1 = 0$; which provides a suitable prediction for the relative performance of the two algorithms.

2.- Basic results and algorithms

Most of the results of this section concern the problem of finding critical points of quadratic functions $f(x) = (1/2)x^tGx + b^tx + c$, where G is a symmetric, not necessarily definite, matrix. Under this hypothesis:

Theorem 2.1.

Let $\delta_1, \ldots, \delta_n$ be linearly independent in R^n, with

$$\gamma_i = G\delta_i, \quad i = 1, \ldots, n.$$

Then, if $x \in R^n$ is arbitrary, the set of critical points of f is the set of points x + z where z stands for the solutions of

(2.1) $$\gamma_i^t z = -\delta_i^t g(x), \quad i = 1, \ldots, n.$$

Proof.

Easy, considering that (2.1) is equivalent to $Gz = -g(x)$.∎

Theorem 2.2.

If (2.1) is compatible, then there exists only one solution z which belongs to the subspace spanned by $\gamma_1, \ldots, \gamma_n$; moreover, this is the minimum norm solution of (2.1). (Thus x + z is the critical point of f which is closest to x).

Proof.

It follows from standard arguments of linear algebra.∎

Let us now give an algorithm for computing the minimum norm solution of (2.1), based in modified Gram-Schmidt orthogonalization (2). We write $b_i = -\delta_i^t g(x)$, $b = (b_1,\ldots,b_n)^t$.

Algorithm 2.1.

Step 1.- $i = 1$, $e_i = \gamma_i$, $c_i = b_i$.

Step 2.- If $e_i = 0$ (this implies $c_i = 0$ because we suppose that the system is compatible), go to Step 3, otherwise $e_i = e_i/\|e_i\|$, $c_i = c_i/\|e_i\|$.

Step 3.- If $i = n$, go to Step 4. Otherwise $i = i + 1$, $e_i = \gamma_i$, $c_i = b_i$. For $j = 1,\ldots, i-1$; perform steps 3.a and 3.b.

Step 3.a.- $e_i = e_i - e_j e_j^t e_i$.

Step 3.b.- $c_i = c_i - c_j e_j^t e_i$.

Go to Step 2.

Step 4.- $z = c_1 e_1 + \ldots + c_n e_n$.

Theorem 2.3.

If (2.1) is compatible, then Algorithm 2.1 computes its minimum norm solution z. In addition, the set of all the solutions of the system is:

$$\{z + (I - \sum_{j=1}^{n} e_j e_j^t)w\}_{w \in R^n}$$

Proof.

By construction, z satisfies (2.1) and is a combination of the γ_i's. The second part of the thesis follows because $\sum_{j=1}^{n} e_j e_j^t$ is the orthogonal projector over the range of G.∎

We have given a way of calculating the minimum norm solution, which is an important distinguished solution of (2.1). Then $z = -G^{\dagger}g(x)$, where $G^{\dagger}$ is the Moore-Penrose pseudoinverse of G. (See [1]). Other important solutions are basic solutions, which are given by $z = -G^{-}g(x)$, where G^{-} is a reflexive generalized inverse of G (see [1]). Scolnik ([22]) gave an algorithm for computing recursively all the basic solutions, which we describe here for the case rank $G = n - 1$.

Algorithm 2.2.

Given z and $A = I - \sum_{j=1}^{n} e_j e_j^t$, which rank $G = n - 1$, compute the basic solutions of (2.1).

Step 1.- Find a_j the first column of A such that

$$\|a_j\| \neq 0.$$

Step 2.- Find a_{ij} the first entry of a_j such that $a_{ij} \neq 0$.

Step 3.- Compute

$v_i = z + \alpha a_j$ where $\alpha = -z_i/a_{ij}$. (Therefore v_i is a basic solution of (2.1) where the i-th component is zero).

Step 4.- If $i = n$, stop.
Otherwise, $i = i + 1$.
Go to Step 3.

We can now define a minimization algorithm for a general function as follows:

Algorithm 2.3.

Given f as in Section 1, $x^k \in R^n$, the steps to calculate x^{k+1} are the following (we suppose $g(x^k) \neq 0$).

Step 1.- Choose $\delta_1^k, \ldots, \delta_n^k$ linearly independent increments, and
$\gamma_i^k = g(x^k + \delta_i^k) - g(x^k)$; $i = 1,\ldots,n$. (Note that if f is quadratic, then $\gamma = G\delta$ as in the theorems before).

Step 2.- Use Algorithm 2.1 to compute the minimum norm solution z of (2.1). If rank $A = n$, go to Step 4.

Step 3.- Use Scolnik's algorithm to compute the basic solutions of (2.1). Let $\{z_1,\ldots,z_p\}$ be the set of all the computed solutions (including minimum norm), and $z = z_j$ where

$$f(x^k + z_j) \leq f(x^k + z_i), \text{ for all } i = 1,\ldots,p.$$

Step 4.- If $z^t g(x^k) > 0$; $z = -z$.
If $|z^t g(x^k)|/\|z\|\,\|g(x^k)\| \leq$ Tol (a small positive number), then $z = -g(x^k)$.

Step 5.- $x^{k+1} = x^k + 2^{-j}z$, where $j \geq 0$ is the first integer which verifies $f(x^k + 2^{-j}z) < f(x^k)$.

Remarks.

1.- There exist several ways to choose $\delta_1^k,\ldots,\delta_n^k$. One of them is to choose $\delta_i^k = (0,\ldots,\underset{i}{1},\ldots,0)^t h_i^k$, where h_k^i is a small number. With this choice Algorithm 2.3 turns to be a stable implementation of the discretized Newton's method. Other possible choosing is $\delta_i^k =$ $= x^{k-i+1} - x^{k-i}$ (using the previous iterations). If so, we may use a less expensive algorithm than 2.1 for computing z (standard methods for modifying matrix factorizations). Another way of choosing the increments (in such a way that the γ_i's are pairwise orthogonal in (2.1) for quadratic functions) is given by Scolnik ([19]). In the numerical experiences we chose δ_i^k along the coordinate axes.

2.- The scheme derived from (2.1) for finding critical points is related with the usual one of the secant methods:

$$(\gamma_1^k,\ldots,\gamma_n^k)w = -g(x^k),$$

$$z = (\delta_1^k,\ldots,\delta_n^k)w,$$

but in the latter the minimum norm solution cannot be easily obtained. This solution is important because it leads to a point $x^k + z$ which is the closest to x^k among all possible choices in the linear manifold of solutions. Thus we may expect that f is better represented by its Taylor development in this point than in the other points of the manifold. On the other hand, basic solutions provide a wider exploration of the space of solutions in order to obtain $f(x^{k+1}) < f(x^k)$. Further justifications of the utility of basic solutions may be found in [18].

3.- The system (2.1), as formed in Step 2 of Algorithm 2.3, may be incompatible. However this is unlikely to occur in a neighborhood of a local minimum of f. To be precise, if x^* is a local minimum of f, then

$$g(x) = G(x^*)(x - x^*) + O(\|x - x^*\|^2),$$

and hence, if we consider negligible the term $O(\|x - x^*\|^2)$, the system $G(x^*)z = -g(x)$ remains compatible. But this is precisely the system that we approximate with equation (2.1). So, in Step 2 of Algorithm 2.3, we ignore the possible incompatibility of the system. (Note that Algorithm 2.1 may run even if $c_i \neq 0$ when $e_i = 0$).

4.- The weakest point of the Algorithm 2.3 is Step 4. If the z computed in Steps 2 and 3 is not a descent direction, then all the work of these steps is wasted. However, this feature is unlikely to occur in most practical cases.

When $G(x^*)$ is positive definite, the usual convergence theorems for secant methods hold:

Theorem 2.4.

Let $f: S \subset R^n \rightarrow R$ be such that f is of class C^2 in the open set S, $x^* \in S$ is a local minimum of f; G(x), the hessian matrix, is such that $G(x^*)$ is positive definite, and for all $x, y \in S$:

$$\|G(x) - G(y)\| \leq K\|x - y\| \quad (K > 0).$$

Suppose that x^k is generated by Algorithm 2.3. Then, there exists $\varepsilon > 0$ such that if $\|x^o - x^*\| \leq \varepsilon$, then $\lim x^k = x^*$ and if $\|\delta_i^k\| \leq O(\|F(x^k)\|)$ for all i, k, then the Q-order of convergence of x^k is 2.

Proof.

The consistence of the scheme is proved taking into account the symmetry of G and following the lines of Section 9.2 of [15]. Then the theorem is proved for the undamped version of the algorithm. But it is easy to prove that the undamped and the damped version of the algorithm are the same in a neighborhood of x^*. So the theorem is proved.∎

3.- Numerical experiences.

We shall denote by GIMIN the method described in Algorithm 2.3.

Its performance is compared in the following with the one of the Harwell subroutine VA13A by using test problems of different sort.

In order to carry out a fair comparison of the algorithms, we shall not limit ourselves to use only the concept of equivalent evaluations. The reason is that excepting when the components of the gradient are approximated by finite differences, the computational cost of evaluating the objective function and its gradient at a certain point, is usually less than the cost of performing n + 1 function evaluations. With this purpose in mind, we shall give the relative computational costs in terms of u units of CPU. For an IBM/145, one u unit is equivalent to 1/38400 seconds of CPU time, and the cost of the double precision operations are the following: Addition, 0.51 u; Multiplication, 1.92 u; Sine, 26.11 u; Cosine, 25.73 u; Tangent, 26.50 u; Square root, 16.51 u; Log, 28.03 u; Sin^{-1}, 33.66 u; Cos^{-1}, 33.92 u; Tan^{-1}, 28.29 u; Exponential, 27.01 u. Of course in different computers the relative costs are more or less the same.

We will consider the following three versions of GIMIN:

Version 0: The user provides separate subroutines for evaluating the function and its gradient.

Version 1: The user provides a subroutine FUN for evaluating the function, with an entry GRAD for computing the gradient, in such a way that if it has to be calculated at a point where the function was already evaluated, all the relevant information for saving computations is used. Separately, the user furnishes a subroutine for calculating the gradient at the points where the function has not been evaluated.

Version 2: The user writes a subroutine FUN for computing the function, with an entry GRH for evaluating the gradient and the matrix (γ_i^k) in such a way that all the common information is used for avoiding unnecessary calculations.

Of course, the results in terms of number of iterations and number of function and gradient evaluations are the same for the three versions but the required CPU time may drastically change.

We define, for GIMIN and VA13A (G and V from hereafter):

C_o: Cost of one undamped iteration in terms of equivalent evaluations; that is, number of function evaluations + n . number of gradient evaluations. Thus, C_o is equal to $(n + 1)n + 1$ in G and $n + 1$ in V.

C_1: Cost of one undamped iteration in u units, taking into account only function and gradient evaluations (without the intrinsic work required by the algorithm) and with the version 1 of G.

C_2: Idem with the version 2 of G. Of course, $C_1(V) = C_2(V)$.

C_3: Computational cost of the intrinsic work of one undamped iteration in u units. Approximately, the intrinsic work is $n^3 + 5n^2/2 + n/2$ additions and products and n square roots for G, and $3n^2 + 13n$ additions and products for V.

We also define:

$C_4 = C_1 + C_3$ and $C_5 = C_2 + C_3$.

The values $C_o, \ldots, C_5$ can be computed for a given function before applying the minimization algorithms. On the contrary, the cost associated with the choice of the relaxation parameter cannot be determined in advance.

We can also calculate beforehand the following:

O: the convergence order of the method, which is 2 for G and the positive root of $t^{n+1} - t^n - 1 = 0$ for V (see [17]).

PQI = log O(V)/log O(G), which is a prediction of the quotient between the number of iterations of G over the number of iterations of V.

P_o: C_o(G).PQI/C_o(V): Prediction of the quotient between the computational cost of G and the one of V in terms of equivalent evaluations.

P_1: C_1(G).PQI/C_1(V): The same as above in u units with the version 1 of G and without taking into account the intrinsic work of the algorithms.

P_2: C_2(G).PQI/C_2(V): Idem with the version 2 of G.

P_3: C_3(G).PQI/C_3(V): Idem but considering only intrinsic work.

P_4: C_4(G).PQI/C_4(V): Prediction of the quotient between the total computational cost required for G(version 1) over the one of V.

P_5: C_5(G).PQI/C_5(V): The same as P_4 but with the version 2 of G.

In each case the user is advised to choose G if $P_j < 1$ and V otherwise, according with the criterion he wishes to adopt. Of course, to employ this kind of predictions has only sense when weak conditions are required in the algorithms for selecting the relaxation step, that is when in normal conditions an iteration does not require a costly linear search. This is the case of GIMIN and VA13A.

We define, as results of the running tests:

NI: Number of iterations.

FE: Number of function evaluations.

PREC: $-\log_{10} \|x - x^*\|$, where x^* stands for the computed optimum.

F: $-\log_{10} |f(x) - f(x^*)|$.

T_o: Total work in terms of equivalent evaluations.

T_1: Total work in u units with the version 1 of G without intrinsic costs.

T_2: Idem with the version 2 of G.

T_3: Total intrinsic work in u units.

T_4: $T_1 + T_3$.

T_5: $T_2 + T_3$.

In connection with these definitions we will give $Q_j = T_j(G)/T_j(V)$, $j = 0, 1,\ldots,5$. Hence, the P_j's are a prediction of the Q_j's. Likewise, PQI is a prediction of EQI (effective quotient of the number of iterations).

h_i^k was taken in G as $10^{-6}\|x^k\|$ in the nonquadratic and $10^{-3}\|x^k\|$ in the quadratic cases. The vector SCALE in VA13A was taken as recommended by its authors. Each run of G or V was stopped when the algorithm reached the precision of the one of the routine FMFP (Davidon-Fletcher-Powell method) of the IBM-SSP with a stopping-criterion of 10^{-6}.

The first nine tests were runned in a Bull with 11 digits of precision and the last three in an IBM 370 in double precision.

Tests functions:

1) (Rosenbrock) $n = 2$; $f(x) = 100(x_2 - x_1^2)^2 + (1 - x_1)^2$; $x_o = (-1.2\ ,\ 1)$; $x^* = (1\ ,\ 1)$; $f(x^*) = 0$; $O(V) = 1.466$.

2) (Brown) $n = 4$; $f(x) = (x_1^2(x_2 - 1)^2 + 50)^2 + (x_2 - 1)^2x_3^2 + (x_3 - 5)^2x_4^2 + ((x_4 - 10)^2 + 50)^2 + (x_1^2 + (x_2 - 1)^2 + (x_3 - 5)^2 + 10)^2 + ((x_4 - 10)^2 + (x_3 - 5)(x_2 - 1))^2$; $x_o = (2.5\ ,\ 4.5\ ,\ 10.75\ ,\ 21)$; $x^* = (0\ ,\ 1\ ,\ 5\ ,\ 10)$; $f(x^*) = 5100$; $O(V) = 1.325$.

3) (Wood's banana) $n = 4$; $f(x) = 100(x_2 - x_1^2)^2 + (1 - x_1)^2 + 90(x_4 - x_3^2)^2 + (1 - x_3)^2 + 10.1((x_2 - 1)^2 + (x_4 - 1)^2) + 19.8(x_2 - 1)(x_4 - 1)$; $x_o = (-3\ ,\ -1\ ,\ -3\ ,\ -1)$; $x^* = (1\ ,\ 1\ ,\ 1\ ,\ 1)$; $f(x^*) = 0$; $O(V) = 1.325$.

4) (Fletcher-Powell's helical valley) $n = 3$; $f(x) = 100(x_3 - 10\phi(x_1,x_2))^2 + (R(x_1,x_2) - 1)^2 + x_3^2$, with $2\pi\phi(x_1,x_2) = \tan^{-1}(x_2/x_1)$ if $x_1 > 0$, $2\pi\phi(x_1,x_2) = \pi + \tan^{-1}(x_2/x_1)$ if $x_1 < 0$, $R(x_1,x_2) = (x_1^2 + x_2^2)^{1/2}$; $x_o = (-1\ ,\ 0\ ,\ 0)$; $x^* = (1\ ,\ 0\ ,\ 0)$; $f(x^*) = 0$; $O(V) = 1.380$.

5) (Powell's quartic) $n = 4$; $f(x) = (x_1 + 10x_2)^2 + 5(x_2 - x_4)^2 + (x_2 - 2x_3)^4 + 10(x_1 - x_4)^4$; $x_o = (3\ ,\ -1\ ,\ 0\ ,\ 1)$; $x^* = (0\ ,\ 0\ ,\ 0\ ,\ 0)$ $f(x^*) = 0$; $O(V) = 1.325$.

6) (Cragg and Levy) $n = 4$; $f(x) = (\exp(x_1) - x_2)^4 + 100(x_2 - x_3)^6 + \tan^4(x_3 - x_4) + x_1^8 + (x_4 - 1)$; $x_o = (1\ ,\ 2\ ,\ 2\ ,\ 2)$; $x^* = (0\ ,\ 1\ ,\ 1\ ,\ 1)$; $f(x^*) = 0$; $n = 4$; $O(V) = 1.325$.

7) (Hilbert's quadratic) $n = 8$; $f(x) = x^tHx$ with $H_{ij} = (1/(i + j - 1)$; $x_o = (0.57\ ,\ 3.45\ ,\ -1.83\ ,\ 8.42\ ,\ 6.9\ ,\ -3.78\ ,\ 0.37\ ,\ 6.12)$; $x^* = (0,\ldots,0)$; $f(x^*) = 0$; $O(V) = 1.213$.

8) $n = 3$; $f(x) = x_1^4 + (x_1 + x_2)^4 + \ldots + (x_1 + \ldots + x_n)^4$; x_o = random $(0\ ,\ 1)$; $x^* = (0,\ldots,0)$; $O(V) = 1.380$.

9) Idem 8, with n = 10 ; O(V) = 1.184.

10) Idem 8, with n = 30; O(V) = 1.085.

11) (Fletcher-Powell's trigonometric) n = 10, m = 20; $f(x) = \sum_{i=1}^{m} (E_i - \sum_{j=1}^{n} A_{ij} \sin x_j + B_{ij} \cos x_j)^2$; $x^* = \text{random}\ (-\pi\ ,\ \pi)$; A_{ij}, B_{ij} random integers between-100 and 100; $E_i = \sum_{j=1}^{n} A_{ij} \sin x_j^* + B_{ij} \cos x_j^*$; $x_{o,i} = x_i^* + \varepsilon_I$, with ε_I a random number between -0.1π and 0.1π; $f(x^*) = 0$; O(V) = 1.184.

12) Idem 11 , with n = 15, m = 30; O(V) = 1.140.

Function Nº		Results NI	PREC	F	FE		Comparison with predictions 0	1	2	3	4	5	
1	G	20	6	10	28	P	1.28	1.24	1.03	0.47	0.61	0.57	PQI=0.55
	V	34	6	12	44	Q	1.12	1.11	1.00	0.51	0.64	0.61	EQI=0.59
2	G	9	10	20	11	P	1.72	1.90	0.96	0.55	0.87	0.65	PQI=0.41
	V	33	10	20	72	Q	0.53	0.59	0.42	0.35	0.45	0.38	EQI=0.27
3	G	39	8	16	52	P	1.72	1.35	0.76	0.55	0.69	0.32	PQI=0.41
	V	75	8	17	95	Q	1.75	1.43	0.84	0.69	0.85	0.72	EQI=0.52
4	G	13	8	16	18	P	1.53	1.73	0.90	0.51	0.93	0.64	PQI=0.47
	V	22	9	17	32	Q	1.36	1.61	0.78	0.64	1.06	0.70	EQI=0.59
5	G	23	4	13	39	P	1.72	1.44	0.99	0.55	0.67	0.61	PQI=0.41
	V	52	4	15	63	Q	1.58	1.45	0.88	0.60	0.74	0.64	EQI=0.44
6	G	30	3	16	34	P	1.72	2.15	0.94	0.55	1.15	0.69	PQI=0.41
	V	82	3	16	98	Q	1.29	1.63	0.70	0.49	0.96	0.57	EQI=0.37
7	G	3	2	6	4	P	2.27	2.11	0.51	0.69	0.88	0.67	PQI=0.28
	V	Didn't converge to the correct minimum.											
8	G	20	3	14	21	P	1.53	1.46	1.15	0.51	0.64	0.60	PQI=0.47
	V	60	3	14	76	Q	0.48	0.25	0.20	0.34	0.30	0.28	EQI=0.33
9	G	19	3	11	20	P	2.42	1.72	1.20	0.74	0.83	0.77	PQI=0.24
	V	197	2	11	253	Q	0.76	0.52	0.38	0.29	0.32	0.30	EQI=0.10
10	G	19	4	12	20	P	3.60	2.60	1.48	1.14	1.19	1.16	PQI=0.12
	V	Stopped because of singularity of the approximate hessian.											
11	G	5	10	21	6	P	2.42	2.60	0.45	0.74	2.11	0.52	PQI=0.24
	V	19	7	19	27	Q	1.87	2.03	0.36	0.80	1.78	0.45	EQI=0.26
12	G	6	10	16	7	P	2.86	3.10	0.34	0.88	2.53	0.48	PQI=0.19
	V	24	6	17	31	Q	2.95	3.11	0.36	1.16	2.69	0.53	EQI=0.25

4.- Final remarks.

In general the predictions are qualitatively correct ($P_j < 1$ iff $Q_j < 1$) even in the cases where the hessian at the solution is singular and thus the methods have not, in fact, the postulated order. In these cases (functions 5, 6, 8, 9 and 10) and in the cases where the hessian is ill conditioned (function 7), G tends to behave better than predicted, also being more efficient than V. In the cases where the trajectory to the optimal point goes through indefinite hessians (functions 3 and 4), G behaves worse than predicted. Both situations were expected; the first because algorithm G was built to take account of singular or ill-conditioned situations; the second because of its weak feature pointed out in a previous remark. However , in all tests , $Q_5 < 1$, that is, the computation time used by the version 2 of G was less than that used by V.

Acknowledgement

To Harald K. Solberg for his programming assistance.

References.

[1] A. BEN ISRAEL and D.N.E. GREVILLE - Generalized inverses: Theory and applications, John Wiley and Sons, 1974.

[2] A. BJORCK - Solving linear least squares problems by Gram-Schmidt orthogonalization, BIT, 7, 1967, 1-21.

[3] C.G. BROYDEN - Quasi Newton methods and their application to function minimization, Math. Comp., 21, 1967, 363-381.

[4] J.E. DENNIS and J.J. MORE - Quasi Newton methods: motivation and theory, Report TR 74-217, Dept. of Computer Science, Cornell University, 1974.

[5] L.C.W. DIXON - Quasi Newton methods generate identical points, Math. Programming, 2, 1972, 388-397.

[6] R. FLETCHER - A new approach to variable metric algorithms, Comput. J., 13, 1970, 317-322.

[7] R. FLETCHER and M.J.D. POWELL - A rapidly convergent descent method for minimization, Comput. J., 6, 1963, 163-168.

[8] R. FLETCHER and M.J.D. POWELL - On the modification of LDL^t factorizations, A.E.R.E. Report TP519, Harwell, 1973.

[9] P.E. GILL and W. MURRAY - Quasi Newton methods for unconstrained optimization, J. Inst. Math. Appl., 9, 1972, 91-108.

[10] P.E. GILL, W. MURRAY and R.A. PITFIELD - The implementation of two revised Quasi Newton algorithms for unconstrained optimization, Report Nac 11/71, National Physical Laboratory, England, 1971.

[11] R. GOLDFARB - Sufficient conditions for the convergence of a variable metric algorithm, in Optimization (R. Fletcher, editor), Academic Press, London-New York, 1969.

[12] D. LUENBERGER - Introduction to linear and nonlinear program-

ming. Addison Wesley, 1973.

[13] J.M. ORTEGA and W.C. RHEINBOLDT - Iterative solution of non-linear equations in several variables, Academic Press , New York, 1970.

[14] M.J.D. POWELL - A view of unconstrained optimization,C.S.S. 14, A.E.R.E., Harwell, 1975.

[15] M.J.D. POWELL - Subroutine VA13A, Harwell Subroutine Library, 1974.

[16] R.W.H. SARGENT and D.J. SEBASTIAN - Numerical experience with algorithms for unconstrained minimization, in Numerical methods for nonlinear optimization (F.A. Lootsma, editor), Academic Press, 1972.

[17] G. SCHULLER - On the order of convergence of certain quasi-Newton methods, Numer. Math., 23, 1974, 181-192.

[18] H.D. SCOLNIK - Reflexive generalized inverses and the solution of nonlinear algebraic equations, 1976, to appear.

[19] H.D. SCOLNIK - Working paper, Candido Mendes University, 1977.

[20] D. SHANNO - Conditioning of quasi-Newton methods for function minimization, Math. Comp., 24, 1970, 647-656.

A CONJUGATE DIRECTIONS METHOD AND ITS APPLICATION

Fridrich Sloboda
Institute of Technical Cybernetics
Dúbravská cesta 1, 809 31 Bratislava, Czechoslovakia

Abstract

A new method of conjugate directions for the minimization of strictly convex functions is described. The method does not use gradient vectors and can be considered as a generalization of a projection method for linear algebraic systems of equations. The main advantage of the method is that for band occurence matrices it requires $O(n)$ function evaluations for the determination of n linearly independent vectors. The matrix of these vectors is upper triangular with positive elements on the diagonal. The conjugate gradient methods require $O(n^2)$ function evaluations in each iteration and the structure of the occurrence matrix has no influence on the total number of function evaluations.

Introduction

In [6],[7] a direct projection method for linear algebraic systems is described. The method is closely related also to other direct methods as the elimination method, the orthogonalization method and the method of conjugate directions. The algorithm is in such a form that it may be used for nonlinear problems and also some of its properties can be transferred to nonlinear problems. Let us consider the system

(1) $$Ax = b$$

where A is a regular n by n matrix and b is an n-vector. Let us consider the system (1) in the form

(2) $$r_i = b_i - \langle a_i, x\rangle = 0 \qquad i=1,2,\dots,n$$

where r_i represent n hyperplanes in the n-dimensionel Euclidean space E_n, $\langle a_i,x\rangle = \sum_{j-1}^{n} a_{ij} x_j$, $a_i = (a_{i1}, a_{i2},\dots,a_{in})$ is the i-th normal vector of the hyperplane r_i and b_i is the i-th component of the vector b.

Let $x_0^{(0)}, x_0^{(1)},\dots,x_0^{(n)}$ be n+1 linearly independent points of the space E_n. The algorithm for solution of (1) is defined by the recurrent relation [6]

(3) $$x_i^{(k)} = x_{i-1}^{(k)} + \frac{b_i - \langle a_i, x_{i-1}^{(k)}\rangle}{(a_i, v_{i-1}^{(i)})} v_{i-1}^{(i)}$$

where $$v_{i-1}^{(i)} = x_{i-1}^{(i)} - x_{i-1}^{(i-1)} \qquad i=1,2,\dots,n \quad k=i,i+1,\dots,n.$$

Let $x_0^{(0)}, x_0^{(1)},\dots,x_0^{(n)}$ be such points that $(a_i, v_{i-1}^{(i)}) \neq 0, i=1,2,\dots,n$. Then the point $x_n^{(n)}$ defined by the algorithm (3) is the solution of (1). Let $x_0^{(0)} = (0,\dots,0)^T$ and let $x_0^{(i)} = (0,\dots,t_i,\dots,0)^T$ where $t_i=1$. Let A be a strictly regular matrix. Then the matrix defined by columns of the vectors $v_0^{(1)}, v_1^{(2)},\dots, v_{n-1}^{(n)}$ is upper triangular with unit elements on the diagonal and

$$(a_i, v_{i-1}^{(i)}) \neq 0 \qquad i=1,2,\dots,n.$$

Let A be a strictly regular symmetric matrix. Let $v_0^{(i)} = x_0^{(i)} - x_0^{(0)}$ be vectors of the form $v_0^{(i)} = (0,\dots,t_i,\dots,0)^T$ where $t_i=1$, $i=1,2,\dots,n$. Then $(Av_{i-1}^{(i)}, v_{j-1}^{(j)}) = 0$ for $i\neq j$. Let A be a strictly regular, q-diagonal band matrix. Let $v_0^{(k)} = x_0^{(k)} - x_0^{(0)}$ be vectors of the form

$$v_0^{(k)} = (0,\dots,t_k,\dots,0)^T \qquad k=1,2,\dots,n$$

where $t_k=1$. Then

$$x_i^{(k)} = x_i^{(i)} + v_0^{(k)} \qquad k>(q-1)/2+i$$

for $i=1,2,\dots,n$ and $k=i+1,\dots,n$. The algorithm (3) provides the same LU decomposition of the matrix A as the elimination method and other related methods and requires $O(1/3n^3)$ arithmetic operations. The total storage requirements are less than $n^2/4+n+2$. Input of data is very convenient for the algorithm since single rows of the matrix are required on each iteration. For a q-diagonal band matrix it is necessary to store $(q-1)/2+1$ vectors. The algorithm can be applied to systems with strictly diagonally dominant or symmetric positive definite matrices.

Let A be a symmetric positive definite matrix. Let $f:E_n \longrightarrow E_1$ be a quadratic function in the form $f(x) = (Ax,x) - 2(b,x) + c$. In [7] it is shown how algorithm (3) can be applied for minimization of the corresponding quadratic function. Let $x_0^{(0)}, x_0^{(1)},\dots, x_0^{(n)}$ be n+1 linearly independent points of the space E_n. Then the algorithm for minimization of the quadratic function $f(x)$ is defined as follows:

$$x_i^{(k)} = x_i^{(i)} + \alpha_{i-1}^{(k)} v_{i-1}^{(i)}$$

where
$$v_{i-1}^{(i)} = x_{i-1}^{(i)} - x_{i-1}^{(i-1)}$$

and $\alpha_{i-1}^{(k)}$ are scalar coefficients such that

$$f(x_{i-1}^{(k)} + \alpha v_{i-1}^{(i)}) = \min! \quad \text{for } \alpha = \alpha_{i-1}^{(k)} \qquad i=1,2,\dots,n$$
$$k=i,i+1,\dots,n.$$

At the point $x_n^{(n)}$ the quadratic function $f(x)$ achieves its minimum. Let us denote $g(x) = f'(x) = (\partial f/\partial x_1, \partial f/\partial x_2,\dots,\partial f/\partial x_n)$. In [7] the following theorems are proved:

Theorem 1. Let $f:E_n \longrightarrow E_1$ be a strictly convex quadratic function. Let $v \in E_n$ be a non-zero vector. Let $x_0, y_0 \in E_n$, $x_0 \neq y_0$ be points that $(g(x_0),v) = 0$ and $(g(y_0),v) = 0$. Then $(Av, x_0 - y_0) = 0$.

Theorem 2. The vectors $v_0^{(1)}, v_1^{(2)},\dots, v_{n-1}^{(n)}$ defined by the above described algorithm are mutually conjugate.

The purpose of the next part of this paper is to describe an algorithm for the minimization of strictly convex functions which for band occurrence matrices requires $O(n)$ function evaluations for the determination of n linearly independent vectors and which can be considered a generalization of (3).

Description of the algorithm

Let $f:E_n \longrightarrow E_1$ be a twice continuously differentiable strictly convex function. Let $x_0^{(0)}, x_0^{(1)}, \ldots, x_0^{(n)}$ be n+1 linearly independent points of the space E_n. Let $x_0^{(0)} \in E_n$ be an initial point and let $x_0^{(k)} = x_0^{(0)} + v_0^{(k)}$, $v_0^{(k)} = (0, \ldots, t_k, \ldots, 0)^T$ where $t_k = \lambda$ is a suitable positive real number. The algorithm for minimization of f(x) is defined as follows:

(4) Algorithm.

Step (0) : Define λ and set $x_0^{(k)} = x_0^{(0)} + v_0^{(k)}$, k=1,2,...,n, set i=1.

Step (1) : Compute

$$x_i^{(i)} = x_{i-1}^{(i-1)} + \alpha_{i-1}^{(i-1)} w_{i-1}^{(i)}$$

where
$$w_{i-1}^{(i)} = v_{i-1}^{(i)} / \|v_{i-1}^{(i)}\|, \quad v_{i-1}^{(i)} = x_{i-1}^{(i)} - x_{i-1}^{(i-1)}$$

and $\alpha_{i-1}^{(i-1)}$ is a scalar coefficient such that

$$f(x_{i-1}^{(i-1)} + \alpha w_{i-1}^{(i)}) = \min! \quad \text{for } \alpha = \alpha_{i-1}^{(i-1)}.$$

Step (2) : Compute

$$x_{i-1}^{(k)} = x_{i-1}^{(k)} - (x_{i-1}^{(k)} - x_i^{(i)}, w_{i-1}^{(i)})\, w_{i-1}^{(i)} \qquad k=i+1,\ldots,n.$$

Step (3) : Compute

$$x_i^{(k)} = x_{i-1}^{(k)} + \alpha_{i-1}^{(k)} w_{i-1}^{(i)} \qquad k=i+1,\ldots,n$$

where $\alpha_{i-1}^{(k)}$ is defined as

$$\alpha_{i-1}^{(k)} = -\frac{c\lambda}{2} \; \frac{f(x_{i-1}^{(k)} + c\lambda w_{i-1}^{(i)}) - f(x_{i-1}^{(k)} - c\lambda w_{i-1}^{(i)})}{f(x_{i-1}^{(k)} + c\lambda w_{i-1}^{(i)}) - 2f(x_{i-1}^{(k)}) + f(x_{i-1}^{(k)} - c\lambda w_{i-1}^{(i)})}$$

and $c \in (0,1\rangle$.

Step (4) : Set i=i+1. If $i \leq n$ then go to Step (1), else go to Step(5).

Step (5) : Replace $x_0^{(0)}$ by $x_n^{(n)}$ and go to Step (0).

According to the choice of λ we obtain the following algorithms:

Algorithm I : $\lambda = \min(h, \|x_n^{(n)} - x_0^{(0)}\|)$, where h is a constant, usually h=0.5(1.0).

Algorithm II: $\lambda = \min(h, |f(x_n^{(n)}) - f(x_0^{(0)})|^p)$, $p \in (0,1\rangle$, usually p=1/2.

In Step (2) we define the orthogonal projections of the point $x_i^{(i)}$, defined in Step (1), on the corresponding parallel directions using the fact that the vectors $w_{i-1}^{(i)}$ are normalized. $\|\cdot\|$ denotes the Euclidean norm. In Step (3) $\alpha_{i-1}^{(k)}$ denotes one discretized Newton step. Step (2) may be considered a predictor and Step (3) a corrector of the local minimizers. The matrix of the vectors $w_{i-1}^{(i)}$, i=1,2,...,n is upper triangular with positive elements on the diagonal which follows from the fact that $v_{i-1}^{(i)}$ is a linear combination of the vectors $v_0^{(k)}$, k=1,...,i, (see [6]). Step (5) indicates that the algorithm after n steps is restarted. The restart ensures that the algorithm in each cycle generates n linearly independent vectors $w_{i-1}^{(i)}$, i=1,2,...,n which span the space E_n. The matrix of these vectors is in each cycle upper triangular with positive elements on the diagonal. The restart enables also to follow the structure of the occurrence matrix (see Theorem 4).

Theorem 3. Let $f:E_n \longrightarrow E_1$ be a twice continuously differentiable strictly convex function which satisfies

(5) $$\lim_{\|x\| \longrightarrow \infty} f(x) = +\infty \; .$$

Then for any $x_0^{(0)} \in E_n$ the Algorithm (4) is well defined and $\{x_n^{(n)}\}$ converges to the unique minimizer of $f(x)$.

P r o o f. Let $w_0^{(1)}$, $w_1^{(2)}$, ..., $w_{n-1}^{(n)}$ be vectors defined by the Algorithm (4). In each cycle we obtain a new set of these vectors. Let us denote

(6) $$p^{(s+1)} = q^{(s)}, \; q^{(s)} = w_{s(\mathrm{mod}\ n)}^{(s)(\mathrm{mod}\ n)+1} \qquad s=0,1,\ldots \; .$$

Let $k_0 \geq 0$ be an index. The sequence $\{p^{(k)}\}$ k=1,2,..., $p^{(k)} \neq 0$ defined by (6) has the property that for $k \geq k_0$, $p^{(j)}$, j=k+1,...,k+n are linearly independent vectors and after a rearrangement the matrix of these vectors is upper triangular with $\lambda / \|v_{i-1}^{(i)}\|$ on the diagonal. Since the vector $v_{i-1}^{(i)}$ is a linear combination of the vectors $v_0^{(k)}$, k=1,2,...,i and the level set $L_0 = L(f(x_0^{(0)}))$ is according to (5) compact, it follows that $\|v_{i-1}^{(i)}\| \leq K\lambda$ where $\lambda = \|v_0^{(k)}\|$, $K > 0$, so that $\lambda / \|v_{i-1}^{(i)}\| \geq 1/K = c > 0$. According to the last statements the sequence $\{p^{(k)}\}$, k=1,2,... defined by (6) is uniformly linearly independent (see Definition 14.6.2 [4], Theorem 14.6.3 [4]).

The condition (5) ensures that $x_n^{(n)}$ is well defined and the level set $L_0 = L(f(x_0^{(0)}))$ is compact (see Definition 4.2.1 [4]). By virtue of Theorem 14.2.10 [4], Theorem 14.1.3 [4] and whith regard to the fact that the sequence $\{p^{(k)}\}$, k=1,2,... is uniformly linearly independent Theorem 14.6.4 [4] applies and Theorem 14.1.5 [4] shows the convergence of $\{x_n^{(n)}\}$.

Let us consider the system

(7) $$\partial f / \partial x_i = f_i(x_1, x_2, \ldots, x_n) = 0 \qquad i=1,2,\ldots,n.$$

<u>Definition</u>. The occurrence matrix of the system (7) is a Boolean matrix associated with the system (7) as follows:
An element of the matrix, s_{ij}, is either a Boolean 1 or 0 according to the rule

$$s_{ij} = \begin{cases} 1 & \text{if the j-th variable appears in the i-th equation} \\ 0 & \text{otherwise.} \end{cases}$$

This occurrence matrix influences the structure of the Algorithm (4).

<u>Theorem 4</u>. Let $f: E_n \longrightarrow E_1$ be a twice continuously differentiable strictly convex function satisfying

$$(8) \qquad \lim_{\|x\| \longrightarrow \infty} f(x) = +\infty \quad .$$

Let the occurrence matrix corresponding to the function $f(x)$ be a q-diagonal band matrix. Let $x_0^{(0)} \in E_n$ be an arbitrary initial point and let $v_0^{(k)} = x_0^{(k)} - x_0^{(0)}$ have the form $v_0^{(k)} = (0,\ldots,t_k,\ldots,0)^T$ where $t_k = \lambda$, $k=1,2,\ldots,n$. Then $x_i^{(k)}$ defined by the Algorithm (4) fulfils

$$x_i^{(k)} = x_i^{(i)} + v_0^{(k)} \qquad \begin{array}{l} k > (q-1)/2+i \\ i=1,2,\ldots,n \\ k=i+1,\ldots,n. \end{array}$$

P r o o f. For $i=1$ according to the Algorithm (4) we have

$$x_1^{(1)} = x_0^{(0)} + \alpha_0^{(0)} w_0^{(1)} .$$

For $x_1^{(1)}$ it holds

$$(9) \qquad (f'(x_1^{(1)}) \ , \ w_0^{(1)}) = 0$$

where $w_0^{(1)} = (1,0,\ldots,0)^T$, i.e., according to the form of $w_0^{(1)}$ and $f'(x_1^{(1)})$ we have

$$(10) \qquad f_1(x_{11}^{(1)}, x_{12}^{(1)}, \ldots, x_{1p}^{(1)}) = 0$$

whereby $p=(q-1)/2+1$. Let us consider the points $x_1^{(k)} = x_0^{(k)} + \alpha_0^{(0)} w_0^{(1)}$ where $x_0^{(k)} = x_0^{(0)} + v_0^{(k)}$, $k > (q-1)/2+1$. For $x_1^{(k)}$ we obtain

$$(11) \quad x_1^{(k)} = x_0^{(k)} + \alpha_0^{(0)} w_0^{(1)} = x_0^{(0)} + \alpha_0^{(0)} w_0^{(1)} + v_0^{(k)} = x_1^{(1)} + v_0^{(k)}$$

for $k > (q-1)/2+1$. According to (10), (11) we obtain

$$(12) \qquad (f'(x_1^{(k)}) \ , \ w_0^{(1)}) = 0 \qquad k > (q-1)/2+1.$$

The assumption (8) ensures the existence and the uniqueness of $x_1^{(k)}$ for which (12) holds. From (11) we obtain

$$v_1^{(k)} = x_1^{(k)} - x_1^{(1)} = v_0^{(k)}, \ (v_1^{(k)}, \ w_0^{(1)}) = 0 \quad k > (q-1)/2+1$$

so that $\alpha_0^{(k)}=0, k>(q-1)/2+1$ and for $x_1^{(k)}$ defined by the Algorithm (4) we have

(13) $$x_1^{(k)}=x_1^{(1)}+v_0^{(k)} \qquad k>(q-1)/2+1.$$

For i=2 according to the Algorithm (4) we obtain

(14) $$x_2^{(2)}=x_1^{(1)}+\alpha_1^{(1)}w_1^{(2)}.$$

For $x_2^{(2)}$ it holds that

(15) $$(f(x_2^{(2)}),\ w_1^{(2)})=0$$

where $w_1^{(2)}=(w_{11}^{(2)},\ w_{12}^{(2)},\ 0,\dots,0)^T$, i.e., according to the form of $w_1^{(2)}$, $f(x_2^{(2)})$ we have

$$f_1(x_{21}^{(2)},\ x_{22}^{(2)},\dots,\ x_{2p-1}^{(2)})\,w_{11}^{(2)}+f_2(x_{21}^{(2)},\ x_{22}^{(2)},\dots,x_{2p}^{(2)})w_{12}^{(2)}=0$$

where $p=(q-1)/2+2$. Let us consider the points $x_2^{(k)}=x_1^{(1)}+\alpha_1^{(1)}w_1^{(2)}$, $k>(q-1)/2+2$. From (13),(14) we have

(16) $$x_2^{(k)}=x_1^{(k)}+\alpha_1^{(1)}w_1^{(2)}=x_1^{(1)}+\alpha_1^{(1)}w_1^{(2)}+v_0^{(k)}=x_2^{(2)}+v_0^{(k)}$$

for $k>(q-1)/2+2$. According to (15),(16) we obtain

(17) $$(f(x_2^{(k)}),\ w_1^{(2)})=0 \qquad k>(q-1)/2+2.$$

The assumption (8) ensures the existence and the uniqueness of $x_2^{(k)}$ for which (17) holds. From (16) it follows

$$v_2^{(k)}=x_2^{(k)}-x_2^{(2)}=v_0^{(k)},(v_2^{(k)},\ w_1^{(2)})=0 \qquad k>(q-1)/2+2$$

so that $\alpha_1^{(k)}=0, k>(q-1)/2+2$ and for $x_2^{(k)}$ defined by the Algorithm (4) we have

$$x_2^{(k)}=x_2^{(2)}+v_0^{(k)} \qquad k>(q-1)/2+2.$$

Let for i=j

(18) $$x_j^{(k)}=x_j^{(j)}+v_0^{(k)} \qquad k>(q-1)/\ 2+j.$$

For i=j+1 we obtain

$$x_{j+1}^{(j+1)}=x_j^{(j)}+\alpha_j^{(j)}w_j^{(j+1)}$$

whereby

(19) $$(f(x_{j+1}^{(j+1)}),\ w_j^{(j+1)})=0$$

and $w_j^{(j+1)}=(w_{j1}^{(j+1)},\ w_{j2}^{(j+1)},\dots,\ w_{jj+1}^{(j+1)},\ 0,\dots,0)^T$, i.e., according to the form of $f(x^{(j+1)})$ and $w_j^{(j+1)}$

$$\sum_{i=1}^{j+1} f_i(x_{j+1}^{(j+1)})\,w_{ji}^{(j+1)}=0$$

where $p=(q-1)/2+j+1$ denotes the last component $x_{j+1p}^{(j+1)}$ of $x_{j+1}^{(j+1)}$ in $f_{j+1}(x_{j+1}^{(j+1)})$. Let us consider the points $x_{j+1}^{(k)}=x_j^{(k)}+\alpha_j^{(j)}w_j^{(j+1)}$ for

$k>(q-1)/2+j+1$. According to (18) we have

$$(20)\quad x_{j+1}^{(k)} = x_j^{(k)} + \alpha_j^{(j)} w_j^{(j+1)} = x_j^{(j)} + \alpha_j^{(j)} w_j^{(j+1)} + v_0^{(k)} = x_{j+1}^{(j+1)} + v_0^{(k)}$$

for $k>(q-1)/2+j+1$. By virtue of (19),(20) we obtain

$$v_{j+1}^{(k)} = x_{j+1}^{(k)} - x_{j+1}^{(j+1)} = v_0^{(k)}, \quad (v_{j+1}^{(k)}, w_j^{(j+1)}) = 0 \qquad k>(q-1)/2+j+1$$

so that $\alpha_j^{(k)} = 0, k>(q-1)/2+j+1$ and for $x_{j+1}^{(k)}$ defined by the Algorithm (4) we have

$$x_{j+1}^{(k)} = x_{j+1}^{(j+1)} + v_0^{(k)} \qquad k>(q-1)/2+j+1$$

which is the assertion of the theorem.

The Algorithm (4) requires in each cycle n linear minimizations and the following number of function evaluations:

	Algorithm I-II	CG
full matrix	$3/2n^2-3/2n$	$2n^2$
5-diagonal band matrix	$6n - 9$	$2n^2$
3-diagonal band matrix	$3n - 3$	$2n^2$.

The conjugate gradient methods [1],[2],[3],[4],[5], require $2n^2$ function evaluations and the structure of the occurrence matrix has no influence on the number of function evaluations. We have considered that the gradient vector is estimated by the symmetric difference formulas. The above described algorithm has the quadratic termination property and in the case of a quadratic function is equivalent to the algorithm (3). For a q-diagonal band matrix it is necessary to store $(q-1)/2+1$ vectors. For $q=1$ we obtain the nonlinear Gauss-Seidel iteration.

Nonlinear systems

Let $f:E_n \longrightarrow E_1$ be a twice continuously differentiable strictly convex function which satisfies

$$\lim_{\|x\| \to \infty} f(x) = +\infty .$$

Let $x_0^{(0)}, x_0^{(1)}, \ldots, x_0^{(n)}$ be n+1 linearly independent points of the space E_n. Let $x_0^{(0)} \in E_n$ be an initial point and let $x_0^{(k)} = x_0^{(0)} + v_0^{(k)}$ where $v_0^{(k)} = (0,\ldots,t_k,\ldots,0)^T$ and $t_k = \lambda$ is a suitable positive real number. Let us denote

$$g(x) = f'(x) = (\partial f/\partial x_1, \partial f/\partial x_2, \ldots, \partial f/\partial x_n) .$$

Then a modification of the Algorithm (4) for minimization of $f(x)$ is defined as follows:

(21) Algorithm.

Step (0): Define λ and set $x_0^{(k)} = x_0^{(0)} + v_0^{(k)}$, $k=1,2,\ldots,n$, set $i=1$.

Step (1): Compute

$$x_i^{(i)} = x_{i-1}^{(i-1)} + \alpha_{i-1}^{(i-1)} w_{i-1}^{(i)}$$

where $w_{i-1}^{(i)} = v_{i-1}^{(i)} / \| v_{i-1}^{(i)} \|$, $v_{i-1}^{(i)} = x_{i-1}^{(i)} - x_{i-1}^{(i-1)}$

and $\alpha_{i-1}^{(k)}$ is a scalar coefficient such that

$$(g(x_{i-1}^{(i-1)} + \alpha w_{i-1}^{(i)}), w_{i-1}^{(i)}) = 0 \qquad \text{for} \quad \alpha = \alpha_{i-1}^{(i-1)} .$$

Step (2): Compute

$$x_{i-1}^{(k)} = x_{i-1}^{(k)} - (x_{i-1}^{(k)} - x_i^{(i)}, w_{i-1}^{(i)}) w_{i-1}^{(i)} \qquad k=i+1,\ldots,n.$$

Step (3): Compute

$$x_i^{(k)} = x_{i-1}^{(k)} + \alpha_{i-1}^{(k)} w_{i-1}^{(i)} \qquad k=i+1,\ldots,n$$

where $\alpha_{i-1}^{(k)}$ is defined as

$$\alpha_{i-1}^{(k)} = -c\lambda \frac{(g(x_{i-1}^{(k)}), w_{i-1}^{(i)})}{(g(x_{i-1}^{(k)} + c\lambda w_{i-1}^{(i)}) - g(x_{i-1}^{(k)}), w_{i-1}^{(i)})}$$

and $c \in (0,1\rangle$.

Step (4): Set $i=i+1$. If $i \leq n$ then go to Step (1) , else go to Step (5).

Step (5): Replace $x_0^{(0)}$ by $x_n^{(n)}$ and go to Step (0).

According to the choice of λ we obtain the following algorithms:

Algorithm III: $\lambda = \min(h, \| x_n^{(n)} - x_0^{(0)} \|)$, where h is a constant, usually $h=0.5(1.0)$.

Algorithm IV: $\lambda = \min(h, \|g(x_0^{(0)})\|)$.

Theorem 3 and Theorem 4 holds also for the Algorithm (21). The Algorithm (21) has the quadratic termination property and is suitable for band occurrence matrices. In this case it is necessary to store only $(q-1)/2+1$ vectors. The matrix of the vectors $w_{i-1}^{(i)}$, $i=1,2,\ldots,n$ is upper triangular which influences the number of arithmetic operations required in Step (1) and Step (3). For $q=3,5,7$ the algorithm requires less arithmetic operations per iteration than the conjugate gradient method.

Let $F: E_n \longrightarrow E_n$ be continuously differentiable and suppose that $F'(x)$ is symmetric and satisfies for some $c > 0$

$$(F'(x)h,h) \geq c(h,h) \qquad \forall\, x,h \in E_n.$$

Then the function $f: E_n \to E_1$

$$f(x) = \int_0^1 x^T F(tx)dt$$

according to Theorem 4.1.6 [4], Theorem 4.3.6 [4] and Theorem 3.4.6 [4] is uniformly convex, $g(x) = f'(x) = (Fx)^T$ and $\lim f(x) = +\infty$ for $\|x\| \longrightarrow \infty$, so that the unique minimizer of $f(x)$ is the unique solution of $Fx = 0$ and the Algorithm (21) can be applied for the solution of $Fx = 0$.

A modification of the above mentioned algorithm with an imperfect step size and a suitable corrector which approximates the vectors defined by the linear minimization principle will be published in a separate paper.

Example. Let $Fx = Ax + \phi x$ where A is a symmetric positive definite n by n matrix and $\phi: E_n \longrightarrow E_n$ is either continuously differentiable on E_n and $\phi'(x)$ is symmetric positive semidefinite for all x, or continuous, diagonal and isotone on E_n. The occurrence matrix corresponding to Fx is usually a band matrix (see [4]).

References

1. Fletcher, R., Reeves, C.M.: Function minimization by conjugate gradients, Comp. J.,2,1964, 149-154.
2. Hestenes, M.R., Stiefel, E.: The method of conjugate gradients for solving linear systems, J. Res. Nat. Bur. Standards, 49, 1952, 409-436.
3. Jacoby, S.L.S., Kowalik, J.S., Pizzo, J.T.: Iterative methods for nonlinear optimization problems, Prentice-Hall, Englewood Cliffs, 1972.
4. Ortega, J.M., Rheinboldt, W.C.: Iterative solutions of nonlinear equations in several variables, AP, New York, 1970.
5. Polak, E.: Computational methods in optimization, AP,New York, 1971.
6. Sloboda, F.: A parallel projection method for linear algebraic systems, to appear in Aplikace matematiky, 23, 1978.
7. Sloboda, F.: Parallel method of conjugate directions for minimization, Apl. mat., 20, 1975, 436-446.

THE DEVELOPMENT OF AN EFFICIENT OPTIMAL CONTROL PACKAGE

R.W.H. Sargent and G.R. Sullivan
Imperial College
London

1. Introduction

Many methods have been proposed for the numerical solution of deterministic optimal control problems (cf. Bryson and Ho, 1969). Early methods attempted, with limited success, to solve the two-point boundary value problem arising from Pontryagin's necessary conditions for an optimum by a shooting method, and the more recent "multiple shooting method" of Bulirsch et al. (1977) is designed to deal with the inherent instability in this technique. Miele and his co-workers (Miele, 1973; Heidemann and Levy, 1975) developed the "sequential gradient-restoration algorithm" which by-passes the stability problems by solving a sequence of linear two-point boundary value problems, using an extension of the gradient or conjugate-gradient method to a function-space. For the unconstrained case, extension of the variable-metric method (Tokumaru et al., 1970) to function-spaces has been made.

An alternative approach is to use a finite-dimensional representation of the control, and hence reduce the problem to a finite-dimensional optimization problem. Piecewise constant control functions were used for unconstrained problems by Horwitz and Sarachik (1968), and for problems without path constraints by Pollard and Sargent (1970). However, the latter authors used a simple projection technique to deal with control constraints and a penalty function to deal with terminal state constraints. In this paper we use a finite-dimensional representation of the control to formulate the general optimal control problem as a nonlinear programme, which can then be solved by a standard algorithm. Objective and constraint functions are evaluated by forward integration of the system equations, while their gradients with respect to the decision variables are obtained via backward integration of an adjoint system; the special structure of these equations makes possible the efficient implementation of integration procedures suitable for stiff systems, yielding an effective general-purpose programme.

2. Optimal Control Problem Formulation

The system is described by a set of state variables $x(t) \varepsilon E^n$, which evolve under the influence of controls $u(t) \varepsilon E^m$ according to the equation

$$\dot{x}(t) = f(t, x(t), u(t), v) \quad , \quad t \varepsilon [t_o, t_f] \tag{1}$$

where $[t_o, t_f]$ is the time interval of interest, and $v \varepsilon E^r$ is a vector of design parameters for the system which satisfy constraints

$$a^v \leq v \leq b^v \quad . \tag{2}$$

The class of admissible controls $u(t)$ consists of piecewise-continuous functions of t which satisfy

$$a^u \leq u(t) \leq b^u \quad , \quad t \varepsilon [t_o, t_f]. \tag{3}$$

The initial state $x(t_o)$ satisfies the conditions

$$a^o \leq x(t_o) \leq b^o \tag{4}$$

For almost all $t \varepsilon [t_o, t_f]$, and for all possible values of v, $x(t)$ and $u(t)$, the vector-valued function $f(t, x, u, v)$ satisfies the conditions:

(i) It is a piecewise-continuous function of t and a differentiable function of x, u and v.

(ii) There exists a function $S(t)$ summable on $[t_o, t_f]$, and a function $\phi(z)$ positive and continuous for $z \geq 0$ but not summable on $[0, \infty)$, such that $||f(t, x, u, v|| \leq S(t)/\phi(||x||)$.

(iii) The derivatives $f_u(t, x, u, v)$, $f_v(t, x, u, v)$ and $f_x(t, x, u, v)$ are bounded and $f_x(t, x, u, v)$ is Lipschitz continuous in x.

These conditions ensure that for each admissible set of design parameters, initial state and control there is a unique solution $x(t)$, $t \varepsilon [t_o, t_f]$, to equation (1), and further that these solutions are absolutely continuous in t and uniformly bounded over all admissible choices of design parameters, initial states and controls.

The state and controls may also be subject to path and terminal constraints of the form

$$a^g(t) \leq g(t, x(t), u(t), v) \leq b^g(t) \quad , \quad t \varepsilon [t_o, t_f] \tag{5}$$

$$a^f \leq F(t_f, x(t_f), v) \leq b^f \tag{6}$$

where $g(t, x, u, v) \varepsilon E^p$ and $F(t, x, v) \varepsilon E^q$.

System performance is measured in terms of the scalar objective function

$$J = F_o(t_f, x(t_f), v) \tag{7}$$

Henceforth $F_o(t, x, v)$ will be taken as the first element of the $(q+1)$-dimensional vector $F(t, x, v)$, which is assumed to be continuously differentiable with respect to its arguments, with the derivative $F_x(t, x, v)$ Lipschitz continuous in x. The function $g(t, x, u, v)$ satisfies the conditions (i), (ii) and (iii) given above.

The optimal control problem is to choose an admissible set of design parameters, initial state and controls, and possibly also the final time, to minimize J, subject to the conditions (1)-(6).

This formulation provides for considerable flexibility. Equality constraints may be imposed by setting the relevant upper and lower bounds to the same value, while constraints may be relaxed by setting the relevant bound to a very small or very large value as appropriate. Path constraints (5) allow for constraints on the state variables themselves, and for system and state dependent constraints on the controls. More general problems can also be treated by defining extra state variables; for example the more general objective function

$$J = F_o(t_f, x(t_f), v) + \int_{t_o}^{t_f} f_o(t, x(t), u(t), v)dt \tag{8}$$

may be treated by defining the additional state variable

$$x_o(t_o) = 0 , \quad \dot{x}_o(t) = f_o(t, x(t), u(t), v) , \quad t\varepsilon[t_o, t_f] , \tag{9}$$

which allows J to be written in the standard form (7).

3. Formulation as a Nonlinear Programme

To treat the optimal control problem by nonlinear programming it is necessary to use a finite number of parameters to define the control function $u(t)$, $t\varepsilon[t_o, t_f]$. For this purpose we divide the interval $[t_o, t_f]$ into a finite number of subintervals and use given basis functions involving a finite set of parameters to represent the control in each subinterval:

$$u(t) = \phi^j(t, z^j) , \quad t\varepsilon(t_{j-1}, t_j] , \quad z^j \varepsilon E^{S_j} , \quad j = 1, 2 \dots J , \tag{10}$$

where $t_J \equiv t_f$. The control is then defined by the parameters z^j and switching times t_j, $j = 1, 2 \dots J$. It is assumed that the functions $\phi^j(t, z^j)$ are continuously differentiable with respect to z^j. Note that this formulation allows for a discontinuity and change of functional form for every element of $u(t)$ at each switching time, which is not necessarily required. However, this formulation is simpler to handle than one allowing each element of $u(t)$ to switch independently.

It is also necessary to deal with the infinite-dimensional path constraints (5), which have always been a stumbling block in solving optimal control problems. Time discretization of these constraints is at best only an approximation, and use of a small time interval generates a very large number of nonlinear constraints. Some investigators (e.g. Bryson and Ho, 1969; Lapidus and Luus, 1970) have suggested a penalty function approach, adding the integral of the path constraint violations to the objective function with a suitable weighting parameter. A solution which avoids the computational difficulties of this approach is to use this integral to convert the path constraints into a terminal constraint by defining a new state variable:

$$\left.\begin{aligned} \dot{x}_{n+1}(t) &= \sum_{i=1}^{p} \{\max\{0, a_i^g(t) - g_i(t, x(t), u(t), v)\}\}^2 \\ &+ \sum_{i=1}^{p} \{\max\{0, g_i(t, x(t), u(t), v) - b_i^g(t)\}\}^2, \\ x_{n+1}(t_o) &= 0 \end{aligned}\right) \tag{11}$$

We then impose the constraint $x_{n+1}(t_f) = 0$ and include this in the terminal constraints (6). It is clear that $x_{n+1}(t_f)$ as calculated from (11) is zero if, and only if, the path constraints (5) are satisfied for all $t \in [t_o, t_f]$ so the two formulations are equivalent.

If the basis functions $\phi^j(t, z^j)$ in (10) are complicated, the constraints (3) on the controls u(t) are of the same type as (5) and must be treated similarly. However, it is frequently possible to express these constraints as functions of the decision variables t_j and z^j, and they take a particularly simple form if a piecewise constant or piecewise linear representation of the control is used.

Thus the set of decision variables is

$$y = (x(t_o), v, z^1, z^2 \,.\,.\, z^J, t_1, t_2 \,.\,.\, t_f) , \tag{12}$$

and we define
$$F(t_f, x(t_f), v) = \psi(y) . \tag{13}$$

Finally, we need the gradient of $\psi(y)$ with respect to y. The appropriate formulae have been derived in a companion paper (Sargent and Sullivan, 1977), and are given here without proof:

$$\psi_x(t_o) = \Lambda(t_o) \qquad , (14)$$

$$\psi_v = \int_{t_o}^{t_f} \Lambda(t).f_v(t, x(t), u(t), v)dt + F_v(t_f), x(t_f), v) \qquad , (15)$$

$$\psi_{z^j} = \int_{t_o}^{t_j} \Lambda(t).f_u(t, x(t), u(t), v)\phi^j_{z^j}(t, z^j)dt \qquad , (16)$$

$$\psi_{t_j} = \Lambda(t_j)\{f(t_j, x(t_j), u(t_j^-), v) - f(t_j, x(t_j), u(t_j^+), v)\}, \ 1 \leqslant j < J, \ (17)$$

$$\psi_{t_f} = \Lambda(t_f).f(t_f, x(t_f), u(t_f), v) + F_t(t_f, x(t_f), v) \qquad , (18)$$

where the adjoint variables $\Lambda(t)$ are given by

$$\dot{\Lambda}(t) = -\Lambda(t).f_x(t, x(t), u(t), v) \quad , \quad \Lambda(t_f) = F_x(t_f, x(t_f), v) . \tag{19}$$

In the present implementation a piecewise constant control is used, yielding

$$\phi^j(t, z^j) = z^j \quad , \quad \phi^j_{z^j}(t, z^j) = I , \tag{20}$$

and the optimal control problem can thus be formulated as the nonlinear programme:

$$\left.\begin{array}{lll} \text{Minimize} & \psi_o(y) & , \\ \text{Subject to} & a_i^f \leqslant \psi_i(y) \leqslant b_i^f & , \ i = 1, 2 \,.\,.\, q , \\ & a^o \leqslant x(t_o) \leqslant b^o & , \\ & a^v \leqslant v \leqslant b^v & , \\ & a^u \leqslant z^j \leqslant b^u & , \ j = 1, 2 \,.\,.\, J , \\ & 0 \leqslant t_j - t_{j-1} & , \ j = 1, 2 \,.\,.\, J . \end{array}\right) \tag{21}$$

The objective function and terminal constraint function values are obtained by integrating the system equations (1) from t_o to t_f, and the gradients are then obtained from (14)-(18) during the backward integration of the adjoint system (20) from t_f to t_o. The various conditions given in Section 2 ensure that these gradients are continuous in y.

It should be noted that, although this is also true for the terminal constraint arising from the path constraints, $x_{n+1}(t_f)$ from (11) and its gradient with respect to the decision variables are both zero for any feasible trajectory, and this can lead to failure of the "constraint qualification" at the optimal solution. This is discussed more fully in the companion paper cited above, and it suffices to note here that the algorithm we have used to solve the above nonlinear programme, the "variable-metric projection (VMP) method" of Murtagh and Sargent (1973), deals satisfactorily with this problem. Apart from this particular difficulty associated with path constraints, the problem as formulated in (21) is in a form suitable for solution by any nonlinear programming algorithm.

Of course, any of the decision variables in (12) may be held fixed if desired. We note in particular that both fixed and free end-time problems are dealt with naturally, and indeed the time-optimal control problem can be solved simply by taking $J \equiv t_f$ without the need to introduce an extra state variable.

4. Integration Methods

The system and adjoint equations are integrated repeatedly during the solution of the problem, and since many engineering systems are characterized by stiff differential equations it is important for efficiency that an appropriate integration method be used.

Most methods for stiff systems (Gear, 1971) are linear multistep methods, based on the formula

$$\left.\begin{aligned} &\sum_{j=0}^{k} \alpha_j x(t_j) = h \sum_{j=0}^{k} \beta_j f(t_j, x(t_j), u(t_j), v) \\ &\text{where} \quad t_j = t_{j-1} - h \,, \quad j = 1, 2 \,.\,.\, k \,, \end{aligned}\right\} \tag{22}$$

and α_j, β_j, j = 0, 1 . . k, are fixed scalars, with $\alpha_o = 1$. For stability it is essential that $\beta_o \neq 0$, so that (22) represents a system of nonlinear equations which determines $x(t_o)$ at the current time t_o in terms of values at past times $t_1, t_2 \,.\,.\, t_k$.

Since in any case we need the Jacobian matrix $f_x(t, x(t), u(t), v)$ for the adjoint system (20), it seems sensible to choose the same time-steps for this latter system so that the appropriate Jacobian matrix can also be used to solve (22) by Newton's method:

$$\{I - h\beta_o f_x(t_o, x(t_o), u(t_o), v)\}.\delta x = \sum_{j=0}^{k} \{h\beta_j f(t_j, x(t_j), u(t_j), v) - \alpha_j x(t_j)\}, \tag{23}$$

where δx represents the correction to be made to the current estimate $x(t_o)$. Each iteration requires solution of the linear system (23), for which purpose we obtain the LU factorization of the iteration-matrix on the left-hand side, using partial pivoting to avoid numerical instability as described by Wilkinson and Reinsch (1971). It is also possible to take account of sparseness (Duff and Reid, 1977) or special structure of the Jacobian (Tyreus et al., 1975; Robertson, 1976) to obtain further reductions in the computational and storage requirements. Given the LU factorization, equation (23) is easily solved for δx by two sequences of substitutions.

Applying (22) to the adjoint system (12), with integration backward in time, gives the following analogue of (23) (with $t_j = t_{j-1}+h$):

$$\{I-h\beta_o f_x(t_o,x(t_o),\ u(t_o),\ v)\}^T.\delta\Lambda^T = \sum_{j=0}^{k}\{h\beta_j f_x^T(t_j,x(t_j),\ u(t_j),\ v)\Lambda^T(t_j)-\alpha_j\Lambda^T(t_j)\}$$

$$= \sum_{j=0}^{k}\{(\beta_j/\beta_o-\alpha_j)\Lambda^T(t_j) - (\beta_j/\beta_o)\{I-h\beta_o f_x(t_j,x(t_j),\ u(t_j),\ v)\}\Lambda^T(t_j)\} \qquad (23a)$$

Since the iteration-matrix in (23a) is simply the transpose of that in (23) its LU factorization is available as U^TL^T. Further, since (12) is linear in $\Lambda(t)$, equation (23a) is also linear and the correct $\Lambda(t_j)$ is obtained on the first iteration. Thus the adjoint system is solved very cheaply if the LU factors are stored at each step.

In fact, the rate of convergence of (23) and (23a) is not greatly impaired if only an approximate iteration-matrix is used, and since the LU factorization is expensive computationally it is desirable to retain the same LU factors for all iterations at a given step, and indeed over a number of successive steps. Numerical tests showed this policy to be satisfactory, as is illustrated by the example given later. Now the Jacobian matrix does not enter directly into the determination of the state trajectory x(t) or the $\psi_i(y)$, but only in the iteration-matrices in (23), (23a) and in the determination of the gradients. Since previous experience (Sargent and Gaminibandara, 1975) has shown that high accuracy in gradients is often not important for the nonlinear programming algorithm there is further scope for savings by re-evaluating the Jacobian only when required for a new LU factorization of the iteration-matrix. Then, as indicated by the second expression in (23a), only these LU factors need be stored on the forward integration, since they provide all the necessary data for integrating the adjoint system. It remains to choose the number of past values (k) and the parameters α_j, β_j, j=0,1 . . k, in equation (22). Dahlquist (1963) demonstrated the importance of the concept of A-stability for stiff systems, and showed that an A-stable linear multistep method cannot have an order greater than two. The A-stable second-order method of smallest truncation error is the trapezoidal rule, given by k=1, $\alpha_1=1$, $\beta_o=\beta_1=\frac{1}{2}$, which therefore seems an obvious candidate. However Liniger and Willoughby (1970) pointed out that for stiff systems it is worth sacrificing the order of polynomial fit to the solution in order to fit the dominant fast-decaying modes. Their first-order formula corresponds to the choice k=1, $\alpha_1=-1$, $\beta_2=1-\beta_1$, with

$$\beta_1 = |h\lambda|^{-1} - (\exp(|h\lambda|) - 1)^{-1} \qquad (24)$$

where λ is the eigenvalue of the Jacobian corresponding to the fast-decaying mode fitted. This is, of course, usually the most negative eigenvalue, which can be estimated by the Gerschgorin Theorem. This formula yields $\beta_1 \leq \frac{1}{2}$ and is A-stable for all λ.

These two integration formulae were compared with the standard fourth-order Kutta-Simpson rule for integration of both system and adjoint equations for a model of a distillation column. This model, for the splitter column of a crude oil distillation train, is described in detail by Sullivan (1977); it consists of 29 non-linear differential equations, including one for an integral objective function as given in (8) and (9), with three control variables and no terminal or path constraints. Initial values $x(t_0)$ were specified and four equal control intervals were used. The system has a stiffness-ratio of about 40 over the region of interest, and each control interval was approximately equal to the longest time-constant associated with the system. An accurate solution was obtained by repeated integrations using the Kutta-Simpson rule, with successive halving of the step-length until two successive runs agreed to the required accuracy. Final extrapolation to $h = 0$ yielded a value of $J = 0.9398 \times 10^{-3}$. Tests were then made with the Liniger-Willoughby method to determine the effects of step-length, frequency of re-evaluation of the iteration-matrix in equations (23) and (23a), and number of Newton iterations in solving (23) for the system equations. The final results are summarized in Table 1.

TABLE 1 Test of Liniger-Willoughby Algorithm

No. of Newton itera-tions	Iteration-matrix re-evaluation frequency			
	Once per control interval		Twice per control interval	
	J	$J_{u'}$	J	$J_{u'}$
1	$.838139896\times10^{-3}$	$-.265271352\times10^{-3}$	$.839066099\times10^{-3}$	$-.264378264\times10^{-3}$
2	$.915477864\times10^{-3}$	$-.263277440\times10^{-3}$	$.915372519\times10^{-3}$	$-.263322118\times10^{-3}$
3	$.913622656\times10^{-3}$	$-.263386438\times10^{-3}$	$.9136406569\times10^{-3}$	$-.263387959\times10^{-3}$

All integration errors will tend to accumulate in the values of the derivatives of the objective function with respect to control levels in the first control-interval, and the value of the most sensitive of these derivatives is given in Table 1 together with the objective function value. In view of the control-level changes it is necessary to re-evaluate the iteration-matrix at the beginning of each control interval, but Table 1 shows that this gives essentially the same accuracy as re-evaluation at double this frequency. The table also shows that the accuracy is sufficient with two Newton iterations of (23) at each integration-step and it is therefore worth fixing the number of iterations a priori rather than testing for convergence.

With 50 integration-steps per control-interval the accuracy of x(t) and J is only moderate, but sufficient for the application in question. Figure 1 gives a comparison between the three integration methods, in terms of computing time for the forward and backward integrations when the step-length is adjusted to give a specified accuracy.

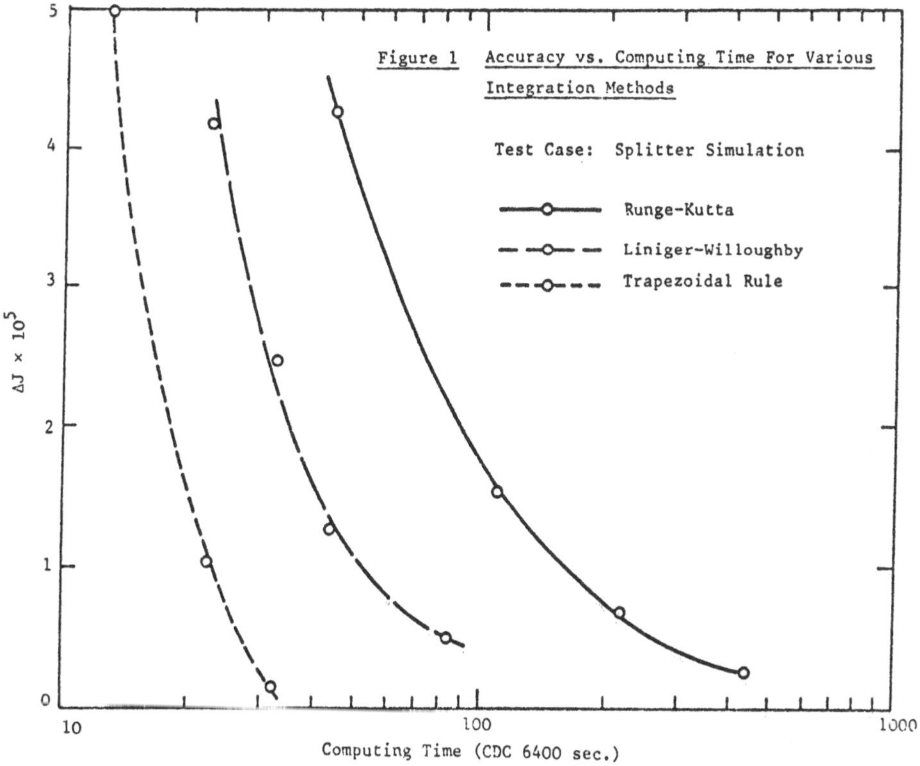

Figure 1 Accuracy vs. Computing Time For Various Integration Methods

The clear advantage of the proposed scheme is evident using both the trapezoidal rule and the Liniger-Willoughby formula, and the advantage would, of course, be greater in problems with terminal constraints. In this example, the trapezoidal rule is about twice as fast as the Liniger-Willoughby method, but this is to be expected with such a low stiffness-ratio; in the computer programme both methods are retained as options, since the Liniger-Willoughby method is likely to prove advantageous for stiffer systems.

5. The Complete Algorithm

The complete algorithm (OPCON) uses the VMP nonlinear programming algorithm, with the integration methods described in Section 4 to evaluate the objective and constraint functions and their gradients as required. The user must supply subroutines to evaluate f(t, x, u, v), g(t, x, u, v), and their derivatives with respect to x, u and v, as well as for F(t, x, v) and its derivatives with respect to t, x and v. He must also specify the bounds in constraints (2)-(6), tolerances on constraints and objective function at the solution, the number of control intervals, the integration step-length, and initial estimates of the decision variables. The user can also select the integration method and change the number of Newton iterations for equation (23) and the frequency of evaluation of the iteration-matrix.

The number of control intervals and the integration step-length must be chosen to give the desired accuracy, which requires judgement and often experimentation. Clearly it is desirable to start with relatively large intervals and step-length and refine them, so that iterations with heavy computing demands are started from relatively good estimates of the solution. The effects of integration step-length are well known, and Horwitz and Sarachik (1963) discuss the effects of piecewise constant approximations to continuously varying controls. It would be easy incorporate a test of integration error on the forward integration and re-evaluate the iteration-matrix automatically when it exceeds a specified tolerance. Similarly the technique proposed by Mellefont and Sargent (1977) for automatic adjustment of the number of control intervals could be added without difficulty.

To illustrate the performance of the algorithm, a simple example concerning the optimal mix of catalyst in a chemical reactor is given. This has been solved analytically by Jackson (1968), who showed that the optimal solution is a piecewise constant control with a singular segment (in the sense of Pontryagin). This is convenient for our purpose, since there are no uncertain inaccuracies resulting from the control discretization. The problem is:

$$\text{Minimize} \qquad J = x_1(t_f) + x_2(t_f)$$

$$\text{Subject to} \qquad \dot{x}_1 = u(k_2x_2 - k_1x_1)$$

$$\dot{x}_2 = u(k_1x_1 - k_2x_2) - (1-u)k_3x_2$$

$$0 \leqslant u(t) \leqslant 1$$

$$x_1(0) = 1 \ , \quad x_2(0) = 0 \ , \quad k_1 = k_3 = 1 \ , \quad k_2 = 10 \ , \quad [t_o, t_f] = [0, 1]$$

The number of control intervals was 5, with initial switching times equally spaced and u = 0.2 in each interval. The final solution is shown in Figure 2, which also shows Jackson's analytical solution and a policy obtained by Nishida et al. (1976) using their modified "Max-H" algorithm.

OPCON took 14 seconds CDC 6400 time, involving 17 iterations with 21 function evaluations, and obtained J = 0.951940 compared with the analytical optimum value of 0.951935. The Max-H algorithm took 15 seconds IBM 370/155 time (≈22 seconds CDC 6400 time), involving 25 iterations, from the same starting trajectory, and

Nishida et al. state that their results compare favourably with gradient and second-variation techniques.

Figure 2 Comparison for Mixed Catalyst Problem

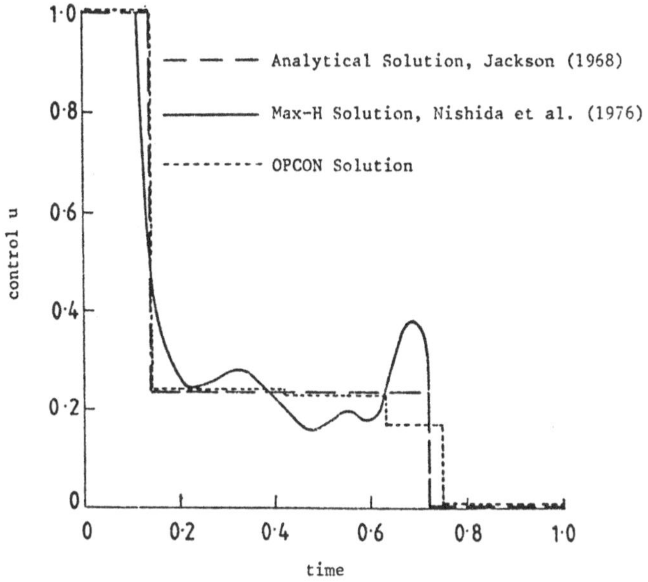

Almost identical performance was obtained with OPCON from an initial control $u(t) = 0.5$, $t\epsilon[0, 1]$, and the analytical solution was attained with a tighter imposed tolerance.

OPCON has also been used successfully on large problems concerned with the optimal control of crude distillation columns, involving up to 50 state variables and 7 controls (Sullivan, 1977), and other smaller problems.

References

1. Bryson, A.E., Jr., and Ho, Y.C., (1969) "Applied Optimal Control", Blaisdell, Waltham, Mass.
2. Bulirsch, R. (1977), "Numerical Computation of Optimal Control Problems in Economy and Engineering", 8th IFIP Conference, Würzburg.
3. Dahlquist, G. (1963) "A Special Stability Problem for Linear Multistep Methods", BIT, 3, 27-43.
4. Duff, I.S., and Reid, J.K. (1977) "Some Features of a Sparse Matrix Code", Report CSS 48, AERE, Harwell.
5. Gear, C.W. (1971) "Numerical Initial Value Problems in Ordinary Differential Equations", Prentice-Hall, New Jersey.
6. Heidemann, J.C., and Levy, A.V. (1975) "Sequential Conjugate-Gradient-Restoration Algorithm for Optimal Control Problems: Part I Theory. J. Optimization Theory and Applications, 15, 203-222.
7. Horwitz, L.B. and Sarachik, P.E. (1968) "A Computational Technique for Calculating the Optimal Control Signal for a Specific Class of Problems", Record of 2nd Asilomar Conference on Circuits and Systems, Pacific Grove, Calif., October 1968, pp. 537-540.
8. Jackson, R. (1968) "Optimal Use of Catalyst for Two Successive Chemical Reactions", Chem. Eng. Sci., 21, 241-260.

9. Lapidus, L., and Luus, R. (1967) "Optimal Control of Engineering Processes", Blaisdell, Waltham, Mass.
10. Liniger, W., and Willoughby, R.A. (1970) "Efficient Integration Methods for Stiff Systems of Ordinary Differential Equations", SIAM. J. Numer. Anal. 7, 46-66.
11. Mellefont, D.J., and Sargent, R.W.H. (1977) "Calculation of Optimal Measurement Policies for Feedback Control of Linear Stochastic Systems", IFIP, Würzburg.
12. Miele, A. (1973), "Gradient Methods in Optimal Control Theory", in Eds. M. Avriel, M.J. Rijckaert and D.J. Wilde "Optimization and Design", Prentice Hall, New Jersey.
13. Nishida, N., Liu, R.A., Lapidus, L., and Hiratsuka, S. (1976) "An Effective Computational Algorithm for Suboptimal Singular and/or Bang-Bang Control, Part I: Theoretical Developments and Applications to Linear Lumped Systems", A.I.Ch.E. J., 22, 503-513.
14. Pollard, G.P., and Sargent, R.W.H. (1970) "Off-line Computation of Optimal Controls for a Plate Distillation Column", Automatica, 6, 59-76.
15. Robertson, H.H. (1976) "Numerical Integration of Systems of Stiff Ordinary Differential Equations with Special Structure", J. Instn. Math. and Applic. 18, 249-263.
16. Sargent, R.W.H. and Gaminibandara, G.G.K.K. (1975) "Optimum Design of Plate Distillation Columns" in Ed. L.C.W. Dixon, "Optimization in Action", Academic Press.
17. Sargent, R.W.H. and Murtagh, B.A. (1973) "Projection Methods for Nonlinear Programming", Math. Prog. 4, 245-268.
18. Sargent, R.W.H., and Sullivan, G.R., (1970), to appear.
19. Sullivan, G.R. (1977) "Development of Feed Changeover Policies for Refinery Distillation Units", Ph.D. Thesis, University of London.
20. Tokumaru, H., Adachi, N., and Goto, K. (1970) "Davidon's Method for Minimization in Hilbert Space with an Application to Control Problems", SIAM. J. Control, 8, 163-178.
21. Tyreus, B.D., Luyben, W.L. and Schiesser, W.E. (1975) "Stiffness in Distillation Models and Use of an Implicit Integration Method to Reduce Computation Times", Ind. Eng. Chem., PDD, 14, 427-433.
22. Wilkinson, J.H. and Reinsch, C. (1971) "Handbook for Automatic Computation - Vol. 2: Linear Algebra", Springer-Verlag, New York.

AN ACCURACY SELECTION ALGORITHM FOR THE MODIFIED GRADIENT PROJECTION METHOD IN MINIMAX PROBLEMS

Jacek Szymanowski
Andrzej Ruszczyński

Institute of Automatic Control
Technical University of Warsaw
00-665 Warszawa, Poland.

1. Introduction

For minimax problems two-iterative methods are often used. In these methods the inner "max" problem is solved by a certain lower-level algorithm and the outer "min" problem is solved by an upper-level algorithm. Usually it is assumed that the lower level solutions are unique and are computed precisely. These assumptions can hardly be satisfied in practice because computation of lower-level solutions requires infinite number of function evaluations. Therefore a certain accuracy selection algorithm has to be used on the lower-level.

One of possible methods for solving minimax problems is the ε-subgradient projection method, as proposed by Auslender in [1]. Short description of this method is contained in Sec.2. In the paper [1] the gradient projection method [2], was modified so as to be applicable to minimax problems in which lower-level solutions are computed with errors. Under general assumptions on the accuracy of lower-level solutions convergence theorem was proved. However, no constructive accuracy selection algorithm has been proposed. Such a proposition was given in [3], where the general problem of accuracy selection for two-level methods was discussed. Short presentation of these ideas is given in sec 3.

A constructive accuracy selection algorithm for the ε-subgradient projection method is proposed in sec.4. In this algorithm stop criteria at the lower-level are tightened only when actually needed. It allows for substantial savings of calculations when compared with the method of establishing maximal accuracy from the beginning of operation. Convergence theorem for the whole method is proved in sec.4.

2. Methods of ε-subgradients

Let X and Y be finite dimensional spaces and let K and D be closed convex subsets of X and Y. Let $f: X \times Y \to R^1$ be a functional which is convex with respect to x and concave with respect to y. Let us consider the minimax problem

$$\min_{x \in K} \max_{y \in D} f(x,y) \tag{1}$$

let us define the functional

$$J(x) = \sup f(x,y), \tag{2}$$

and the sets

$$\hat{Y}(x) = \{ y \in Y: \quad f(x,y) = J(x) \} \tag{3}$$

We assume that $J(x) < \infty$ for $x \in K$. Finally, let level asets of the functional J be compact and let for each $x \in K$ the set $\hat{Y}(x)$ contain a single element $\hat{y}(x)$. Then the solution of (1) exists and

$$J(x) = \frac{\partial f(x, \hat{y}(x))}{\partial x} \tag{4}$$

There are many methods in which the problem (1), or equivalent to it problem of minimization of J over K, may be solved. We shall focus our attention on the gradient projection methods. Levitin and Polyak proposed the gradient projection method which consists of the iteration

$$x_{k+1} = P_K \left(x_k - \tau_k \nabla J(x_k) \right) \tag{5}$$

where $P_k(z)$ denotes the unique projection of a vector $z \in X$ on K and $\tau_k > 0$ denotes step size. Under the Lipschitz assumption

$$\| \nabla J(x_1) - \nabla J(x_2) \| \leq M \| x_1 - x_2 \| , \quad \forall x_1, x_2 \in K \tag{6}$$

convergence of the algorithm has been proved for the case where the step size τ_k satisfies

$$0 < \varepsilon \leq \tau_k \leq \frac{2(1 - \varepsilon)}{M}, \quad \forall k \tag{7}$$

with ε any scalar with $0 < \varepsilon < 2/(2+M)$. Next, Bertsekas [2] observed that it is not necessary to know the constant M for the choice of numbers τ_k which provide convergence.

In the method of Bertsekas τ_n is the first number of the sequence $(s, S \quad , s\beta^2, s\beta^3, \ldots, s\beta^m, \ldots)$ for which

$$J(x_n) - J(x_{n+1}) \geq \frac{\sigma \| x_{n+1} - x_n \|^2}{\tau_n} \tag{8}$$

where $s > 0$, $0 < \beta < 1$ and $0 < \sigma < 1$.

Let us note however that in the case of minimax problems the value of J and the derivative ∇J are computed by the lower-level algorithm which solves the problem

$$\max_{y \in D} f(x,y) \tag{9}$$

In general, solution of such a problem requires infinite number of iterations of a certain algorithm. In practice, computations are truncated after finite number of iterations. Thus neither the exact value of J nor its derivative are available at the upper level. Let us assume, that for a fixed accuracy ε at the lower level we obtain a certain approximate solution $\tilde{y}(x,\varepsilon)$ such that

$$J(x) - \varepsilon \leqslant f(x,\tilde{y}(x,\varepsilon)) \leqslant J(x) \tag{10}$$

and approximation of the gradient of J:

$$g(x,\varepsilon) = \frac{\partial f(x,\tilde{y}(x,\varepsilon))}{\partial x} \tag{11}$$

In such a situation gradient projection methods cannot be applied in the form described above. Along with the sequence $\{x_n\}$ one should define another sequence $\{\varepsilon_n\}$ of the accuracy parameters for the lower-level. Let us observe however that $g(x,\varepsilon)$ defined by (11) is an ε-subgradient of J, i.e.

$$g(x,\varepsilon) \in \partial_\varepsilon J = \{ d\colon f(z) \geqslant f(x) + \langle d, z-x \rangle - \varepsilon \} \tag{12}$$

Basing on this property, Auslender [1] suggested a modification of the step size rule (8) which guarantees convergence provided that the sequence $\{\varepsilon_n\}$ satisfies some additional conditions. In the Auslender's method the step size τ_n is the first number of the sequence $(s, s\beta, s\beta^2, \dots, s\beta^m, \dots)$ for which

$$J_{\varepsilon_n}(x_n) - J_{\varepsilon_n}(z(x_n,\varepsilon_n,\tau_n)) \geqslant \frac{\sigma \| x_n - z(x_n, n, \tau_n) \|^2}{\tau_n} - 2\sqrt{\varepsilon_n} \tag{13}$$

where

$$J(x) - \varepsilon \leqslant J_\varepsilon(x) \leqslant J(x) \tag{14}$$

$$z(x,\varepsilon,\tau) = P_K(x - \tau g(x,\varepsilon)), \tag{15}$$

and $s > 0$, $0 < \beta < 1$, $0 < \sigma < 1$.

For such a τ_n we define $x_{n+1} = z(x_n, \varepsilon_n, \tau_n)$ (16)

Theorem 1 (Auslender [1])

Let $\varepsilon \in (0,1]$. Let $K_\varepsilon = \{ z : \exists\, x \in K,\ \| z-x \|^2 \leq \varepsilon \}$ and let

$$\| \nabla J(x') - \nabla J(x'') \| \leq M \| x' - x'' \| , \quad \forall x', x'' \in K \tag{17}$$

Let the sequence $\{\varepsilon_n\}$ be such that

$$0 < \varepsilon_n \leq \varepsilon \tag{18}$$

and

$$\sum_{n=0}^{\infty} \sqrt{\varepsilon_n} < \infty \tag{19}$$

Then for any $x_0 \in K$ the sequence $\{x_n\}$ is well defined and minimizes J over K, i.e.

$$\lim_{n\to\infty} J(x_n) = \min_{x\in K} J(x) \tag{20}$$

It follows from this theorem that there are many ways in which the sequence $\{\varepsilon_n\}$ may be constructed. One of possible methods is to take $\varepsilon_n = \frac{1}{n^p}$ with $p > 2$. It is doubtful however that such a choice would be satisfactory for any practical problem. It seems reasonable to update ε on-line, on the base of current observations of changes in the objective function.

3. Accuracy selection algorithm for two-level methods

In the paper [3] convergence of two-level methods of mathematical programming was analysed. It was assumed that we have to solve the problem

$$\min_{x\in K} G(x, \hat{y}(x)) \tag{21}$$

in which $\hat{y}(x)$ is the solution of the lower-lever problem

$$\min_{y\in D(x)} Q(x,y)$$

Let us observe that the minimax problem (1), (2) is included in this scheme $(G(x,y) = f(x,y);\ Q(x,y) = -f(x,y))$.

Let us assume that there is a certain algorithm which solves (22) at the lower level and suppose that this algorithm is provided with a stop criterion dependent on the parameter $\delta > 0$. Taken together they define a mapping

$$\tilde{y} : X \times R^1 \to Y \tag{23}$$

such that $\tilde{y}(x,\delta)$ is an approximate solution of (22) computed with the accuracy δ . Moreover, it is postulated that $\tilde{y}(x,\delta) \in D(x)$

and $\lim_{\delta \to 0} \tilde{y}(x, \delta) = \hat{y}(x)$ for all $x \in K$.

Next, let us assume that there is another algorithm at the upper-level which produces a sequence

$$x_{n+1} \in B_{\Gamma}(x_n) \tag{24}$$

where B_{Γ} is a certain point-to-set mapping dependent on the function which is minimized. Let the stop criterion at the upper level define sets $\Omega_{\Gamma}(\varepsilon)$ of approximate solutions of the minimization problems. Finally, let us assume that $\Gamma(x) - \inf_{z \in K} \Gamma(z) \leq \varphi(\varepsilon)$ when $x \in \Omega_{\Gamma}(\varepsilon)$ and Γ is a function from a certain class Ξ_{ε}. It is postulated that $\lim_{\varepsilon \to 0+} \varphi(\varepsilon) = 0$.

Under above assumptions the following algorithm for the solution of two-level problems was proposed in [3] :

Step 0: Select $x_0 \in K$, $\varepsilon_0 > 0$, $\lambda \in (0,1)$

Set $k = 0$

Step 1: Set $u_0 = x_k$

Step 2: Compute $\tilde{y}(u_0, \varepsilon_k)$ and $G_{\varepsilon_k}(u_0) = G(u_0, \tilde{y}(u_0, \varepsilon_k))$

Step 3: Set $j = 0$

Step 4: Perform one iteration of the upper-level algorithm

$$u_{j+1} \in B_{G_{\varepsilon_k}}(U_j)$$

Step 5: If

$$G_{\varepsilon_k}(u_{j+1}) \leq G_{\varepsilon_k}(u_j) - \varepsilon_k$$

then go to step 6; otherwise go to Step 7.

Step 6: Set

$$x_{k+1} = u_{j+1}$$

$$u_0 = u_{j+1}$$

$$\varepsilon_{k+1} = \varepsilon_k$$

$$k = k+1$$

and go to step 3

Step 7: If

$$u_{j+1} \in \Omega_{G_{\varepsilon_k}}(\varepsilon_k)$$

then go to Step 8; otherwise set

$u_j = u_{j+1}$

$j = j+1$

and go to Step 4

Step 8: Set

$x_{k+1} = x_k$

$\varepsilon_{k+1} = \lambda\varepsilon_k$

$k = k+1$

and go to Step 1.

Under rather general assumptions about the properties of functions G and Q, the algorithms and the stop criteria it was proved in [3] that the sequence $\{x_k\}$ minimizes $G(x, \hat{y}(x))$ over K. Important feature of the proposed method of cooperation of the algorithms at upper and lower level is the fact that the inner problem (22) at x distant from the solution are solved roughly, avoiding superfluous computational expences. Accuracy at the lower level is increased when the errors due to inaccurate computations are of the some order as the improvements of the function observed at the upper level. Of course, for each particular method many obvious questions arise such as how to choose the constant λ and how to balance sensitivity of stop criteria to the parameter ε in the algorithms used. Proper choice of the stop criteria may considerable reduce cost of computations. In the next section we shall incorporate the above accuracy selection algorithm to the Auslender's method and prove convergence of the whole two-level algorithm.

4. Accuracy selection algorithm for the ε-subgradient projection method

Firstly, we shall define stop criteria for upper and lower level. Let us assume that for a given accuracy $\delta > 0$ the approximate solution $\tilde{y}(x, \delta)$ of the inner problem

$$\max_{y \in D} f(x,y) \tag{25}$$

satisfies the inequalities

$$J(x) - \mu\delta^2 \leq f(x, \tilde{y}(x,\delta)) \leq J(x) \tag{26}$$

where $\mu > 0$ is a certain fixed constant.

Such a property is typical for stop criteria based on the norm of the gradient.
At the upper level we shall decide that $u_{j+1} \in \Omega(\delta)$ if the last decrease of the function J_{ε} observed by the upper-level (recall that $\varepsilon = \mu\delta^2$, according to (26)) is less than δ.

Under above assumptions we may specify the whole algorithm as follows:

Step 0: Select $x_0 \in X$, $\delta_0 > 0$, $\lambda \in (0,1)$. Set $n = 0$.

Step 1: Compute $\tilde{y}(x_n, \delta_n)$ satisfying (26) and set $\varepsilon_n = \mu\delta_n^2$

Step 2: Compute τ_n and $z(x_n, \varepsilon_n, \tau_n)$ according to the Auslender's method (13)-(15)

Step 3: If

$$J_{\varepsilon_n}(z(x_n, \varepsilon_n, \tau_n)) \leq J_{\varepsilon_n}(x_n) - \delta_n \tag{27}$$

then go to step 4; otherwise go to Step 5.

Step 4: Set

$$x_{n+1} = z(x_n, \varepsilon_n, \tau_n)$$
$$\delta_{n+1} = \delta_n$$
$$\varepsilon_{n+1} = \varepsilon_n$$
$$n = n+1$$

and go to Step 2.

Step 5: Set

$$x_{n+1} = x_n$$
$$\delta_{n+1} = \lambda\delta_n$$
$$n = n+1$$

and go to Step 1.

let us observe that $\varepsilon_n = \mu\delta_n^2$ and only one accuracy parameter is used in the method, as in Sec.3. We have preserved the denotation ε_n for the error in the function observed at the upper level in in order to point out close relations to the Auslender's method.

Theorem 2

Let the assumption (17) of the theorem 1 be satisfied.
Let $\mu\delta_0^2 \leq \varepsilon$. Then for any $x_0 \in K$ the sequence $\{x_n\}$ generated by the above algorithm minimizes J over K, i.e.

$$\lim_{n \to \infty} J(x_n) = \min_{x \in K} J(x)$$

<u>Proof:</u>

A. Let us assume that Step 4 is executed finite number of times. Then there exist $N \geqslant 0$ such that $x_n = x_N$ and

$$J_{\varepsilon_n}(z(x_N, \varepsilon_n, \tau_n)) > J_{\varepsilon_n}(x_N) - \delta_n \tag{28}$$

for all $n \geqslant N$. On the other hand, according to the Step size rule (13):

$$J_{\varepsilon_n}(x_N) - J_{\varepsilon_n}(z(x_N, \varepsilon_n, \tau_n)) \geqslant \frac{\sigma \| x_N - z(x_N, \varepsilon_n, \tau_n) \|^2}{\tau_n} - 2\sqrt{\varepsilon_n} \tag{29}$$

It follows from (28) and (29) that

$$\frac{\sigma \| x_N - z(x_N, \varepsilon_n, \tau_n) \|^2}{\tau_n} \leqslant 2\sqrt{\varepsilon_n} + \delta_n \tag{30}$$

for all $n \geqslant N$. It was proved in [1] that τ_n choosen according to (13) satisfies the inequality

$$\tau_n \geqslant B \cdot \min \left(\frac{2(1-\sigma)}{M}, \frac{1}{(M+1) \| g(x_n, \varepsilon_n) \|} \right)$$

with $B > 0$. Thus, there exists $\gamma > 0$ such that $\tau_n > \gamma$ for $n \geqslant N$. On the other hand $\tau_n \leqslant S$. Consequently we obtain from (30) the inequality

$$\| x_n - P_K(x_N - \gamma g(x_n, \varepsilon_n)) \|^2 \leqslant S(2\sqrt{\varepsilon_n} + \delta_n) \tag{31}$$

Finally, let us observe, taat Step 5 is executed for $n \geqslant N$ and $\delta_n \to 0$, $\varepsilon_n \to 0$. After transition to the limit in (31) we obtain

$$\| x_N - P_K(x_N - \gamma \nabla J(x_N)) \| = 0$$

and x_N is the solution.

B. Let us assume that Step 4 is executed infinite number of times. Let

$$I = \{ n : J_{\varepsilon_n}(z(x_n, \varepsilon_n, \tau_n)) \leqslant J_{\varepsilon_n}(x_n) - \delta_n \}$$

Thus

$$x_{n+1} = z(x_n, \varepsilon_n, \tau_n) \quad \text{for} \quad n \in I$$

$$x_{n+1} = x_n \quad \text{for} \quad n \notin I$$

Consequently, it remains to prove that the subsequence $\{x_n\}_{n\in I}$ minimizes J over K.

Let us define the subsequences $\{x_{k_i}\}$ and $\{x_{n_i}\}$ of the sequence $\{x_n\}$ as follows.

$$K_o = \min I \tag{32}$$

$$n_i = \min \{n > k_i \; ; \; n \notin I\} \tag{33}$$

$$K_{i+1} = \min \{n > n_i ; n \in I\} \tag{34}$$

Note that the choice of n_i according to (33) is always possible. Otherwise there would exist $k_i \geqslant 0$ such that for all $n > k_i$ Step 4 would be executed. Consequently, inequality (27) would be fulfilled infinite number of times with fixed $\delta_{k_i} > 0$, $\varepsilon_{k_i} > 0$. Thus $\lim_{n\to\infty} J_{\varepsilon_{k_i}}(x_n) = -\infty$ which is contradictory to the assumption that J has compact level sets. Let us observe that $\delta_n = \delta_{k_i}$ for $k_i \leqslant n < n_i$ and $\delta_n = \delta_{k_i} \cdot \lambda^{n-n_i}$ for $n_i < n \leqslant k_{i+1}$. Thus $\delta_{k_{i+1}} = \delta_{k_i} \lambda^{k_{i+1}-k_i}$.

We shall estimate the difference $J(x_{k_{i+1}}) - J(x_{k_i})$. From the definition of the set I we obtain

$$J_{\varepsilon_{k_i}}(x_{n_i}) \leqslant J_{\varepsilon_{k_i}}(x_{k_i}) - \sum_{n=k_i}^{n_i-1} \delta_n$$

Since $x_{k_{i+1}} = x_{n_i}$, it follows from the last inequality that

$$J_{\varepsilon_{k_i}}(x_{k_{i+1}}) \leqslant J_{\varepsilon_{k_i}}(x_{k_i}) - \sum_{n=k_i}^{n_i-1} \delta_n$$

Consequenty

$$J(x_{k_{i+1}}) \leqslant J(x_{k_i}) + \varepsilon_{k_i} - \sum_{n=k_i}^{n_i-1} \delta_n$$

Hence

$$J(x_{k_{i+1}}) \leqslant J(x_{k_o}) + \sum_{j=0}^{i} \varepsilon_{k_j} - \sum_{j=0}^{i} \sum_{n=k_j}^{n_j-1} \delta_n =$$

$$= J(x_{k_o}) + \mu \sum_{j=0}^{i} (\delta_{k_j})^2 - \sum_{\substack{n\in I\\ n\leqslant i}} \delta_n \tag{35}$$

Let us note that $\delta_{k_{j+1}} = \delta_{k_j} \lambda^{k_{j+1}-n_j} \leq \lambda \delta_{k_j}$. Thus

$$\mu \sum_{j=0}^{i} (\delta_{k_j})^2 \leq \frac{\mu \delta_{k_o}^2 (1 - \lambda^{2i+2})}{1 - \lambda^2} \tag{36}$$

It follows from (35) and (36) that

$$\sum_{\substack{n \in I \\ n \leq i}} \delta_n \leq J(x_{k_o}) - J(x_{k_{i+1}}) + \frac{\mu \delta_{k_o}^2 (1 - \lambda^{2i+2})}{1 - \lambda^2} \tag{37}$$

Since J is bounded from below on K, it follows from the last inequality that the series $\sum_{n \in I} \delta_n$ converges. Consequently, $\sum_{n \in I} \sqrt{\varepsilon_n} < \infty$ and assumption (19) of theorem 1 is fulfilled for the subsequence $\{x_n\}_{n \in I}$.

Let us note that the accuracy selection algorithm can be incorporated to other methods for minimax problems and to a wide class of two-iterative methods such as augmented Lagrangian methods and two--level methods for large-scale problems.

References:

1. A. Auslender: "Minimization of convex functions with errors", IX International Symposium on Mathematical Programming, Budapest 1976.
2. D. Bertsekas: "On the Goldstein-Levitin-Polyak gradient projection method", IEEE Tran. Aut.Contr., April 1976, vol. AC-21, No-2
3. J. Szymanowski, A. Ruszczyński: "Convergence analysis for two-level algorithms of mathematical programming", International Symposium on Mathematical Programming, Budapest 1976.

SINGLE-ITERATIVE SADDLE-POINT ALGORITHMS FOR SOLVING CONSTRAINED OPTIMIZATION PROBLEMS VIA AUGMENTED LAGRANGIANS

Andrzej Wierzbicki, Andrzej Janiak and Tomasz Kręglewski
INSTITUTE OF AUTOMATIC CONTROL
TECHNICAL UNIVERSITY OF WARSAW

00-665 Warsaw. Poland, Nowowiejska 15/19.

The theory of augmented Lagrange functions makes it possible to unify most of known constrained optimization algorithms as special cases of a general Newton-like algorithms of saddle-point seeking. Penalty methods, gradient projection methods, multiplier methods, quadratic approximation methods etc can be treated uniformly in this framework. Moreover, a new class of single-iterative saddle-point algorithm generalising the previously mentioned methods results from this approach. The particular algorithms are described in detail : a double-variable metric algorithm with variable metrics used for the approximation of the primal and the dual hessian, and a single-variable metric algorithm with a variable metric used for the approximation of the primal hessian and with an inversion of the dual hessian. Both algorithms have finite termination in quadratic case and quadratic convergence in more general case; both are effective as well for nonlinear as for linear constraints. Examples of application and results of preliminary tests are given at the end.

1. Problem setting, notation, fundamentals

Comider the classical nonlinear programming problem

$$/1.1a/ \quad \hat{x} = \arg \min_{x \in X_0} f(x) \ ; \quad X_0 = \left\{ x \in R^n : g(x) \in -D \subset R^m \right\}$$

where the inequality and equality constraints are expressed jointly by the positive cone D in the space of constraints R^m

$$/1.1b/ \quad D = \left\{ d \in R^m : d_i \geqslant 0, \ i = 1,\dots k; \ d_i = 0, \ i = k+1,\dots m \right\}$$

The notation used in the paper is not quite typical but convenient for possible generalizations to infinite-dimensional problems.

Suppose a solution $\hat{x}$ of /1.1a/ does exist and let the functions $f : R^n \to R^1$, $g : R^n \to R^m$ be twice differentiable /at least in a

Hausdorf neighbourhood of X_0/. Assume that the problem /1.1a/ is normal [1] , that is, for each solution $\hat{x} \in X_0$ there exists a normal Lagrange multiplier $\hat{\lambda}$ satisfying the fundamental triple condition

/1.2a/ $g(\hat{x}) \in -D; \quad \langle \hat{\lambda}, g(\hat{x}) \rangle = 0; \ \hat{\lambda} \in D^* = \{\lambda \in R^m : \langle \lambda, d \rangle \geqslant 0 \ \forall d \in D\}$

/where $\langle \cdot, \cdot \rangle$ denotes scalar product and star-duality or transposition/ and such that $\hat{\lambda}, \hat{x}$ is a stationary point

/1.2b/ $L_x(\hat{\lambda}, \hat{x}) = f_x(\hat{x}) + g_x^*(\hat{x})\hat{\lambda} = 0$

of the normal Lagrange - function

/1.2c/ $L(\lambda, x) = f(x) + \langle \lambda, g(x) \rangle$

2. Saddle-point theorems for augmented Lagrange functions.

The basic idea, due to Rockafellar [3] , is that even if the normal Lagrange function /1.2c/ does not have a saddle-point for a non-convex problem, it is nevertheless possible to formulate an augmented Lagrange function

/2.1a/ $\Lambda(\varrho, w, x) = f(x) + \frac{1}{2}\varrho \|(g(x) + w)^{D^*}\|^2 - \frac{1}{2}\varrho \|w\|^2$

which may have a saddle-point at $\hat{w}, \hat{x}$ with $\hat{w} = \frac{1}{\varrho}\hat{\lambda}$. Here the norms are Euclidean, $w \in R^m$ is an additional variable equivalent to Lagrange multiplier under $w = \frac{1}{\varrho}\lambda$, and $\varrho > 0$ is a penalty coefficient. The operation $(\cdot)^{D^*}$ denotes the projection on the cone D^*

/2.1b/ $(d)^{D^*} = \arg \min_{\bar{d} \in D^*} \|d - \bar{d}\| ; \ \|(d)^{D^*}\| = \operatorname{dist}(d, -D)$

If D is defined by /1.1b/, then $D^* = \{d \in R^m : d_i \geqslant 0, \ i = 1, \dots k;$ d_i - arbitrary, $i = k+1, \dots m\}$ and the projection $(d)^{D^*}$ is the vector composed of $\max(0, d_i)$ for $i = 1, \dots k$ and of d_i for $i = k+1, \dots m$. Hence

/2.1c/ $\Lambda(\varrho, w, x) = f(x) + \frac{1}{2}\varrho\left(\sum_{i=1}^{k}(\max(0, g_i(x) + w_i))^2 + \sum_{i=k+1}^{m}(g_i(x) + w_i)^2 - \sum_{i=1}^{m} w_i^2\right)$

If k = 0 and only equality constraints are considered, then the augmented Lagrange function reduces to

/2.1d/ $\Lambda(\varrho, w, x) = f(x) + \sum_{i=1}^{m}\varrho w_i g_i(x) + \frac{1}{2}\varrho \sum_{i=1}^{m} g_i^2(x) = L(\lambda, x) + \frac{1}{2}\varrho \|g(x)\|^2$

This formula, introduced by Hertenes in [5], justifies the name augmented Lagrange function; it is fact the normal Lagrange function /1.2c/ augmented by the penalty term $\frac{1}{2}\varsigma \, \| g(x) \|^2$.

Lemma 2.1. [2], [6] The necessary conditions of optimality /1.2a,b/ are equivalent to the following equations

/2.2a/ $\Lambda_w(\varsigma,\hat{w},\hat{x}) = \varsigma\left((g(\hat{x})+ \hat{w})^{D^*} - \hat{w}\right) = 0 \in R^m$

/2.2b/ $\Lambda_x(\varsigma,\hat{w},\hat{x}) = f_x(\hat{x}) + \varsigma \, g_x^*(\hat{x})\left(g(\hat{x}) + \hat{w}\right)^{D^*} = 0 \in R^n$

for a given $\varsigma > 0$ and $\hat{w} = \frac{1}{\varsigma}\hat{\lambda}$.

The equivalence of the inequality conditions of Kuhn-Tucker type /1.2a/ to the equation /2.2a/ is of extreme importance. One of the conclusions of this equivalence is that the saddle-points of the augmented Lagrange function, if they exist, are unconstrained in the sign of the dual variable w.

Theorem 2.4. [3] , [6] Let the necessary conditions of optimality /1.2a,b/ be satisfied at a given pair $\hat{\lambda},\hat{x}$ for the problem /1.1a/. Then the full complementarity condition is equivalent to the differentability of the projection $\left(g(\hat{x}) + w\right)^{D^*}$ in w at $\hat{w} = \frac{1}{\varsigma}\hat{\lambda}$ for any $\varsigma > 0$. Under full complementarity, the augmented Lagrange function /2.1a/ is twice differentiable in w,x at $\hat{w},\hat{x}$ for any $\varsigma > 0$, if f and g are twice differentiable, and the second-order sufficiency condition is equivalent to the fact that $\langle \Lambda_{xx}(\varsigma,\hat{w},\hat{x})\bar{x},\bar{x}\rangle > 0 \quad \forall \ \bar{x} \neq 0, \ \bar{x} \in R^n$ for $\varsigma > 0$. Under full complementarity, full rank and second-order sufficiency conditions, a saddle-point of the augmented Lagrange function exists :

/2.3a/ $$(\hat{w},\hat{x}) = \arg \min_{x \in R^n} \max_{w \in R^m} \Lambda(\varsigma,w,x) = \arg \max_{w \in R^m} \min_{x \in R^n} \Lambda(\varsigma,w,x)$$

and is /at least locally/ unique; moreover, the Lagrange multipliers $\hat{w}$ or $\hat{\lambda} = \varsigma\,\hat{w}$ are differentiable as the functions of constraint perturbation p at p = 0 in the perturbed problems

/2.3b/ $$\min_{x \in X_p} f(x) \ ; \quad X_p = \left\{ x \in R^n : g(x) \in p - D \subset R^m \right\}$$

The second derivative $\Lambda_{xx}(\varsigma,\hat{w},\hat{x})$ has the form :

/2.4a/ $$\Lambda_{xx}(\varsigma,\hat{w},\hat{x}) = f_{xx}(\hat{x}) + \varsigma \, g_{xx}(\hat{x})(g(\hat{x})+\hat{w})^{D^*} + \varsigma \, \tilde{g}_x^*(\hat{x})\,\tilde{g}_x(\hat{x}) = L_{xx}(\hat{\lambda},\hat{x}) + \varsigma \, \tilde{B}^*\tilde{B}$$

where $\tilde{B} = \tilde{g}_x(\hat{x})$ is the Jacobian matrix for active constraints.

Moreover :

/2.4b/ $\Lambda_{xw}(\varsigma,\hat{w},\hat{x}) = \varsigma\,(\tilde{B}^* \ 0)$

where $\tilde{B}$ was defined above and must be supplemented by a zero matrix for inactive constraints, and

$$/2.4c/ \quad \Lambda_{ww}(\varrho,\hat{w},\hat{x}) = -\varrho(0, \bar{I})$$

where the second derivative is zero for active constraints and the diagonal $-\varrho$ for inactive constraints.

3. A general quasi-Newton scheme for saddle-point seeking.

Almost all algorithms of nonlinear programming solve the system of equations /2.2a,b/, that is, $\Lambda_x(\varrho,\hat{w},\hat{x}) = 0 \in R^n$ and $\Lambda_w(\varrho,\hat{w},\hat{x}) = 0 \in R^m$. If the augmented Lagrange function $\Lambda(\varrho,w,x)$ is twice differentiable in a neighourhood of $(\hat{w},\hat{x})$ and the second-order derivatives are easy to compute and invert, then a generalised Newton algorithm could be applied for finding $(\hat{w},\hat{x})$

$$/3.1a/ \quad (x^{i+1}, w^{i+1}) = (x^i,w^i) - H^{-1}\left(\Lambda_x(\varrho,w^i,x^i),\Lambda_w(\varrho,w^i,x^i)\right)$$

where

$$/3.1b/ \quad H = \begin{bmatrix} \Lambda_{xx} & \Lambda_{xw} \\ \Lambda_{wx} & \Lambda_{ww} \end{bmatrix} \; ; \; H^{-1} = \begin{bmatrix} \Lambda_{xx}^{-1} + \Lambda_{xx}^{-1}\Lambda_{xw}\hat{\Lambda}_{ww}^{-1}\Lambda_{wx}\Lambda_{xx}^{-1} & -\Lambda_{xx}^{-1}\Lambda_{xw}\hat{\Lambda}_{ww}^{-1} \\ -\hat{\Lambda}_{ww}^{-1}\Lambda_{wx}\Lambda_{xx}^{-1} & \hat{\Lambda}_{ww}^{-1} \end{bmatrix}$$

and $\hat{\Lambda}_{ww}^{-1} = \Lambda_{ww} - \Lambda_{wx}\Lambda_{xx}^{-1}\Lambda_{xw}$. In order to increase the radius of convergence and to avoid computing and inverting second-order derivatives /particularly Λ_{xx}/, a quasi-Newton method can be applied with approximations V for Λ_{xx}^{-1} and M for $-\hat{\Lambda}_{ww}^{-1}$

$$/3.2a/ \quad (x^{i+1},w^{i+1}) = (x^i,w^i) - \bar{H}^{-1}\left(\Lambda_x(\varrho,w^i,x^i)\ ,\Lambda_w(\varrho,w^i,x^i)\right)$$

where

$$/3.2b/ \quad H^{-1} = \begin{bmatrix} V - V\Lambda_{xw}M\Lambda_{wx}V & V\Lambda_{xw}M \\ M\Lambda_{wx}V & -M \end{bmatrix}$$

Theoretically, it would be also possible to use an approximation for Λ_{xw}, but it is usually not necessary because of the simple form /2.4b/ of this derivative. Data used in quasi-Newton approximations of second derivatives is related to the finite differences of gradients and of independent variables. Assuming strong approximative properties of H^{-1} /which can be quaranteed, e.g., by special modifications of rank-one variable metric approximations for V and M/ and under other typical assumptions it is possible to prove [8] that the convergence of /3.2a/ is superlinear or even quadratic at each (n+m)-th iteration.

The radius of convergence of the general algorithm /3.2a/ cannot be estimated without specifying more detailed properties of the method. Various assumptions concerning the choice of V^i, M^i and their relation to Λ_x, Λ_w are possible; in fact, these assumptions account for all fundamental classes of constrained optimization methods.

Gradient projection and multiplier function methods.

Gradient projection methods, as introduced first in [10], are obtained from /3.2a,b/ if the current point x^i is admissible, $g(x^i) \in -D$, and $w^i = \frac{1}{\varsigma}\lambda^i \in D^*$ is such that the dual equation

$$/3.3a/ \quad \Lambda_w(\varsigma, w^i, x^i) = \varsigma\left((g(x^i) + w^i)^{D^*} - w^i\right) = 0$$

is satisfied. Moreover, assume that V^i is arbitrary matrix - for example, $V^i = I$ or V^i is a variable metric approximating Λ_{xx}^{-1} - and that M^i is chosen in a strict correlation to V^i

$$/3.3b/ \quad \tilde{M}^i = (\Lambda_{\tilde{w}x}V^i\Lambda_{x\tilde{w}})^{-1} = \varsigma^{-2}\left(\tilde{g}_x(x^i)V^i\,\tilde{g}_x^*(x^i)\right)^{-1}; \quad \bar{M}^i = 0$$

where $\tilde{M}^i$ corresponds to the active constraints and $\bar{M}^i$ to the inactive. Then the quasi-Newton step in $\tilde{w}$ corresponding to the active constraints takes the form

$$/3.3c/ \quad \tilde{w}^{i+1} = \tilde{w}^i - \tilde{M}^i\Lambda_{\tilde{w}x}V^i\Lambda_x(\varsigma, w^i, x^i) = -\varsigma\tilde{M}^i\tilde{g}_x(x^i)V^i\,f_x(x^i)$$

and $\tilde{w}^{i+1}$, being independent of $\tilde{w}^i$, can be interpreted as the Lagrange multiplier corresponding to the projection of the direction $-V^if_x(x^i)$ on the subspace tangent to active constraints. The quasi-Newton step in x determines actually the projected direction

$$\begin{aligned}/3.3d/ \quad x^{i+1} &= x^i - (V^i - V^i\Lambda_{x\tilde{w}}\tilde{M}^i\Lambda_{\tilde{w}x}V^i)\Lambda_x(\varsigma, w^i, x^i) = \\ &= x^i - \left(V^i - V^i\,\tilde{g}_x^*(x^i)(\tilde{g}_x(x^i)V^i\tilde{g}_x^*(x^i))^{-1}\tilde{g}_x(x^i)\,V^i\right)f_x(x^i) = \\ &= x^i - V^i\left(f_x(x^i) + \varsigma\,\tilde{g}_x^*(x^i)\;\tilde{w}^{i+1}\right)\end{aligned}$$

Here $\pi^i = V^i - V^i\Lambda_{x\tilde{w}}(\Lambda_{\tilde{w}x}V^i\Lambda_{x\tilde{w}})^{-1}\Lambda_{\tilde{w}x}V^i$ is a generalized projection operator on the subspace $\tilde{T}(x^i) = \{\bar{x} \in R^n : \Lambda_{\tilde{w}x}\bar{x} = 0\}$ and $\pi^i\Lambda_{x\tilde{w}}\tilde{w}^i = 0$ for all $\tilde{w}^i$.

The interpretation of gradient projection methods in terms of augmented Lagrange functions makes it possible to generalise these methods to the case when x^i is not necessarilly admissible,

$g(x^i) \notin -D$. The quasi-Newton iteration for the dual variables $\tilde{w}$ corresponding to active constraints results [6] in the formula :

$$/3.4a/ \quad \tilde{w}^{i+1} = -\tilde{g}(x^i) - \frac{1}{\varrho}\left(\tilde{g}_x(x^i)V^i\tilde{g}_x^*(x^i)\right)^{-1}\left(\tilde{g}_x(x^i)V^i f_x(x^i) - \tilde{g}(x^i)\right) = \triangleq \hat{\tilde{w}}(\varrho, x^i)$$

This result generalises the expression /3.3c/ for the case when $\tilde{g}(x^i) \neq 0$. It should be checked whether $\tilde{w}^{i+1} \in \tilde{D}^*$; if not, the definition of constraints activity might be changed and $\tilde{w}^{i+1}$ recomputed. The next x^{i+1} is then defined by

$$/3.4b/ \quad x^{i+1} = x^i - V^i\left(f_x(x^i) + \varrho\, \tilde{g}_x^*(x^i)(\tilde{w}^{i+1} + \tilde{g}(x^i))\right)$$

which generalises /3.3d/.

The function $\hat{\tilde{w}}(\varrho, x^i)$ defined by /3.4b/ satisfies the axioms given by Martensson [12] for multiplier functions ; but the multiplier functions given in [12] do not posess all the strong properties of the function $\hat{\tilde{w}}(\varrho, x^i)$. Therefore, Martensson did not use single-iterative procedures and proposed several iterations in x after determining a value of the multipliers. The algorithm /3.4a,b/ can be considered both as a generalised gradient projection and a multiplier function method. But in its essence, it is a single-iterative saddle-point seeking method and as such shall be discussed in further paragraphs.

Quadratic approximation methods.

A large class of constrained optimization methods consists in iterative solving of quadratic programming problems resulting from a quadratic - linear approximation to the orginal problem :

$$/3.5a/ \quad \min_{\bar{x} \in \bar{X}^i} \left(\langle f_x{}^i, \bar{x}\rangle + \frac{1}{2}\langle L_{xx}^i\, \bar{x}, \bar{x}\rangle\right); \; \bar{X}^i = \left\{ \bar{x} \in R^n : g^i + g_x^i\, \bar{x} \in -D \right\}$$

where $f_x{}^i = f_x(x^i)$, $L_{xx}^i = L_{xx}(\lambda^i, x^i)$ with $L(\lambda, x) = f(x) + \langle\lambda, g(x)\rangle$, $g^i = g(x^i)$, $g_x^i = g_x(x^i)$. The solution to this problem, $\hat{\bar{x}}$, and the corresponding Lagrange multiplier, $\hat{\bar{\lambda}}$, generate the next point for example by $x^{i+1} = x^i + \hat{\bar{x}}$, $\lambda^{i+1} = \hat{\bar{\lambda}}$. It can be shown [6] that $\hat{\bar{x}}, \hat{\bar{\lambda}}$ can be found by attempting to solve the equations :

$$/3.5b/ \quad \Lambda_x^i + \Lambda_{xx}^i \hat{\bar{x}} + \Lambda_{xw}^i \hat{\bar{w}} = 0$$

$$/3.5c/ \quad \Lambda_w^i + \Lambda_{wx}^i \hat{\bar{x}} + \Lambda_{ww}^i \hat{\bar{w}} = 0$$

equivalent to the generalized Newton iteration /3.1a,b/. If the activity of constraints does not change from x^i to $x^i + \hat{\hat{x}}$, then the solution $\hat{\hat{x}}, \hat{\hat{w}}$ of /3.5b,c/ yields the solutions $\hat{x}, \hat{\lambda}$ of /3.5a/ by $\hat{x} = \hat{\hat{x}}$, $\hat{\lambda} = \lambda^i + \varrho \hat{\hat{w}}$. If the activity of constraints does change, then these solutions do not coincide and the quadratic approximation method proceeds to find $\hat{x}, \hat{\lambda}$ satisfying the constraint $\hat{x} \in \overline{X}^i$, that is, satisfying approximatively the original constraints, whereas the generalized Newton iteration /3.1a,b/ or even the quasi-Newton iteration /3.2a,b/ admits a violation of constraints but proceeds to find in the next iteration points satisfying approximatively the once violated constraints.

Penalty methods.

Various penalty methods are most closely related to the augmented Lagrange function approach. They consist in minimizing $\Lambda(\varrho, w, x)$ in x to satisfy /usually only approximatively/ the condition $\Lambda_x(\varrho, w, x) = 0$ and then changing w and/or ϱ to satisfy the condition $\Lambda_w(\varrho, w, x) = 0$.

The classical increased penalty method consists in letting $\varrho \to \infty$ which results in bringing $\Lambda_w(\varrho, w, x)$ to zero. The shifted penalty method, as introduced by Powell [13], Hestenes [6] and developed further in many publications [7,11,14], is a special case of the general quasi-Newton iteration /3.2a,b/ and consist in assuming $M^i = \frac{1}{\varrho} I$ which results in the simple gradient step in w

/3.6a/ $$w^{i+1} = w^i + \frac{1}{\varrho}\Lambda_w(\varrho, w^i, \hat{x}(\varrho, w^i)) = \left(g\left(\hat{x}(\varrho, w^i)\right) + w^i\right)^{D^*}$$

where $\hat{x}(\varrho, w^i)$ denotes the minimal point of $\Lambda(\varrho, w^i, x)$ resulting in $\Lambda_x(\varrho, w^i, \hat{x}(\varrho, w^i)) = 0$. The shifted penalty techniques are well substantiated, since $\lim_{\varrho \to \infty} \varrho \hat{\Lambda}_{ww}(\varrho, x, w) = -I$ and $M^i = \frac{1}{\varrho} I$ approximates well $-\hat{\Lambda}_{ww}^{-1}$ for sufficiently large ϱ.

Fletcher [15] introduced so called ideal penalty method, changing /3.6a/ to

/3.6b/ $$\tilde{w}^{i+1} = \tilde{w}^i + \tilde{M}^i \Lambda_{\tilde{w}}(\varrho, w^i, \hat{x}(\varrho, w^i))$$

with $\tilde{M}^i$ defined as in /3.3b/. This algorithm is also a special case of the general quasi-Newton iteration /3.2a,b/, but is double-iterative /first find $\hat{x}(\varrho, w^i)$/ similarly as all penalty methods.

4. Single - iterative saddle-point seeking algorithms for generally constrained optimization problems.

All the methods considered before were shown to be variants of the Newton-like or quasi-Newtons methods of saddle-point seeking for an augmented Lagrangian. One of their disadventages is a double-iterative structure : saddle-point is sought by a multiple minimization of augmented Lagrangian or equivalent function and by corrections of dual variables after each minimization ; moreover, these minimizations are usually rather precise. Such a structure of algorithms increases the numerical effort required for solving a problem. The application of general saddle-point theory of augmented Lagrangians leads to a general, single-iterative algorithm /3.2a/, /3.2b/. Two versions of this algorithm are presented here. The first one uses the variable metric to approximate the primal hessian inversion $\Lambda_{xx}^{-1} \approx V$. The dual hessian $\hat{\Lambda}_{ww}$ is calculated with the use of this approximation

$$\hat{\Lambda}_{ww} = \Lambda_{ww} - \Lambda_{wx} \Lambda_{xx}^{-1} \Lambda_{xw} \approx \Lambda_{ww} - \Lambda_{wx} V \Lambda_{xw}$$

and then its inversion is computed.
The second algorithm uses two variable metric algorithms to approximate Λ_{xx}^{-1} and $\hat{\Lambda}_{ww}^{-1}$ independently [16,17].
New dual variables w^{k+1} are computed in both algorithms not at each iteration, but only if satisfactory convergence occurs. Penalty coeficients are also changed if necessary.

Algorithm A1 /Single iterative algorithm for saddle-point seeking with one variable metric/

/i/ Set initial ϱ, x^0 and w^0 /usually $w^0=0$/, $i=0$, $k=0$.

/ii/ Compute $\tilde{b}_x^i = \Lambda_x(\varrho, w^k, x^i)$.

/iii/ For $i \geqslant 1$ compute $s_x^i = x^i - x^{i-1}$, $r_x^i = \tilde{b}_x^i - b_x^{i-1}$

and $V^i = V\left(s_x^j, r_x^j\right)_{j=\max(1,i-N+1)}^{j=i}$; V^i approximates Λ_{xx}^{-1}.

/iv/ If convergence is not satisfactory increase ϱ and go to/vi/.

/v/ Dual step :

$$M^k = \left(\Lambda_{ww} - \Lambda_{wx} V \Lambda_{xw}\right)^{-1}; \quad M^k \text{ approximates } \hat{\Lambda}_{ww}^{-1}$$

and

$$w^{k+1} = w^k + M^k\left(\Lambda_{wx} V \tilde{b}_x^i - \Lambda_w(\varrho, w^k, x^i)\right)$$

set $k = k+1$

/vi / Compute $b_x^i = \Lambda_x(\varrho, w^k, x^i)$.

/vii/ Primal step :

$$x^{i+1} = x^i - \tau^i V^i b_x^i$$

set i = i+1 and go to /ii/.

<u>Algorithm A2</u> /Single iterative algorithm for saddle-point seeking with two variable metrics/

This algorithm is similar to A1 except the step /v/ which has the form

/v/ For $k \geqslant 1$ compute $s_w^k = w^k - w^{k-1}$,

$$r_w^k = \Lambda_w(\varrho, w^k, x^{i_k+1}) - \Lambda_w(\varrho, w^{k-1}, x^{i_k} - V^{i_k} \Lambda_x(\varrho, w^{k-1}, x^{i_k}))$$

$$M^k = M\left(s_w^j,\ r_w^j\right)_{j=\max(1,k-m+1)}^{j=k}$$

$$w^{k+1} = w^k - M^k \Lambda_w(\varrho, w^k, x^i - V^i \Lambda_x(\varrho, w^k, x^i))$$

set k = k+1 and $i_k = i$.

The algorithm A1 is recommended for solving nonlinear programming problems whereas the algorithm A2 is recommended for solving optimal control problems.

5. <u>Preliminary numerical results</u>.

The algorithm A1 was tested only. The primal hessian Λ_{xx} was approximated by modified rank-one variable metric algorithm [8] and the dual hessian $\hat{\Lambda}_{ww}$ was inverted by the use of LQL^* Cholesky factorization. This algorithm is rather robust : it solves even problems which do not possess optimal Lagrange multipliers, since at each iteration a problem with shifted constraints and finite Lagrange multipliers is solved. The sequence of solutions of such perturbed problems converges to the solution of the orginal one.

For regular problems, the algorithm converges rapidly. The Rosen-Suzuki [18] problem and two problems TP1 and TP3 used in Colville's comparison [19] have been solved by the algorithm. The stopping test required that the norms of the primal and the dual gradients were less than 10^{-5}:

Test problem	Number of variables	Number of constraints	Number of function and constra-ints evalu-ations	Number of gradients evalua-tions	Number of dual hessian inver-sions
Rosen -Suzuki	4	3	61	20	5
Colville TP1	5	15	56	18	4
Colville TP3	5	16	63	21	4

It is hoped that after some modifications the algorithm will be one of the most efective techniques of nonlinear constrained optimization.

Bibliography

1. Findeisen W, J.Szymanowski, A.P.Wierzbicki: Theory and computational methods of optimization /in Polish/. PWN, Warsaw 1977
2. Mangasarian O. : Unconstrained methods in nonlinear programming. SIAM - AMS Proc., 9 /1976/, p.p. 169-184.
3. Rockafellar, R.T.: Augmented Lagrange multiplier functions and duality in non-convex programming. SIAM J.Control 12, 268-285 /1974/.
4. Wierzbicki A.P., St.Kurcyusz : Projection on a cone, penalty functionals and duality theory for problems with inequality constraints in Hilbert space. SIAM J.Control 15, No 1 /1977/
5. Hestenes, M.R.: Multiplier and gradient methods. JOTA 4, 303-320 /1969/.
6. Wierzbicki A.P. Towards unification of nonlinear programming methods via augmented Lagrangians. Int.Conf. on Methods of Mathematical Programming. Zakopane, Poland, September 1977.
7. Wierzbicki A.P.: A penalty shiffting method in constrained optimization and its convergence properties. Archiwum Automatyki i Telemechaniki 16, 395-416 /1971/
8. Kręglewski T, A.P.Wierzbicki: Further properties and modifications of the rank-one variable metric method. Int.Conf. on Methods of Mathematical Programming,Zakopane, Poland, September 1977.
9. Ortega J.M, Rheinboldt W.C. Iterative solution of nonlinear equations in several variables. Academic Press 1970.
10. Rosen J.B.: The gradient projection method in nonlinear programming. Part I, II. J. of SIAM, 8, 181-217 /1961/,9, 514-532 /1962/
11. Janiak A. Effectiveness comparison of some penalty method algorithms for static optimization /in Polish/ MSc Thesis, Institute of Automatic Control, Technical University of Warsaw, 1972.
12. Martensson K. : A new approach to constrained function minimization. JOTA 12, 531-555 /1972/
13. Powell, M.J.D.: A method for nonlinear constraints in minimization problems. In R.Fletcher : Optimization, Ac.Press 1969.
14. Bertsekas D.P.: Multiplier methods - a survey. Automatica 12, 135-146 /1976/
15. Fletcher, R. : An ideal penalty function for constrained optimization. J.Inst.Maths.Applics. 15, 319-342 /1975/.
16. Wierzbicki A.P.: Ein neuer Algorithmus zur Sattelpunktermittlung und ein universaler Strafverschiebungsalgorithmus für beschränkte nichtlineare Optimierungsprobleme in R^n und Hilbertraum, 21 Internationale Wissenschaftliche Kolloquium der T.H.Ilmenau, 1976.
17. Wierzbicki A.P.: A primal-dual large scale optimization method based on augmented Lagrange functions and interaction shift prediction. Richerche di Automatica, 1977.
18. Rosen J.B., Suzuki S. 1965 Communs ACM 8, 113.
19. Colville A.R., A comparative Study of Nonlinear Programming Codes, IBM, New York Scientifie Center, Report No 320-2949, 1968.

IMPROVED LOWER BOUNDS TO 0/1 PROBLEMS VIA LAGRANGEAN RELAXATION

Gianfranco d'Atri
Institut de Programmation
Université de Paris VI, France

ABSTRACT. We exploit linear constraints satisfied by feasible solutions to minimization problems for giving a general procedure finding lower bounds to their optimal value. It provides old and new tight bounds to some well known integer or combinatorial problems.

SECTION 1: Introduction.

Many integer or combinatorial optimization problems are solved by algorithms employing a subroutine which computes a bound to the optimal value of generated subproblems, in order to decide whether some subset of feasible solutions may contain the optimum or not.
An appropriate choice of such subroutine is crucial to the algorithm performance, so it is important to look for "tight" bounds, i.e. very near to the optimal value.
In this paper, we propose a general bounding procedure for problems having a (partial) linear characterization of feasible solutions; it is then specialized to some difficult integer problems, so providing bounds "improved" with respect to the most usual ones.
Given the minimization problem (P)

$$(P) \qquad \min_{x \in X} \; c(x)$$

with X a discrete subset of $\mathbb{R}^n$ and $c(x)$ a real function, a "lower bound" to (P) is a number $v \leq v(P)$; $v(\cdot)$ denoting the optimal value of problem $(\cdot)$.
For obtaining a lower bound we have to solve a "relaxation" of (P), that is a problem (RP) satisfying conditions 1.1 and 1.2

$$(RP) \qquad \min_{x \in X'} \; c'(x)$$

where X' is a relaxation of the feasible set and $c'(x)$ a reduced cost function.

$$X \in X' \qquad 1.1$$

$$c(x) \geq c'(x) \text{ for } x \in X \qquad 1.2$$

Remark that a relaxation of (RP) is still a relaxation of (P).
When X is the set of integral points satisfying a given system of linear inequalities $g_1(x) \geq 0, \ldots\ldots, g_2(x) \geq 0$, a largely used relax-

ation is the continuous one, that is (RP) with $c'(x) = c(x)$ and $X' = \{x \in \mathbb{R}^n / g_1(x) \geq 0, \ldots\ldots, g_2(x) \geq 0\}$. This type of relaxation is mainly employed because we have good algorithms for continuous problems, and, in particular, for linear ones, LP-Symplex Method is a consolidated technique. But, unfortunately, it has a poor performance on continuous problems arising from integer or combinatorial ones.

Really, a good bounding procedure is expected to have the following

-*PROPERTY A : it is performed "efficiently".*

Here and throughout the paper "efficiently" means "in a polynomial number of operations as a function of problem size". This is a general agreement but, it is a theoretical measure of algorithm performance; we put emphasis on the fact that, in practice, an exponential behaviour might be better than an order-10 polynomial one.

Due to the fact that for many combinatorial problems we don't know whether a polynomial algorithm does exist or not, a more appropriate way of treating efficiency should be to decide to consider efficient a family of algorithms,and those built up with them according to a set of rules.

Lagrangean relaxation. Now, let us suppose $c(x) = cx$ a linear function and let be T a relaxation of X, $A\,x \geq_a$ k linear inequalities satisfied by X and u a non-negative multipliers k-vector, then (RL) is a "Lagrangean relaxation" of (P)

$$\text{(RL)} \qquad \min_{x \in T} \; cx + u(a - Ax) \tag{1.3}$$

If the minimization problem 1.3 defined over the set T is efficiently solved, then a bounding procedure based on (RL) has property A, provided that the reduced cost function is efficiently computed.

The last remark isn't a trivial one; for example, the system defining Conv(X), the convex hull of X, provides Lagrangean relaxations but it may be constituted by a non-polynomial number of linear inequalities. Meaningful and efficiently solvable Lagrangean relaxations have been introduced for many integer programming problems (see Geoffrion [8] for reference, and also for a comprehensive study of the subject),and, furthermore, the naif approach "construct one Lagrangean relaxation and solve it" has been improved by iterative schemes; their similar structure is the following.

Let be $T°$ be a special relaxation of X and $A°x \geq a°$ a special set of linear constraints, chosen in such a way that 1.4 holds

$$X = T° \cap \{x \in \mathbb{R}^n / A°x \geq a°\} \tag{1.4}$$

The sequence (RP_i), $i=1\ldots r$, of relaxations is constructed and solved,

$$(RP_i) \qquad \underset{x \in T^\circ}{\text{MIN}} \quad (c - u^i A^\circ)\, x + u^i a^\circ \qquad 1.5$$

where (u^i), $i=1\ldots r$, is a sequence of multipliers vectors, determined according to some stepping rule (subgradient optimization [7], [10], [11]):

The resulting lower bound is $\underset{i \leqslant r}{\text{MAX}}\; v(RP_i)$

The Dual. The previous procedure (efficient whenever r is bounded by a polynome in the problem size) is a sub-optimization technique **for** the particular dual problem: (D_{R°) of (P), where R° is the family of all Lagrangean relaxations constructed with relaxed set T° and using constraints $A^\circ x \geqslant a^\circ$, and a "dual" of (P) is any problem (D_R), with R a family of relaxations.

$$(D_R) \qquad \underset{(RP) \in R}{\text{MAX}} \; v(RP)$$

It may be showed (see [8]) that $v(D_{R^\circ}) = v(\overline{P})$, with

$$(\overline{P}) \qquad \begin{array}{l} \text{MIN} \quad cx \\ A^\circ x \geqslant a^\circ \\ x \in \text{Conv}(T^\circ) \end{array}$$

Though 1.4 holds, generally $\text{Conv}(X) \neq \text{Con}(T^\circ) \cap \{x \in \mathbb{R}^n / A^\circ x \geqslant a^\circ\}$; so, a duality gap $v(P) - v(D_{R^\circ})$ usually exists.

For improving the bound obtainable through (D_{R°), we must work on a larger family R ; this may be done by taking a less relaxed set T° and putting more or tighter constraints in the objective of (RP_i), but this increases the difficulty of solving it.

The different approach employed in this paper uses the family R' of unconstrained Lagrangean relaxations 1.6

$$\begin{array}{l} \text{MIN} \quad c'x + c'_0 \\ L_j \leqslant x_j \leqslant U_j \ \text{ for } j = 1\ldots.n \end{array} \qquad 1.6$$

with $c'x + c'_0$ a reduced cost function obtained efficiently as in 1.3 .

Note that the definition of family R' is based on the set of algorithms accepted as efficient, according to the remarks following property A.

For many problems we show that R° may be imbedded in R', that is we can efficiently associate to each relaxation (RP) of R° a relaxation (RP') of R' in such a way that $v(RP) = v(RP')$.

The dual $(D_{R'})$ isn't optimized, but it's given a general procedure for constructing a sub-optimizing sequence (RP'_i) ; so doing, it isn't guaranteed that max $v(RP'_i) >$ max $v(RP_i)$, but, obviously, the combined sequence, formed by (RP'_i) and the relaxation (RP''_i), images of (RP_i) in R', provides on "improved" lower bound to (P).

SECTION 2 : The method

Let us rewrite problem (P) under the hypothesis that c(x) is a linear function and the variables are bounded, then

$$\text{(P)}\qquad \begin{aligned} &\text{MIN}\ \ cx + c_0 && 2.1\\ &x \in X && 2.2\\ &x \in Z && 2.3 \end{aligned}$$

where $Z = \{x \in \mathbb{R}^n / L_j \leqslant x_j \leqslant U_j\}$, (usually, $0 \leqslant x_j \leqslant 1$).
However, it is easy to see how to extend the analysis which follows to non-bounded variables and, if function c(x) is not linear, we can consider formulation 2.1, 2.2, 2.3, as a linear-objective relaxation of (P). The explicit use of constraint 2.3 will result clear further. Let $c'x + c'_0$ be a reduced cost function for (P) and $Ax \geqslant a$ a(k,n)-linear system of inequalities satisfied by all $x \in X$, and let us denote by $\bar{A}$ the quadruple (A, a, c', c'_0), then consider problem

$$(P_{\bar{A}})\qquad \begin{aligned} &\text{MIN}\ \ c'x + c'_0 &&\\ &Ax \geqslant a && 2.4\\ &x \in Z && 2.5 \end{aligned}$$

Let us now recall the fundamental result from linear programming theory :

THEOREM 1 : There exists a multipliers vector u' such that

$$v(P_{\bar{A}}) = \underset{Z}{\text{MIN}}\ c''x + c''_0$$

with $c'' = c' - u'A$ and $c''_0 = c'_0 + u'a$.

Remark that the set of solutions to $v(P_{\bar{A}})$ is

$$\bar{X} = \{x \in \mathbb{R}^n / x_j = L_j \text{ if } c''_j < 0 \text{ ; } x_j = U_j \text{ if } c''_j > 0;\ L_j \leqslant x_j \leqslant U_j \text{ if } c''_j = 0\} \qquad 2.6$$

(if $c'' \geqslant 0$ and $0 \leqslant x_j \leqslant 1$, then $v(P_{\bar{A}}) = c''_0$).

Theorem 1 shows that a source of meaningful relaxations, belonging to the family R' defined in section 1, is any system of linear inequalities for which the corresponding LP-problem is efficiently solved.
Thus, for devising an effective procedure, we have to identify linear continuous problems for which efficient algorithms are known : a representative list of them is given below, also for subsequent use.
Furthermore, theorem 1 clarifies when R° may be imbedded in R' : for example, if problems 1.5 are of type 1, 2 or 3 in the following list (see [8] for applications).

Efficiently Solved Problems

1) LP-problems with a fixed number of rows k(with respect to problem size = n) may be solved by a non-cycling version of the Symlex Method in polynomial time. A remarkable case is k = 1(continuous knapsack problem) which is solved in O(n lg n) operations.

2) Linear problems defined on the convex hull of integer solutions to 2.4, 2.5 when A is an integer matrix, a a non negative integer vector, L_j and U_j are integer and the following condition holds

$$\sum_i |A_{ij}| \leqslant 2 \qquad 2.7$$

where A_{ij} is the entry of A in the i-th row and j-th column.
This is th General Matching problem and it may be solved in polynomial time by the algorithm in [6],[5].
A special case is the Assignment problem, and a remarkable one is the 2xn problem, which is solved in O(n),[1].

3) Problems defined on the convex hull of 0-1 points of $\mathbb{R}^n$ representing basis of matroids, in which independence may be tested efficiently.
A well known example in the minimum spanning tree [12] or 1-tree [11], for which the Kruskal's algorithm works.
A remarkable case is given by the solutions to 2.8,

$$\sum_{j \in J} x_j = b, \; J \subset \{1,\dots,n\}, b \text{ integer} \qquad 2.8$$

an O(n) algorithm solves minimisation problems under constraint 2.8 .

Bounding Procedure. Now we state the general bounding procedure, that we call BOUND. It employs a subroutine SELECT whose purpose is to select sets of linear inequalities satisfied by feasible points of (P) according to appropriate rules, which will be discussed later. Let us suppose, without loss of generality, that c is nonnegative.

BOUND :

step 0: $i=1$; $c^1 = c$; $c_0^1 = c_0$; $v^1 = \min_Z cx + c_0$

1: SELECT system $A^i x \geqslant a^i$, if none is found then STOP

2: Solve the linear problem

$$(P^i) \qquad \begin{array}{l} \text{MIN } c^i x + c_0^i \\ A^i x \geqslant a^i \\ x \in Z \end{array}$$

and get optimal multipliers vector u^i

3 : Set $c^{i+1} = c^i - u^i A^i$

$c_0^{i+1} = c_0^i + u^i a^i$

$v^{i+1} = v(P^i)$; $i = i + 1$ and GO TO 1

END

BOUND computes the optimal value v^i of the sequence of relaxations $(RP_i^!)$, $i=1\ldots\bar{r}$, where $\bar{r}$ is the value of i at the STOP.

$$(RP_i^!) \qquad \underset{Z}{\text{MIN}}\ c^i x + c_0^i \tag{2.9}$$

Remark that $(RP_{i+1}^!)$ is constructed from (P^i) according to theorem 1. The sequence v^i is non decreasing , indeed

$$v^i = \underset{Z}{\min}\ c^i x + c_0^i \quad \text{(unconstrained)} \tag{2.10}$$

$$v^{i+1} = \underset{Z}{\min}\ c^i x + c_0^i \quad \text{subject to } A^i x \geqslant a^i \tag{2.11}$$

Further, the next theorem specifies when it is increasing.

THEOREM 2 : Let be X^i the set of all solutions to $(RP_i^!)$ as in 2.6 ; if no $x \in X^i$ satisfies $A^i x \geqslant a^i$, then holds the following :

PROPERTY B : $v(RP_{i+1}^!) > v(RP_i^!)$.

Proof. X^i is the subset of Z over which $c(x) = v^i$, then $A^i x \geqslant a^i$ cuts off all points of Z with value v^i.

An example :

$$\begin{aligned} \text{MIN}\ \ 3x + 2y + 5z + 3w & \\ y + z + w &\geqslant 2 \\ x \qquad\quad + w &\geqslant 1 \\ x \qquad + z \qquad &\geqslant 1\ , \quad x,y,z,w = 0 \text{ or } 1. \end{aligned}$$

Let us suppose that SELECT takes the linear constraints in the order, then minimizing the objective $c^1 x$ subject to $y + z + w \geqslant 2$ yields reduced costs $c^2 = (3, -1, 2, 0)$ and $c_0^2 = 6$, $v^2 = 5$; set $X^2 = \{(0, 1, 0, 0), (0, 1, 0, 1)\}$, then $x + w \geqslant 1$ is satisfied by $(0, 1, 0, 1)$: v^3 doesn't improve v^2 and the the reduced costs are unchanged. Lastly solving $c^2 x + c_0^2$ subject to $x + z \geqslant 1$ yields $v^3 = 7$ which is greater than $v^2 = 5$.

<u>Comments</u>. If the explicit constraint 2.3 is not considered and the Lagrangean relaxations MIN $c'(x)$, $x \geqslant 0$,are employed, then, solving $(P_{\bar{A}})$ should provide nonnegative reduced costs, so a larger set $\bar{X}$, as defined in 2.6.

If problem $(P_{\bar{A}})$ is 0/1 with integer extremal points then the cardinality of X^i is

$2^{(\text{number of zero reduced costs})}$, (e.g., type 2 or 3 of the previous list).
A different approach, equivalent for the results, but providing further insights in the connection with subgradient optimization, is to build up an enormous LP-problem with all known inequalities satisfied by set X and arising from $(P_{\bar{A}})$ type problems, writing down its usual dual and to work on a restricted set of non-zero dual variables, i.e. Lagrangean multipliers (see [4]).

SECTION 3 : Applications

In this section, we propose the application of BOUND to a collection of integer problems for which no efficient solving algorithm is known ; for each of them we describe the family of linear continuous systems to be chosen by SELECT, and we discuss conditions essuring property A or B.
Iterative schemes similar to this one have already appeared in the literature : Christofides [3] and Camerini-Maffioli [2] devised bounding algorithms for the Travelling Salesman problem (TS), and Gondran-Lauriere for the Set Covering (SC)[9] and the Set Partitioning (SP) problems ; and really, we should go back to the well known work of Little et al. [13].
The common aspect of the cited techniques is that they only exploit the linear constraints which are explicit in the integer programming formulation, so the best obtainable bound is not better than the one given by the usual continuous relaxation. We avoid this limitation by adding known inequalities.
Furthermore, the result of theorem 2 allows to analyse the behaviour of the procedure and suggests general strategies for the iterative constraints selection.
In the following pages, we briefly examine Covering (CP) and Partitioning (PP), Multicommodity Disconnecting (MD), 0/1-Linear (IP) and 0/1 Quadratic Problems (QP) and (TS).
To describe complete specialized algorithms the order for the selection should be given, but this isn't done : we refer to the previously cited papers or to further researches.

Covering and Partitioning. Let be A a (k,n)-matrix with 0/1 entries and a positive, then

(CP) $$\text{MIN } cx \qquad Ax \geqslant a \;;\; x_j = 0 \text{ or } 1$$

and

(PP) $$\text{MIN } cx \qquad Ax = a \;;\; x_j = 0 \text{ or } 1$$

(SC) and (SP) are the special cases with a = (1, 1, ..., 1).

Let be $t \leq k$, fixed, and I a subset of t row-indices, suppose that the (t,n)-submatrix A_I has at most two ones in each column, then $A_I x \geq a_I$ satisfies condition 2.7 : thus, any linear problem defined over 3.3 is efficiently solved

$$\mathrm{Conv}\{x \in \mathbb{R}^n / A_I x \geq a_I, x_j = 0 \text{ or } 1\} \qquad 3.3$$

For any $t \leq k$, SELECT_t chooses among submatrices of t rows and satisfying 2.7, and gives BOUND the convex hull defined in 3.3 ; BOUND with SELECT_t has Property A, whenever k is polynomial in n (taken as problem size).

A special case is SELECT_1 which provides k matroidal problems solved in O(n) and gives for BOUND the overall behaviour O(kn) ; it is easy to see that $\min\{c^i_j / a^i_j = 1\} > 0$, where $c^i x$ is the reduced cost function and $a^i x \geq a'$ is the constraint considered at the i-th iteration, assurers that property B holds at the i-th iteration.

Furthermore, while for (CP) there is no gain in continuing to perform BOUND with SELECT_1, $t \neq 1$, once SELECT_1 has exhausted all the constraints, for (PP) an improved bound may be obtained by considering some constraints twice or more, as showed in the example.

Example :

c ⇒	3	3	2	2	6	6	6	6	a
	1		1		1		1		2
A		1		1		1		1	2
	1	1	1	1					1

Let us perform BOUND with SELECT_1

i = 1 ; c^2 = 0 3 -1 2 3 6 3 6 ; $v^2 = 5$

i = 2 ; c^3 = 0 0 -1 -1 3 3 3 3 ; $v^3 = 10$

i = 3 ; c^4 = 1 1 0 0 3 3 3 3 ; $v^4 = 11$ for (PP), but

no improvement is possible for (CP). For (PP), considering together constraints 1 and 3, the bound is improved to $v^5 = 11 + 3 = 14$.

<u>Multicommodity Disconnecting</u>. Given a directed graph G = (M;E) whose arcs are assigned a cost, let be S^+ and S^- two special subsets of t corresponding vertices, called respectively "sources" and "sinks" : we have to find a minimum cost subset D of E, such that in the graph (M;E-D) there is no path from any source to the corresponding sink. Denoting by $\mathbb{P}$ the set of all possible paths from sources to sinks, its integer programming formulation [12] is

$$\text{(MD)} \qquad \begin{aligned} &\text{MIN} \quad cx \\ &\sum_{e \in P} x_e \geq 1 \quad \text{for } P \in \mathbb{P} \ ; \\ &x_e = 0 \text{ or } 1 \end{aligned} \qquad 3.5$$

(MD) is a covering problem with $O(2^m)$ constraints, where $m = |M|$.
Let be $\mathbb{P}'$ a subset of paths sources-sinks such that each arc e belongs to at most two paths in $\mathbb{P}'$, then the constraints 3.5 for $P \in \mathbb{P}'$ satisfy 2.7 and the associated linear problem is efficiently solved.
Properties A and B hold if, at each iteration, SELECT takes a subset $\mathbb{P}'$ of paths in (M;E-D'), where D' is the set of arcs having zero or negative reduced costs.

__The Travelling Salesman__ : Given the directed graph G = (M;E) whose arcs are assigned a cost, we have to find a cricuit passing through each vertex exactly once, with minimal cost.
Denoting the set of arcs leaving vertices of $S \subset M$ by $w^+(S)$ and of those entering by $w^-(S)$, one integer programming formulation is the following (see d'Atri [4] for the symetric case)

$$\text{(TS)} \qquad \begin{array}{ll} \text{MIN } cx & \\ \sum_{w^+(S)} x_e = 1 + w_S \text{ , for any } S \subset M & 3.6 \\ \sum_{w^-(S)} x_e = 1 + w_S \text{, for any } S \subset M & 3.7 \\ x_e = 0 \text{ or } 1 \text{ ; } 0 \leqslant w_S \leqslant |S| - 1 \text{ and integer} & \end{array}$$

where there is a variable w_S for any subset S of M.

Let be $\mathbb{P} = \{S_1,\ldots,S_k\}$ a partition of M, the constraints 3.6 and 3.7 for S_i belonging to $\mathbb{P}$ satisfy 2.7 : they define a matching problem over the "contracted" graph G' = (M';E'), where M' = {1,...,k} and there is an arc e' from i to j for any arc e of G from some vertex of S_i to some vertex of S_j. The matching problem on G' is efficiently solved ($|E|$ as a measure of problem size), furthermore property A is assured for BOUND, if at each iteration SELECT takes a partition smaller than the previous one (see cited references for different ways of doing it).
Property B holds if the arcs leaving, or those entering at least one element of the partition have all positive reduced costs.

__0-1 Linear Problem.__ Let be A an integer (k,n)-matric and a an integer k-vector, then

$$\text{(IP)} \qquad \begin{array}{l} \text{MIN } cx \\ Ax = a \\ x_j = 0 \text{ or } 1 \end{array}$$

Let $t \leq k$ be fixed, and I a subset of t row indices, any problem defined over 3.8 may be efficiently solved.

$$\mathrm{Conv}^r(x_j = 0 \text{ or } 1 / A_I x = a_I) \qquad 3.8$$

where $\mathrm{Conv}^r(Y)$ denotes a partial linear characterization of Conv (Y)given by $A_I x = a_I$ and $0 \leq x_j \leq 1$ plus r cutting planes generated according to some rule.
A special case is r = 0, then BOUND with SELECT choosing among systems of inequalities as in 3.8 provides a lower bound not better than the usual continuous relaxation of (IP).
Nevertheless, the remarkable case r = 0, t = 1 requires only O(k.n.lgn) steps.

0-1 Quadratic Problem. Let be C and $A^h = a^h$, h= 1...k, upper triangular (n,n)-matrices and $a = (a^1,\dots,a^k)$ a k-vector, then

$$(QP) \qquad \begin{array}{l} \mathrm{MIN}\ x C x \\ x A^h x \geq a^h, \ h = 1\dots k \\ x_j = 0 \text{ or } 1 \end{array}$$

If we set $x_{ij} = x_i . x_j$ for $i \leq j$, (QP) can be formulated as an integer programming program (QP') :

$$(QP') \qquad \begin{array}{ll} \mathrm{MIN}\ \Sigma\ C_{ij} x_{ij} & 3.10 \\ \Sigma\ A^h_{ij} x_{ij} \geq a^h, \ h = 1\dots k & 3.11 \\ x_{ij} = x_i . x_j, \text{ for } i \leq j & 3.12 \\ x_j = 0 \text{ or } 1 & \end{array}$$

where the summations in 3.10 and 3.11must be intended for $i \leq j \leq n$.
Let us suppose to have obtained a reduced cost matrix C', exploiting constraints 3.11, then problem (QP_{rs}), $r<s$, is a relaxation of (QP') solvable in a finite number of steps

$$(QP_{rs}) \qquad \begin{array}{ll} \mathrm{MIN}\ \Sigma\ C'_{ij}\ x_{ij} + c'_0 & \\ x_{rs} \leq x_{rr} \ ; \ x_{rs} \leq x_{ss} \ ; \ x_{rs} \geq x_{ss} + x_{rr} - 1 & 3.13 \\ 0 \leq x_{rr}, \ x_{ss}, \ x_{rs} \leq 1 & 3.14 \end{array}$$

indeed constraints 3.13 and 3.14 define the convex hull of 0-1 solutions to 3.12 for i = r and j = s.
SELECT has property A if it explores all couples (r,s), $r \leq s$,

Property B holds if, at the iteration we select the couple (r,s) one of the following conditions is satisfied

$$C'_{rs} < 0 \quad \text{and} \quad (C'_{rr} > 0 \text{ or } C'_{ss} \geq 0) \qquad 3.15$$

$$C'_{rs} > 0 \quad \text{and} \quad (C'_{rr} < 0 \text{ and } C'_{ss} \leq 0) \qquad 3.16$$

REFERENCES

1 D.L.Adolphson-G.N.Thomas,*"A linear time algorithm for a 2xn transportation problem"*, SIAM J.of Comp., 6 (September 1977), 481-486

2 P.M.Camerini-F.Maffioli,*"Polynomial bounding for NP-hard problems"*, Proceedings of the IXth Symposium on Mathematical Programming, (1976), Budapest, (to appear)

3 N.Christofides,*"Bounds for the Traveling Salesman problem"*, Opns.Res., 20 (1973), 1044-1056

4 G.d'Atri,*"Improved lower bounds to the Travelling Salesman problem"* R.A.I.R.O. serie verte, (to appear)

5 J.Edmonds-E.L.Johnson,*"Matching, Euler tours and the Chinese Postman"*, Mathematical Programming, 5 (1973), 88-124

6 J.Edmonds-E.L.Johnson,*"Matching:a well solved class of integer programs"*, in Combinatorial Structures and their Applications, (1970), Gordon and Breach, N.Y., 89-92

7 M.L.Fisher-W.D.Northup-J.F.Shapiro,*"Using duality to solve discrete optimization problems: theory and computational experience"*,Math.Prog.Study, 3 (1975), 56-94

8 A.M.Geoffrion,*"Lagrangean relaxation for Integer Programming"*,Math.Prog.Study, 2 (1974), 82-114

9 M.Gondran-J.L.Laurière,*"Un algorithme pour le problème de recouvrement"*, R.A.I.R.O. serie verte, 2 (1975), 33-51

10 M.Held-H.P.Crowder-P.Wolfe,*"Validation of subgradient optimization"*, Mathematical Programming, 6 (1974), 62-88.

11 M.Held-R.M.Karp,*"The Travelling Salesman problem and minimum spanning trees:II"* Mathematical Programming, 1 (1971), 6-25

12 J.BKruskal,*"On the shortest spanning subtree of a graph and the Travelling Salesman problem"*, Proc.of the Amer.Math.Soc., 2 (1956), 48-50

13 J.D.Little et al. ,*"An algorithm for the Travelling Salesman problem"*, Opns.Res., 11 (1963), 972-989

14 M.Bellmore-H.J.Greenberg-J.J.Jarvis,*""Multi-commodity diconnecting sets"*, Manag.Sc., 16 (1970), B/427-433

ACKNOWLEDGEMENT. This work was written while the author was at the "Institut de Programmation" with a fellowship from the Italian National Council of Researches.

A UNIFIED APPROACH TO RESHIPMENT, OVERSHIPMENT AND POST-OPTIMIZATION PROBLEMS*

G. Finke
Nova Scotia Technical College
Halifax, Canada

Abstract

In the past, various different solution methods have been proposed for certain modifications of the classical transportation problem, such as reshipment, overshipment and post-optimization problems. It will be shown in this article that the three preceding problems may be solved by a modified version of the primal transportation algorithm. This unified approach leads to a very efficient and direct solution process, which neither requires a transformation of the problem nor an augmentation of the number of variables.

Introduction

The survey articles [4] and [10] summarize the intensive computation studies which have been performed on transportation and transshipment problems over the last years. As a consequence, one has to rank the existing algorithms with respect to their efficiency as follows:

(1) primal simplex method;

(2) primal-dual (Ford-Fulkerson) method;

(3) dual simplex method.

Due to this development, there is a renewed interest in the primal approach to transportation and transportation-type problems. However, there are certain problems associated with transportation problems which seem to require non-primal methods. For instance, the dual method (3) is used for post-optimization problems, and Dwyer's reshipment problems [6] are solved by the Hungarian method (2). But it is relatively easy to see that one can tackle these problems in a primal framework, provided one considers more general transshipment networks. One simply has to interpret negative amounts as reverse shipments from destinations to origins. We shall show in this article how to modify directly the classical transportation algorithm so that it can handle variables which are unrestricted in sign. This extension requires surprisingly few alterations and, in addition to the two problems already mentioned, also leads to an efficient solution method for overshipment problems, i.e. for Charnes' lower bounded and upper unbounded distribution model [3].

---

* This research was supported in part under NRC grant A4117.

The Variant of the Primal Transportation Algorithm

Let us consider the classical linear transportation problem in the following form:

$$\min \sum_{i=1}^{m} \sum_{j=1}^{n} c_{ij} x_{ij}$$

$$(A) \qquad \text{s.t.} \sum_{j=1}^{n} x_{ij} = a_i \qquad (1 \leq i \leq m) ,$$

$$\sum_{i=1}^{m} x_{ij} = b_j \qquad (1 \leq j \leq n) ,$$

$$x_{ij} \geq 0 \quad \text{for all } i,j \;;$$

where the stocks a_i and demands b_j are non-negative numbers satisfying $\sum a_i = \sum b_j$.

By allowing reverse shipments from the destinations to the origins, one obtains variables that are unrestricted in sign. Setting

$$x_{ij} = x_{ij}^{+} - x_{ij}^{-} \quad \text{with} \quad x_{ij}^{+}, x_{ij}^{-} \geq 0 \;,$$

we assign the unit costs c_{ij}^{+} to a shipment x_{ij}^{+} from i to j and the unit costs c_{ij}^{-} to a reverse shipment from j to i ($c_{ij}^{+}, c_{ij}^{-} \geq 0$). Consequently, we have the following special transshipment problem:

$$\min \sum c_{ij}^{+} x_{ij}^{+} + \sum c_{ij}^{-} x_{ij}^{-}$$

$$(B) \qquad \text{s.t.} \sum_{j} (x_{ij}^{+} - x_{ij}^{-}) = a_i ,$$

$$\sum_{i} (x_{ij}^{+} - x_{ij}^{-}) = b_j ,$$

$$x_{ij}^{+} \geq 0 \,,\; x_{ij}^{-} \geq 0 \;.$$

Problem (B) is a direct generalization of problem (P3) in [9]. In strict analogy, one can therefore derive the complementary slackness conditions

$$x_{ij}^{+} (c_{ij}^{+} - u_i - v_j) = 0 \;,$$

$$x_{ij}^{-} (c_{ij}^{-} + u_i + v_j) = 0 \;;$$

and obtain the following variant of the primal algorithm (for details see [9]).

Modified Algorithm: Start with an initial basic solution for problem (B) and generate a sequence of basic solutions using operations 1. and 2. below, until the exit criterion is satisfied.

1. Let I be the set of subscripts which belong to the basic variables. Define costs c^*_{ij} , $(i,j) \in I$, as follows

$$c^*_{ij} = \begin{cases} c^+_{ij} & \text{if} \quad x_{ij} \geq 0 \\ -c^-_{ij} & \text{if} \quad x_{ij} < 0 \end{cases} .$$

Then determine shadow costs u_i and v_j which satisfy $u_i + v_j = c^*_{ij}$, $(i,j) \in I$. Exit if $-c^-_{ij} \leq u_i + v_j \leq c^+_{ij}$ for all i,j (then the present program is optimal).

2. Select a square (h,k) with $u_h + v_k < -c^-_{hk}$ or $u_h + v_k > c^+_{hk}$. The variable x_{hk} enters the basis.

In order to determine the variable which leaves the basis consider the set C of all basic variables in the cycle generated by x_{hk} . Then partition C into C^+ and C^- where C^+ is defined as follows: Enter x_{hk} at a (small) positive level. Move around the cycle C , thereby alternately subtracting and adding this amount. Now C^+ is the set of all variables that decrease their absolute values.

a) If $u_h + v_k > c^+_{hk}$, then $C^+ \neq \emptyset$. Eliminate the variable $x^+ \in C^+$ which has the smallest absolute value.

b) If $u_h + v_k < -c^-_{hk}$, then $C^- \neq \emptyset$. Eliminate the variable $x^- \in C^-$ which has the smallest absolute value.

Let an optimal basic solution of problem (A) for the costs $c_{ij} = c^+_{ij}$ be given so that the corresponding shadow costs u_i and v_j satisfy $u_i + v_j \leq c^+_{ij}$. Then the modified optimality criterion implies the following

<u>Theorem 1</u>. If $u_i + v_j \geq -c^-_{ij}$ for all i,j , then the problems (A) and (B) have the same optimal values.

A computer code for the modified algorithm is described in [9]. We found that, depending on the range of the costs c^-_{ij} , the computation time for a complete basis exchange only took 2-5% more than for the ordinary primal algorithm.

Reshipment Problems

Reshipment problems have been introduced by Dwyer [6]. They correspond to the following special case of problem (B):

$$c^+_{ij} = c^-_{ij} = c_{ij} \quad .$$

Dwyer's solution process requires three steps.

(1) Replace nonoptimal costs $c_{\alpha\beta}$, i.e. costs such that

$$c_{\alpha\beta} > c_{\alpha j} + c_{ij} + c_{i\beta} \quad \text{for some } i,j \quad ;$$

with the cost $c_{\alpha\beta} + c_{ij} + c_{i\beta}$ of the indirect route $(\alpha,j),(j,i),(i,\beta)$ from α to β .

(2) Solve the transportation problem with respect to the transformed costs.

(3) Retrace the shipments $x_{\alpha\beta} > 0$ associated with nonoptimal costs to the original routes.

As an immediate consequence of Theorem 1, one has the following

Corollary 1. If $u_i + v_j \geq -c_{ij}$ for all i,j , then reverse shipments cannot improve the optimal value of the ordinary transportation problem (A).

If the condition of Corollary 1 is not satisfied, the modified algorithm continues directly with the final tableau. (Illustrative examples can be found in the Appendix.)

Solving the 100 x 100 random examples of [8] and [9] as reshipment problems resulted in an average improvement of 5.2% . The optimal solutions contained a mean number of 18 reverse shipments.

Post-Optimization Problems

If the stocks a_i and demands b_j are changed for the ordinary problem (A), an optimal solution may become infeasible. However, the dual feasibility is preserved and the dual method is therefore proposed to reoptimize the program (compare [1] and [2]).

There are several possibilities to use the modified algorithm for post-optimization problems. But the obvious way, namely $c^+_{ij} = c_{ij}$ and $c^-_{ij} = k$ with k large, does not seem to be the best approach. The following cost structure is proposed in [9]:

$$c^+_{ij} = c^-_{ij} = c_{ij} + k \quad ; \text{ k large enough .}$$

If the costs satisfy $0 \leq a \leq c_{ij} \leq b$, one may set $k = b - a$. It is relatively easy to see that, for these costs, the optimal solution to problem (B) is automatically non-negative and hence solves the post-optimization problem (see Lemma 1 in [9]).

Overshipment Problems

Overshipment problems are defined as the following relaxation of problem (A):

$$\text{(C)} \qquad \begin{aligned} &\min \sum_i \sum_j c_{ij} x_{ij} \\ &\text{s.t.} \sum_j x_{ij} \geq a_i \ , \quad \sum_i x_{ij} \geq b_j \ ; \\ &x_{ij} \geq 0 \ . \end{aligned}$$

An optimal solution of (C) usually satisfies the supply and demand constraints as strict inequalities ('Transportation Paradox'). Several solution methods have been proposed [3, 5, 11] all of which require a reformulation and augmentation of the size of the problem. The following theorem is restated from [8].

Theorem 2. Problem (C) is equivalent to problem (B) with costs $c_{ij}^{+} = c_{ij}$ and $c_{ij}^{-} = 0$. If $\{x_{ij} = x_{ij}^{+} - x_{ij}^{-}\}$ is an optimal solution of (B), then the set $\{x_{ij}^{+}\}$ is optimal for (C) and the total overshipment is given by $\sum\sum x_{ij}^{-}$.

Theorems 1 and 2 imply the following (see also [11]).

Corollary 2. If $u_i + v_j \geq 0$ for all i, j, then overshipments cannot improve the optimal value of the ordinary problem (A).

The computational results of 100 x 100 random examples are rather unexpected. By allowing additional shipments, one can obtain an average cost reduction of 18.8%, and the total overshipment amounts to 20.5%.

It should be noted that the amount $\sum\sum x_{ij}^{-}$ varies for alternate optimal solutions of (C). The modified algorithm may be used to find its minimum (except for certain cases of degeneracy). Suppose an optimal solution is given with shadow costs satisfying $0 \leq u_i + v_j \leq c_{ij}$. Then define

$$c_{ij}^{+} = \begin{cases} 0 & \text{if } u_i + v_j = c_{ij} \\ \infty & \text{otherwise} \end{cases} \quad \text{and} \quad c_{ij}^{-} = \begin{cases} 1 & \text{if } u_i + v_j = 0 \\ \infty & \text{otherwise} \end{cases} .$$

The average minimum overshipments for the random examples worked out to 17.9%.

Appendix

1. Reshipment. Consider the 3 x 4 transportation problem from [7].

	4	5	3	7
6	3	6	1	5
11	7	9	2	7
2	2	4	2	1

One obtains the following optimal solution and corresponding shadow costs

x_{ij}:

4	2		
	3	3	5
			2

u_i, v_j:

	6	9	2	7
-3	3	6	-1	4
0	6	9	2	7
-6	0	3	(-4)	1

$u_i + v_j \leq c_{ij}$

Since $u_3 + v_3 = -4 < -c_{33}$, one can continue the algorithm. The partition is $C^{+} = \{x_{23} = 3, x_{34} = 2\}$ and $C^{-} = \{x_{24} = 5\}$. Hence x_{33} enters the basis at the negative level -5.

4	2		
	3	8	
		-5	7

x_{ij}

	3	6	-1	2
0	3	6	-1	2
3	6	9	2	5
-1	2	(5)	-2	1

u_i, v_j

In the next step, we have $u_3 + v_2 = 5 > c_{32}$. The corresponding partition is $C^+ = \{x_{22} = 3\}$, $C^- = \{x_{23} = 8 , x_{33} = -5\}$, and therefore x_{32} enters the basis at the positive level 3 .

4	2		
		11	
	3	-8	7

x_{ij}

	1	4	-2	1
2	3	6	0	3
4	5	8	2	5
0	1	4	-2	1

u_i, v_j

$|u_i + v_j| \leq c_{ij}$

The last tableau displays the optimal reshipment solution. The previous optimal value \$94 for the transportation problem has been reduced to \$81 .

<u>2. Overshipment.</u> Let us solve the 4 x 5 problem of [8, 11]

	4	11	12	8	11
7	14	15	6	13	14
18	16	9	22	13	16
6	8	5	11	4	5
15	12	4	18	9	10

The transportation problem has the following optimal solution and corresponding shadow costs

		7		
4	6		8	
		5		1
	5			10

x_{ij}

	16	9	21	13	15
-15	1	(-6)	6	-2	0
0	16	9	21	13	15
-10	6	-1	11	3	5
-5	11	4	16	8	10

u_i, v_j

$u_i + v_j \leq c_{ij}$

This solution is not optimal for problem (C). Since $u_1 + v_2 = -6 < 0$, the variable x_{12} may be introduced into the basis at a negative level according to the algorithm. The partition of the cycle generated by x_{12} is given by $C^+ = \{x_{13} = 7 , x_{35} = 1 , x_{42} = 5\}$ and $C^- = \{x_{33} = 5 , x_{45} = 10\}$. Consequently, x_{33} leaves the basis and x_{12} enters at level -5 .

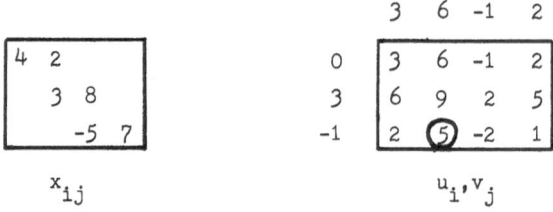

	-5	12		
4	6		8	
				6
	10			5

	16	9	15	13	15
-9	7	0	6	4	6
0	16	9	15	13	15
-10	6	(-1)	5	3	5
-5	11	4	10	8	10

The algorithm terminates with the next iteration.

	-5	12		
4	6		8	
	-5			11
	15			

	7	0	6	4	5
0	7	0	6	4	5
9	16	9	15	13	14
0	7	0	6	4	5
4	11	4	10	8	9

$$0 \le u_i + v_j \le c_{ij}$$

The optimal solution for problem (C) is obtained by replacing negative variables with zero (Theorem 2).

	4	21	12	8	11
12			12		
18	4	6		8	
11					11
15		15			

We have an overshipment of 10 units, and the optimal value \$444 for the transportation problem has been reduced to \$409 .

<u>Remark</u>: If $\sum a_i \neq \sum b_j$, a fictitious row (column) is added with costs that are minimal in the corresponding column (row) (see [5, 8]).

3. <u>Post-Optimization.</u> The transportation problem

	5	3	4	3
3	3	6	3	5
8	7	9	4	7
4	2	4	4	1

has the following optimal solution

3			
1	3	4	
1			3

Suppose the stocks and demands are changed to (6, 11, 2) and (4, 5, 3, 7). Replace the costs c_{ij} with $c_{ij} + 10$ and the modified algorithm reoptimizes as follows:

	4	5	3	7
6	6			
11	3	5	3	
2	-5			7

	0	2	-3	23
13	13	15	10	36
17	17	19	14	(40)
-12	-12	-10	-15	11

6			
	5	3	3
-2			4

	-6	19	14	17
19	13	(38)	33	36
0	-6	19	14	17
-6	-12	13	8	11

4	2		
	3	3	5
			2

	16	19	14	17
-3	13	16	11	14
0	16	19	14	17
-6	10	13	8	11

$$|u_i + v_j| \leq (c_{ij} + 10)$$

References

[1] E. Balaş and P. Ivănescu, "On the Transportation Problem - Part I and Part II", Cahiers du Centre d'Etudes de Recherche Opérationnelle, 4, (1962) 98-116 and 131-160.

[2] A. Charnes and M. Kirby, "The Dual Method and the Method of Balaş and Ivănescu for the Transportation Problem", Cahiers du Centre d'Etudes de Recherche Opérationnelle, 6, (1964) 55-18.

[3] A. Charnes, F. Glover and D. Klingman, "The Lower Bounded and Partial Upper Bounded Distribution Model", Nav. Res. Log. Quart., 18, (1971) 277-281.

[4] A. Charnes, D. Karney, D. Klingman, J. Stutz and F. Glover, "Past, Present and Future Large Scale Transshipment Computer Codes and Applications", Comput. and Ops. Res., 2, (1975) 71-81.

[5] P. S. Dwyer, "A General Treatment of Upper Unbounded and Bounded Hitchcock Problems", Nav. Res. Log. Quart., 21, (1974) 445-464.

[6] P. S. Dwyer, "Transportation Problems with Some x_{ij} Negative and Transshipment Problems", Nav. Res. Log. Quart., 22, (1975) 751-776.

[7] G. Finke, "Une méthode de post-optimisation pour le problème de transport", Proc. CAAS Conference 1976, Université Laval, Québec.

[8] G. Finke, "The Lower Bounded and Upper Unbounded Distribution Model", Proc. 8th Southeastern Conf. on Combinatorics, Graph Theory, and Computing (1977).

[9] G. Finke and J. H. Ahrens, "A Variant of the Primal Transportation Algorithm", to appear in INFOR.

[10] F. Glover, D. Karney and D. Klingman, "Implementation and Computational Comparisons of Primal, Dual and Primal-Dual Computer Codes for Minimum Cost Network Flow Problems", Networks, 4, (1974) 191-212.

[11] W. Szwarc, "The Transportation Paradox", Nav. Res. Log. Quart., 18, (1971) 185-202.

SOLVING AN INTEGER PROGRAMMING PROBLEM

Hoang Hai Hoc
Department of Electrical Engineering
Ecole Polytechnique
Montreal, Canada

ABSTRACT

We consider a sequence of master (zero-one integer programming) problems arising from the application of Benders decomposition to mixed integer programming models. This sequence of related problems was solved by reoptimization using implicit enumeration based algorithms for postoptimizing zero-one programs. Both computational experience and results are reported.

INTRODUCTION

We consider the following family of zero-one integer programming problems (IP)

$$\underset{x_1,\ldots,x_n}{\text{Minimize}} \quad \underset{i=1,\ldots,m}{\text{maximize}} \quad (d_i - \sum_{j=1}^{n} a_{ij}x_j)$$

subject to

$$\sum_{j=1}^{n} c_j x_j \leq b \; ; \quad x_j = 0 \text{ or } 1, \; j = 1,\ldots,n$$

where $m = 1,2,\ldots$

These problems can be viewed as nonlinear zero-one knapsack problems with nonadditive objective functions. This sequence of problems is to be considered when one solves a nonlinear mixed integer programming model for the optimization of network topologies (10) using the generalized Benders' decomposition (3). However this case is not a simple coïncidence. In fact, the application of generalized Benders' decomposition to any linear (nonlinear) mixed integer programming model always gives rise to a similar sequence of master problems, each of which is obtained from the previous one by adding one or more constraints. Moreover, although Benders' decomposition principle is not a complete algorithm, in the sense that it does not specify how to solve either master programs or subprograms, it was successfully used to solve practical problems such as electric power generation planning (11), multicommodity distribution system design (6), etc.

The major conclusion arising from these previous studies (6,10,11), is the remarkable effectiveness of Benders' decomposition as a computational strategy for the classes of mixed integer programming models reported. The numerical experience shows that only a few (less than a dozen) cuts are needed to find and verify a solution within one percent of the global optimum. Computational time is then determined completely by the efficiency of the algorithms used to solve the sequences of master problems and subproblems. For this reason, we are presently interested in investigating computational techniques suitable to the solution of a sequence of master problems, in particular the family of problems (IP). To be more specific, we shall restrict ourselves to the class of enumerative techniques for its effectiveness as a practical tool to solve combinatorial problems.

We note that the members of the family of problems (IP) are closely related. For $m = 2,3...$, the m-th problem is actually obtained by adding a new Benders' cut to the (m-1)-th problem. One can then examine the family of problems (IP) from the viewpoint of parametric and postoptimality analysis in integer linear programming (7). We propose to solve the family of problems (IP) by employing implicit enumeration based algorithms for postoptimizing zero-one programs (12,14). Computational experience on the family of problems (IP) should then shed more insight into postoptimality analysis by implicit enumeration.

IMPLICIT ENUMERATION

This section presents an implicit enumeration which solves a problem (IP) and gathers the information required for the postoptimization. For this purpose let us define the equivalent problem (IPE) as follows.

$$\text{minimize } y$$

subject to

$$\sum_{j=1}^{n} -c_j x_j \leq B \tag{1}$$

$$D_i + \sum_{j=1}^{m} a_{ij} x_j \leq y, \; i = 1,2,...,m \tag{2}$$

$$x_j = 0 \text{ or } 1, \; j = 1,2,...,n \tag{3}$$

where

$$B = b - \sum_{j=1}^{n} c_j < 0$$

$$D_i = d_i - \sum_{j=1}^{n} a_{ij} > 0, \; i = 1,...,m$$

and all the coefficients a_{ij}, c_j, and d_i are positive.

If we take only constraints (3) into consideration there are 2^n possible assignment of values to $(x_1, x_2, ..., x_n)$. Let an assignment of values to a subset of these variables be called a partial solution. Let the variables x_j without assigned values be called free variables. Any assignment of values to the free variables is called a completion of the associated partial solution. In particular, each partial solution has its zero-completion.

As the implicit enumeration search proceeds partial solutions are generated in an attempt to find a feasible zero-completion, since nonzero-completion of a feasible partial solution (i.e. a partial solution with feasible zero-completion) cannot have a better objective function value.

A partial solution is fathomed (a) if its zero-completion is feasible, or (b) if it can be shown that none of its completions can yield a feasible solution better than the best feasible solution found to date. Each fathomed partial solution of problem (IPE) with infeasible zero-completion will then be classified in two ways: (a) by the single resource constraint (1), and (b) by the objective function constraints (2). For the purpose of this paper, only partial solutions obtained during the search, which are feasible or fathomed by objective function constraints need be saved for postoptimization. Consequently, let us define

$A_o = \{ s |$ the zero-completion of s is infeasible, and one or more constraints (2) fathom s $\}$

$A_f = \{ s |$ the zero-completion of s is feasible $\}$

and associate the following index sets with each partial solution s obtained during the search

$$J_s^d = \{ j \in N : x_j(s) = d \} , \; d = 0,1$$

$$N_s = N - J_s^o - J_s^1$$

where $N = \{ 1,2,\ldots n \}$ and $x_j(s) = d$ implies that the j variable is assigned value d in the partial solution s. Hence, J_s^o and J_s^1 are the index sets for the fixed variables and N_s is the index set of the free variables with respect to s.

Algorithm I, which follows immediately, may be used to classify and collect the partial solutions while solving the problem (IPE). Certain details are omitted such as the termination criteria used, the rules for generating new partial solutions, as well as for backtracking from old ones. Any methods in (1,4,5,9) can be used. The value $\overline{Z}$ which appears in the algorithm is valid upper bound on the optimal solution. This bound is continually updated during the search.

ALGORITHM I

1. Initialize $N_s = N$, $p = o$, $A_f = A_o = J_s^o = J_s^1 = \emptyset$

2. Compute $t_o(s) = B + \sum_{j \in J_s^1} c_j$

$$v_o(s) = t_o(s) + \sum_{j \in N_s} c_j = b - \sum_{j \in J_s^o} c_j$$

$$v_i(s) = \overline{Z} - D_i - \sum_{j \in J_s^1} a_{ij}, \; i = 1,2,\ldots,m$$

3. If $t_o(s) < o$, go to 4. Otherwise, s is feasible with objective function value $Z(s) = \overline{Z}$ - maximum $(v_i(s), i = 1,\ldots,m)$ Set $p = p + 1$, add s to A_f. If $Z(s) < \overline{Z}$, set $\overline{Z} = Z(s)$. Backtrack, and go to 2.

4. For each free variable j in N_s define $G^o(j)$ and $G^1(j)$ as follows:

(i) If $v_o(s) < c_j$ then $o \in G^1(j)$

(ii) For $i = 1, 1\ldots,m$, if $v_i(s) < a_{ij}$ then $i \in G^o(j)$

and perform the following steps:

a) If $G^o(\bar{j})$ and $G^1(\bar{j})$, $\bar{j} \in N_s$, are nonempty then there are no better feasible completions of s. Add $\bar{s}$ to A_o, where $J^o_{\bar{s}} = J^o_s$ and $J^1_{\bar{s}} = J^1_s u\{\bar{j}\}$. Backtrack, and go to 2.

b) If $G^1(j)$ is nonempty then skip to partial solutions $\bar{s}$, where $J^o_{\bar{s}} = J^o_s$ and $J^1_{\bar{s}} = J^1_s u\{\bar{j}\}$, and go to 2.

c) If $G^1(\bar{j})$ is nonempty then add $\bar{s}$ to A_o, where $J^o_{\bar{s}} = J^o_s$ and $J^1_{\bar{s}} = J^1_s u \{\bar{j}\}$. Skip to partial solution $s*$, where

$J^o_{s*} = J^o_s u \{\bar{j}\}$, $J^1_{s*} = J^1_s$, and go to 2.

5. Generate, and go to 2.

POSTOPTIMIZATION USING IMPLICIT ENUMERATION

It has been shown by Roodman (14) that useful postoptimization capabilities for the zero-one integer programming problem can be obtained from an implicit enumeration algorithm modified in such way as to classify and collect fathomed partial solutions as obtained through the use of algorithm I of the previous section. The underlying principle is as follows: whenever a search along a branch of the enumeration tree is terminated, the fathomed partial solution s can be attributed to a constraint k. Unless constraint k is somehow relaxed, s and its completions will remain infeasible, regardless of other changes to the 0-1 integer program. By considering the set A_k of all partial solutions attributed to k, one can obtain the minimum relaxation in k before any partial solution in A_k becomes potentially feasible. Only partial solutions in A_k need be examined if one relaxes constraint k. More recently, Piper and Zoltners (12) have presented a storage structure to cope with the difficult data collection task inherent to the approach, as well as a collection of algorithms using this storage structure to do the postoptimization after one or more problem parameter changes.

Let us consider now the family of problems (IPE) as a sequence of closely related problems. Each problem in this sequence is obtained from the previous one by adding a new constraint (Benders' cut). This added constraint is actually not verified by the previous optimal solution, and it is always necessary to reoptimize the problem after adding a constraint. To carry out this reoptimization, we used the approach suggested by Roodman, as refined by Piper and Zoltners. Conceptually, we obtained the following procedure:

ALGORITHM II

1. Update $z(s)$ for all partial solutions s in A_f and A_o.
Determine the best solution $\bar{s}$ in A_f.
Let $\bar{x} = x(\bar{s})$, $\bar{z} = z(\bar{s})$.

2. For each partial solution s in A_o with $z(s) < \bar{z}$, perform the following steps:

a) Let $A_o = A_o - \{ s \}$.

b) Examine the completions of s using algorithm I.

We remark that when algorithm I is used, during postoptimization, to examine the completions of partial solution s then J^o_s , J^1_s , N_s describe s; and p, A_o , and A_f describe the current state of the postoptimization. Step 1 of the algorithm I should then be modified appropriately.

ALGORITHM IMPLEMENTATION AND COMPUTATIONAL RESULTS

Algorithms I and II were implemented in order to study the computational behavior of implicit enumeration based algorithms for reoptimizing zero-one programs. No particular attention was given to the efficiency of the resulting computer programs.

The computer program implementing algorithm I represents essentially a variation of the additive algorithm (1) specialized to problem (IPE) with an efficient bookeeping schema (4,5) which keeps track of the enumeration. This variation includes only ceiling test and cancellation tests. As a branching strategy it is chosen to fix at 1 the free variable j for which the ratio a_{mj}/c_j is minimum.

Althouth algorithm II is very simple comceptually, its implementation presents some interesting problems from the viewpoint of the storage structure. First, to accomodate problems of about 30 variables, each partial solution is packed into two 32 bit words. Two bits are attributed to each variable: one bit is used to indicate that the variable is free or fixed, and the other represents the fixed value of the variable. Secondly due to the fact that the set of partial solutions A_o and A_f are too large to be stored in core, random access disc files are used to save A_o and A_f. Moreover, the set A_o is actually subdivided into A^1_o, A^2_o ,...,A^q_o according to the values of the objective function of the partial solutions collected. For all partial solutions belonging to A^i_o the objective function values fall inside the i-th predetermined interval. Partial solutions in A_o and A_f are stored sequentially in records which are chained by pointers.

In order to facilitate file processing, a sufficiently large array is used as buffer, and organized into LIFO sublists. Each sublist contains partial solutions belonging to a subset A^i_o or to the set A_f , and residing temporarily in core. When this buffer array is full, and at the end of the reoptimization procedure, all partial solutions are transferred on disc files. The output file obtained at the end of a reoptimization is used as input file for the next reoptimization.

All programs were written in FORTRAN IV Level G, and executed under the control of OS/360 MVT, on a system IBM/360 Model 75. Several families of test problems with size ranging from 10 to 30 variables were solved. First we solved separately each problem in a family. Then we solved each family of problems as a sequence of related problems using reoptimization technique. Each family consisted of 5-8 problems. Computational results indicate that the time required to solve a family of problems using reoptimization technique is actually 2-3 times longer than the total time required to solve the problems separately. This is partially due to the time consuming operations of packing and unpacking partial solutions (by means of integer arithmetic in FORTRAN IV) as well as to the inefficiency of the file processing procedure presently used. Furthermore, an important explanatory factor is the continually increasing number of partial solutions collected and re-examined while reoptimizing. In fact, as reoptimization proceeds, one is going down the enumeration tree deeper and deeper, and generating more and more nodes which will eventually be fathomed, saved and re-examined. Hence, one should be prevented from going down the tree too deeply by means of efficient bounding process using embedded linear programs surrogate constraints, etc. This represents a difficulty common to all implicit enumeration, and branch bound algorithms. Last, the reoptimization technique considered will be efficient only if the part of the enumeration tree to be explored for solving the modified problems is not very different from the part of the tree explored while solving the original problem. This may be the case, if only a few coefficients of the zero-one integer program are subject to changes. Extensive computational experience is planned in the future to study further these aspects.

REFERENCES

1. E. Balas, "An Additive Algorithm for Solving Linear Programs with Zero-One Variables", Operations Research, Vol. 13, pp. 517-546 (1965).

2. J.F. Benders, "Partioning Procedures for Solving Mixed Variables Programming Problems", Numerische Mathematik, Vol. 4, pp. 238-252 (1962).

3. A.M. Geoffrion, "Generalized Benders Decomposition", Journal of Optimization Theory and its Applications, Vol. 10, pp. 237-260 (1972).

4. __________, "Integer Programming by Implicit Enumeration and Balas Method", SIAM Review, Vol. 9, pp. 178-190 (1967).

5. __________, "An Improved Implicit Enumeration Approach for Integer Programming", Operations Research, Vol. 17, pp. 437-454 (1969).

6. A.M. Geoffrion & G.W. Graves, "Multicommodity Distribution System Design by Benders Decomposition ", Management Science, Vol. 20, pp. 822-844 (1974).

7. A.M. Geoffrion & R. Nauss, "Parametric and Postoptimality Analysis in Integer Linear Programming", Management Science, Vol. 23, pp. 453-466 (1977).

8. F. Glover, "Surrogate Constraints", Operations Research, Vol. 16, pp. 741-749 (1968).

9. F. Glover, "A Multiphase Dual Algorithm for the Zero-One Integer Programming Problem", Operations Research, Vol. 13, pp. 879-919 (1965).

10. Hoang Hai Hoc, "Network Improvements via Mathematical Programming", Paper presented at the IX th International Symposium on Mathematical Programming, Budapest August 23-27, 1976.

11. F. Noonan & R.J. Giglio, "Planning Electric Power Generation: A Nonlinear Mixed Integer Programming Model Employing Benders Decomposition", Management Science, Vol. 23, pp. 946-956 (1977).

12. C.J. Piper & A.A. Zoltners, "Implicit Enumeration Based Algorithms for Post-optimizing Zero-One Programs", Naval Research Logistics Quarterly, Vol. 22, pp. 791-809 (1975).

13. C.J. Piper & A.A. Zoltners, "Some Easy Postoptimality Analysis for Zero-One Programming", Management Science, Vol. 22, pp. 759-765 (1976).

14. G.M. Roodman, "Postoptimality Analysis in Zero-One Programming by Implicit Enumeration", Naval Research Logistics Quarterly, Vol. 19, pp. 435-447 (1972).

ACKNOWLEDGEMENT

This work was supported in part by National Research Council Canada - Grant No. A-8816.

WORST-CASE ANALYSIS FOR A CLASS OF COMBINATORIAL OPTIMIZATION ALGORITHMS

Dirk Hausmann and Bernhard Korte
Institut für Ökonometrie und Operations Research
Universität Bonn

1. Introduction

We want to present a method to measure the worst-case complexity for some very general problems. As the precise, set-theoretical definitions given in the next section are quite abstract, we will first introduce our ideas in a rather intuitive language.

The input of a problem we have in mind does not only consist of easily encodeable informations such as integers, matrices, and graphs - we will call them "side information" later on - but also of general structures, such as systems of sets, real functions, or set functions. In particular instances of the problem, these general structures are usually not explicitly given in the input but must be retrieved by a special procedure for example by checking a specified property or by evaluating a formula. Since this effort for retrieving information which belongs implicitly to the input depends strongly on the particular instance of the problem, we do not want to consider it as a part of the complexity of the general problem but we take the view that this is taken care of by some oracle. This oracle may be thought of as a kind of subroutine R which can be called on by an algorithm (main program) for some argument X and which then gives back a specified information $R(X)$ about the general structure. For the reason explained above we do not want to count the number of steps which the oracle needs for providing the information $R(X)$, but we will count every call on the oracle as a single step instead. Moreover, as we are only interested in lower bounds on the complexity, we will count only these calls on the oracle. The consequence of these reflections is that we do not need the Turing machine concept (cf. e.g. [1]) which is by far the most important tool in the theory of complexity but which is rather complicated because it was conceived to count every step in the computation which we do not want to do.

For this reason we define here another algorithmic concept, the oracle algorithm. From our formal definitions in section 2 it will follow immediately that every Turing machine with an oracle tape (in the sense of [2] or [3]) can also be considered as an oracle algorithm. Hence if we prove that every oracle algorithm for some problem P needs at least k calls on the oracle, this result implies immediately that every Turing machine (with an oracle tape) for P needs at least k steps.

We think that for the kind of problems we consider here a proof of complexity results in our terminology of oracle algorithms seems to be more to the point than in the classical terminology of Turing machines.

2. Oracle algorithms

In this section we want to give formal definitions of an oracle problem (i.e. a problem the input of which contains a "general structure" represented by an "oracle") and of oracle algorithms solving it. As we mentioned in the introduction, the only steps of an oracle algorithm which we want to count are the calls on the oracle. Hence we need not divide the procedure which leads from one to the next call on the oracle into single steps - this is done in the Turing machine concept - but we formalize this procedure as a mapping. To all "side informations" and informations given by the oracle so far, this mapping assigns the next oracle argument or the output. In order to clearly distinguish between oracle arguments and outputs we allow as outputs only pairs having the symbol STOP in the first component.

Now we define an <u>oracle problem</u> to be a pair $P = (\text{INPUT},\text{SOL})$ where

INPUT is a set of pairs (e,R) where e, called the <u>side information</u>, is an element of some set E and where R, called the <u>oracle</u>, is a mapping $R : DOM_e \to REG_e$, (note that domain and region of the oracle may depend only on the side information e).

SOL is a mapping which to every $(e,R) \in \text{INPUT}$ assigns a subset of OUTPUT where OUTPUT is a set of pairs having the symbol STOP in the first component.

For any $(e,R) \in \text{INPUT}$, $\text{SOL}(e,R)$ can be interpreted as the set of all solutions to the problem instance specified by (e,R). Now let $P = (\text{INPUT},\text{SOL})$ be an oracle problem, let N be the set of all nonnegative integers, $N=\{0,1,2,...\}$, and let A be a mapping which assigns to every $e \in E$ a sequence $(A_{e,i})_{i\in N}$ where any $A_{e,i}$ is a mapping from $DOM_e^i \times REG_e^i$ into $DOM_e \cup \text{OUTPUT}$. Here DOM_e^i (resp. REG_e^i) denotes the i-fold Cartesian product of DOM_e (resp. REG_e). Thus for $i=0$, $A_{e,i} = A_{e,0}$ can be identified with a single element in $DOM_e \cup \text{OUTPUT}$.

Let $e \in E$ be any side information and $R:DOM_e \to REG_e$ any oracle, not necessarily $(e,R) \in \text{INPUT}$. Then we define the corresponding <u>sequence of oracle arguments</u> $(\overline{X}_i)_{i\in N} = (\overline{X}_i(e,R,A))_{i\in N}$ by the following recursion:

$$\overline{X}_o := A_{e,o}$$

$$\overline{X}_i := \begin{cases} A_{e,i}(\overline{X}_o,\ldots,\overline{X}_{i-1}, R(\overline{X}_o),\ldots,R(\overline{X}_{i-1})), & \text{if } \overline{X}_{i-1} \in DOM_e \\ \overline{X}_{i-1} & \text{, if } \overline{X}_{i-1} \in OUTPUT. \end{cases}$$

If for any $(e,R) \in$ INPUT some element $\overline{X}_i(e,R,A)$ of the corresponding sequence of oracle arguments is in OUTPUT and moreover in SOL(e,R), then A is called an <u>oracle algorithm</u> for problem P. For any $(e,R) \in$ INPUT let $\overline{k}(e,R,A)$ designate the smallest integer i such that $\overline{X}_i(e,R,A) \in$ OUTPUT, it is called the <u>number of calls on the oracle</u> for the input (e,R).

Every mapping $A_{e,i}$ can be interpreted as the part of the (oracle) algorithm between the i-th and the (i+1)st call on the oracle. This part can exploit all informations about the input obtained so far and produces either the output or the argument $\overline{X}_i \in DOM_e$ for the next call on the oracle. As long as $\overline{X}_i \in DOM_e$, the next element $\overline{X}_{i+1}$ is determined by the mapping $A_{e,i}$ and the algorithm may be considered intuitively as "running". But for $i \geq \overline{k}(e,R,A)$, the algorithm may be considered intuitively as "having stopped and provided the output $\overline{X}_{\overline{k}(e,R,A)}$", for these indices i, the sequence $(\overline{X}_i)_{i\in N}$ is continued constantly.

3. The maximization problem over a general independence system

In this section we want to consider a very important "general problem" and apply our formal definitions and methods to it. Let n be a positive integer, c a vector in $\mathbb{R}^n_+$ and S an <u>independence system</u> over $\{1,\ldots,n\}$, i.e. a non-empty system of subsets of $\{1,\ldots,n\}$ such that any subset of a set in S, belongs to S, too. The subsets belonging to S are called <u>independent</u> (with respect to S). The following problem is known as the <u>maximization problem over a general independence system</u>

$$(1) \qquad c(F) = \sum_{i\in F} c_i = \text{Max!} \qquad \text{s.t. } F \in S.$$

A lot of well-known combinatorial optimization problems are instances of this general problem. It is easy to see that the (informally stated) problem (1) can be formulated as an oracle problem with side information e = (n,c), an oracle R giving relevant informations about S and SOL(n,c,R) being the set of all optimal solutions of (1). As there are several ways to define an independence system S, we might consider also several types of oracles specifying S. We propose to use the (perhaps) most natural one which checks the independence of a subset, i.e. an oracle $R: 2^{\{1,\dots,n\}} \to \{YES, NO\}$ such that $R(F) = YES$ iff $F \in S$. (Note that domain and region of this oracle depend only on the side information $e = (n,c)$.) Other possible oracles for an independence system S were considered in [5].

A well-known approximative algorithm for problem (1) is the greedy algorithm which first orders the integers $1,\dots,n$ according to non-increasing components of c, then starts with the independent set $G = \emptyset$ and for each integer $i \in \{1,\dots,n\}$ in turn adds i to the current independent set G if the enlarged set $G \cup \{i\}$ belongs again to the independence system S.

To illustrate our formal definitions given in section 2, we briefly formulate the greedy algorithm as an oracle algorithm. This formulation is given here only for illustrative purposes and may be skipped by the reader.

Consider the following oracle problem $P = (INPUT, SOL)$ where

$$INPUT = \{(n,c,R) \mid n \geq 1,\ c \in \mathbb{R}^n_+,\ R: 2^{\{1,\dots,n\}} \to \{YES, NO\},\ R^{-1}(YES) \text{ is an independence system}\}$$

$$SOL(n,c,R) = \{(STOP, X) \mid X \in R^{-1}(YES)\}.$$

Now let (n,c) be a given side information. Note that $DOM_{(n,c)} = 2^{\{1,\dots,n\}}$, $REG_{(n,c)} = \{YES, NO\}$. Let $(j_o,\dots,j_{n-1})$ be an ordering of $\{1,\dots,n\}$ such that $c_{j_i} \geq c_{j_{i+1}}$. Clearly this ordering depends on the side information (n,c). Now we define the mappings $A_{n,c,i}$ for $i \in N$ by

$$A_{n,c,o} = \{j_o\}$$

$$A_{n,c,i}(X_o,\dots,X_{i-1}, Y_o,\dots,Y_{i-1}) = \begin{cases} (X_{i-1} \setminus \{j_{i-1}\}) \cup \{j_i\}, & \text{if } i < n,\ Y_{i-1} = NO \\ X_{i-1} \cup \{j_i\}, & \text{if } i < n,\ Y_{i-1} = YES \\ (STOP,\ X_{n-1} \setminus \{j_{n-1}\}), & \text{if } i \geq n,\ Y_{n-1} = NO \\ (STOP,\ X_{n-1}), & \text{if } i \geq n,\ Y_{n-1} = YES \end{cases}$$

Then the greedy algorithm for problem P can formally be defined as the mapping which assigns to any side information (n,c) the sequence of mappings $(A_{n,c,i})_{i \in N}$ defined above. The interpretation of the mappings $A_{n,c,i}$ should be obvious now. For $i < n$ each of these mappings provides the next set the independence of which is to be checked; this set equals the last set X_{i-1} -

with or without the last enlargement j_{i-1} depending on the last reply of the oracle - enlarged by the next integer j_i. It is also obvious that the number of calls on the oracle for input (n,c,R) is $\overline{k}(n,c,R,A)=n$.

In a former paper [7], we characterized a performance guarantee for the greedy algorithm. Let (n,c,S) be any instance of the problem (1) and G an approximate solution yielded by the greedy algorithm. Moreover we define the <u>rank quotient</u>

$$q(n,S) = \min\{lr(H)/ur(H) : H \subseteq \{1,\dots,n\}\}$$

where $lr(H)$ $(ur(H))$ is the minimum (maximum) cardinality of a maximal independent subset (with respect to S). Then we proved the following performance guarantee of the greedy algorithm:

$$(2) \qquad c(G) \geq q(n,S)\cdot\max\{c(F) : F \in S\}$$

If S is a <u>matroid</u>, i.e. if $q(n,S)=1$, then it is well-known (see [4]) and follows immediately from (2) that no algorithm can have a better performance guarantee than the greedy algorithm. But is there perhaps a polynomial algorithm that for any problem instance (n,c,S) where S is not a matroid, has a globally better performance guarantee than (2)? Using our general oracle approach we will show that the answer to this question is in the negative. For this purpose we define the following oracle problem

$P = (\text{INPUT},\text{SOL})$ where

$$\text{INPUT} = \{(n,c,R) \mid n \geq 1,\ c \in \mathbb{R}^n_+,\ R:2^{\{1,\dots,n\}} \to \{\text{YES},\text{NO}\},$$
$$R^{-1}(\text{YES}) \text{ is an independence system - but no matroid - on } \{1,\dots,n\}\}$$

$$\text{SOL}(n,c,R) = \{(\text{STOP},X) \mid X\in R^{-1}(\text{YES}),\ c(X)>q(n,R^{-1}(\text{YES}))\cdot\max\{c(F) \mid F\in R^{-1}(\text{YES})\}\}$$

The following theorem is a precise formulation of our intuitive statement that no polynomial oracle algorithm for the maximization problem over a general independence system has a globally better performance guarantee than the greedy algorithm.

<u>Theorem 1:</u> There is no oracle algorithm A for problem P above such that the number of calls on the oracle $\overline{k}(n,c,R,A)$ is bounded by a polynomial in n.

Proof: Suppose A is such an algorithm and p is a polynomial such that $\overline{k}(n,c,R,A) \leq p(n)$ for any $(n,c,R) \in$ INPUT. Then there is an integer n such that $\binom{n}{[n/2]} -2 \geq p(n)$. Let n be a fixed integer with this property and let $m = [n/2]$, $r=\binom{n}{m}-2$ and $c = (1,\dots,1) \in \mathbb{R}^n$. Moreover let $R':2^{\{1,\dots,n\}} \to \{\text{YES},\text{NO}\}$ be defined by

$$R'(F) = \begin{cases} \text{YES, if } |F| < m \\ \text{NO, otherwise.} \end{cases}$$

Let $(\overline{X}_i(n,c,R',A))_{i\in N}$ be the corresponding sequence of oracle arguments. (Note that the independence system $R'^{-1}(\text{YES})$ is a matroid, hence $(n,c,R') \notin$ INPUT and we do not know whether $\overline{k}(n,c,R',A) \leq p(n)$.)
Let $(X_i)_{i\in N}$ be the same sequence "without the indicator (STOP,·)", i.e. a sequence with the property

$$\overline{X}_i(n,c,R',A) = \begin{cases} (\text{STOP},X_i), \text{ if } \overline{X}_i(n,c,R',A) \in \text{OUTPUT} \\ X_i \quad\quad , \text{ otherwise} \end{cases}$$

Since $\{1,\dots,n\}$ has $r+2$ subsets of cardinality m, there is a subset $F_o \subseteq \{1,\dots,n\}$ with $|F_o| = m$ which is different from $X_o,\dots,X_r$. Now let $R:2^{\{1,\dots,n\}} \to \{\text{YES},\text{NO}\}$ be defined by

$$R(F) = \begin{cases} \text{YES, if } |F| < m \text{ or } F = F_o \\ \text{NO, otherwise.} \end{cases}$$

As $X_i \neq F_o$, $0 \leq i \leq r$, an easy induction on i yields

$$\overline{X}_i(n,c,R,A) = \overline{X}_i(n,c,R',A) \quad \text{for } 0 \leq i \leq r.$$

But $(n,c,R) \in$ INPUT, thus for $k = \overline{k}(n,c,R,A) \leq p(n) \leq r$ we have:

$$\overline{X}_k(n,c,R,A) = \overline{X}_k(n,c,R',A) = (\text{STOP},X_k) \in \text{SOL}(n,c,R),$$

hence

(3) $c(X_k) > q(n,R^{-1}(\text{YES}))\max\{c(F)\,|\,F \in R^{-1}(\text{YES})\}$.

and $X_k \in R^{-1}(YES)$. On the other hand, $X_k \nsubseteq F_o$, hence $|X_k| < m = |F_o| = \max\{c(F) | F \in R^{-1}(YES)\}$. Thus we get the following contradiction to (3):

$$c(X_k) \leq \frac{m-1}{m} \cdot \max\{c(F) | F \in R^{-1}(YES)\} = q(n,R^{-1}(YES)) \cdot \max\{c(F) | F \in R^{-1}(YES)\}.$$

4. Other applications of the oracle approach

We hope to have made clear our point that the oracle approach presented in this paper can be used as a general method to examine in a precise manner the complexity of problems which are too general for the usual Turing machine approach. For example in [6] , we have applied our formal definitions of oracle problems and oracle algorithms to show that three problems which can be interpreted as the approximation of a fixed point of a continuous function cannot be solved by any (bounded) oracle algorithm without informations about the continuous function that exceed all those informations which can be derived by arbitrary many function evaluations.

We want to finish this paper with another but similar application of the oracle approach, viz. the proof that for the minimization problem over a general clutter

$$(4) \qquad \begin{array}{l} c(F) = \text{Min!} \\ \text{s.t. } F \in B \end{array}$$

there is no polynomial approximation algorithm with a constant (non-trivial) performance guarantee. In (4), B is a clutter over $\{1,\ldots,n\}$, i.e. a system of subsets of $\{1,\ldots,n\}$ such that no set $F \in B$ is a subset of a different set $F' \in B$. Note that a system of subsets of $\{1,\ldots,n\}$ is a clutter iff it is the set of all bases (i.e. maximal independent sets) of an independence system over $\{1,\ldots,n\}$.

To make our claim above more precise, we define for any real number $\alpha > 0$

$P_\alpha = (\text{INPUT}, \text{SOL}_\alpha)$ where

$$\text{INPUT} = \{(n,c,R) | n \geq 1,\ c \in \mathbb{R}^n_+,\ R : 2^{\{1,\ldots,n\}} \to \{\text{YES},\text{NO}\},\ R^{-1}(\text{YES}) \text{ is a clutter}\}$$

$$\text{SOL}_\alpha(n,c,R) = \{(\text{STOP},X) | X \in R^{-1}(\text{YES}), c(X) \leq \alpha \cdot \min\{c(F) | F \in R^{-1}(\text{YES})\}\}.$$

Theorem 2: For any real number $\alpha > 0$, there is no algorithm A for problem P_α such that the number of calls on the oracle $\bar{k}(n,c,R,A)$ is bounded by a polynomial in n.

Proof: Let $\alpha > 0$ be given and suppose that A is such an algorithm and p is a polynomial such that $\bar{k}(n,c,R,A) \leq p(n)$ for any $(n,c,R) \in \text{INPUT}$. Then there is an integer n such that

$$\binom{n-1}{[(n-1)/2]} - 2 \geq p(n) \quad \text{and} \quad [(n-1)/2] - 1 > 0.$$

Let $m = [(n-1)/2]$, $r = \binom{n-1}{m} - 2$ and $c = (\alpha m, 1,1,\ldots,1) \in \mathbb{R}^n$.
Let $R' : 2^{\{1,\ldots,n\}} \to \{\text{YES},\text{NO}\}$ be defined by

$$R'(F) = \begin{cases} \text{YES, if } 1 \in F \text{ and } |F| = m \\ \text{NO, otherwise.} \end{cases}$$

Let $(\bar{X}_i(n,c,R',A))_{i\in N}$ be the corresponding sequence of oracle arguments and $(X_i)_{i\in N}$ the same sequence without the indicator (STOP,$\cdot$). Obviously there exists a subset F_o different from $X_o,\ldots,X_r$ such that $1 \notin F_o$ and $|F_o| = m$.
Let R be defined by

$$R(F) = \begin{cases} \text{YES, if } (1 \in F,\ |F| = m) \text{ or } F = F_o \\ \text{NO, otherwise.} \end{cases}$$

By arguments similar to those of the proof of Theorem 1, we get for $k = \bar{k}(n,c,R,A) \leq p(n) \leq r$:

$$\bar{X}_k(n,c,R,A) = \bar{X}_k(n,c,R',A) = (\text{STOP},X_k) \in \text{SOL}(n,c,R)$$

hence

$$(5) \quad c(X_k) \leq \alpha \cdot \min\{c(F)\,|\,F \in R^{-1}(\text{YES})\} \leq \alpha \cdot c(F_o) = \alpha \cdot |F_o| = \alpha m$$

and $X_k \in R^{-1}(\text{YES})$ which together with $X_k \neq F_o$ implies $c(X_k) = \alpha m + m-1$, a contradiction to (5).

References

[1] A.V. Aho, J.E. Hopcroft, J.D. Ullmann, The Design and Analysis of Computer Algorithms, Addison Wesley, Reading, Mass. (1974).

[2] Th. Baker, J. Gill, R. Solovay, "Relativizations of the P=? NP question", SIAM Journal on Computing 4 (1975) 431-442.

[3] S.A. Cook, "The complexity of theorem-proving procedures", Proceedings of the Third Annual ACM Symposium on the Theory of Computing, Shaker Heights, Ohio (1971) 151-158.

[4] J. Edmonds, "Matroids and the greedy algorithm", Mathematical Programming 1 (1971) 127-136.

[5] D. Hausmann, B. Korte, "Lower bounds on the worst-case complexity of some oracle algorithms", Working Paper No. 7757-OR, Institut für Ökonometrie und Operations Research, University of Bonn (1977).

[6] D. Hausmann, B. Korte, "Oracle algorithms for fixed-point problems - an axiomatic approach", Working Paper No. 7766-OR, Institut für Ökonometrie und Operations Research, University of Bonn (1977), to be published in: R. Henn, B. Korte, W. Oettli (ed.), Operations Research und Optimierung, Springer-Verlag, Berlin, Heidelberg, New York (1978).

[7] B. Korte, D. Hausmann, "An analysis of the greedy heuristic for independence systems", to be published in: Annals of Discrete Mathematics, Vol 2, Proceedings of the Conference on Algorithmic Aspects of Combinatorics, Qualicum Beach, B.C. (1976).

AN IMPROVED METHOD OF SUCCESSIVE OPTIMA FOR THE ASSIGNMENT PROBLEM

László Mihályffy,
SZÁMKI Research Institute for Applied Computer Sciences,
P. O. B. 227, H-1536 Budapest, Hungary

1. Introduction. The problem of the optimal assignment (abridged AP) is one of the best known combinatorial optimization problems; among its usual formulations the following one is perhaps the simplest:

Find an $n \times n$ permutation matrix $X = (x_{ij})$

minimizing the linear function $\sum_{i=1}^{n} \sum_{j=1}^{n} c_{ij} x_{ij}$

where $C = (c_{ij})$ is a given $n \times n$ matrix with nonnegative entries. There are many efficient algorithms for solving AP (see e. g. [1], [2], [3] and [5]); in this paper a further one called the method of successive optima (abridged MSO) is given. (An earlier version of MSO was presented at the IXth International Symposium on Mathematical Programming, Budapest, August 1976; hence the attribute "improved" in the title.) The most conspicuous feature of MSO is its simplicity; its computational efficiency is also quite agreeable.

MSO is based on the simple notions of approximate solutions of AP and chain matrices; these are introduced in Definitions 2.2 and 2.5 below, respectively. Starting from a suitable initial approximate solution $X^o = (x^o_{ij})$ the method produces a sequence

$$(1.1) \qquad X^o, X^{(1)}, \ldots, X^{(q)}$$

of approximate solutions such that $q \leq n-1$, $X^{(i)} - X^{(i-1)}$ is a chain matrix for $i = 1, 2, \ldots, q$, and $X^{(q)}$ is an optimal solution of AP. The device for determining any member $X^{(i)}$ of this sequence is a simple routine called the scheme of reduced minors of the cost matrix C; this is described in Definition 2.7.

Owing to space limitations no complete verification of MSO is given in the paper. In place of this only a brief sketch of the verification is provided, which is composed partly of some lemmas and theorems concerning the properties of approximate solutions, chain matrices and reduced minors, partly of some additional remarks in the last section of the paper. The proofs of the statements are not reproduced here; they can be found in [4].

2. Terminology, definitions. For the sake of brevity the notation $F(X) = \sum_{i=1}^{n} \sum_{j=1}^{n} c_{ij}x_{ij}$ will be used throughout the paper.

2.1. Definition. An $n \times n$ matrix $X = (x_{ij})$ is called a degenerate permutation matrix if it can be obtained fron an $n \times n$ permutation matrix by replacing certain positive entries, i. e. certain 1's by zeroes. If X is a degenerate permutation matrix with ℓ positive entries $(0 \leq \ell \leq n)$, the degree of degeneracy is defined as $p = n - \ell$.

2.2. Definition. A degenerate permutation matrix $X = (x_{ij})$ with degree p is an approximate solution of AP if $F(X) \leq F(X')$ holds for any degenerate permutation matrix $X' = (x'_{ij})$ with degree p.

2.3. Corollary. If X and Y are approximate solutions of AP, p and q are the corresponding degrees of degeneracy, and $p < q$, then, because of the nonnegativity of c_{ij} we have $F(X) \geq F(Y)$.

The sequence (1.1) produced by MSO is such that the degrees of degeneracy belonging to the approximate solutions constitute a strictly decreasing sequence. At any stage of the process the current approximate solution is supposed to be of the form

$$X^{(i)} = \left(\begin{array}{c|c} 0 & 0 \\ \hline 0 & I \end{array}\right) \tag{2.1}$$

where I denotes unit matrix of order $n-p$ and p is the degree of degeneracy of $X^{(i)}$; this involves in practice suitable rearrangements of rows and columns of the cost matrix C.

2.4. <u>Definition</u>. Let $N = \{1, 2, \ldots, n\}$ and let t be a non-negative integer. The sequence

$$(i_1, j_1), (i_2, j_2), \ldots, (i_{2t}, j_{2t}), (i_{2t+1}, j_{2t+1})$$

of pairs from $N \times N$ represent a chain if the following conditions are fulfilled:

- either $i_1 = i_2$, $i_3 = i_4$, $\ldots$, $i_{2t-1} = i_{2t}$ and $j_2 = j_3$, $j_4 = j_5$, $\ldots$, $j_{2t} = j_{2t+1}$, or $i_2 = i_3$, $i_4 = i_5$, $\ldots$, $i_{2t} = i_{2t+1}$ and $j_1 = j_2$, $j_3 = j_4$, $\ldots$, $j_{2t-1} = j_{2t}$;
- for an arbitrary k from N there are at most two elements of the sequence $\{i_1, i_2, \ldots, i_{2t+1}\}$ that are equal to k, and the same applies to the sequence $\{j_1, j_2, \ldots, j_{2t+1}\}$.

2.5. <u>Definition</u>. An $n \times n$ matrix $U = (u_{ij})$ is a <u>chain matrix</u> if $(i_1, j_1), \ldots, (i_{2t+1}, j_{2t+1})$ is a chain in the above sense, and

$$u_{ij} = \begin{cases} 1 & \text{if } (i, j) = (i_k, j_k) \text{ for some odd } k, \; 1 \le k \le 2t+1, \\ -1 & \text{if } (i, j) = (i_k, j_k) \text{ for some even } k, \; 2 \le k \le 2t, \\ 0 & \text{otherwise.} \end{cases}$$

By this definition for any chain matrix U there are positive integers i and j such that the ith row total and the jth column total of U are equal to 1, and any row total of U other than the ith as well as any column total of this matrix other than the jth vanishes. For the 6×6 chain matrices shown in Figure 1 $i = 1$, $j = 2$.

	1				

(a)

		1			
	1	-1			

(b)

		1			
		-1		1	
	1			-1	

(c)

Figure 1

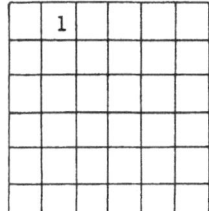

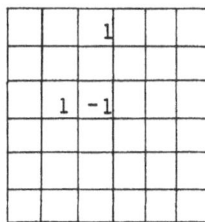

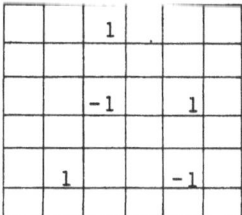

With any approximate solution $X = (x_{ij})$ of AP, or, in general, with any degenerate permutation matrix X a set M consisting of all chain matrices U such that $X + U$ is a degenerate permutation matrix is associated. It is easy to see that M is not empty, and for $U \in M$ the degree of degeneracy of $X + U$ is one less than that of X.

2.6. Lemma. Let X and M be an approximate solution of AP and the corresponding set of chain matrices, respectively. Then there is a $U_o \in M$ such that $X + U_o$ is an approximate solution of AP. (For $U \in M$ $X + U$ is, of course, in general no approximate solution of AP.)

Given an approximate solution X of AP, and thereby the corresponding set M of chain matrices, the following subproblem (SP) is formulated:

Find $U \in M$ that minimizes the linear function $F(U)$.

This subproblem is solved by means of the so-called reduced minors of the cost matrix C. In the following definition p denotes the degree of degeneracy of the current approximate solution X, which is supposed to be of the form (2.1), and $c_{ij}^{(o)} = c_{ij}$ for all $(i, j) \in N \times N$.

2.7. Definition. For $k = 1, 2, \ldots, n-p$ $C^{(k)}$, the kth reduced minor of C is an $(n-k)\times(n-k)$ matrix defined as follows:

$$(2.2) \qquad c_{ii}^{(k)} = c_{ii}^{(k-1)}, \quad i = p+1, p+2, \ldots, n-k;$$

$$(2.3) \qquad c_{ij}^{(k)} = \min\{c_{ij}^{(k-1)}, \; c_{i,n-k+1}^{(k-1)} - c_{n-k+1,n-k+1}^{(k-1)} + c_{n-k+1,j}^{(k-1)}\}$$

for $1 \le i, j \le n-k$, $(i, j) \neq (p+1, p+1), \ldots, (n-k, n-k)$.
(In case of $k = n-p$ (2.2) is to be ignored.)

2.8. Theorem. The value of the optimum of the above subproblem is equal to the minimum of the entries of $C^{(n-p)}$, the last reduced minor of the cost matrix C, i. e.

$$(2.4) \qquad c_{k\ell}^{(n-p)} = F(U^o)$$

where $c_{k\ell}^{(n-p)} = \min_{1\le i,\, j\le p} c_{ij}^{(n-p)}$, and U^o is an optimal solution of SP.

In what follows it is shown how to find U^o provided that the left-hand side of (2.4) is already known; this is done by means of a numerical example.

19	20	16	18	16	17
19	20	16	16	14	17
17	18	(14)	14	11	19
18	18	15	(13)	14	16
16	15	14	14	(10)	13
16	16	12	17	11	(12)

(a)

19	20	16	18	16
19	20	16	16	14
17	18	14	14	11
18	18	15	13	14
16	15	13*	14	10

(b)

19	20	16	18
19	19*	16	16
17	16*	14	14
18	18	15	13

(c)

19	20	16
19	19	16
17	16	14

(d)

19	18*
19	18*

(e)

Figure 2

The encircled entries of the cost matrix C in Fig. 2a represent an approximate solution of the corresponding AP; thus in the present case $n = 6$, $p = 2$. In Figs. 2b - 2e the reduced minors C', C'', C''' and C^{IV} of C are displayed; the asterisks indicate the fact that the application of (2.3) implied $c_{ij}^{(k)} < c_{ij}^{(k-1)}$ for the corresponding cells. The minimum of the entries of the last reduced minor, that is now C^{IV}, is 18; it is attained for c_{12}^{IV} and c_{22}^{IV}. For the sake of unambiguity let us consider c_{12}^{IV}. The asterisk shows that - by virtue of (2.3) - $c_{12}^{IV} = c_{13}''' - c_{33}''' + c_{32}'''$. In C''' (see Fig. 2d) there are no asterisks at all, and therefore $c_{12}^{IV} = c_{13}'' - c_{33}'' + c_{32}''$. Considering C'' (i. e. Fig. 2c) we see that c_{32}'' (= 16) has an asterisk, while c_{13}'' and c_{33}'' are not marked. Again with reference to (2.3), one concludes that $c_{12}^{IV} = c_{13}' - c_{33}' + c_{35}' - c_{55}' + c_{52}'$, and finally, by similar reasoning the result $c_{12}^{IV} = c_{13} - c_{33} + c_{35} -$

$c_{55} + c_{52}$ is obtained. The latter expression is the objective function value belonging to the matrix U of Fig. 1c, and, by Theorem 2.8, this matrix is a solution of the subproblem under the circumstances.

2.9. Lemma. The result of the procedure shown in the example is always a chain matrix.

3. The algorithm. In what follows the observations made in the preceding section are summarized. The result is an algorithm for solving AP; however, owing to certain technical considerations this algorithm needs some further refinement. The result of this improvement is a "revised" algorithm, and only this is recommended for practical purposes.

To describe the algorithm it is useful to introduce the notation $S = \{(p+1, p+1), (p+2, p+2), \ldots, (n, n)\}$. This set represents obviously the current approximate solution (2.1) and it is very advantageous to use S also to register the individual stages that U (i. e. the solution of SP) passes through. This means that after finishing the construction of U immediately the new approximate solution X + U is obtained. It should also be noted that there are many possibilities to choose a suitable initial approximate solution to AP; the particular one described in Step 1 of the algorithm is far from being the best, but by all means it is one of the simplest.

ALGORITHM:

1. Find a minimal entry c_{ij} of the cost matrix. Interchange rows i and n, as well as columns j and n. Set $p = n-1$.
2. Using (2.2) - (2.3) compute the reduced minors $C', C'', \ldots, C^{(n-1)}$. Give a mark to $c_{ij}^{(r)}$ if it is strictly smaller than $c_{ij}^{(r-1)}$.
3. Find a minimal entry $c_{ij}^{(n-p)}$ of $C^{(n-p)}$. Set $S = S \cup \{(i,j)\}$ and $r = n-p$.
4. Check S for a (k, ℓ) such that $c_{k\ell}^{(k)}$ has a mark. (None of the pairs $(p+1, p+1), (p+2, p+2), \ldots, (n, n)$ can have a mark.) If there is no such (k,ℓ), diminish r by 1, and repeat this step unless $r = 0$. If $r = 0$, go to step 6.
5. (A marked entry $c_{k\ell}^{(r)}$ has been found.) Update S in the fol-

lowing manner: $S = S \cup [(k, n-r+1), (n-r+1, \ell)\} \setminus \{(k, \ell), (n-r+1, n-r+1)\}$ (see the above example), and diminish r by 1. Go to step 4 if $r \neq 0$, and go to step 6 if $r = 0$.

6. S represents now $X + U$, the new approximate solution. The rows and columns of C should be rearranged in order that $X + U$ may be of the form (2.1). Interchange first rows i and p as well as columns j and p, and replace simultaneously (i, ℓ) and (k, j) in S by (p, ℓ) and (k, p), respectively. After this interchange rows k and ℓ for every $(k,\ell) \in S$ such that $k \neq \ell$. Diminish p by 1, and go to step 2 unless $p = 0$. (The original form of S should, of course, be restored.) If $p = 0$, an optimal solution of AP has been found.

This algorithm has a serious drawback, namely, it works only if C and its reduced minors $C', C'', \ldots, C^{(n-p)}$ are stored simultaneously in the memory of the computer. However, at the cost of some extra computation the reduced minors of C can be replaced by a single $n \times n$ matrix $\hat{C} = (\hat{c}_{ij})$ that is defined as follows: for $i, j = 1, 2, \ldots, n$

$$\hat{c}_{ij} = c_{ij}^{(n-t)} \qquad \text{where} \quad t = \max\{i, j, p\}.$$

Provided that both C and $\hat{C}$ are available, the above algorithm can be modified as follows:

(i) in step 2 the reduced minors should be computed and stored in such a manner that $C^{(r-1)}$ may partly be rewritten by $C^{(r)}$ $(r = 1, 2, \ldots, n-p)$. No marking is needed;

(ii) the term "$c_{k\ell}^{(r)}$ has a mark" should be replaced by the condition

$$c_{k\ell}^{(r)} < c_{k\ell}^{(r-1)};$$

(iii) for $(k, \ell) \in S$ the comparison of $c_{k\ell}^{(r)}$ and $c_{k\ell}^{(r-1)}$ should be accomplished by means of repeated application of (2.3): $c'_{k\ell} = \min\{c_{k\ell}, c_{kn} - c_{nn} + c_{n\ell}\}$, $c''_{k\ell} = \min\{c'_{k\ell}, c'_{k,n-1} - c'_{n-1,n-1} + c'_{n-1,\ell}\}$, etc. (All what is needed to this computation is available from $\hat{C}$.)

The memory requirement of the above algorithm becomes obviously reasonable if these improvements are taken into account. An easy counting shows that $7n^4/8$ is an upper bound of the arithmetic operations necessary to solve AP by the modified algorithm.

4. Sketch of verification. The correctness of the above algorithm depends obviously in the first place on Theorem 2.8 and on Lemma 2.9. To give a general view of the arguments used to verify MSO it is sufficient to consider here only Theorem 2.8; this will be reduced below to two simpler statements.

4.1. Lemma. Suppose that the current approximate solution of AP is of the form (2.1), and let U be an arbitrary chain matrix from M. Then there are indices $1 \le i, j \le p$ and there is an ordered set S of integers $i_1, i_2, \ldots, i_s$ belonging to the closed interval $(p+1, n)$ such that $u_{k\ell} = 1$ for $(k, \ell) = (i, i_1), (i_1, i_2), (i_2, i_3), \ldots, (i_{s-1}, i_s), (i_s, j)$, $u_{k\ell} = -1$ for $(k, \ell) = (i_1, i_1), (i_2, i_2), \ldots, (i_s, i_s)$, and $u_{k\ell} = 0$ for all other pairs of indices. The case $S = \emptyset$ is not excluded.

4.2. Definition. Let $T = \{i\ , i_2, \ldots, i_t\}$ be a nonempty ordered set of integers not exceeding n. A matrix $V = (v_{ij})$ is called a "loop" matrix if $v_{k\ell} = 1$ for $(k, \ell) = (i_1, i_2), (i_2, i_3), \ldots, (i_{t-1}, i_t), (i_t, i_1)$, $v_{k\ell} = -1$ for $(k, \ell) = (i_1, i_1), (i_2, i_2), \ldots, (i_t, i_t)$, and otherwise $v_{k\ell} = 0$.

In what follows H denotes the set of all loop matrices such that $p+1 \le i_1, i_2, \ldots, i_t \le n$. In the following two auxiliary theorems the assumptions of Lemma 4.1 are supposed to be valid, and the notations of this lemma are used. $c_{ij}^{(n-p)}$ denotes a proper entry of the (n-p)th reduced minor of C (see Def. 2.7).

4.3. Lemma. Let U be an arbitrary element of M, that is determinded by the pair (i, j) and the ordered set $S = \{i_1, \ldots, i_s\}$. U satisfies the following inequality: $c_{ij}^{(n-p)} \le F(U)$.

4.4. Lemma. For $i, j = 1, 2, \ldots, p$ there is a $U \in M$ and there are $V_1, V_2, \ldots, V_r \in H$ such that

$$c_{ij}^{(n-p)} = F(U) + F(V_1) + \ldots + F(V_r).$$

Considering that under the circumstances $F(U) \ge 0$, $F(V_1) \ge 0$, $\ldots$, $F(V_r) \ge 0$ (cf. Def. 2.2 and Cor. 2.3) these lemmas imply Theorem 2.8. The following notion is the key to their proof.

4.5. Definition. Let $U \in M$ determined by (i, j) and $S = \{i_1, \ldots, i_s\}$ in the sense of Lemma 4.1. Let $S \neq \emptyset$ and let i_k be the greatest element of S. Finally, let U' be the chain matrix determined by (i, j) and $S' = \{i_1, \ldots, i_{k-1}, i_{k+1}, \ldots, i_s\}$; U' is obviously an element of M. The assignment $U \rightarrow U'$ is called contraction.

Example: the chain matrix of Fig. 1b is obtained by contraction from the chain matrix of Fig. 1c, and the same is true for the chain matrices of Figs. 1a and 1b.

Between contraction and the reduced minors of the cost matrix there is a close connection, that can easily be discovered and can be exploited to prove Lemmas 4.3, 4.4 and 2.9. In the course of the verification a certain "inverse operation" of contraction is also needed; this is, however, somewhat complex and is not described here.

REFERENCES

[1] Balinski, M. L. - R. E. Gomory: A primal method for the assignment and transportation problems. Management Science 10 (1964) 578-597.

[2] Barr, R. S., F. Glover, D. Klingman: The alternating basis algorithm for assignment problems. Math. Programming 13 (1977) 1-13.

[3] Komáromi, É.: A finite primal method for the assignment problem. Problems of Control and Information Theory 3 (1974) 157-166.

[4] Mihályffy, L.: An improved method of successive optima for solving the assignment problem (in Hungarian). SZÁMKI Tanulmányok (Studies published by the Research Institute for Applied Computer Sciences), Budapest. To appear in 1977.

[5] Munkres, J.: Algorithms for the assignment and transportation problems. J. Soc. Indust. Appl. Math. 5 (1957) 32-38.

ACCELERATED GREEDY ALGORITHMS
FOR MAXIMIZING SUBMODULAR SET FUNCTIONS

Michel MINOUX
Ecole Nationale Supérieure de
Techniques Avancées - PARIS - France

Abstract :

Given a finite set E and a real valued function f on $\mathcal{P}(E)$ (the power set of E) the optimal subset problem (P) is to find $S \subseteq E$ maximizing f over $\mathcal{P}(E)$. Many combinatorial optimization problems can be formulated in these terms. Here, a family of approximate solution methods is studied : the greedy algorithms.

After having described the standard greedy algorithm (SG) it is shown that, under certain assumptions (namely : submodularity of f) the computational complexity of (SG) can often be significantly reduced, thus leading to an accelerated greedy algorithm (AG). This allows treatment of large scale combinatorial problems of the (P) type. The accelerated greedy algorithm is shown to be optimal (interms of computational complexity) over a wide class of algorithms, and the submodularity assumption is used to derive bounds on the difference between the greedy solution and the optimum solution.

I - INTRODUCTION

Among the great variety of existing methods for the solution of combinatorial optimization problems, the so-called "greedy algorithms" are conceptually the simplest and perhaps the most widely used in practise. One of the first algorithm of this type seems to be the famous KRUSKAL'S algorithm for the minimum spanning tree problem [4] . More generally, EDMONDS [5] has shown that these methods provide optimal solutions for the class of problems involving the search for a maximum (minimum) weight basis of a matroïd.

Most of real-world problems, of course, do not belong to this class, but it has long been noticed by many authors [1][10][11] that greedy algorithms often lead to very good approximate solutions. These observations had not been explained until a recent publication [15] where it is shown that the quality of greedy solutions are strongly related to a specific property of set functions called : submodularity.

Our purpose, here, is to carry on the study of combinatorial problems involving the maximization of submodular set functions. In particular we show that, when the submodularity property holds, the number of computations necessary to get a greedy solution can often be significicatly reduced, and we describe an "accelerated" greedy algorithm which allows the treatment of large scale combinatorial problems of type (P).

Computational experiments illustrate the efficiency of the procedure. For example, optimum network problems on graphs involving about 180 nodes and 500 edges can be solved within a few minutes, using the accelerated greedy algorithms ; for the same result a standard greedy algorithms would require several hours of computations on the same computer (see [1]).

These experimental results are confirmed theoretically by proving that the accelerated greedy algorithms is optimal, in terms of computational complexity, among a wide class of algorithms (see section IV).

Finally, it is shown how submodularity can be used to derive bounds for the difference between the greedy solution and the optimum solution of a given problem. These bounds, being specific of each particular problem, are usually tighter than those in [15] .

II - THE OPTIMAL SUBSET PROBLEM : A GENERAL FORMULATION OF COMBINATORIAL OPTIMIZATION PROBLEMS.

Let $E = (e_1 , e_2 \dots e_n)$ be a finite set with n elements and $f : \mathcal{P}(E) \to \mathbb{R}$ a real function on $\mathcal{P}(E)$ the power set of E. Such a function will be called a set function.

The optimal subset problem (called problem (P)) is to determine a subset S^* of E such that :

$$f(S^*) = \max_{S \subset E} f(S)$$

It is easily shown that (P) is equivalent to the general mixed or pure integer program with bounded variables.

In view of this fact we shall restrict ourselves to problem (P). First, let us take a few examples :

Example 1 : minimum spanning tree [4]

$G = [X, U]$ is a (unoriented) connected graph with node set X and edge set U.

Associated with each edge $u \in U$ of G is a real number w(u) called the weight of u.

Let $S \subset U$ be a subset of edges, and $w(S) = \sum_{u \in S} w(u)$.

We want to determine a minimum spanning tree of G. If we take E as the set of edges U, and :

. $f(S) = -w(S)$ if $G_S = [X, S]$ is connected

. $f(S) = -\infty$ otherwise,

the problem is converted into (P).

Example 2 : center location problem : [8] [9] [10] .

This type of problem is frequently encountered in practise. In its simplest form, it may be stated as follows.

A set $I = \{1, 2, \ldots M\}$ of M customers must be supplied from a certain (unknown) set of centers. A (finite) list of all possible locations for the centers : $E = \{e_1, e_2 \ldots . e_n\}$ is given.

For $j \in J$, $\alpha_j > 0$ is the cost of installation of a center at location e_j .

For every $i \in I$, and for $e_j \in E$, the cost $\gamma_{ij} > 0$ of assignment of customer i to location e_j is given.

If the subset $S \subset E$ of location is chosen, the total cost of the solution is the sum of :

. the cost of the centers : $\alpha(S) = \sum_{e_j \in S} \alpha_j$

. the cost $\gamma(S)$ of the assignment of customers to centers :

$$\gamma(S) = \sum_{i \in I} \underset{e_j \in S}{\text{Min}} \{\gamma_{ij}\}$$

The problem is then to find an optimal subset of centers $S \subset E$ such that the cost is minimal.

This problem is easily recognized as a problem (P) with function f defined by :

$$f(S) = -[\alpha(S) + \gamma(S)] \qquad \forall S \subset E$$

Example 3 : the optimum network problem [1][7]

Let $G = [X, U]$ be a connected (unoriented) graph with node set X and edge set U.

With each edge $u \in U$ we associate two numbers :

$\delta(u)$ = fixed cost of link u

$\rho(u)$ = cost of link u per unit capacity.

$(\rho(u) \geq 0 \ \forall u)$.

For any $S \subset U$, the cost of subgraph : $G_S = [X, S]$ of G is the sum of two costs.

. A fixed cost $\delta(S) = \sum_{u \in S} \delta(u)$

. A variable cost $\rho(S) = \sum_{i \neq j} \lambda_{ij} \cdot l_{ij}(S)$

where $l_{ij}(S)$ is the length of the shortest chain between nodes i and j in G_S, with lengths $\rho(u)$ on the edges. ($l_{ij}(S) = +\infty$ if i and j are not connected in sub-graph G_S). The λ_{ij} are given non negative point to point capacity requirements. The problem is to determine a subgraph $G_{S^*} = [X, S^*]$ for which the cost is minimum. Many problems of network engineering (computer networks, telephone networks, transportation networks ...) may be formulated in this way.

It reduces to problem (P) by taking E = U (the edge set of G) and by defining function f by :

$$f(S) = -\left[\delta(S) + \rho(S)\right] \qquad \forall S \subset U$$

The optimum network problem belongs to the class of the so-called : fixed cost linear programming problems [11] [12] [13] ; these may be similarly re-formulated as type (P) problems.

III - THE STANDARD GREEDY ALGORITHM

In this section we show how greedy algorithms can be used to provide exact or approximate solutions to problems (P).

Starting with solution $S = \emptyset$, for instance, a subset S of E is grown, little by little, by successively adding some elements of E which do not belong to the current S. At each step, that element e_i is added for which the increase in cost :

$f(S + \{e_i\}) - f(S)$ is maximum.

This may be summarized by the following procedure.

Standard greedy algorithm (SG)

(a) Take $S^\circ = \emptyset$; itération k = 0

(b) at step k, S^k is the current solution of cost $f(S^k)$ and $|S^k| = k$

(c) for all $e_i \in E - S^k$, compute :

$\Delta^k(e_i) = f(S^k + \{e_i\}) - f(S^k)$.

(d) select e_{io} such that $\Delta^k(e_{io}) = \underset{e_i \in E-S^k}{\text{Max}} \Delta^k(e_i)$.

if $\Delta^k(e_{io}) \leqslant 0$ Terminated = the current solution S^k is (loccaly) optimal.

otherwise :

(e) Set $S^{k+1} \leftarrow S^k + \{e_{io}\}$

$k \leftarrow k + 1$

and return to (b).

The solution determined by (SG) will be called a greedy solution of (P). Notice that, unless the minimum $\Delta^k(e_{io})$ is unique at each step k of the algorithm, the greedy solution is not necessarily unique.

For the minimum spanning tree problem, [4] (SG) is recognized as KRUSKAL II algorithm.Thus there exists a class of problems for which (SG) produces an optimum solution : these are one-matroid optimization problems (involving the search for a maximum or minimum weight basis of a matroid).

Though most combinatorial problems arising in practise do not belong to this class, it has long been recognized that greedy algorithms often lead to very good approximate solutions. This is the case for example 2 above (locations of centers [8] [9] [10] [12]), example 3 (the optimum network problem [1] [7]) and more generally, for some classes of fixed charge linear programming problems [11] [12] [13] (example 3). See [1] for some computational results.

The computational complexity of (SG) can easily be evaluated, in terms of the number of calculations of f.

At the first step of the algorithm, $S^\circ = \emptyset$ and the number of such calculations is : n . At the second step : n - 1 . At the kth step , n - k + 1 . Hence, the total number of calculation may be bounded above by =

$$n + (n - 1) + (n - 2) \dots + 1 = \frac{n\,(n + 1)}{2}$$

And the computational complexity of (SG) is $\mathcal{O}(n^2)\ .\ \mathcal{O}(f)$ where $\mathcal{O}(f)$ is the complexity of one f(S) calculation.

For the center location problem (example 2) with N centers and M customers :

$\mathcal{O}(f) = M.N$, hence (SG) is $\mathcal{O}(M\ N^3)$

For the optimum network problem (example 3) the calculation of f(S) for any given S involves the determination of all shorst paths in a N-node unoriented graph, thus : $\mathcal{O}(f) = \mathcal{O}(N^3)$. Then, if M is the number of edges, (SG) is : $\mathcal{O}(M^2\ N^3)$.

These examples show that, even though greedy algorithms are usually much faster than enumerative methods (like Branch and Bound), they may prove inefficient in solving large scale problems of type (P).

IV - AN ACCELERATED GREEDY ALGORITHM

In this section, we show how the efficiency of the standard greedy algorithm can be improved thus leading to an "accelerated greedy algorithm" (AG). When function f is <u>submodular</u> , we prove that the solutions given by (SG) and (AG) are identical.

A set function $f : \mathcal{P}(E) \rightarrow \mathbb{R}$ is said to be <u>submodular</u> if and only if, for every $S \subset E$, $T \subset E$:

$$f(S) + f(T) \geqslant f(S \cup T) + f(S \cap T). \qquad (1)$$

(see [20])

We now give another more useful characterization of submodular set functions in terms of the differences :

$$\delta(A,\ e_i) = f(A + \{e_i\}) - f(A).$$

<u>Theorem 1</u> :

the two following properties are equivalent :

- function $f : \mathcal{P}(E) \rightarrow \mathbb{R}$ is <u>submodular</u> ;
- for every $A \subset E$, $B \subset A$; $e_i \notin A$:

$$\delta(A,\ e_i) \leqslant \delta(B,\ e_i) \qquad (2)$$

Proof : see [2] or [15] for instance.

For set functions, submodularity plays a role which is, in a sense, similar to that of concavity for ordinary functions.

In fact, relation (2) can be interpreted by saying that the benefit incurred by adding element e_i to a given solution A can only decrease if others elements are first added to A.

It can be shown (cf [2]) that submodularity holds for example 1 (minimum spanning tree) and example 2 (optimal location of centers) above, and for a number of other important combinatorial problems (see ref [15] for a list of such examples).

It is easily seen that it does not hold for general fixed charge linear programming problems. However for the optimum network problem it can be shown [1][2] that there exists a submodular function f' which closely approximates f.

The accelerated greedy algorithm (AG)

(a) Take $S^\circ = \emptyset$ as a starting solution
step $k = 0$

(b) for every $e_i \in E$, compute :

$$\Delta(e_i) = f(\{e_i\}) - f(\emptyset)$$

(c) At step k, let S^k be the current solution, of cost $f(S^k)$

Select $e_{io} \in E - S^k$ such that :

$$\Delta(e_{io}) = \operatorname*{Max}_{e_i \in E - S^k} \{\Delta(e_i)\}$$

If e_{io} has already been selected once at step k set : $\delta = \Delta(e_{io})$ and go to (e)

(d) compute :

$$\delta = f(S^k + \{e_{io}\}) - f(S^k)$$

and set : $\Delta(e_{io}) \leftarrow \delta$

if $\delta < \operatorname*{Max}_{\substack{e_i \in E - S^k \\ e_i \neq e_{io}}} \{\Delta(e_i)\}$

return to (c)

otherwise :

(e) if $\delta \leq 0$ STOP : solution S^k is (locally) optimal - Otherwise ($\delta > 0$) :

(f) Set : $S^{k+1} \leftarrow S^k + \{e_{io}\}$

$\Delta(e_{io}) = 0$

$k \leftarrow k + 1$

and return to (c).

Theorem 2 (convergence of (AG))

When function f is submodular, the accelerated greedy algorithm (AG) produces a greedy solution.

Proof : refer to [2].

Remark : if the greedy solution is unique, then the solutions produced by (SG) and (AG) are identical.

Applied to the optimum network problem (example 3) (AG) requires on the average only 2 or 3 calculations of f at each level. In this case the computational complexity : $\frac{n^2}{2}\mathcal{O}(f)$ reduces to about $3n.\mathcal{O}(f)$, which shows an improvement by a factor n/6 (on the average). It follows that the accelerated greedy algorithm can be used to solve large scale combinatorial problems of the (P) type, for which even standard greedy algorithms would prove inefficient. For instance, examples of optimum network problems on graphs with about 180 nodes and 500 edges have been solved within a few minutes by (AG) (*). (See [1][2] for computational results). On the same problems, (SG) would have been 50 to 100 times slower !

The above results provide an experimental confirmation of the efficiency of (AG) when applied to some real problems of the (P) type. Unfortunately,these cannot be demonstrated theoretically and it is easy to build worst-case examples for which (AG) requires the same number of computations as (SG).

However, what will be shown next is that these are, in a sense, the best possible results, by proving that, among a wide class of algorithms, (AG) is optimal in terms of number of calculations.

For any $S \subset E$, define a neighbourhood of S as a set of solutions $S' \subset E$ such that : $S' \supset S$ and $|S'| = |S| + 1$, and let's design by $\mathcal{L}$ the class of local optimization algorithms (LOCO algorithms as termed by EDMONDS [5]) :

(a) step $k = 0$, $S^\circ = \emptyset$ (starting solution)

(b) at step k , S^k is the current solution. Determine S^* such that :

$$f(S^*) = \underset{S \in \mathcal{N}(S^k)}{\text{Max}} f(S)$$

where $\mathcal{N}(S^k)$ is any neighbourhood of S^k.

(c) if $f(S^*) \leqslant f(S^k)$ STOP. Otherwise :

set : $S^{k+1} \leftarrow S^*$, $k \leftarrow k + 1$ and return to (b).

We now state :

Theorem 3 :

Algorithm (AG) is optimal in the class of LOCO algorithms ie : no other algorithm in $\mathcal{L}$ can provide a greedy solution to any submodular maximization problem with fewer computations of f.

<u>Proof</u> : Let (P1) be any submodular maximization problem, and suppose that (AG) requires p computations of f for getting a greedy solution. Consider an algorithm (A') $\in \mathcal{L}$ which computes a greedy solution of (P1) within a number $q < p$ of f computations and let's show that there exists at least one submodular maximization problem for which (A') doesn't provide a greedy solution. Since (A') performs fewer computation than (AG), there is at least one level at which the number of computations is lower ; let k be the first level at which this occurs. At this level, (A') will add an element e_{io} such that :

$\Delta^k(e_{io})$ is not the maximum of all $\Delta(e_i)$. Hence, there exists $e_{i1} \neq e_{io}$, with $\Delta(e_{i1}) > \Delta^k(e_{io})$.

Since the submodularity condition only requires that $\forall e_i$: $\Delta(e_i) \geqslant \Delta^{\ell}(e_i)$ at each step $t \geqslant \ell$ we can build a new problem (P'1) which consists in maximizing a submodular function f' such that :

(a) $\Delta'^{\ell}(e_i) = \Delta^{\ell}(e_i)$

$\forall \ell = 1 \ldots k, \quad \forall e_i \notin S^{\ell} + \{e_{i1}\}$

(b) $\forall \ell = 1 \ldots k : \Delta'^{\ell}(e_{i1}) = \Delta(e_{i1})$ at step ℓ of (AG) applied to problem (P1); this implies in particular :

$$\Delta'^k(e_{i1}) > \Delta'^k(e_{io}) = \Delta^k(e_{io})$$

(c) $f'(S^k + \{e_{io}\} + \{e_j\}) = -\infty \quad \forall e_j \notin S^k + \{e_{io}\}$

Applied to (P'1), algorithm (A') would stop at solution $S^k + \{e_{io}\}$, when at the same time (AG) would get a greedy solution $S^* \supset S^k + \{e_i\}$, hence such that : $S^* \neq S^k + \{e_{io}\}$ and :

$$f'(S^*) \geqslant f'(S^k + \{e_{i1}\}) = f'(S^k) + \Delta'^k(e_{i1}) > f'(S^k) + \Delta'^k(e_{io}) = f'(S^k + \{e_{io}\})$$

This clearly demonstrates that no algorithm in $\mathcal{L}$ performing fewer calculations than (AG) can insure a greedy solution for any submodular maximization problem. □

As a consequence of the above result, it is seen :

(a) that the number of f computations required by (AG) may be taken as a measure of the difficulty of submodular maximization problems ;

(b) the worst-case examples mentionned above (on which (AG) achieves a maximum number of calculations) must be actually recognized as problems of intrinsical difficulty.

Our last concern is to show how submodularity can be used to derive bounds on the difference between the optimal solution and the greedy solution of (P)

Suppose that f is submodular, and let $\bar{S}$ ($|\bar{S}| = p$) be the greedy solution produced by (SG) or (AG).

For i = 1 .... n, consider the quantites :

$$\delta_i = \delta(\emptyset, e_i) = f(\{e_i\}) - f(\emptyset)$$

and suppose that the elements e_i of E are considered in decreasing order :

$$\delta_1 \geqslant \delta_2 \geqslant \ldots\ldots \geqslant \delta_n$$

then :

<u>Theorem 4</u> :

1) $f(\emptyset) + \sum_{i=1}^{p} \delta_i$

is a upper bound on the value of an optimal solution of cardinality p.

2) $f(\emptyset) + \sum_{i/\delta_i \geqslant 0} \delta_i$

is a upper bound on the value of the true optimum of (P)

<u>Proof</u> : see [2] .

In [15] submodularity was used to derive general worst-case bounds between an optimal solution and a greedy solution. Theorem 3 above is however interesting in that it provides bounds which, being specific of each particular problem, are generally tighter than those given in [15].

It is clear, that for combinatorial problems of complex structure, it will be often impossible to prove, by theoretical considerations, that the submodularity assumption holds. However, submodularity may be tested, experimentally, by comparing the solution produced by (SG) and (AG), on medium size examples. For large scale problems, anyway, it is likely that only the accelerated greedy algorithm (AG) will be praticable.

REFERENCES

[1] MINOUX (M.) "Multiflots de coût minimum avec fonctions de coût concaves". Annales des télécommunications, 31, n° 3-4, (1976)

[2] MINOUX (M.) "Algorithmes gloutons et algorithmes gloutons accélérés pour la résolution des grands problèmes combinatoires". Bulletin de la Direction des Etudes et Recherches - EDF (France) Série C N° 1 (1977) pp.41-58

[3] LAURIERE (J.L.) "Un langage et un programme pour énoncer et résoudre des problèmes combinatoires" - Thèse, doc.ès sciences. Université PARIS VI - Mai 1976

[4] KRUSKAL (J.B.) "On the shortest spanning subtree of a graph and the travelling salesman problem" - Proc. Am. Math. Soc. 2 (1956) pp. 48-50

[5] EDMONDS (J.) "Matroïds and the greedy algorithm" - Mathematical programming 1, (1971), pp. 127-136.

[6] GONDRAN (M.) "L'algorithme glouton dans les algèbres de chemins" Bulletin Dir. Et. Rech. EDF Série C N° 1 (1975) pp.25-32

[7] BILLHEIMER (J.W.) GRAY (P.) "Network design with fixed and variable cost elements".
Transp. Science 7, n° 1 (1973) pp. 49-74

[8] LEGROS (J.F.) MINOUX (M.) OUSSET (A.) "Local networks optimization" Proc. ISSLS Conference London (May 1976)

[9] COOPER (L.) "The transportation location problem" Ops. Res. 20, n° 1 (1972) pp. 94-108

[10] KUENNE (R.E.) SOLAND (R.M.) "Exact and approximate solutions to the multisource weber problem". Mathematical programming 3 (1972) pp.193-209.

[11] STEINBERG (D.I.) "The fixed charge problem" - Nav. Res. Log. Quart. 17 (1970) pp. 217-236.

[12] BALINSKY (M.L.) "Fixed Cost Transportation Problems" - Nav. Res. Log. Quart. 8 (1961) pp. 41-54

[13] MALEK-ZAVAREI (M.), FRISCH (I.T.) "On the fixed cost flow problem" Intern.Journal Control 16, n° 5, (1972), pp. 897-902

[14] EDMONDS (J.) "Submodular functions, matroïds, and certain polyhedra" in : Combinatorial structures and their applications, R. Guy ed. pp.69-87 Gordon and Breach 1971.

[15] FISCHER (M.L.) NEMHAUSER (G.L.) WOLSEY (L.A.) "An analysis of approximations for maximizing submodular set functions" IX Internat. Symp. on Mathematical Programming BUDAPEST Hungary (1976).

[16] SAVAGE (S.L.) "Some theoretical implications of local optimization" Mathematical Programming 10 (1976) pp. 354-366.

RESOURCE ALLOCATION IN A SET OF NETWORKS UNDER MULTIPLE OBJECTIVES

R. Petrović
Mihailo Pupin Institute
Belgrade, Yugoslavia

A b s t r a c t

In this paper a class of dynamic resource allocation in a set of networks is treated. The paper is concerned with the construction of formal models of resource allocations under multiple objectives. We formalize the basic concepts of the partial resource exchange among networks, such as optimization of the partial resource exchange, singular resource points, Pareto optimal resource points. Theorems concerning resource exchanges are given, which provide a constructive way of deriving cooperative solution to resource allocation under multiple objectives.

INTRODUCTION

The subject of this paper will be resource allocation in a set of networks. The resource allocation problems in networks treated thus far are those of a single decision-maker with only one optimization criterion for the whole set of networks/1/. We introduce the possibility of more than one decision-maker, in which case the value of the criterion function for any decision-maker, responsible for resource allocation in one networks in a given set, depends not only on his own allocation, but also on the allocations of the others.

DESCRIPTION OF THE PROBLEM

The problem stated is the following:

(a) Given a set $\pi = \{G_1,\dots,G_s,\dots,G_S\}$ of directed networks (oriented, antisymetric graphs; such network is defined by $G_s = \{N_s, L_s\}$, $s = 1,\dots,S$, with proper numbering of nodes, i.e., if two arcs $(i,j) \in L_s$, $(j,k) \in L_s$ then the elements of N_s are indexed so that $i < j < k$, for all $i,j,k \in N_s$, $s = 1,\dots,S$

(b) To each arc $(i,j) \in L_s$, of the network G_s, $s = 1,\dots,S$ is associated a discrete multistage linear transformation:

$$x^{t+1}_{ij,s} = x^t_{ij,s} - u^t_{ij,s}(\Delta t); \qquad x^o_{ij,s} > 0 \text{ are given, terminal state is zero} \tag{1}$$

The meaning of the functions $x_{ij,s}$ and $u_{ij,s}$ is the following: $x^t_{ij,s}$ is nonincreasing function describing the state of the arc $(i,j) \in L_s$ at stage (time) t; $u^t_{ij,s} \geq 0$ is stepwise control variable representing the resource allocated to the arc $(i,j) \in L_s$ at stage t; Δt is the length of the stage t. The boundary conditions are posed: the terminal state of each arc in all $G_s \in \pi$ has to be zero.

(c) The resources allocated to arcs are subject to constraints:

$$\begin{aligned} u_{jk,s} &\equiv 0 \quad \text{for} \quad x_{ij,s} \neq 0 \\ u_{jk,s} &\equiv 0 \quad \text{for} \quad x_{jk,s} = 0 \qquad \text{for all } i,j,k \in N_s \\ & \qquad\qquad\qquad\qquad\qquad s = 1,\dots,S \\ & \qquad\qquad\qquad\qquad\qquad t = 1,2,\dots \end{aligned} \tag{2}$$

In addition, the integral constraints are:

$$\sum_{(i,j)\in L_s} u^t_{ij,s} \leq u^o_s > 0, \qquad t = 1,2,... \tag{3}$$

where u^o_s, s = 1,...,S are given in advance.

(d) Corresponding to resource allocation in $G_s \in \pi$, assume that there is given a local performance criterion F_s of the following form:

$$F_s : U_s \times X_s \to R, \qquad s = 1,...,S \tag{4}$$

where U_s is a set of resource allocations $u_{ij,s}$ and X_s is a set of arcs states $x_{ij,s}$ and R is real line. Sets U_s and X_s are to be thougth as the resource control space and network space, correspondingly:

$$\begin{aligned} U_s &= \{u_{ij,s} \mid \text{for } (i,j) \in L_s, \text{ subject to (2) and (3)}\} \\ X_s &= \{x_{ij,s} \mid \text{for } (i,j) \in L_s, \text{ subject to (1)}\} \end{aligned} \tag{5}$$

Fig. 1. shows arcs state trajectory in network space.

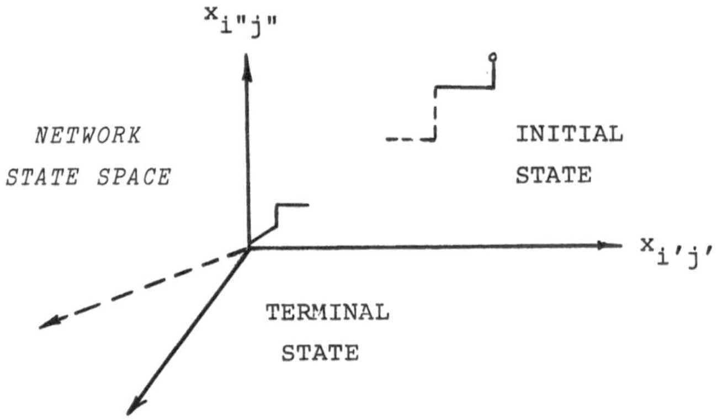

Fig. 1.

It is supposed that at t = 0 for each network $G_s \in \pi$ an initial amount of resource u^o_s is available, so the total amount of resource for the set π is limited and equal $\sum_s u^o_s$.

LOCAL OPTIMIZATION PROBLEMS

For each $G_s \in \pi$ the following local optimization problem can be stated and solved by using standard optimization techniques, (see for example /2/):

"Find $u^t_{ij,s}$ for all $(i,j) \in L_s$, and all $t = 1,2,\ldots,$ which transfer the initial state point from the position $x^o_{ij,s}$ to the final zero state, satisfying constraints (2) and (3), for which the objective function $F_s = F_s(\{u^t_{ij,s}\}) = f_s(u^o_s)$ takes on the minimum value". As it is well known, within this setup many interesting optimization results have been obtained[/3/].

CENTRALIZED OPTIMIZATION PROBLEM

In centralized decision making one can talk about resource reallocation of initially available total resource amount $\sum_s u^o_s$, among all networks $G_s \in \pi$. The centralized optimization problem can be stated as follows:

Find minimum of F over all u_s,

$$F = \sum_{s=1}^{S} f_s(u_s) \tag{6}$$

subject to:

$$\sum_{s=1}^{S} u_s = \sum_{s=1}^{S} u^o_s \tag{7}$$

and constraints (1)-(3).

The problem given in (6) and (7), including constraints (1)-(3) can be approached in two different ways:

(1) One-level optimization, where all networks $G_s \in \pi$ have to be aggregated into one complex network, forming the collection of all N_s and the union of all L_s. Then, all available resource $\sum_s u^o_s$ allocate in optimal way with respect to the overall performance criterion in (6). As it is known, the main computational difficulties in solving the above stated problem arise due to pure fact that all f_s in (6) are implicit in nature and depend on network $G_s \in \pi$.

(2) Another solution concept is two-level approach where the function of the first level decision making units are local optimal resource allocations of given amounts of u_s, $s = 1,\dots,S$ with respect to f_s, and the coordination level function is to allocate u_s, $s = 1,\dots,S$ so that F in (6) which is an explicit function of the first level performances f_s, $s = 1,\dots,S$ takes minimum.

OPTIMAL PARTIAL RESOURCE REALLOCATION

The resource may be reallocated among networks in a subset $\sigma \subset \pi$, where σ has any number of elements between 2 and S. Define system of subsets of networks $\Sigma = \{\sigma\}$. In the following we will focuss our attention on subsets σ with two elements only: $\sigma = \{G_{s'}, G_{s''}\}$ with criterion $|f_{s'}(u_{s'}) + f_{s''}(u_{s''})|$, where s' and s" are any two elements of N_s.

Denote $u = |u_1,\dots,u_s,\dots,u_S|$. For each $\sigma \subset \pi$ there exists a set $T_\sigma(u)$:

$$T_\sigma(u) = \{v \mid v = |v_1,\dots,v_s,\dots,v_S|, \sum_{G_s \in \sigma} u_s = \sum_{G_s \in \sigma} v_s , \quad u_s = v_s \quad \text{for} \quad G_s \notin \sigma\} \tag{8}$$

Find the mapping $\tau_\sigma(u)$:

$$\tau_\sigma(u) = \{v \mid v \in T_\sigma(u), F(v) = \max_{w \in T_\sigma} \sum f_s(w_s)\} \tag{9}$$

we say: τ_σ is the effective mapping at resource point u if $F|\tau_\sigma(u)| < F(u)$. Consequently, the most effective partial resource reallocation is for such a subset σ^* that:

$$F|\tau_{\sigma^*}(u)| = \max_{\sigma \in \Sigma} F|\tau_\sigma(u)| \tag{10}$$

In addition, resource point $\bar{u}$ is singular if there is no effective mapping at $\bar{u}$. Accordingly, $\bar{u}$ is the solution of the problem:

$$\text{minimize} \quad \sum_{G_s \in \sigma} f_s(u_s), \text{ for any } \sigma \in \Sigma$$

$$\text{subject to:} \quad \sum_{G_s \in \sigma} u_s = \sum_{G_s \in \sigma} u_s^o$$

Define a sequence of subsets $\sigma^1,\sigma^2,\ldots,\sigma^n,\ldots$ A corresponding sequence of the resource points $u^o,u^1,\ldots,u^n,\ldots$ is allowable if there is τ_σ so that $u^n \in \tau_{\sigma^n}(u^{n-1})$, $n = 1,2,\ldots$ It is proved /4,5/ that if all $f_s(u_s)$, $s = 1,\ldots,S$ are concave and $\bar{u}$ is a singular resource point and resource reallocation between any two networks in π are permited, then $\bar{u}$ is an optimal resource point. Further, for any initial resource point u^o, there exists an optimal sequence of resource reallocation among networks in subsets of networks in π, so that $F(u^n) \to \min \sum_s f_s(u_s)$, for $n \to \infty$.

In the following an efficient way for optimal partial resource reallocation between two networks G_s; $s = s',s''$ will be explained. We introduce the pair:

$|f_s^*(u_s),K_s|$: local optimal criterion value for G_s; $s = s',s''$, and the corresponding terminal stage (time).

The optimal resource reallocation between any two networks $G_{s'} \leftrightarrow G_{s''}$ can be obtained from the solution of the functional equation:

$$F_{s's''}(u^o_{s'}+u^o_{s''}) = \max_{u_{s'}} |f^*_{s'}(u_{s'}) + f^*_{s''}(\tilde{u}_{s''})| \tag{11}$$

where $F_{s's''}$ is the optimal criterion value for the pair of networks $G_{s'} \leftrightarrow G_{s''}$ and $\tilde{u}_{s''}$ is defined:

$$\tilde{u}_{s''} = u^o_{s'} + u^o_{s''} - u_{s'}\,, \qquad 0 \leq t \leq K_{s'} \tag{12}$$

$$u_{s''} = u^o_{s'} + u^o_{s''} \qquad t > K_{s'}$$

The solution of (11) for variable $(u^o_{s'} + u^o_{s''})$ is a simple process by using a straightforward one dimensional search technique.

According to this model of the partial resource reallocation, initially available resource for one network, for example $G_{s'}$, may be transferred to some other network $G_{s''}$, where it could be used more efficiently. The question arises how to share the benefits gained. In principle there are two possible mays:

(a) There is a pair of positive coefficients $a_{s'} \geq 0$, $a_{s''} = 1-a_{s'}$; any network $G_{s'}$ from which the amount of resource $u^o_{s'} - u^1_{s'}$ is transferred to some other network $G_{s''}$ has to obtain the part of the benefit produced by resource reallocation:

$$a_{s'}\{|(f_{s'}(u^1_{s'}) + f_{s''}(u^1_{s''})| - |f_{s'}(u^o_{s'}) + f_{s''}(u^o_{s''})|\}$$

(b) Second possible way of sharing the benefit of resource reallocation is based on establishing an internal pricing mechanism. For example, let from $G_{s'}$ an amount of resource $u^o_{s'} - u^1_{s'}$ be transfered to some other network $G_{s''}$. Then, the part $p(u^o_{s'} - u^1_{s'})$ of benefit has to be reimburced to $G_{s'}$ where p is an internal price of the resource. Obviously, resource exchange between two networks is useful for both networks, if the internal price p is chosen in such a way that the following inequality holds:

$$f_{s''}(u^1_{s''}) - f_{s''}(u^o_{s''}) \leq p(u^o_{s'} - u^1_{s'}) \quad \text{for} \quad s = s', s''$$

PARETO OPTIMAL RESOURCE POINTS

Let us consider the possibility of cooperation in resource reallocation between pairs of networks in π, having in mind Pareto optimal set for the problem. Denote $\Phi(u) = |f_1(u_1),\dots,f_s(u_s),\dots,f_S(u_S)|$. Then, for any $\sigma = \{G_{s'}, G_{s''}\} \subset \pi$ the criterion is $|f_{s'}(u_{s'}), f_{s''}(u_{s''})|$. We find T_σ such that the following holds:

$$T_\sigma(u) = \{v | v = |v_1,\dots,v_s,\dots,v_S|, \sum_{G_s \in \sigma} u_s = \sum_{G_s \in \sigma} v_s$$

$$u_s = v_s, \text{ for } G_s \notin \sigma\}$$

$$\Phi(v) \leq \Phi(u)$$

$$\text{if } u^* \in T_\sigma(u) \text{ and } \Phi(u^*) \leq \Phi(v), \text{ then } \Phi(u^*) = \Phi^* \tag{13}$$

The mapping $\tau_\sigma(u)$ defined by (13) is called alowable if $\sigma \subset \Sigma$. We say that the resource point $u \geq 0$ is Pareto singular if $u \in \tau_\sigma(u)$ for all $\sigma \in \Sigma$. In addition, the resource point $u \geq 0$ which satisfies the condition $u \in \tau_\pi(u)$ is Pareto optimal. Similarly as in the previous section a sequence of resource points $u^o, u^1, \dots, u^n, \dots$ is Pareto optimal sequence, if for $n = 1,2,\dots$ there exists τ^n_σ such that $u^n \in \tau^n_\sigma(u^{n-1})$. Any limit point of this sequence is Pareto optimal point. It is proved /4,5/ that if all f_s, $s = 1,\dots,S$ are concave and if resource reallocationsbetween any two networks in π are permitted, then Pareto singular resource point is Pareto optimal. Therefore a negotiated solution of Pareto type can be reached using the sequential application of the partial resource reallocation between pairs of networks in π.

The following general scheme for the derivation of the cooperative solutions of Pareto type is conceived.

Step 1. Solve local optimization problems for given u_s^o for all $G_s \varepsilon \pi$.

Step 2. Exchange resource between $G_{s'} \leftrightarrow G_{s''}, u_{s'}^o \pm \delta u$ and $u_{s''}^o \mp \delta u$.

Step 3. If $f_{s'}(u_{s'}^o \pm \delta u) \leq f_{s'}(u_{s'}^o)$

$f_{s''}(u_{s''}^o \mp \delta u) \leq f_{s''}(u_{s''}^o)$

resource exchange is effective (go on to Step 1.), if not it is ineffective (go on to Step 5.).

Step 4. Apply effective exchange, select another pair of networks and begin the computation process over again by returning to Step 2.

Step 5. Do not apply ineffective exchange, select another pair of networks and return to Step 2.
If there is no effective exchange the obtained $u \geq 0$ is Pareto singular, and consequently Pareto optimal.

CONCLUSION

Various formulations of the unified and the cooperative solutions of resource allocation in a set of networks under multiple objectives are given. It is shown that in the case of multiple objectives the resource allocation problem is essentially different in many respects from that of a single criterion case. The formalization of the basic concepts of the partial resource exchange, as for example singular resource point, Pareto optimal resource point and sequences of resource exchange, provides a constructive way of deriving cooperative solution to resource allocation under multiple objectives.

REFERENCES

/1/ Herroelen, W.S.: RESOURCE - CONSTRAINED PROJECT SCHEDULING - THE STATE OF THE ART, Opl. Res. Q. Vol. 23, pp. 261, (1973).

/2/ Woodgate, H.S: PLANNING NETWORKS AND RESOURCE ALLOCATION, Datamation, Vol. 14, pp. 36, (1968).

/3/ Petrović R.: OPTIMIZATION OF RESOURCE ALLOCATION IN PROJECT PLANNING, Ops. Res. Vol. 16, pp. 259, (1968).

/4/ Petrović R.: THE COOPERATIVE SOLUTION OF MULTIPROJECT RESOURCE ALLOCATION IN PLANNING NETWORKS, Proc. V INTERNET, Birmingham, U,K. (1976).

/5/ Polterovich V.: BLOCK METHODS IN CONCAVE PROGRAMMING AND ITS ECONOMIC INTERPRETATION (in russian) Econ. and Math. Methods, Vol. 5, No. 2 (1969).

AN ALGORITHM FOR SOLVING THE GENERALIZED TRANSPORTATION PROBLEM

Zsuzsa Pogány
Computing Centre for Universities

H-1093 Budapest Dimitrov tér 8

Introduction

In this paper we present an algorithm for solving the generalized transportation problem (GTP) which can be regarded ad the generalization of the predecessor indexing or triple-label method of the classical transportation problem published by Glover, Karney and Klingman ([1], [2]). In Hungary, Dömölky and Frivaldszky published the same method in 1966 ([3]), but their results have not become known in other countries. As it was reported in [4], transportation problems can be solved at least 100 times faster by this algorithm than by a general-purpose linear programming code utilizing sophisticated procedures for exploiting sparse matrices.

The predecessor indexes may be interpreted as the creation of links from 'fathers' to 'sons' to impose a consistent 'ancestry relation' on the spanning tree of the transportation network. The algorithm can not be applied for solving the GTP in its original form, because the rank of an m-row n-column GTP may be m+n or m+n-1, which involves different graph structures. In case of full rank the graph may be disconnected having several loops, the structure of which can not be described using three labels. In 1967 Frivaldszky developed a highly efficient algorithm for solving the GTP using three labels and some supplementary information about the subgraphs and loops ([5]). In our paper we present an algorithm which differs from that of Frivaldszky in defining four labels instead of three which enables a unified treatment of the different graph structures and simplifies the algorithm. In the case of rank m+n-1 and all weighting factors equal to 1, the algorithm reduces itself to that of the classical transportation problem.

Description of the algorithm

Let us consider the following problem:

$$(1)\quad \text{minimize } z = \sum_{i=1}^{m} \sum_{j=1}^{n} c_{ij}x_{ij}$$

subject to

$$(2)\quad \sum_{j=1}^{n} x_{ij} = a_i \qquad a_i \geq 0 \;,\; i = 1,2,\ldots,m$$

$$(3)\quad \sum_{i=1}^{m} p_{ij}x_{ij} = b_j \qquad p_{ij} \neq 0 \;,\; b_j \geq 0 \;,\; j = 1,2,\ldots,n$$

$$(4)\quad x_{ij} \geq 0 \qquad i = 1,2,\ldots,m \quad j = 1,2,\ldots,n$$

Any of the equations in (2) or (3) may be given as inequality as well.

This is a linear programming problem to be solved by a special-purpose algorithm exploiting the special structure of the problem. Let us suppose now that we know a feasible basic solution. Then an iteration of the algorithm consists of the following steps:

1. Determination of the u_i and v_j simplex multipliers or potentials corresponding to the current basic solution. In the course of this we have to solve a system of equations given by $u_i + p_{ij}v_j = c_{ij}$ for all basic elements.
2. Test of optimality: if $c_{ij} - u_i - p_{ij}v_j \geq 0$ for all nonbasic variable the current basic solution is optimal, otherwise chose a variable x'_{ij} for which $c_{ij} - u_i - p_{ij}v_j < 0$ to enter the basis.
3. Determination of the new basic solution after the basis change: let us increase the value of x'_{ij} from zero to Θ and simultaneously change the values of some 'old' basic variables so that the constraint equations be kept satisfied. We calculate these compensatory changes in terms of Θ and then drop a variable from the basis. In the course of this we have to solve a system of equations which is the transpose of that of step 1.

The efficiency of the algorithm mostly depends on the efficiency of solving these systems of equations. Therefore let us examine the special features of the systems of equations. In order to do this let us consider the usual tableau-formulation of a GTP problem (Fig. 1).

x_{11} p_{11} c_{11}	x_{12} p_{12} c_{12}					a_1 u_1
x_{21} p_{21} c_{21}	x_{22} p_{22} c_{22}	x_{23} p_{23} c_{23}				a_2 u_2
	x_{32} p_{32} c_{32}					a_3 u_3
	x_{42} p_{42} c_{42}		x_{44} p_{44} c_{44}			a_4 u_4
				x_{55} p_{55} c_{55}	x_{56} p_{56} c_{56}	a_5 u_5
				x_{65} p_{65} c_{65}	x_{66} p_{66} c_{66}	a_6 u_6
b_1 v_1	b_2 v_2	b_3 v_3	b_4 v_4	b_5 v_5	b_6 v_6	

Fig. 1 The tableau-formulation of the GTP.
We filled only the basic cells.

It is well-known that each basic solution has one of the following properties:

Property 1: there is at least one row or column which contains exactly one basic entry. Deleting this row or column, the reduced system also has property 1 or property 2.

Property 2: each row and column contains exactly two basic elements and the number of rows is equal to the number of columns.

It follows from this that the matrices of the systems of equations to be solved are triangular or near-triangular. Such systems of equations can be solved very fast easily if a precedence ordering of the equations is known so that solving the equations in the order indicated by the precedence ordering, each equation contains exactly one undetermined variable. In the case of near-triangularity (a sub-system having property 2) let us chose one of the variables as a parameter and determine the other variables in terms of the parameter.

This required precedence ordering may be given by defining an appropriate orientation of the graph corresponding to the current basic solution. Let us make the graph directed the following way:

At each node corresponding to a row or column having property 1, let us direct the only joined edge towards this node. Then let us drop the node and edge and repeat the procedure until all the nodes and the joined edges having property 1 have been dropped. It may happen that a subsystem having property 2 remained. This means that the graph contains one or more loop. Let us direct the loops in a consequent way. In Fig. 2 can be seen the direct graph corresponding to the basic solution in Fig. 1. For convenience we denoted the column indices by 7, 8, ..., 12.

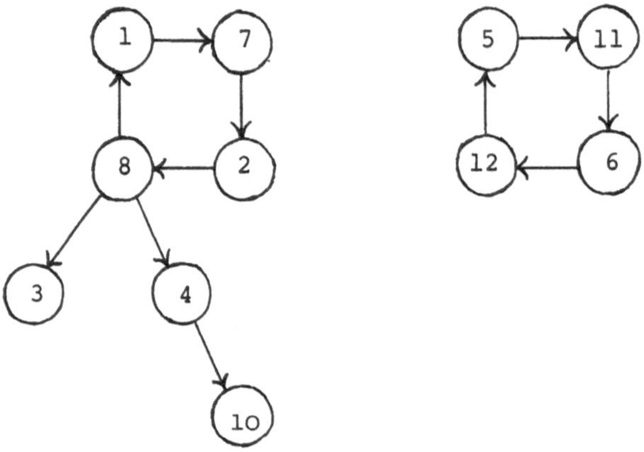

Fig. 2 The directed graph

Now let us define four functions to describe the sturcture of this directed graph. The functions will be given by assigning to each node the indeces of some other nodes.

1. Loop

L(i) = 0 for all nodes that are not in a loop.
L(i) = j if the edge (i,j) in a loop points towards i.

The functions Predecessor, Successor and Brother are the same as defined by Glover. It is important that only edges that are not in a loop are considered here.

2. Predecessor

P(i) = j if the only edge pointing to node i originates from node j.
P(i) = 0 if there is no edge pointing towards node i (considering only the edges that are not in a loop).

3. Successor

Let us chose any of the edges originated from node i and define S(i) = j the node towards which this chosen edge points. S(i) = 0 if there is no edge originated from node i.

4. Brother

Let us determine a cyclic order of the edges which originate from the same node. The nodes, at which these edges terminate, are brothers of each other. We register them by the index of the cyclic following node.

In Fig. 3 we augmented the tableau of Fig. 1 to represent the functions Loop, Predecessor, Successor and Brother for each row and column.

Now let us solve the systems of equations making use of these functions. Let us consider first the determination of the new basic solution after a basic change.

1	x	x					8			
2	x	x	x				7		9	
3		x						8		4
4		x		x				8	10	3
5					x	x	12			
6					x	x	11			
L	1	2			5	6				
P			2	4						
S		4								
B			9	10						

Fig. 3 The augmented tableau. We denote the basic cells with x.

Determination of the new basic solution

When a former zero variable enters the basis, two sequences of compensatory changes in the value of some old basic variables are necessary to keep the row and column equations satisfied. We call them row path and column path. In the course of their determination we use only the functions P and L.

Let us denote variable to enter the basis (i,j), its value Θ. Then the row path is given by the sequence of basis elements

i, P(i)
P(i), P(P(i))
.
.
.

and so on until we have for some row or column the value of function P zero. Then let us examine the value of function L of this row or column. If it is zero as well, the row path is finished; otherwise let us attach the elements of the loop to it. The values of the compensatory changes in terms of Θ can be calculated simultaneously. It is either the negative of the previously calculated value or the negative multiplied by the ratio of the p_{ij} coefficients, while we are not in a loop. Entering a loop, let us chose a loop element as a parameter and determine the compensatory changes in terms of Θ and in terms of the parameter. After going through the loop the value of the parameter can be determined and the two parts of the compensatory changes added. The column path is quite similar to the row path with the exception that it begins with the basic element (j, P(j)).

It may happen that the row and column path have a junction. In this case we add up the separately calculated values of the compensatory changes in the common part of the row and column paths. Then we can determine the variable to leave the basis. In Fig. 4 we present the row and column paths of our illustrative example, supposing that the variable to enter basis was (5, 10).

(1, 8) } (1, 7)	loop broke, direction unchanged
(2, 7) →	left the basis
(2, 8) }	loop broke, direction reversed
(4, 8) } (4,10)	reverse the direction
→ (5,10)	attach to row path
(5,12) (6,12) (6,11) (5,11)	row path unchanged

Fig. 4 The row and column paths and modifications after the basis change

In consequence of the basis change the structure of the graph changes and some of the functions describing the structure of the graph have to be modified. The modified function are needed for the calculation of the potentials. It is important that the modifications are usually confined to a small part of the graph.

Modification of the functions

In the course of the modification of the functions three cases are to be distinguished, depending on the relation of the row and column paths.

Case 1: There was no junction.

In this case, according to the place of the variable which left the basis, either the functions corresponding to the elements of the row path or the functions corresponding to the elements of the column path remain unchanged. The direction of the edges between the entered and left variables have to be reversed. If the variable which left the basis was in a loop, some additional modifications are needed.

Case 2: There was a junction at a node that was not in a loop.

In this case a new loop has formed. The modifications are rather similar to that of Case 1.

Case 3: There was a junction at a node in a loop.

In this case a 'double-loop' has formed. According to the place of the variable which left the basis, three sub-cases are possible.

Determination of the potentials

After the modification of the functions the potentials can be determined systematically. Here two cases are to be distinguished:

Case 1: If the new variable is in a loop, a system of equations have to be solved. Then, starting from the elements of the loop we can determine the values of the potentials which change in consequence of the change of the potentials in the loop. We determine the potentials in the order indicated by functions S, P and B, quite similar to the classical transportation problem.

Case 2: The new variable is not in a loop. Then let one of the potentials associated with the row or column of the new basic variable be unchanged. The other potential can be determined and starting from it, the values of the potentials which have to be modified can be determined systematically using functions S, P and B.

In our illustrative example let (2, 7) be the variable which left the basis. Then the changes in the values of the potentials are as follows:

The new basic variable was (5, 10). Let u_5 be unchanged. The value of v_{10} can be determined. The order of determination of potentials is shown by functions S and B. The values of the functions after the basis change is given in Fig. 5.

We start from v_{10}.

S(10) = 4, determine u_4
S (4) = 8, " v_8
S (8) = 1, " u_1
S (1) = 7, " v_7
S (7) = 0, B (7) = 7 already determined, go back to P (7) = 1
B (1) = 2, determine u_2 (after backward steps we consider the function B)
S (2) = 9, determine v_9
S (9) = 0, B (9) = 9, step back to P(9) = 2
B (2) = 3, determine u_3
S (3) = 0, B (3) = 1 already determined.

The other potentials given by the functions are already determined. They are: P(3)=8, B(8)=8, P(8)=4, B(4)=4, P(4)=10. We have returned back to the starting point, the procedure is finished.

row	L	P	S	B	column	L	P	S	B
1		8	7	2	7		1		7
2		8	9	3	8		4	1	8
3		8		1	9		2		9
4		10	8	4	10		5	4	10
5	12				11	5			
6	11				12	6			

Fig. 5 The functions after the basis change

The 'reduced-cost variant' of the algorithm:
In some cases it may be useful to store the $\hat{c}_{ij} = c_{ij} - u_i - p_{ij}v_j$ reduced costs instead of the original c_{ij} data. This accelerates and also simplifies the algorithm because only the functions Loop and Predecessor are needed. The row and column paths can be determined as above. To update the reduced costs after a basis change, let mark somehow the place of the basis variables in the matrix $[\hat{c}_{ij}]$. (The

reduced costs associated with the basic variables are zero. If we write there an appropriate number instead of zero, which can not occur among the reduced costs, the places of the basic cells are identified.) When we calculated the value of a particular Δu_i or Δv_j change of potential, let us substract it from the corresponding row or column of the reduced cost matrix. If we find in this row or column a basic element, determine next the Δu_i or Δv_j associated with this basic element and then substract it from the appropriate column or row. In this way all the modifications in the values of the reduced costs can be determined. This variant of the algorithm may be useful especially when $m \ll n$.

Determination of a feasible basic solution and the rank of the basis

Contrary to the classical transportation problem, an initial solution of the GTP, constructed by some heuristic procedure, is usually not feasible. Let us enlarge the original system with a row and column of artificial variables having $c_{ij} = M$ (a very large number), construct an initial solution of the augmented problem and then minimize it using the described algorithm. It may happen that after some iterative step artifical variable with zero value remains in the basis. This means either degeneracy or that the rank of the basis is m+n-1. To decide this we have to solve a system of equations given by $\hat{u}_i + p_{ij}\hat{v}_j = a_{ij}$ where a_{ij} denotes the coefficients of x_{ij} in the equation corresponding to the artifical variable with zero value. This system of equations differs from that of the potentials only in the right-hand-side and can be solved by the same procedure.

Computational experience

The testing of the variants of the algorithm on problems of different size and rank has not been completed, it will be reported later.

References

[1] F. Glover and D. Klingman, "Locating stepping-stone paths in distribution problems via the predecessor index method", Transportation Science 4(2)(1970) 220-225

[2] F. Glover, D. Karney and D. Klingman, "The augmented predecessor index method for locating stepping-stone paths and assigning dual prices in distribution problems", Transportation Science 6[2] (1972) 171-180

[3] B. Dömölky, S. Frivaldszky, "Solving the transportation problem by means of graph representation" (in Hungarian), Információ-Elektronika 1966. №1. 47-49

[4] F. Glover, D. Karney and A. Napier, "A comparison of computation times for various starting procedures, basis change criterion and solution algorithms for transportation problems", Management Science 20 (1974) 793-813

[5] S. Frivaldszky, "Solving the generalized transportation problem by means of graph representation" (in Hungarian), Információ-Elektronika 1967. №4. 300-302

[6] J. Lourie, "Topology and computation of the generalized transportation problem", Management Science 11(1) (1964) 177-187

[7] G. Dantzig, Linear programming and extensious (Princeton Univ. Press, Princeton, N.J. 1963).

AN EFFICIENT ALGORITHM FOR SOLVING A STOCHASTIC,

INTEGER PROGRAMMING PROBLEM ARISING IN RADIO NAVIGATION

Thomas M. Simundich *
Hoffman Electronics Corp.
El Monte, California

INTRODUCTION

A problem that arises in many in radio navigation systems that fix position from phase measurements is the so-called lane resolution problem. The problem is one of determining a unique phase of a fundamental frequency from noisy, time varying phase measurements of two harmonic frequencies. This paper formulates the lane resolution problem as a stochastic, integer programming problem and then gives an effective algortihm for solving the programming problem. The problem formulation and algorithm development will be addressed to the Omega navigation system [1].

Omega is a world wide radio navigation system. Each Omega station broadcasts in time slots 3 frequencies, 10.2, 11 1/3 and 13.6 KHz. The phases of these 3 frequencies vary in space as a function of distance from the transmitter. A single frequency will yield ambiguous positions radially from a transmitter λ_i apart, where λ_i is the frequency wave length. The distance λ_i is referred to as a lane. The Omega frequencies yield lanes of 15.87, 14.28 and 11.9 nautical miles (nm) respectively. However when utilizing a differential navigation scheme it is beneficial to work with the broader lane widths of the fundamental frequencies of 1.1 1/3, 2.2 2/3 and 3.6 KHz. These frequencies yeild lane widths of 142.83, 71.42 and 47.61 nautical miles respectively.

The straightforward method of determining the phases of these fundamental frequencies is to difference the phases of the higher frequencies. Unfortunately the phase measurements are often corrupted by noise and small phase errors in the differencing frequencies may result in large distance errors in the broader width lanes. As an example consider an error of 0.1 cycle (referred to as 10 centi cycles or 10 cec.) in the 10.2 KHz phase measurement. This corresponds to a distance error of 1.59 nm. Further consider an error of - 10 cec in the 11 1/3 KHz phase measurement. This distance error is - 1.43 nm. The resulting error in the phase of the difference of these two frequencies 20 cec or in distance 28.56 nm. This motivates the lane resolution approach given in this paper.

* Currently with the System Science Department, University of California at Los Angeles, California 90024.

STATEMENT OF PROBLEM

A relationship between the phase difference frequency and the frequency wavelengths exists as follows

$$(\emptyset_{1-2*}\lambda_{1-2})/\lambda_1 = \lambda_2/\lambda_1((\emptyset_{1-2*}\lambda_{1-2})/\lambda_2)) \tag{1}$$

where

$\emptyset_{1-2}$ is the phase of the difference frequency

λ_i are the frequency wave lengths

the above relationship can be expressed in terms of cycle count, N_i, and phase of the differencing frequencies, namely

$$(N_1 + \emptyset_1) = \lambda_2/\lambda_1(N_2 + \emptyset_2) \tag{2}$$

the N_i are positive integers modulo 9, 12, and 10 for 10.2, 11 1/3 and 13.6 KHz respectively. By determining either N_1 or N_2 the broad lane is given by

$$\emptyset_{1-2} = \lambda_i/\lambda_{1-2}\ (N_i + \emptyset_i) \tag{3}$$

In reference to the initial example if there exists no error in the cycle count, N_2, then the error in the broad lane phase is 1 cec or 1.43 nm instead of 28.56 nm.

From EQ (2) the lane resolution problem can be stated as follows:

Given: phase measurements $\emptyset_{m_i}$ and $\emptyset_{m_2}$

where

$$\emptyset_{m_i} = \emptyset_i + \eta_i \tag{4}$$

and η_i is a non zero mean Gaussian noise

Minimize: E{J}

$$J = N_1 - \alpha N_2 + \emptyset_{m_1} - \alpha\emptyset_{m_2} \tag{5}$$

subject to N_i being an integer and

$$N_i \geq 0 \tag{6}$$

$$N_1 < \alpha_1 \tag{7}$$

$$N_2 < \alpha_2 \tag{8}$$

$$N_2 - \alpha N_2 = \alpha\emptyset_2 - \emptyset_1 \tag{9}$$

where

$$\eta = \alpha\eta_2 - \eta_1 \tag{10}$$

$$\alpha = \alpha_1/\alpha_2 = \lambda_1/\lambda_2 \qquad (11)$$

for Omega α is either 10/9, 6/5 or 4/3.

The following algorithm development shows how the N_i are determined to yield a braod lane resolution.

ALGORITHM DEVELOPMENT

When the constraint (9) in the programming problem is met the expectation of the criterion function becomes

$$\alpha_2 N_1 - \alpha_1 N_2 = \alpha_1 \emptyset_2 - \alpha_2 \emptyset_1 + \alpha_2 E(\eta) \qquad (12)$$

since the left half side (Lhs) of EQ (12) is an integer then the right hand side (Rhs) of EQ (12) is an integer. This implies that $\alpha_1 \emptyset_{m_2} - \alpha_2 \emptyset_{m_1}$ should be an integer, K. In general it will not be, because of the measurement noise, propagation errors, computer rounding and etc. However if a time ensemble of $\alpha_1 \emptyset_{m_2} - \alpha_2 \emptyset_{m_1}$ is recursively averaged then

$$\mathrm{AVG}\{\alpha_1 \emptyset_{m_2} - \alpha_2 \emptyset_{m_1}\} \to \alpha_1 \emptyset_2 - \alpha_2 \emptyset_1 + \alpha_2 E(\eta) \qquad (13)$$

therefore if $E(\eta) < .5/\alpha_2$ then rounding the Lhs of (13) to an integer K eliminates the effects of the noise provided that the averaging process is sufficient to yield a small variance in the estimate of the Rhs of (13). $E(\eta) < .5/\alpha_2$ corresponds to errors of 5.55 cec, 10 cec, and 16.67 cec for the 142.83 nm, 71.42 nm and 47.61 nm lanes respectively.

In practice, the computation of K gives rise to a problem since the phase measurements change in time due to various non noise sources such as clock drifts in the signal tracking, signal propagation errors and position change. These effects, however, are proportionally additive to each frequency and will have no effect on the scaled phase difference provided that the effect of lane boundary crossings is considered during the averaging process. This is illustrated in the numerical example.

After the integer K is computed EQ (12) may be written as

$$\alpha_2 N_1 - \alpha_1 N_2 = K \qquad (14)$$

EQ (14) is a linear Diophantine equation [2] which can be solved for a unique N_1 and N_2 by the classical rule of virgins [3]. The methodology of the rule of virgins can be written as a simple algorithm. The algorithm development is as follows. First solve for N_1 since it has the smaller coefficient α_2

$$N_1 = (\alpha_1/\alpha_2) * N_2 + K/\alpha_2 \qquad (15)$$

EQ (15) may be rewritten as

$$N_1 = N_2 + X \tag{16}$$

$$X = N_2/\alpha_2 + K/\alpha_2 \tag{17}$$

since N_1 and N_2 are integers X must also be an integer. Consider the case where $0 \leq |K| \leq \alpha_2 - 1$. Since N_2 is greater than or equal to zero and N_2/α_2 is less than 1 therefore if K is positive then X must be equal to 1 and if K is negative then X must equal zero. Thus in the case where K is positive

$$N_2 = \alpha_2 - K \tag{18}$$

and

$$N_1 = N_2 + 1 \tag{19}$$

and if K is negative

$$N_1 = N_2 = - K \tag{20}$$

now consider the case where $K = - \alpha_2$. In this case EQ (15) can be written as

$$N_1 = N_2 - 1 + X \tag{21}$$

$$X = N_2/\alpha_2 \tag{22}$$

or

$$N_1 = (1 + \alpha_2)X - 1 \tag{23}$$

$$N_2 = \alpha_2 X \tag{24}$$

since N_1 must be positive X must be equal to 1 or N_1 equal to α_2. This forces N_2 to equal α_2 but since N_2 is modulo α_2 then $N_2 = 0$. Finally, consider the case where K equals α_1. In this case EQ (15) can be written as

$$N_1 = N_2 + 1 + X \tag{25}$$

$$X = (N_2 + 1)/\alpha_2 \tag{26}$$

or

$$N_1 = (\alpha_2 + 1)X \tag{27}$$

$$N_2 = \alpha_2 X - 1 \tag{28}$$

since N_2 must be positive X must be equal to 1 or N_2 equal to α_{2-1}. This forces N_1 equal to $\alpha_2 - 1$ or α_1 but since N_1 is modulo α_1 then $N_1 = 0$. These last two cases correspond to a situation where the differencing phases straddle a broad lane boundary.

The above can be summarized in an algorithm as follows:

Step 1: compute K

Step 2: if K is negative then $N_1 = - K$ and $N_2 = (-K) \bmod \alpha_2$ otherwise go to 3

Step 3: if $K = \alpha_1$ set $N_1 = 0$ and $N_2 = \alpha_2 - 1$ otherwise $N_1 = (\alpha_1 - K)_{\bmod \alpha}$ and $N_2 = (\alpha_2 - K)_{\bmod \alpha_2}$

NUMERICAL EXAMPLE

Consider the noisy 11 1/3 and 10.2 KHz phase measurements of columns 2 and 3 of Table 1. The instaneous phase differences are shown in column 8 along with the instaneous true phase difference in column 7. Column 9 shows the error in natural miles. The recursive average after the 6th sample of the scaled phase difference is shown in column 6. A lane count is maintained during the averaging and is shown in columns 4 and 5. A change in lane count can be easily detected provided that the phase never changes more than a half cycle between phase samples.

After the 10th sample a K = 6 is obtained. Applying the lane resoltuion algorithm to this K yields $N_1 = 3$ and $N_2 = 4$. In order to determine the broad lane phase at the 10th sample, the maintained 10.2 KHz cycle count and resolved number of cycles are added to yield $N_1 = 5$. Therefore the 1.1 1/3 KHz broad lane phase can be computed from (3) with $N_i = 5$, $\emptyset_i = .13$ $\lambda_i/\lambda_{1-2} = 1/9$ to yield $\emptyset_{1-2} = .57$. The error in this broad lane phase computation is - 1.4 nm compared to 21.4 nmiles. Similarly a broad lane phase computation from the 11 1/3 KHz yields a phase of .586. The error in this phase computation corresponds to a distance error of .86 nm.

CONCLUSIONS

In this paper the lane resolution problem of radio navigation has been formulated as a stochastic, integer programming problem and an algorithm has been presented which efficiently solves the programming problem. The advantages of this algorithm are:

1) it is easily implemented

2) it works in real time

3) under certain assumptions it eliminates much of the error due to noise.

Presently the algorithm is part of the software package of the Hoffman Electronics HON 360 Omega System.

REFERENCES

[1] SWANSON, E. R. "Omega" Navigation, Vol. 18 No. 2, pp. 168-175, 1971.

[2] NAGELL, T. Number Theory. New York: Chelsea Publishing Co., 1964.

[3] SAATy, T. L. Optimization in Integers and Related Extremal Problems. New York: McGraw-Hill, 1970.

Table 1

Sample	11 1/3 Phase $\emptyset_1$	10.2 KHz Phase $\emptyset_2$	11 1/3 KHz N_1	10.2 KHz N_2	Averages $10\ \emptyset_2 - 9\ \emptyset_1$	1.1 1/3 KHz True Phase $\emptyset_1 - \emptyset_2$	1.1 1/3 KHz Instaneous Phase	Error in Nautical Miles
1	0	.64	0	0	6.40	.4	.36	- 5.7
2	.28	.84	0	0	6.14	.42	.44	2.8
3	.47	.03	0	1	6.12	.44	.44	0
4	.52	.21	0	1	6.44	.46	.31	-21.3
5	.81	.41	0	1	6.52	.48	.4	-11.3
6	.92	.59	0	1	6.70	.5	.33	-24.1
7	.14	.74	1	1	6.76	.52	.40	-17.1
8	.47	.77	1	1	6.47	.54	.70	22.9
9	.64	.09	1	2	6.43	.56	.55	- 1.4
10	.86	.13	1	2	6.25	.58	.73	21.4

USING PSEUDOBOOLEAN PROGRAMMING IN DECOMPOSITION METHOD

Stanisław Walukiewicz
Systems Research Institute
Polish Academy of Science
ul. Newelska 6, 01-447 Warsaw POLAND

1. Introduction

Consider a mathematical programming problem

$$W:\ f(x^*,y^*) = \min f(x,y) \quad \text{subject to} \quad G(x,y) \geqslant 0,\ x \in X,\ y \in Y.$$

If after fixing $y = \bar{y}$, we have more tractable problem, than one may solve the problem W by the Benders' decomposition algorithm [1], [4]. For this reason we call y <u>complicating variables</u> and the problem

$$\min f(x,\bar{y}), \quad \text{subject to} \quad G(x,\bar{y}) \geqslant 0,\ x \in X$$

is called an <u>easy - to - solve subproblem</u>.

The main idea of the Benders' approach is the following. Instead of looking for x^* and y^* in the product $X \times Y$, we look for y in the set Y, and this set changes from one iteration to the other. Analitical discription of it is obtained using the duality theory [7].

So far the stronest results have been obtained for linear mixed integer programming (MIP) problems when complicating variables are integer and easy - to - solve subproblem are linear programming problems, in many cases-classical transportation ones, therefore we consider first the application of Benders' decomposition to such problems. In the next section we consider computational aspects of the Benders' decomposition method. In the last section we discuss how the pseudoboolean programming may be used in the decomposition method.

2. Linear MIP

The linear mixed integer programming problem, denoted as problem W, may be stated as follows:

$$W:\ z^* = f(x^*,y^*) = \min(cx + dy)$$

$$\text{subject to} \quad Ax + Dy \geqslant b, \quad x,y \geqslant 0 \ \text{and integer}$$

We assume that all dimensions of vectors and matricies are such that all multiplications are well-defined. We not distingvish the transposition of vectors and matricies.

For a given $\bar{y}$ the problem W reduces to the ordinary linear programming problem

$$P:\ z^* = d\bar{y} = \min cx$$

$$\text{subject to} \quad Ax \geqslant b - D\bar{y}, \quad x \geqslant 0,$$

and its dual is

$$D:\ g(u^*) = d\bar{y} + \max u(b - D\bar{y})$$

$$uA \leqslant c, \quad u \geqslant 0.$$

The convex polyhedron $U = \{u \geqslant 0 : uA \leqslant c\}$ is independent of $\bar{y}$. There are only two possibilities: $U = \emptyset$ and $U \neq \emptyset$. If $U = \emptyset$, then by the duality theorem of linear programming the problem P is infeasible or unbounded for any choise of $\bar{y}$. The same is true for the problem W. If $U \neq \emptyset$, then it has a finite number of extreme points u^p, $p = 1,2,\ldots,P$ and/or a finite number of extreme rays ϑ^r, $r = 1,2,\ldots,R$. If there exist a direction ϑ^r such that $\vartheta^r(b - D\bar{y}) > 0$, then by the duality theorem the problem P and therefore the problem W is infeasible for that choise of $\bar{y}$. Therefore

$$\vartheta^r(b - D\bar{y}) \leqslant 0, \quad r = 1,2,\ldots,R$$

provides necessary and sufficient conditions on $\bar{y}$ to permit feasible solutions x to the mixed integer programming problem. As we are only interested in this case, the problem D is

$$D: \quad g(u^*) = d\bar{y} - \max_{p=1,\dots,P} u^p(b - D\bar{y})$$

subject to

$$\vartheta^r(b - D\bar{y}) \leqslant 0, \quad r = 1,2,\dots,R.$$

Now, we can write the problem W as

$$W: \quad f(y^*) = \min(dy + \max_{p=1,\dots,P} u^p(b - Dy))$$

subject to

$$\vartheta^r(b - Dy) \leqslant 0, \quad r = 1,2,\dots,R$$

$$y \geqslant 0, \quad \text{integer.}$$

Let

$$z = dy + \max_{p=1,\dots,P} u^p(b - Dy),$$

then $z \geqslant dy + u^p(b - Dy)$ for any $p = 1,2,\dots,P$ and problem W is equivalent to the parametric integer programming problem

$$I: \quad z^* = \min z$$

subject to
$$z \geqslant dy + u^p(b - Dy), \quad p = 1,2,\dots,P,$$
$$0 \geqslant \vartheta^r(b - Dy), \quad r = 1,2,\dots,R,$$
$$y \geqslant 0, \quad \text{integer.}$$

This equivalence states that a mixed integer programming problem is equivalent to the problem I, which can be considered as a special integer problem. We call the constraints in the problem I, a point (ray) inequalities. Usually, the polyhedron U, has a vast number extreeme points and/or rays and therefore it is virtually impossible to generate all point and ray inequalities. Benders sugests in [1] to add these inequalites if they are needed in the solution process.

To control the solution process we need a lower bound $\underline{z}$ and an upper bound $\bar{z}$ on the optimal solution to the mixed integer programming problem. Any incumbent - the best feasible solution found so for - may serve as an upper bound on $f(x^*,y^*)$. Also $g(u^*)$ - the optimal solution to the problem D is an upper bound on $f(x^*,y^*)$, since the problem P is more restricted than the mixed integer problem. A lower bound $\underline{z}$ may be also calculated in two ways. First, the optimal solution to the problem I with not all point and ray inequalities gives a lower bound on $f(x^*,y^*)$. Second, we may obtain $\underline{z}$ form the following relations

$$z^* = \min_{y \in Y} (dy = \max_{u \in U} u(b - Dy)) \geqslant \min_{y \in Y} (dy + \max_{u \in S} u(b - Dy)) = \underline{z},$$

where $S \subseteq U$ is an easy constructed set e.g. parallelepiped.

The Benders' decomposition algorithm can be discribed in the following way:

STEP1 (Initialization) Select a nonnegative integer vector $\bar{y}$ and set $\underline{z} = -\infty$, $\bar{z} = +\infty$, $P = R = 0$.

STEP2 (Linear Programming Phase). Solve the linear problem

D: $g(u^*) = d\bar{y} + \max u(b - D\bar{y})$

subject to

$$uA \leqslant c,$$
$$u \geqslant 0$$

This gives an optimal extreeme point u^p or an extreeme ray direction ϑ^r. Increase P or R by one and replace $\bar{z}$ by $d\bar{y} + u^p(b - D\bar{y}) = g(u^*)$, whenever $\bar{z} \geqslant g(u^*)$.

STEP3 (Integer Programming Phase) Solve the integer programming problem

I: $f(\bar{\bar{y}}) = \max z$

subject to

(1) $$z \geqslant dy + u^p(b - Dy), \quad p = 1,2,\dots,P,$$
$$0 \geqslant \vartheta^r(b - Dy), \quad r = 1,2,\dots,R,$$
$$y \geqslant 0 \text{ integer.}$$

Set $\underline{z} = f(\bar{y})$.

STEP4 (Termination) If $\underline{z} < \bar{z}$, then set $\bar{y} = \bar{y}$ and go to STEP2. Otherwise $(\underline{z} = \bar{z})$ $\bar{\bar{y}}$ is an optimal solution to the problem W, i.e. $y^* = \bar{\bar{y}}$ and x^* can be calculated as an optimal solution to the problem P for $\bar{y} = y^*$. Stop.

3. Computational Remarks

From the discription of the Benders' decomposition algorithm one can see, that near - optimal solutions to the problem I are sufficient for the convergence of this procedure. Possibly the most developed form of this statemant is in the paper [2], and now we follow the main of this paper.

If $(\bar{x},\bar{y})$ is a feasible solution to our initial problem W, hen looking for the optimal solution to it is equivalent to solving a system of linear in equalities

$$cx + dy < \bar{z} \tag{2}$$
$$Ax + Dy \geqslant b \tag{3}$$
$$x \geqslant 0, \quad y \geqslant 0 \quad \text{and integer,} \tag{4}$$

where now $\bar{z} = c\bar{x} + d\bar{y}$. Fixing $y = \bar{y}$, we obtain the system

$$-cx > d\bar{y} - \bar{z} \tag{5}$$
$$-Ax \leqslant D\bar{y} - b \tag{6}$$
$$-x \leqslant 0. \tag{7}$$

The solution to the problem W better than $\bar{z}$ exists if and only if the system (5)-(7) is consistent. By the theorem of alternative [7] we have that the system (5)-(7) is inconsistent if and only if at least one of the systems

$$uA \leqslant c \tag{8}$$
$$u(Dy - b) \leqslant dy - \bar{z} \tag{9}$$
$$u \geqslant 0 \tag{10}$$

or the system

$$uA = 0 \tag{11}$$
$$u(Dy - b) < 0 \tag{12}$$
$$u \geqslant 0 \tag{13}$$

is consistent.

It is easy to see that solutions of the system (8)-(10) are the optimal ones to the problem D. And the system (11)-(13) has a solution if the problem W is infeasible.

By Y we denote the set of complicating variables, i.e.

$Y = \{y \geqslant 0, \text{ integer}\}$. In [2] the following theorem is proved.

If W is feasible and $\bar{z} = g(u^*)$ is an optimal solution to the problem D, then for every $y \in Y$, which satisfies

$$u^*A \leqslant c$$
$$u^*(Dy - b) \leqslant dy - \bar{z}$$

does not correspond a solution of W with $z < g(u\ )$, i.e. the better solutions if any are in the halfspace

(14) $$dy + u\ (b - Dy) \leqslant \bar{z}.$$

We note that (14) is a point inequality (1) in the problem I. Since this inequality cuts off some part of Y, we will call (14) <u>Benders´ cut</u>. Now we have the second termination rule of the Benders´ decomposition algorithm, namely, $Y = \emptyset$.

In [2] a more general system of the following form

$$c(y)x + dy < \bar{z}$$
$$A(y)x + D(y) \geqslant 0$$
$$y \in Y.$$

was considered. This paper contains also generalization of the above results for systems of nonlinear inequalities.

The next computational remark concerns with the problem D. Since in every interation of the Benders´ decomposition algorithm we tray increase the lower bound and decrease the upper one, we may avoid solving the problem D up to the optimality.

Let $\underline{z}_k$ $(\bar{z}_k)$ be a value of the lower (upper) bound at the k-th iteration. Then instead of looking for the optimal solution to the problem D it is sufficient to solve the system

(15) $$uA \leqslant c$$
(16) $$u \geqslant 0$$
(17) $$\bar{z}_k \leqslant \bar{z}_{k-1}, \quad k = 2.3, \ldots,$$

where $z_k = d\bar{y} - \max u(b - D\bar{y})$.

Adding the upper bound constraint to the problem I, we obtain the problem

(18) $\underline{z}_k = \min z$

(19) $z \geqslant dy + u^p(b - Dy)$

(20) $\underline{z}_k \geqslant \underline{z}_{k-1}, \quad k = 2,3,\ldots$

(21) $y \geqslant 0$ and integer

where u^p is current extreeme point of the convex polyhedrom U. This system can be considered as a knapsack problem, for which, specially for the case $y_j = 0$ or 1, many reduction techniques proved their efficiency [10].

4. Pseudoboolean Programming in Benders´ Decomposition

Geoffrin and Graves [5] successfully applied the Benders´ decomposition method to solve real multicommodity transportation and intermediate facilities location problem. Form this paper one can conclude that the efficiency of the Benders´ decomposition depends havily on the thightnes of the formulation of an integer programming problem, i.e., on the fact how close the system of constraints describes the convex hull of the fealisible integer points. This is one more evidence that obtaininig thighter equivalent formulation of an integer programming problem is important [13, 11]. The comparison of the bounds obtained from different cuts given in [8] shows that, the Benders´ cuts are not so strong as surrogate constraints.

The main drawback of the Benders´ decomposition method is poor convergence of it approching the optimal solution. If we use the algorithm described in section 2, than near (x^*,y^*) we have to solve up to the optimality large integer programming problems. If we use the theorem of alternative, than near (x^*,y^*) the problem o feasibility, i.e. finding $\bar{y}$ in each interation becomes hard since the set Y is almost empty.

To improve the convergence of the Benders´ decomposition algorithm we suggest using the pseudoboolean programming in it. The detailed discription of the pseudoboolean programming is given in [6] and the implementationon and some imporments are given in [9].

We use the characteristic function to describe the set Y and it is easy to see that characteristic function for smaller set is more tractable.

Some optimization problems of elastic trusses may be formulated as a mixed integer nonlinear problems with nonlinear descrete part and linear continuous part (see e.g. [3] for one of such formulations). So far there is no algorithm for solving such problems. We note that the transformation of nonlinear integer (boolean) part to linear problem as it was shown in [12] is inefficient. Therefore incorporating the pseudoboolean programming in the Benders' decomposition offers a new algorithm for solving problems with linear continuous part and nonlinear integer part.

REFERENCES

[1] Benders J.F.: Partitioning Procedures for Solving Mixed-Variables Programming Problems. Numerische Mathematik, 4 (1962) 238-252.

[2] Castellani G., Giannessi F.: Decomposition of Mathematical Programs by Means of Theorem of Alternative for Linear and Nonlinear Systems. Proceedings of the IX International Symposium on Mathematical Programming, Budapest 1976.

[3] Faner M., Niemierko A.: On the Possibility of Optimization of Elastic Prusses by Pseudoboolean Programming. Proceedings of the IX International Symposium on Mathematical Programming, Budapest 1976.

[4] Geoffrion A.M.: Generalized Benders Decomposition. JOTA, 10 (1972) 237-260.

[5] Geoffrion A.M., Graves G.W.: Multicommodity Distribution System Design by Benders Decomposition. Management Science, 20 (1974) 822-844.

[6] Hammer P.L., Rudeanu S.: Boolean Methods in Operation Research and Related Areas. Springer-Verlag, 1968.

[7] Mangasarian O.L.: Nonlinear Programming, McGraw-Hill, 1969.

[8] Rardin R.L., Unger V.E.: Surrogate Constraints and the Strenght of Bounds Derived from 0-1 Benders' Partitioning Procedures. Operations Research, 24 (1976) 1169-1175.

[9] Walukiewicz S., Słomiński L., Faner M.: An Improved Algorithm for Pseudo-Boolean Programming. Proceedings of 5-th IFIP Conference on Optimization Techniques. Roma May, 5-8, 1973.

[10] Walukiewicz S.: Prawie liniowe zadania programowania dyskretnego. Prace IOK, Seria B, Zeszyt Nr 23, 1975.

[11] Walukiewicz S., Kaliszewski I.: Thighter Equivalent Formulat-

ion of Integer Programming Problems. Proceedings of the IX International Symposium on Mathematical Programming, Budapest 1976.

[12] Watters L.J.: Reduction of Integer Polynomial Programming Problems to Zero-One Linear Programming Problems. Operations Research, 15 (1967) 1171-1174.

[13] Williams H.P.: Experiments in the Formulation of Integer Programming Problems. Mathematical Programming Study, 2 (1974) 180-197.

SOLVING THE GENERAL PROJECT SCHEDULING PROBLEM WITH MULTIPLE CONSTRAINED RESOURCES BY MATHEMATICAL PROGRAMMING

Roman Słowiński, Jan Węglarz

Institute of Control Engineering, Technical University of Poznań, 60-965 Poznań, Poland

ABSTRACT The problem of the time-optimal allocation of multiple-constrained resources among dependent operations is considered for the case when operation models are given in the form of differential equations relating operation performance speed to amounts of resources allotted. An approach is presented which permits the problem to be reduced to a mathematical programming one and enables a proper insight to be obtained into the properties of optimal solutions.

1. PROBLEM FORMULATION

Let us consider the following project scheduling problem. There are given:

- a set of n operations $O = \{O_1, O_2, \ldots, O_n\}$, which is partially ordered by a binary relation $\prec$ defined as follows: if $O_i \prec O_j$, then the execution of O_i must be completed before the execution of O_j can begin; this set will be denoted by $(O, \prec)$;
- a set of continuous /i.e. homogeneous, arbitrarily splittable/ resources $\mathcal{R} = \{\mathcal{R}_1, \mathcal{R}_2, \ldots, \mathcal{R}_p\}$ containing p resource types whose available amounts are correspondingly equal to $N_1, N_2, \ldots, N_p$ units.

For operation O_i, $i=1,2,\ldots,n$, there are known:

- the intervals including feasible amounts of particular resource types which may take part in the performance of O_i at moment t:

$$\langle a_{ik}, b_{ik} \rangle, \qquad 0 \le a_{ik} \le b_{ik} \le \infty, \quad k=1,2,\ldots,p;$$

- the proportions of particular resource types required in the performance of O_i;
- the mathematical model

$$d\, x_i(t)/dt = f_i[\bar{r}_i(t)] \qquad /1/$$

where $x_i(t)$ is the state of O_i at moment t; $x_i(0) = 0$; $x_i(T_i) = w_i$ /T_i is the finishing time of O_i and is unknown in general a priori, w_i is a known final state which will be called the size of O_i/; $\bar{r}_i^T(t) = (r_{i1}(t), r_{i2}(t), \ldots, r_{ip}(t))$, $r_{ik}(t)$ denotes the amount of resource $\mathcal{R}_k$ taking part in the performance of O_i at moment t; $f_i(\cdot)$ is a continuous, nondecreasing function, $f_i(0) = 0$. The performance of each operation may be arbitrarily split. We will assume that O_i is not being performed at moment t if and only if $\bar{r}_i^T(t) = \bar{0}$. Note that the equation $x_i(T_i) = w_i$ denotes a <u>performance condition</u> of O_i which may also be presented in this form:

$$\int_0^{T_i} f_i[\bar{r}_i(t)]\,dt = w_i \qquad /2/$$

We will be searching for the matrix $R^*(t) = [\bar{r}_1^{*T}(t), \bar{r}_2^{*T}(t), \ldots, \bar{r}_n^{*T}(t)]$ for which the <u>completion time of the whole project</u>, T, is minimized, i.e., $T = T_{min} = T^*$, subject to the resource constraints:

$$\sum_{i \in F_t} r_{ik}(t) \le N_k, \quad \text{for every } t \ge 0, \quad k=1,2,\ldots,p, \qquad /3/$$

and

$$r_{ik} \in \langle a_{ik}, b_{ik} \rangle \cup \{0\}, \quad \text{for every } t \ge 0, \text{ and every } i \in F_t, \quad k=1,2,\ldots,p, \qquad /4/$$

where F_t denotes the set of all operations which may be performed simultaneously at moment t while fulfilling precedence and resource constraints / the problem of generating sets F_t will be considered later/. The set $\{0\}$ in /4/ corresponds to the possibility of splitting an operation.

The above problem has been considered by several authors, for example in [1,2,4,5,6], but only for the case of one resource type. In this paper, in order to solve the multiple resource case under the given assumptions, we will adapt the approach which has been elaborated for the case of one resource type. Special attention will be paid to the properties of optimal solutions.

Before passing to the solution of the problem, let us comment briefly on its practical and mathematical nature. First of all, we consider resources which may be allotted to operations in amounts belonging to certain intervals. Such resources will be called <u>continuous resources</u>. These are, for example, such resources as fuel flow, power, approximately manpower or primary memory pages in computer systems. Mathematical models of operations, given in the form /1/, are more general than those usually considered in project scheduling problems. They allow, in a natural way, the variation of the amount of resources allotted to operations during their performance. The state of an operation, $x_i(t)$, may represent the number of objective work units performed up to moment t /for example the number of standarised computer

instructions or cubic metres of soil/ or, in the case of linear function f_i, the number of resource-hours /for example, man-hours/ performed up to moment t.
From the mathematical point of view, the problem resembles the classical, time-optimal control problem. There are however in our problem some specific points which have to be taken into account.
The principal points follow from the consideration of precedence relationships among operations. This leads to specific constraints in the optimization problem, and to the objective function becoming a function of the min-max type. It is sufficient to consider the set of independent operations /i.e. operations among which there are no precedence constraints - that means - they may be performed simultaneously/ to notice that $T = \max_i \{T_i\}$ has to be minimized.

These points together mean that the application of known methods from optimal control theory seems to be unsuitable in this case. Moreover, we would like to have an approach which would allow for a clear insight into the properties of optimal solutions in particular cases. For the above reasons we will apply an approach which gives such possibilities and makes use of all specific points of our problem.

2. APPROACH TO THE SOLUTION

Before passing to the solution of the problem, let us make use of the assumption concerning the knowledge of the proportions of particular resource types required in the performance of particular operations. On the basis of this assumption we may write:

$$\bar{r}_i(t) = u_i(t)\bar{\alpha}_i \,, \qquad i=1,2,\ldots,n,$$

where $u_i(t) \in \langle 0, 1\rangle$ for every $t \geq 0$, and $\bar{\alpha}_i = \{\alpha_{ik}\}_{k=1}^{p}$ is the vector of parameters for which the modulus $|\bar{\alpha}_i|$ is maximized subject to the known proportions and to constraints /4/.
Hence, we may reduce the mathematical model of operation O_i to:

$$d\, x_i(t)/dt = g_i[u_i(t)] \quad \text{where} \quad g_i(1) = f_i(\bar{\alpha}_i)\,, \qquad /1'/$$

the performance condition to:

$$\int_0^{T_i} g_i[u_i(t)]\, dt = w_i, \qquad /2'/$$

and resource constraints to:

$$\sum_{i \in F_t} \alpha_{ik} u_i(t) \leq N_k \,, \qquad k=1,2,\ldots,p, \qquad /3'/$$

$$u_i(t) \in \langle a_{ik}/\alpha_{ik}, b_{ik}/\alpha_{ik}\rangle \cup \{0\} \quad \text{for every } t \geq 0, \qquad /4'/$$
$$\text{and every } i \in F_t,\ k=1,2,\ldots,p.$$

Thus, instead of $R^*(t)$ we have to find the vector function $\bar{u}^*(t) =$

$= \left(u_1^*(t),\ u_2^*(t),\ldots,\ u_n^*(t)\right)$, for which $T = T^*$ subject to /2'/-/4'/.

Now let us introduce some definitions. Let $\bar{u}$ denote a value of the vector function $\bar{u}(t)$, called a <u>normalized resource allocation</u>. The set of all points $\bar{u}$ fulfilling /3'/, $U^1_{F_t}$, is equal to:

$$U^1_{F_t} = \bigcap_{k=1}^{p} U_k ,$$

where U_k is the set of all points $\bar{u}$ fulfilling /3'/ for a given k. The maximum value of coordinate i of the elements from $U^1_{F_t}$ will be denoted by u_i^N. As an example, for $F_t = \{1,2\}$ and p=2, $U^1_{F_t}$ may have the form shown in Fig. 1. Let P_i denote an interval of the following form:

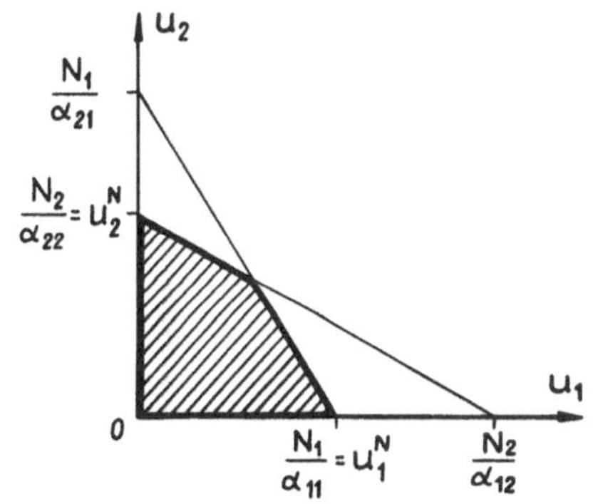

Fig.1. The set $U^1_{F_t}$ for $F_t = \{1,2\}$ and p=2.

$$P_i = \bigcap_{k=1}^{p} P_{ik} \cup \{\bar{u}:\ u_i=0\}, \quad i \in F_t,$$

where

$$P_{ik} = \langle 0,\ldots,0,a_{ik}/\alpha_{ik},0,\ldots,0;\ \infty,\ldots,\infty,b_{ik}/\alpha_{ik},\infty,\ldots,\infty\rangle$$

is the generalized $|F_t|$ - dimensional interval /$|F_t|$ denotes the power of the set F_t/ corresponding to the relation /4'/ for a given k; the set $\{\bar{u}:\ u_i=0\}$ follows from the possibility of splitting O_i. Thus, the set $\bigcap_{i \in F_t} P_i$ corresponds to the set of relations /4'/. For the interval P_i we shall define $u_i^L = \max_k \{a_{ik}/\alpha_{ik}\}$ and $u_i^R = \min_k \{b_{ik}/\alpha_{ik}\}$. Let us note that for a given $\bar{\alpha}_i$, $u_i^R = 1$. An example of sets P_{ik}, P_i, for $F_t=\{1,2\}$ and p=2, is shown in Fig. 2.

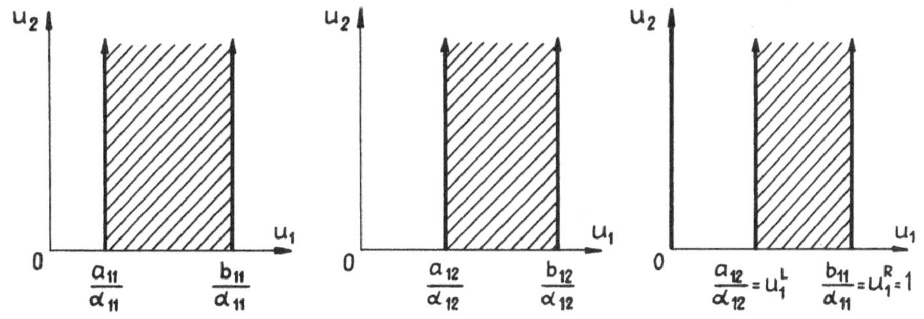

Fig.2. a/ The set P_{11}, b/ The set P_{12}, c/ The set P_1.

The set of <u>feasible normalized resource allocations</u> among operations from F_t will be denoted by

$$U_{F_t} = U^1_{F_t} \cap \bigcap_{i \in F_t} P_i .$$

For an example, see Fig. 3. Let us also define the set V_{F_t} in the following way:

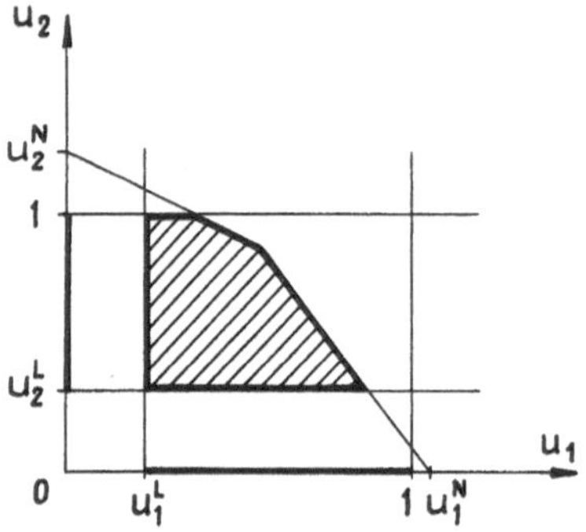

Fig.3. The set U_{F_t} for $F_t=\{1,2\}$ and p=2.

$\bar{v} \in V_{F_t}$ if and only if $\bar{u} \in U_{F_t}$, /5/

where $v_i=g_i(u_i)$, for every $i \in F_t$. Analogically we may define the sets $V^1_{F_t}$, V_k, P^v_i, P^v_{ik} which correspond to the sets $U^1_{F_t}$, U_k, P_i, P_{ik}, as in /5/. Because /5/ determines a univalent mapping, the concept of normalized resource allocation refers to the coordinates $\bar{u}$ as well as to $\bar{v}$.

Now, let us return to the problem of generating sets F_t. For this purpose let the set $(O,\prec)$ be described by an "operation-on-edge" graph G, and assume that the set of nodes in G /i.e. events/ is also partially ordered; this means that node i occurs not later than node j if $i<j$. Such an ordering is always possible and may be obtained, for example in the following way. The node without any input edges receives number 1. Next we eliminate from G all edges starting from node 1, and include the remaining nodes, which have no input edges, into the first "front". The edges being eliminated and nodes of the front, receive separately the following numbers in the arbitrary way. The next fronts are created by the successive elimination of all edges starting from the nodes of the preceding front. As can be seen, the ordering of nodes is unique if and only if each front only contains one node. In the following, we shall consider the sequence $\mathcal{F}$ of the sets F_t, having the following properties:

1° $F_t \neq \emptyset$ for each $F_t \in \mathcal{F}$.

2° There is no set $F_t \in \mathcal{F}$ which is a subset of another set in $\mathcal{F}$.

According to the definition of F_t, sequence $\mathcal{F}$ follows from both precedence and resource constraints. Let us assume, for the moment, that there are no resource constraints. Then, the consecutive sets $F_t \in \mathcal{F}$ include all operations which may be performed between the occurence of successive pairs of nodes. Of course, for a given graph G there exists one and only one sequence $\mathcal{F}$ if and only if the ordering of nodes in G is unique. Otherwise, there exist as many sequences $\mathcal{F}$ as feasible orderings of graph nodes. For example, in the graph presented in Fig. 4, front 1 contains node 2, front 2 - node 3, and front 3 - node 4, thus the sequence $\mathcal{F}$ has the form: $\{1,2\}$, $\{2,3,4\}$, $\{4,5\}$. If, however, the graph does not have the edge corresponding to operation O_3, front 2 would contain nodes 2 and 3 which could be numbered inversely. In consequence, we would obtain two sequences $\mathcal{F}$: $\{1,2\}, \{2,3\}, \{3,4\}$ and $\{1,2\}, \{1,4\}, \{3,4\}$.

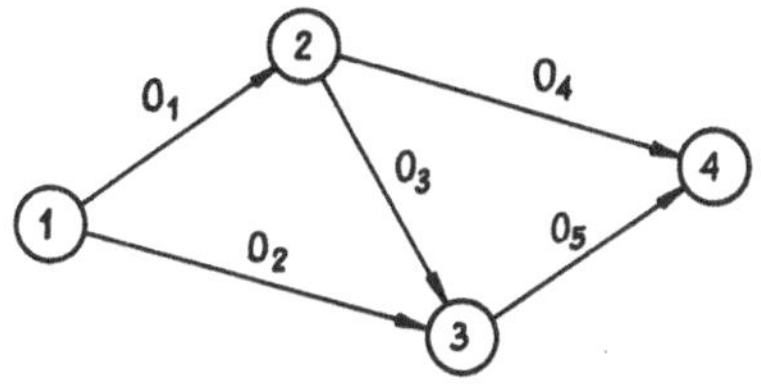

Fig.4. Example of the graph G.

Now let us take into account the influence of resource constraints on the construction of sequences $\mathcal{F}$. It is easy to note that from the viewpoint of these constraints, operations from any set F_t may be performed simultaneously if

$$\sum_{i \in F_t} \alpha_{ik} u_i^L \le N_k, \quad k=1,2,\dots,p. \qquad /6/$$

Thus in order to construct a sequence $\mathcal{F}$ in the general case, we start with a sequence $\mathcal{F}$ obtained under the assumption that there are no resource constraints, and we generate the subsets of the sets F_t, which fulfill condition /6/ and observe the properties of sequence $\mathcal{F}$. This construction is of course unique. Let us notice that for a given project and a given ordering of nodes of the graph G, the sets from the obtained sequence $\mathcal{F}$, exhaust all the possibilities of the parallel performance of operations.

Let us introduce some new definitions. First, let the sets from the sequence $\mathcal{F}$ be numbered from 1 to m taking into account the corresponding ordering of nodes in G. Let Q_i denote the set of the numbers of these sets from $\mathcal{F}$, which include operation O_i, x_{ij} denote a part of the size of operation O_i /unknown a priori/ performed in a time interval connected with $F_j \in \mathcal{F}$, and $T_j\left(\{x_{ij}\}_{i \in F_j}\right)$ denote the performing time of the parts of operations in the set F_j.

Assume that for a given project there exists one and only one sequence $\mathcal{F}$. Then the following theorem holds.

THEOREM 1

The completion time of the project is at the minimum if the function

$$T = \sum_{j=1}^{m} T_j\left(\{x_{ij}\}_{i \in F_j}\right) \qquad /7/$$

reaches a global minimum subject to

$$\sum_{j \in Q_i} x_{ij} = w_i, \quad \text{for } x_{ij} \ge 0, \text{ and every } i \in F_j, \quad j=1,2,\dots,m, \qquad /8/$$

where

$$T_j\left(\{x_{ij}\}_{i \in F_j}\right) = \inf\left\{T_j > 0: \bar{x}_j/T_j \in \text{conv } V_{F_j}\right\}, \; j=1,2,\dots,m, \qquad /9/$$

$\bar{x}_j = \{x_{ij}\}_{i \in F_j}$, conv V_{F_j} is the smallest convex set containing set V_{F_j} /see /5//.

PROOF

In [5], it has been proved that for the case of independent operations

/i.e. such that there are no precedence constraints among them/ the completion time of the project is at the minimum if and only if it may be presented in the form $T = \inf\{T > 0: \bar{w}/T \in \text{conv } V\}$, where $\bar{w} = (w_1, w_2, \dots, w_n)$ is the vector of operation sizes. According to the definition of the sequence $\mathcal{F}$, the parts of operations performed in every set F_j are independent and thus the theorem for independent operations is true for every set F_j. /By a "part of operation" O_i we understand an operation differing from O_i only by its size which is a part of w_i/. It is also easy to note that the completion time of the project is the sum of the performing times of consecutive sets of operation parts in the sequence $\mathcal{F}$. Thus finding the global minimum of /7/, subject to /8/, we find such a division of operation sizes into parts, which will ensure the attainment of the minimum completion time of the project. ◇

It follows from Theorem 1 that $T_j(\{x_{ij}\}_{i \in F_j})$ is obtained by using the normalized resource allocation /which need not be feasible/ determined in the intersection point of the straight line, described by parametric equations:

$$v_i = x_{ij}/T_j, \quad \text{for every } i \in F_j, \qquad /10/$$

with the boundary of the set conv V_{F_j}. This boundary is identical with the set of all convex combinations of the elements of V_{F_j}. Because the form of the set V_{F_j} is determined by the form of the function $g_i(\cdot)$ and the resource constraints /see /5//, we may establish a priori many important properties of optimal solutions to our problem, as well as the properties of the methods of their computation.

In the next section we shall formulate some corollaries concerning these properties. To save space we shall not give the proofs of the corollaries, they follow however almost immediately from Theorem 1 and several simple facts among which, the most important is that for $g_i(\cdot)$, and all $i \in F_j$, being concave, the set $V^1_{F_j}$ is convex. We hope that the presented illustrative material will be sufficient to help the reader in proving these corollaries.

3. SOME PROPERTIES OF THE SOLUTIONS

For simplicity, we shall consider the case of independent operations, which is justifiable essentially by Theorem 1. Let us assume moreover that $\sum_{i=1}^{n} \alpha_{ik} u_i^L \le N_k$, $k=1,2,\dots,p$, which means that the sequence $\mathcal{F}$ contains only one set $F_j = \{1,2,\ldots,n\}$. Thanks to this, we may drop the index F_j. Let us introduce some further definitions which will be useful in the following [3].

Let $\bar{v}^{\bullet *}$ denote the intersection point[1/] of the straight line /10/ with the boundary of the set conv V. $T^{*}_{\mathrm{conv}\,V^1}$ will denote the minimum performing time of the set of operations, under the assumption that $V = V^1$. A solution is said to be <u>unique</u> if $\bar{v}^{\bullet *}$ is a convex combination of at most one n-tuple of elements of the set V. For example, the solution presented in Fig. 6b[2/], in the light of the above definition, is unique because the point $\bar{v}^{\bullet *}$ is a convex combination of only one pair of elements of the set V: $A = (0, g_1(u_1^N))$ and $B = (g_2(u_2^N), 0)$ which correspond to the serial performance of the operations, with maximum

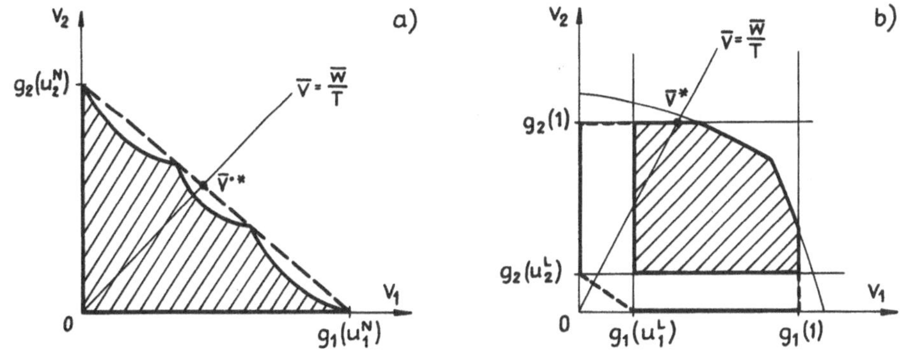

Fig.5. Examples of non-unique solutions:
a/ $V = V^1$ and g_i, i=1,2, are convex; b/ g_i, i=1,2, are concave.

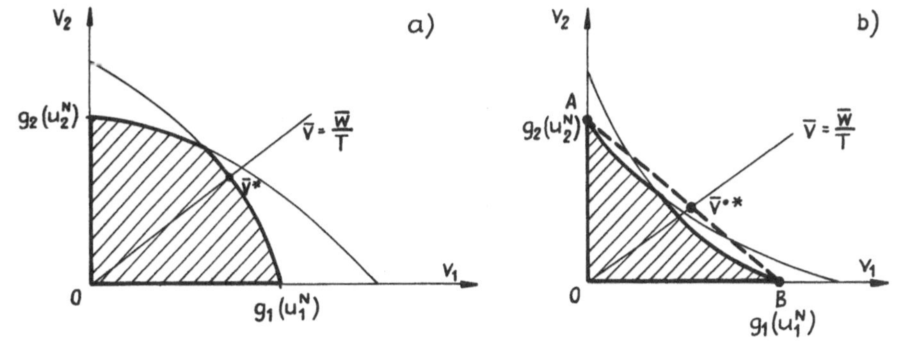

Fig.6. Examples of unique solutions:
a/ conv $V = V^1$, n=2, p=2; b/ $V = V^1$ and g_i, i=1,2, are convex.

1/ Point $\bar{v}^{\bullet *}$ need not determine a feasible normalized resource allocation.

2/ In the following figures, the boundary of the set V is marked by a continous heavy line, and the part of the boundary of the set convV which is different from the boundary of V - by a dotted line.

amounts of resources. Note that the definition of a unique solution does not mean that there only exists one function $\bar{u}^*(t)$ minimizing T. When applying linearly dependent combinations of the n-tuple making $\bar{u}^*(t)$, one obtains different functions $\bar{u}^*(t)$ for the same T. However, the linearly dependent combinations involve more splitting of operations which is unsatisfactory in practice. Examples of non--uniques solutions are shown in Fig.5, and examples of unique solutions - in Fig. 6.

COROLLARY 1

$T_j(\{x_{ij}\}_{i\in F_j})$ given by /9/ is a convex function.

COROLLARY 2

For any $\bar{v}^{\cdot *}$ the optimal solution has the form:

$$\bar{v}^*(t) = \bar{v}^{*l}, \quad t \in \langle t_l - \Delta t_l, t_l), ^{1/} \; l=1,2,\dots,r\leq n, \quad t_r = T^*, \tag{11}$$

where $\bar{v}^{*l}$, $l=1,2,\dots,r$ fulfill the equation:

$$\sum_{l=1}^{r} \lambda_l \, \bar{v}^{*l} = \bar{v}^{\cdot *}, \quad \lambda_l \geq 0, \quad l=1,2,\dots,r,$$
$$\sum_{l=1}^{r} \lambda_l = 1, \qquad \Delta t_l = \lambda_l T^*, \; l=1,2,\dots,r, \tag{12}$$

and

$$T^* = w_i/v_i^{\cdot *}, \qquad i=1,2,\dots,n. \tag{13}$$

COROLLARY 3

If $V = V^1$ and $g_i(\cdot)$, $i=1,2,\dots,n$, are concave, then

$$T^*_{\mathrm{conv}\,V1} = \max_k \left\{T^*_{\mathrm{conv}\,V_k}\right\} \tag{14}$$

where $T^*_{\mathrm{conv}\,V_k}$ is equal to the positive root of the equation:

$$\sum_{i=1}^{n} \alpha_{ik} \, g_i^{-1}(w_i/T) = N_k, \quad k=1,2,\dots,p. \tag{15}$$

Let us note, that under the assumptions made in the last corollary, conv $V = V^1$ and in the optimal solution, the operations are performed with fixed amounts of resources, and end simultaneously. An example is shown in Fig. 6a.

COROLLARY 4

The optimal solution is unique, if conv $V = V^1$ and at most one function $g_i(\cdot)$, $i=1,2,\dots,n$, is linear in any interval of values of u_i.

COROLLARY 5

If $V = V^1$ and $g_i(\cdot)$, $i=1,2,\dots,n$, are convex, two situations may arise:

1° For any pair of operations i,j, the "external" part of the boundary of the set conv V^1, in the coordinate system (v_i,v_j),

---

1/ The last interval in /11/ is closed on both sides.

is a straight line of the form /Fig. 5a,6b/:

$$v_i = v_j \, g_j(u_j^N)/g_i(u_i^N) + g_j(u_j^N), \quad i,j=1,2,\dots,n; \quad i \neq j. \qquad /16/$$

Then

$$T^*_{conv} V^1 = \sum_{i=1}^{n} w_i/g_i(u_i^N). \qquad /17/$$

2° For at least one pair of operations i,j, the "external" part of the boundary of the set $conv\ V^1$, in the coordinate system (v_i, v_j), is a broken line. Then $\bar{v}^{\bullet *}$ /and thus $T^*_{conv} V^1 = \bar{w}/\bar{v}^{\bullet *}$/ are determined generally by formula /12/ /Fig. 7/.

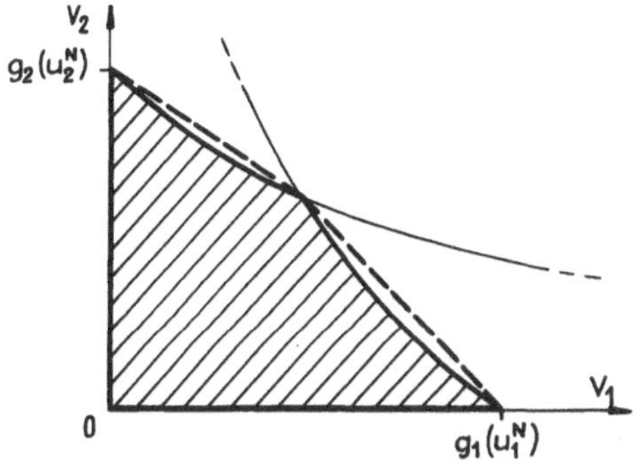

Fig.7. Example of sets V^1 and $convV^1$ for convex g_i, $i=1,2$.

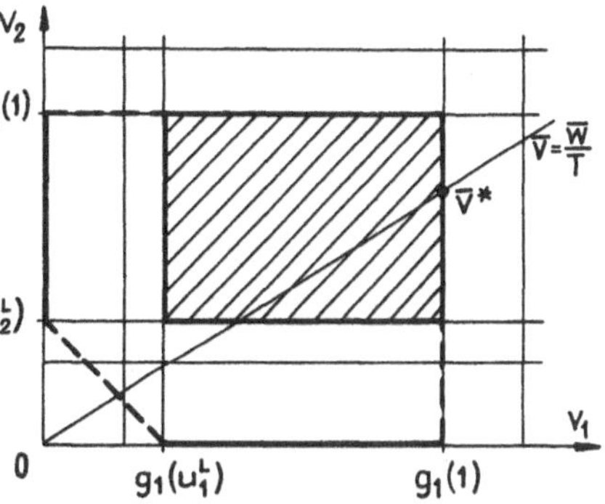

Fig.8. The optimal solution for $V = P_1^V \cap P_2^V$.

COROLLARY 6

If $V = \sum_{i=1}^{n} P_i^V$, the minimum performing time of the set of operations may be calculated from /Fig. 8/:

$$T^* = \max_i \left\{ w_i/g_i(1) \right\}. \qquad /18/$$

COROLLARY 7

If $conv\ V = conv\ V^1 \cap conv \bigcap_{i=1}^{n} P_i^V$, then the minimum performing time of the set of operations may be calculated from the formula:

$$T^* = \max \left\{ T^*_{conv} V^1, \left\{ w_i/g_i(1) \right\}_{i=1}^{n} \right\}. \qquad /19/$$

As a result of proving the above corollaries, it is possible to calculate the optimal solution for particular cases.

4. FINAL REMARKS

The problem considered in this paper belongs to the class of problems lying across the border of operations research and control theory. It emphasises the integration of system sciences both in mathematical model and optimization techniques, and in the range of practical applications.

The approach to a solution, presented here, is not the only possible approach; however it seems that it does give the broadest possibilities of analysing the properties of optimal solutions a priori, and thus, leads to the right choice of an appropriate technique for solving the problem.

REFERENCES

1. Bubnicki, Z.: Optimal control of a complex of operations with random parameters, Podstawy Sterowania 1 /1971/,3-10.
2. Burkov, V.N.: Optimal control of a complex of operations,/Russian/ IV Congress of IFAC, Technical Session 35, Warsaw, 1969, 46-57.
3. Słowiński, R.: Optimal Control of Multiple-Resource Allocation in a Complex of Operations, /Polish/ unpublished doctoral dissertation, Technical University of Poznan, 1977.
4. Węglarz, J., Słowiński, R.: Computational aspects of a certain class of time-optimal resource allocation problems in a complex of operations, Foundations of Control Engineering 1 /1976/,123-133.
5. Węglarz, J.: Application of the convex sets theory in a certain problem of time-optimal control of a complex of operations, Systems Science 1 /1975/,67-74.
6. Węglarz, J.: Time-optimal control of resource allocation in a complex of operations framework, IEEE Trans. Systems, Man and Cybernet., SMC-6, 11/1976/,783-788.

THRESHOLD METHODS FOR BOOLEAN OPTIMIZATION PROBLEMS WITH SEPARABLE OBJECTIVES

U. Zimmermann
Mathematisches Institut
Universität zu Köln
Weyertal 86
5000 Köln 41

Abstract

Let P denote the set of all subsets of $N:=\{1,2,...,n\}$. Let $(H,*,\leq)$ be a negatively ordered commutative semigroup with internal composition '*' and order relation '$\leq$'. Separable objectives $f : P \to H$ have the general form $f(X) = \mathop{*}_{i \in X} c_i$ with coefficients $c_i \in H$ for $i \in N$. The separable objective shall be maximized over $B_k \subseteq P$ which consists only of sets of cardinality $k \in N$. Especially the set of all intersections of maximal cardinality for two matroids is considered. From the threshold method of Edmonds and Fulkerson for bottleneck extrema we derive a class of suboptimal algorithms for the general problem. During the algorithm lower and upper bounds for the optimal objective value are determined.

1. Introduction

In 1974 assignment problems in certain ordered commutative semigroups were considered by BURKARD, HAHN and ZIMMERMANN [1]. Thereafter different types of Boolean (combinatorial) optimization problems together with suitable algorithms have been studied by means of the algebraic approach. The algebraic structure is always given in terms of a special *ordered commutative semigroup* (o.c.s) $(H,*,\leq)$ with internal composition '*' and order relation '$\leq$'. $(H,*,\leq)$ is called o.c.s if the following three axioms hold

(1) $(H,\leq)$ is a nonempty totally ordered set

(2) $(H,*)$ is a commutative semigroup

(3) $a \leq b \Rightarrow a * c \leq b * c$ for all $a,b,c \in H$.

The neutral element in H is denoted by e. Typical simple examples are the o.c.s $(\mathbb{R},+,\leq)$ and $(\mathbb{R},\min,\leq)$. For convenience we define

$$\mathop{\ast}_{i\in N} c_i := c_1 * c_2 * \ldots * c_n$$

for $N := \{1,2,\ldots,n\}$. Let P denote the set of all subsets of N. Then an objective function $f : P \to H$ is called *separable* if

$$f(X) := \begin{cases} e & X = \emptyset \\ \mathop{\ast}_{i\in X} c_i & \emptyset \neq X \in P \end{cases}$$

holds with $c_1, c_2, \ldots, c_n \in H$.

The combinatorial structure of the problem is given by the set $B_k \subseteq P$ of all feasible sets. We assume that B_k consists only of sets of identical cardinality $k \in \mathbb{N}$. Then we consider the optimization problem

(BOP) $$\max_{X\in B_k} f(X) \quad .$$

For the special semigroup $(\mathbb{R},\min,\leq)$ this is the bottleneck problem

$$\max_{X\in B_k} \min\{c_i \mid i\in X\}$$

which can be solved optimally and efficiently by the threshold method of EDMONDS and FULKERSON [3]. In general there is no efficient algorithm known, but the threshold method can be applied to find efficiently suboptimal solutions and certain bounds, if we assume w.l.o.g.

(4) $$c_i \leq e \text{ for all } i \in N.$$

This assumption is trivial if $(H,*,\leq)$ is *negatively ordered*, that is if

$$a * b \leq a$$

holds for all $a,b \in H$. An example is $(\mathbb{R},\min,\leq)$. In order to achieve (4) we can define a new BOP equivalent to the original one if

(5) $$a < b \Rightarrow \exists c \in H : \; a = b * c$$

holds for all $a,b \in H$. We define the coefficients c_i' of a new separable function f' by a certain transformation

$$(T) \quad \begin{array}{ll} c * c_i' := c_i & \text{for } c_i < c \\ c_i' := e & \text{else} \end{array}$$

with $c := \max\{c_i \mid i \in N\}$. Then every solution of the (BOP) with respect to f' is a solution of the original problem. At the end of section 2 we show that axiom (4) is not necessary but convenient for our investigations.

2. The General Threshold Method

We consider the *independence system* corresponding to

$$S: = \{Y \mid \exists X \in B_k : Y \subseteq X\}$$

and the *rank function* $r : P \to N$ corresponding to B_k

$$r(N'): = \max\{|Y| \mid Y \subseteq N', Y \in S\}$$

for all $N' \subseteq N$. A set $N' \subseteq N$ is called *closed* if there is no set $N'' \subseteq N$ with $N' \subsetneq N''$ and $r(N') = r(N'')$. The system of all closed sets is denoted by $C(B_k)$. If $N' \subseteq N'' \in C(B_k)$ and $r(N') = r(N'')$, then N" is called a *minimal closed cover* of N'. Clearly, every set N' has at least one minimal closed cover but in general it is not unique.

Without loss of generality we assume that

$$(6) \qquad c_n \leq c_{n-1} \leq \cdots \leq c_1 \leq e$$

holds in the following. Let $N' \subseteq N$ and $N \setminus N' = \{i_1, i_2, \ldots, i_r\}$ with $i_1 < i_2 < \cdots < i_r$. With $s: = k - r(N')$ we define the *best remainder set* to N'

$$(7) \qquad \Delta X(N') := \begin{cases} \emptyset & s = 0 \\ \{i_1, i_2, \ldots, i_s\} & \text{else} \end{cases}$$

A first consequence is the following lemma.

(2.1) Lemma

Let $N' \subseteq N$ and $X \in B_k$. Then

$$f(X) \leq f(\Delta X(N')) .$$

Proof. By (6) we find $f(X) \leq f(X \cap (N \setminus N'))$. As $s \leq |X \cap (N \setminus N')|$ holds we find by (3) $f(X \cap (N \setminus N')) \leq f(\Delta X(N'))$.

An immediate corollary of this lemma is

(8) $$\max_{X \in B_k} f(X) \leq \min_{N' \in P} f(\Delta X(N')) .$$

To get the minimal value of the right hand bound it is enough to consider only the closed sets.

(2.2) Lemma

$$\min_{N' \in P} f(\Delta X(N')) = \min_{N' \in C(B_k)} f(\Delta X(N'))$$

Proof. As $C(B_k) \subseteq P$ it suffices to show "$\geq$". Let $N' \subseteq N$. Choose a minimal closed cover N" of N'. Then $N'' \in C(B_k)$ and $f(N'') \leq f(N')$.

For bottleneck objectives we find equality in (8). Consider the set

$$C_{k-1} := \{N' \in C(B_k) \mid r(N') = k - 1\} \quad .$$

Then from [2] we may conclude without difficulties

(9) $$z := \max_{X \in B_k} \min \{c_j \mid j \in X\} = \min_{N' \in C_{k-1}} \max \{c_j \mid j \in N \setminus N'\} .$$

Together with (2.1), (2.2) this implies the following theorem.

(2.3) Theorem

Let z be defined by (9). Then

$$\max_{X \in B_k} f(X) \leq \min_{N' \in C(B_k)} f(\Delta X(N')) \leq z$$

holds for an arbitrary separable function $f: P \to H$.

In the case of $f(X) = \min\{c_j \mid j \in X\}$ we find equalities in (2.3).

The following threshold algorithm is a systematic tool to find upper and lower bounds for the optimal value during the computation of a feasible set. For bottleneck objectives the algorithm terminates with the optimal solution.

General threshold method

We assume $c_n \leq c_{n-1} \leq \dots c_1 \leq e$.

(1) $N' := \{c_1, c_2, \dots, c_k\}$; $z := f(N')$.

(2) Find an independent set $X \subseteq N'$ of maximal cardinality and a minimal closed cover N'' of N';
If $|X| = k$ stop.

(3) $z := \min\{z, f(\Delta X(N''))\}$.

(4) Choose $\emptyset \neq Y \subseteq \Delta X(N'')$;
$N' := N' \cup Y$;
Go to (2).

As in each step the set N' is enlarged by at least one element the algorithm has the complexity $O(n)$ times the complexity of step (2). This is obviously an efficient method if there are efficient methods for determining maximal independent sets and minimal closed sets with respect to the special combinatorial structure. For example this is true for matroid intersection and matching problems (cf. LAWLER [3]). At the termination of the algorithm we find a suboptimal solution $X \in B_k$ and an upper bound z for the optimal value z_{opt} of the (BOP):

$$f(X) \leq z_{opt} \leq z .$$

With respect to the algebraic approach the method is a rather general one, since one has usually to assume further axioms in the semigroup $(H, *, \leq)$ fo find optimal solutions (cf. [1],[4],[5] and [6]).

Finally we consider the special case, that the coefficients of the separable objective function have been transformed according to (T). We can use the original coefficients c_i instead of the transformed c'_i if we modify step (3) in the right way. As the transformation does not change the ordering of the coefficients all other steps can be performed using the original coefficients without modification.

(2.4) Lemma

Let $N'' \subseteq N$, $s := k - r(N'')$. Then according to (T)

$$f'(\Delta X(N'')) * c^{s} = f(\Delta X(N'')) .$$

Proof. According to (T) we find immediately

$$f'(X) * c^{|X|} = f(X)$$

for every $X \in P$. ($|X|$ denotes the cardinality of X). Then the lemma follows from $s = |\Delta X(N'')|$.

If we calculate the bound z for the original problem using the bound z' of the transformed one by means of (2.4) we find

$$z' * c^{k} = z .$$

Therefore the modified step ③ is

③' $$z := \min \{z, f(\Delta X(N'')) * c^{r(N'')}\} .$$

It is possible in all our theoretical investigations to use

$$g(\Delta X(N'')) := f(\Delta X(N'')) * c^{r(N'')}$$

instead of f. Thus axiom (4) may not be fulfilled in general.

3. A Threshold Method For Matroid Intersection Problems

For the special combinatorial structure of the matroid intersection problem the general algorithm can be improved. Most of the results of this section can be found in ZIMMERMANN [4]. Algorithms for the optimal solution of the matroid intersection problem in certain ordered semigroups have been described in [4],[5] and [6].

Let $\emptyset \neq S \subseteq P$. Then $M = M(N,S)$ is called a *matroid* if the following two properties hold:

(10) $Y \in S \wedge X \subseteq Y \Rightarrow X \in S$

(11) $X,Y \in S \wedge |X| < |Y| \Rightarrow \exists j \in Y \setminus X : X \cup \{j\} \in S$.

The elements of S are called the *independent sets* of the matroid M(N,S). S is called the independence system of M. Let us consider two matroids M_B, M_R with independence systems S_B, S_R and rank functions r_B, r_R as defined at the beginning of section 2. The elements of $S: = S_B \cap S_R$ are called the *intersections* of M_B and M_R. Let

$$B_k: = \{X \in S \mid |X| = k\}$$

for $k \in N$. Let r denote the rank function corresponding to B_k. Then we consider the *algebraic k-intersection problem*

$$\max_{X \in B_k} f(X) .$$

In step ③ of the general threshold algorithm we can apply a special method for matroid intersection problems. For a given set $N' \subseteq N$ the cardinality intersection algorithm (cf. LAWLER [3]) determines an intersection of maximal cardinality in N' and a minimal cover $N'' \subseteq N$ of N'. The minimal cover consists of two subsets N_B, N_R which are closed with respect to M_B, M_R. They are in general not disjoint. Using this modification we find the following special threshold method.

<u>Threshold method for Matroid Intersection</u>

① $N': = \{c_1, c_2, \ldots, c_k\}$; $z: = f(N')$.

② Find an intersection $X \subseteq N'$ of maximal cardinality and a minimal closed cover $N_B \cup N_R$ of N';
If $|X| = k$ stop.

③ $z: = \min \{z, f(\Delta X(N_B \cup N_R))\}$.

④ Choose Y,Y' with
$\emptyset \neq Y \subseteq \Delta X(N_B \cup N_R)$,
$Y' \subseteq N_B \cap N_R$;
$N': = (N' \cup Y) \setminus Y'$;
Go to ② .

The proof of the finiteness of this algorithm is not so easy as in the general case. This is due to the possible nonempty choice of Y' in step (4) . In [4] the following theorem can be found.

(3.1) Theorem

Step (2) of the algorithm can be performed at most $\frac{1}{2}k\ (k+1)$ times regardless of the choice in step (4) .

Using a special choice in step (4) it can be shown that the Hungarian method for partition-matroid intersection problems (cf. ZIMMERMANN[4]) is in a certain sense a special case of the threshold method. The sequence of sets N' determined in the Hungarian method coincides with the sequence due to the special choice in (4) .

A final example shall illustrate the threshold method. For convenience we consider the well known sum assignment problem

$$\max_{\varphi \in B_n} \Sigma c_{i\varphi(i)}$$

where B_n is the set of all permutations $\varphi : N \rightarrow N$.

The cost coefficients $c_{ij} \in \mathbb{N}$ are given by the following matrix C

$$C := \begin{bmatrix} 11 & 8 & 4 & 5 \\ 10 & 9 & 3 & 1 \\ 8 & 7 & 4 & 5 \\ 1 & 8 & 2 & 3 \end{bmatrix}$$

The intersections of the matroid problem are those sets of entries in C whose elements have no row or column in common. The closed sets N_B (N_R) are the unions of some columns (rows). We choose $Y' \equiv \emptyset$ and $Y \equiv \Delta X(N_B \cup N_R)$ in step (4) of the threshold method. In the following figures we indicate

the elements of N' by ◯

the elements of the intersection X by ◯*

the minimal cover $N_B \cup N_R$ by ////

the elements of $X(N_B \cup N_R)$ by □ .

After the first performance of step ② we find

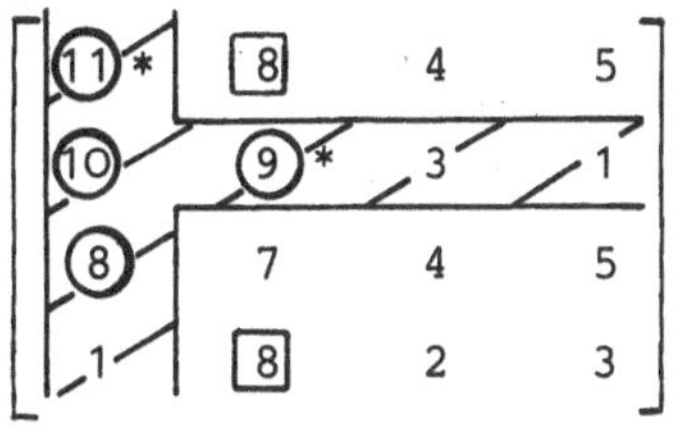

To derive the upper bound we use the modification of step ③ and
find z = 38. In the next performance of step ② we cannot enlarge
the intersection

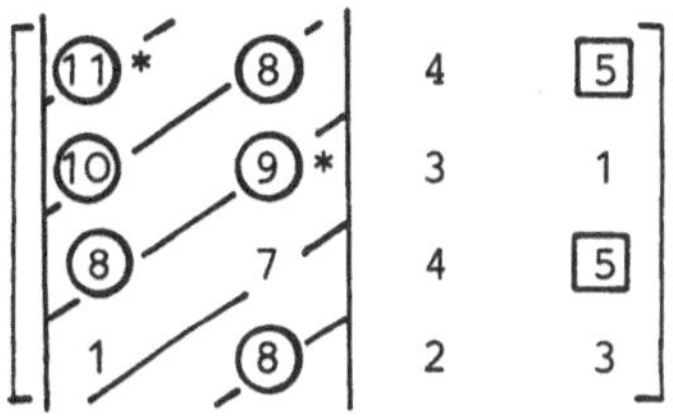

and find z = min {38,32} = 32. Now we determine an intersection of
cardinality 3. The upper bound is not

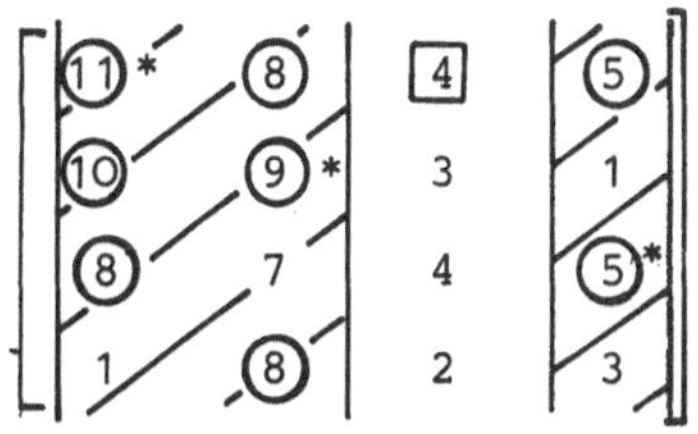

improved : z = min {37,32} = 32, but in the next performance of
step ② we find an intersection of maximal cardinality

$$
\begin{bmatrix}
⑪ & ⑧ & ④* & ⑤ \\
⑩* & ⑨ & 3 & 1 \\
⑧ & 7 & 4 & ⑤* \\
1 & ⑧* & 2 & 3
\end{bmatrix}
$$

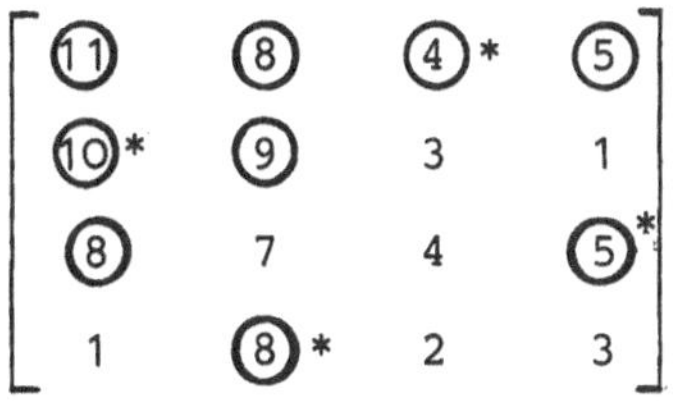

with value f(x) = 27. Although the upper bound is 32 it is easy to
see that X is an optimal solution of the problem.

References

[1] BURKARD, R.E.; HAHN, W.; ZIMMERMANN, U.: An algebraic approach to assignment problems, Math. Prog. 12 (1977) 318-327.

[2] EDMONDS, J.; FULKERSON, D.R.: Bottleneck extrema, J. Comb. Th. 8 (1970) 229-306.

[3] LAWLER, E.: Combinatorial optimization: networks and matroids, Holt, Rinehart and Winston, New York (1976).

[4] ZIMMERMANN, U.: Boole'sche Optimierungsprobleme mit separabler Zielfunktion und matroidalen Restriktionen, Thesis, Mathematisches Institut, Universität zu Köln (1976).

[5] ZIMMERMANN, U.: Matroid intersection problems with generalized objectives, to appear in the Proceedings of the IX International Symposium on Mathematical Programming, North Holland.

[6] ZIMMERMANN, U.: Duality and the algebraic matroid intersection problem, submitted for publication.

COMPARISON OF SOME EDUCATIONAL PLANNING MODELS

A. Lukka
Lappeenranta University of Technology
Box 57, 53101 Lappeenranta 10, Finland

This study is centered on the problem of educational investments, especially in higher education. One aim in educational planning is to give the opportunity to as many students as possible to have the education they want. This demands very wide and flexible educational capacities. The preferences of the students are determined by many different factors such as ambition, interest, general attitudes, opportunities for getting the education, future status and future income. The education the students finally have may not lead to very satisfactory results. For instance, there may be unemployment or underemployment in their particular fields. The employment situation will affect the educational choices of future students and adjustments in the educational system will be needed.

In an economical system, unemployment, underemployment, and the provision of education all cost money. To save expence, the preferences of the students might be manipulated, or restrictions may be forced upon the educational opportunities.

Attempts have been made to solve this problem by using the latter method, for instance Balinsky and Reisman [1], Burckhardt [2], Correa [3], Thonstad [6] and the author [4].

The purpose of this paper is to consider the applicapability of these models with special attention to conditions in Finland. The selected models seemed to be the most advanced of the educational planning models which also considered the demand and supply of manpower.

Balinsky and Reisman [1] represent in their paper a hierarchy of models where the educational system consists of one educational specialty with various stages of educational attainment. Manpower demand is projected, as well as the enrollment pool. The latter is the only restriction, but assumed not to be active. Graduation time is a constant, no class repeaters, and educational capacities are not included. The only exit-flows from the system are to employment through the labor pool, mortality or transfers to other specialties are not included. Objective function consists of the costs of students and costs of the labor pool summed up over time. The decision variables are the entrants to the first level, though the transition rates may be manipulated. The system is depicted in figure 1.

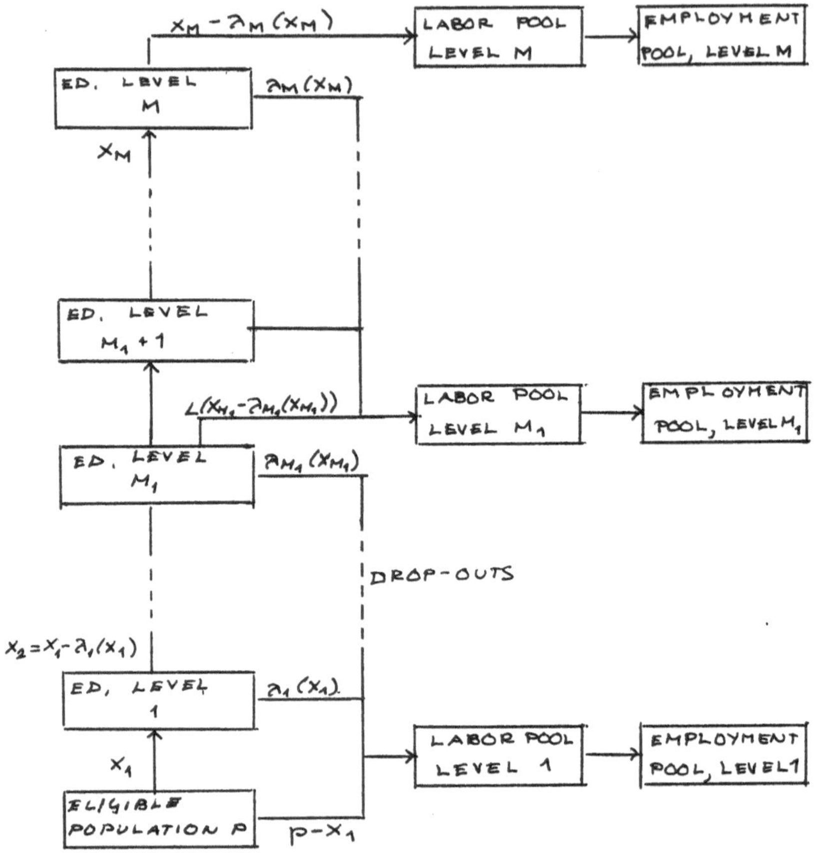

Figure 1. Generalized Balinsky-Reisman model.

Burckhardt [2] considers a different system consisting of one educational attainment with various lines of specialization, the enrollment pool, success of studies (= graduation rates), retirements from the work force and total manpower demands are projected. The only restrictions come from the preferences of the students or, in other words, the willingness of the students to study. Educational capacities and potential entrants are included in the objective, not in the restrictions. All specialties have the same constant graduation time. The objective function consists of the weighted squared differences between total manpower demand and supply, number of entrants and educational capacities, the total number of entrants and the enrollment pool, summed up over specialties and time. The system is shown in figure 2.

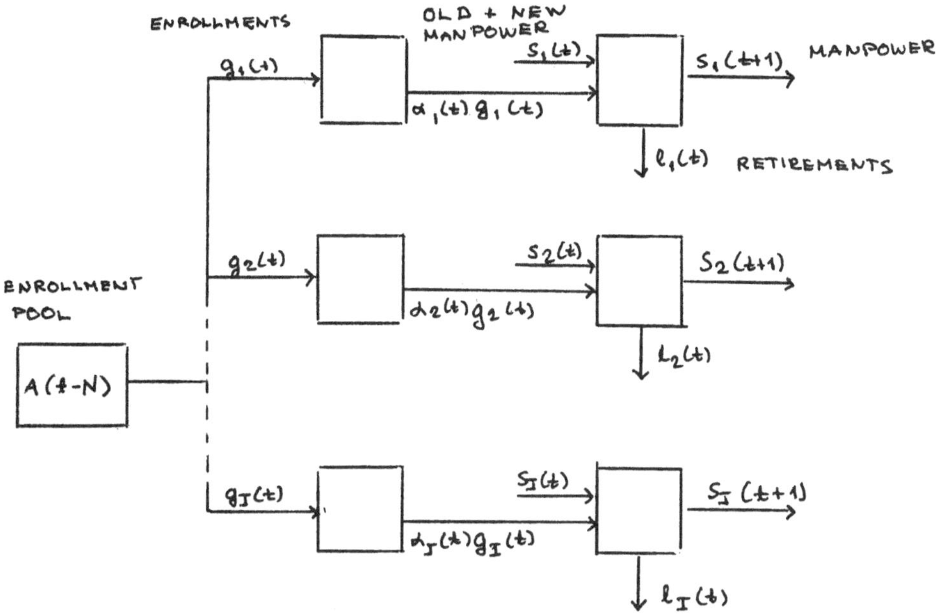

Figure 2. Burckhardt's model.

Correa [3] presents a survey. He considers many aspects in educational planning problems, and suggest several methods of connecting these aspects into models. All his models are linear. Most of the aspects are shown in table 1. Both he and Thonstad [6] pay special attention to the number of teachers required by the educational system. Both also suggest some method of meeting manpower demands, but exclude all constraints (enrollment pool, educational capacities) in this context.

The model by the author [4] is a multilevel, multispecialty model, where educational capacities, enrollment pool and enrollment restrictions (for instance political) are projected and given in constraints.

The specific features of these models are summarized in table 1. Table 1 is not completely conclusive. The x's have different meanings in different models. The (x)'s imply that the corresponding feature is implied, but not explicitly used in the model, or that the feature will be found in other models by the same author or authors, indicating thus that the author is aware of the problem.

FEATURES OF THE MODELS

	Balinsky-Reisman	Burckhardt	Correa	Thonstad	Lukka
Multilevel	X		X	X	X
Multispecialty		X	X	X	X
Ed. costs in constraints			X		(X)
Ed. costs in the objective	X				(X)
Ed. capacity in constraints			X		X
Ed. capacity in the objective		X			
Use of labor pool	X	X			
Enr. wishes	(X)	X	(X)	(X)	(X)
Special constraints on enr.		(X)			X
Graduation time constant	X	X			
Graduation time variable			X	X	X
Drop outs included	X		X		X
Differences in schools					X
Previous enr. included			X	(X)	X
Manpower problem	X	X	(X)	(X)	X

Table 1.

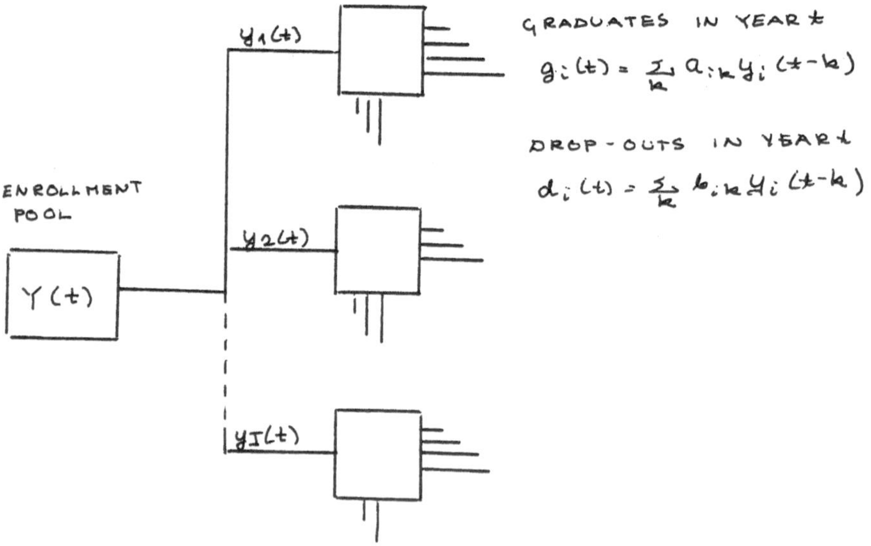

Figure 3. Lukkas one-level model.

It turned out that the models by Correa [3] and Thonstad [6] were not adaptable to the problem of combining educational and manpower planning. However, with some modifications, this could have been accomplished. They both have some features that the others are lacking (f.i. including the need for educational capacities, how to obtain them and their cost).

The model by Balinsky and Reisman [1] is very attractive, clear and simple. The objective function was given by costfunction in a general form, which was not difficult to modify for the comparison.

The difficulties with this model were the lack of any educational capacity constraints,mortality,and transitions out of the system.

In Finland the educational capacities are very active constraints. Also the enrollment pool can be an active constraint especially if enrollment wishes are taken into account. If these constraints are added to the model, it resembles closely a one-specialty version by the author.

The final comparison between the models by Burckhardt [2] and a one-level version by the author is now outlined.

Educational capacities in both models have to be projected. In the author's model they are in the constraints and consider the total number of students in school using these capacities.

According to Burckhardt they are in the objective and concern only entrants. Since Burckhardt has constraints in his model, this is not a real simplification. Also weights for corresponding deviations have to be determined. The same remarks can be made about the enrollment pool.

In Burckhardt's model the labor pool is cumulative, where as the author makes the error of considering only new graduates and current demand for new manpower.

Mortality, including transitions out of the system due to illness etc., is not included explicitly in Burckhardt's model, but may be implicit in the coefficient of the success of studies. It is explicit in the author's model, but it may be of slight importance.

Graduation time, e.g. time needed to complete studies, is in Burckhardt's model constant, and equal in all specialties. Graduation time statistics show significant variation in Finland eg to obtain the degree of M.A. or M.Sc. takes 3-10 years. Moreover, the graduation time distributions are also different in different universities. However, if enrollments are steady, constant graduation times would simplify the model.

Success of studies, the coefficient is dependent on time in Burckhardt's model. The author was satisfied with a constant, in order to simplify the model. This constant is obtained from the graduation time and drop-out time distributions. The differences in schools is another complicating factor, and used only in the author's model. In the planning model the entrants are distributed into different schools according to the capacities of the schools, graduation and drop-out time distributions are applied and the results are combined, so it actually only means a few more calculations in the beginning in the hope of improving accuracy.

Burckhardt has formulated his model as a control problem, but the solution methods were not very straightforward in general multi-specialty cases. It was transformed rather easily into a quadratic programming problem, enrollments as decision variables. The same, not so easy transformation was done with the author's model. The resulting problems were of the same nagnitude. The number of decision variables is the number of specialties (P) multiplied by the number of planning period time units (T). The quadratic problems were solved by a computer routine by Ravindran [5].

Many weights were needed in Burckhardt's model, and if they were not carefully chosen, the capacities and enrollment pool supply were very easily exceeded.

If the manpower demand is strongly fluctuating, Burckhardt's model will result in stronger enrollment fluctuations than the model by the author, especially if the labor pool is not sufficient. The fluctuations can be handled by smoothing restrictions on the consecutive enrollment numbers, but they enlarge the models. Final analysis of the effect of the constant graduation time is not yet finished.

REFERENCES

[1] BALINSKY, W., and A. REISMAN, "Some Manpower Planning Models Based on Levels of Educational Attainment", Management Science, 18:2, 1972.

[2] BURCKHARDT, W., "Bildungs- und Arbeitsmarktplanung unter Einsatz von Verfahren der Kontrolltheorie", Zeitschrift für Operations Research, Band 17, 1973.

[3] CORREA, H., "A Survey of Mathematical Models in Educational Planning", Mathematical Models in Educational Planning, OECD, Paris 1967.

[4] LUKKA, A., "Models for Planning Educational Structures", NOAK -74, Helsinki 1974.

[5] RAVINDRAN, A., "A Computer Routine for Quadratic and Linear Programming Problems", Communications of the ACM, 15:9, 1972.

[6] THONSTAD, T., Education and Manpower Theoretical Models and Empirical Application , Oliver & Boyd Ltd., Edinburg and London, 1969.

MATHEMATICAL PROGRAMMING IN HEALTH-CARE PLANNING

R.Minciardi, P.P.Puliafito, R.Zoppoli

Istituto di Elettrotecnica
University of Genoa
Viale Causa, 13
16145 Genoa, Italy

1. Introduction and statement of the problem

The problem we are dealing with in this paper refers to health-care districting already defined in [1]. Since a survey of mathematical programming in health-care planning is beyond the scope of the paper, some general comments on possible methodological approaches in this field are worth mentioning.

More specifically, we want to enhance some aspects which are strictly related to the planning process itself and turn out to be very important both from a theoretical and an applicative point of view. Whenever a formally stated planning problem is faced, three distinct phases may be considered:

a) Decomposition. It arises in large-scale problems, when we may take some operational advantage in decomposing or simplifying the problem by exploiting its physical structure.

b) Generation of alternatives. It is the usual central problem of mathematical programming, and consists in finding the set of feasible solutions.

c) Decision. This phase requires not only the "optimizing" step, but also that proper values be assigned to the feasible solutions.

If it is possible to define a precise performance criterion, this phase consists only of the optimizing procedure.

Phases b) and c) are generally solved by means of a single operating step (i.e., the optimizing algorithm). In health-care planning, however, the possibility of assigning precise values to alternatives, and then to decisions, is quite rare. Actually, assigning values to alternatives constitutes by itself a very complicated process, which involves several entities (decision makers, groups of people , technical operators, etc.). Therefore, it is difficult to express the performance function in a well defined analytical form. Moreover, the values may be assigned by taking into account several attributes. This may lead to pose the problem within a multigoal framework, or to introduce the familiar artifice of transforming a vector-valued criterion into a scalar-valued one.

For these reasons, it is in general convenient to adopt as flexible methods as possible, and then to let steps b) and c) remain separated, that is to prefer enumerative algorithms (thus obtaining all feasible solutions) to enumerative ones.It it also to be remembered that the three steps are interrelated. Steps b) and c), for instance, may be strongly influenced by the kind of decomposition.

The planning problem described in the paper arises whenever, for socio-administrative reasons, a geographical region must be partitioned into an unknown number of districts so that, within every district, the total amount of service supplied by the existing hospitals of known location and capacity may satisfy a given demand. This districting problem takes on the form of a set partitioning problem (see [1] also for references). Then, we want to find an optimal zero-one vector x^o which solves the problem

$$\min c^T x$$
$$Ax = \underline{1} \qquad (1)$$

where the element a_{ij} takes on the value 1 if commune i belongs to district j, and the value 0 otherwise. $\underline{1}$ is a vector of all 1's, and c is a suitable cost vector.

Each column of A corresponds to a feasible district: the i-th component x_i of x is 1 if a given collection of feasible districts includes district i, otherwise it is 0. To classify a district as "feasible", one must verify that the population of the district is below a given level, which depends on the district kind. More specifically, urban districts may admit larger populations than rural districts.

The construction of matrix A may require too large an amount of computer time due to its dimensions, which depend on the number of communes that constitute the region. Therefore, we reduce the problem to another partitioning problem, characterized by a smaller constraints matrix $\tilde{A}_{\alpha,\theta}$. This matrix is obtained from a reduced graph which is the result of the following linear program (transportation problem)

$$\min \sum_{i=1}^{n} \sum_{j=1}^{t} f(d_{ij}, \theta)\, z_{ij} \qquad , z_{ij} \geq 0$$

$$\sum_{j=1}^{t} z_{ij} = a_i, \quad i = 1,\ldots, n$$

$$\sum_{i=1}^{n} z_{ij} \leq \alpha b_j \ , \quad j = 1,\ldots, t$$

where $\alpha \geq \bar{\alpha} = \sum_{i=1}^{n} a_i / \sum_{j=1}^{t} b_i$. a_i is the population of commune i; b_j is the number of beds of hospital j; z_{ij} is the population of commune i assigned to hospital j; d_{ij} is the distance between i and j; α and θ are given parameters, and the coefficient $f(d_{ij}, \theta)$ is plotted in Fig. 1.

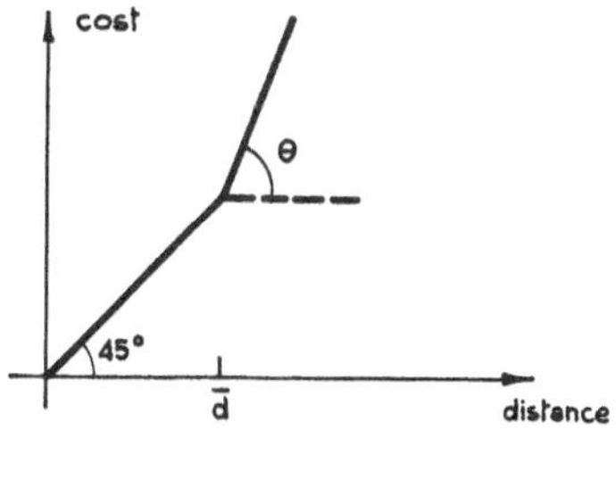

Fig. 1 - Cost function of distance

Although the districting problem may be formally stated in the terms of problem (1), our approach consists in looking for a set of "reasonable" solutions. In other words, we will renounce to find out a unique partition of the region which optimizes some (necessarily) questionable index of quality. Such an approach, however , suffers from at least two serious shortcomings.

The first one is related to the computational procedure itself, whenever the number of feasible solutions turns out to be so large as to give the planner (who is the final user of the algorithm) too poor indications on the overall districting structure.

The second shortcoming stems from the fact that, to avoid the use of a scalar-valued criterion, or possibly of a vector-valued criterion describing the several requirements of a health-care system, we must try to meet these requirements by introducing a certain number of constraints. It is well known, however, that there is a conceptual symmetry between criterion functions and constraints.

A promising way of reducing, at least in part, the effects of these two shortcomings consists in stating the problem within the framework of fuzzy sets theory (see, for instance, [2]). This approach may enable us to avoid too stiff interpretations of the planner's qualitative issues, still providing the decision maker with an efficient tool for ranking the admissible solutions generated by the algorithm described in the following. A discussion of a fuzzy sets approach will be the subject of a forthcoming paper.

2. A method for generating all feasible districts

Let us now consider the problem of finding the columns of matrix $\widetilde{A}_{\alpha,\theta}$, that is, the feasible districts. The solution to the transportation problem allows us to reduce the number of problem variables, by aggregating groups of communes. All communes univocally assigned to the same hospital are grouped together to form a new unit (called Kernel), in general not geometrically connected. If not connected, a Kernel k_j is made up of several components (each of which is connected): a main component m_j (including the commune with a hospital) and c_j dependent components $(s_j^1, \ldots, s_j^{c_j})$.

The territory is now partitioned into kernels and not univocally assigned (n.u.a.) communes. From T.P. output, a graph may now be drawn, which has nodes corresponding to kernel components and to n.u.a. communes, and links corresponding to T.P. assignations. Namely, a link exists between a dependent component and its main component, as well as between a n.u.a. commune n_j and each main component (that is, a hospital) to which it is assigned.

Besides the T.P. graph, the neighbouring reduced graph must also be considered, which has the same nodes as the T.P. one, and links corresponding to neighbouring relationships.

Our problem is to find all feasible districts, defined as follows:

a) a feasible district u_i is a connected set of Kernels and n.u.a. communes, with at least one kernel (i.e., a hospital), i.e.

$$u_i = \{k_1, k_2, \ldots, k_{h_1}, n_1, n_2, \ldots, n_{h_2}\}, \; h_1 \geq 1 \;;$$

b) a feasible district u_i must include all the n.u.a. communes assigned only to kernels included in u_i, and can include a n.u.a. commune only if it is assigned to (at least) one Kernel included in u_i:

$$\Gamma(n_j) \subseteq u_i \Rightarrow n_j \in u_i$$

$$n_j \in u_i \Rightarrow \Gamma(n_j) \cap u_i \neq \emptyset$$

(in the above notation, the n.u.a. commune n_j is assigned to the set of kernels $\Gamma(n_j)$);

c) the population p of a feasible district is bounded by the following constraints:

$$p \geq p_{MIN}; \; p \leq p_{MAX}; \; p \leq \alpha_{MAX} \cdot q$$

(q is the number of beds in the district).

From the above definition, one can see that both above mentioned graphs are necessary to generate all feasible districts. Then, a decomposition of the problem into single-graph subproblems may be profitable.

Each of these subproblems requires that all feasible districts be found which include a certain combination of kernels. To generate these combinations efficiently, the t kernels are ordered starting from the smallest to the largest population. Thus, the following algorithm may be used.

Algorithm 1.

Step 1 (Starting). Start with the null combination $X = \emptyset$ and set $j = 0$.

Step 2 (Population test). Increment j and try to join the j-th kernel to the combination X. If the population exceeds p_{MAX}, go to Step 3. Otherwise, go to Step 4.

Step 3 (Backtrack). Remove from X the last kernel included, say k_p. Set $j = p$ and go to Step 2.

Step 4 (Generation of a new combination). Include the kernel k_j in X, to form the new combination.

Step 5 If $j = t$, go to Step 6; otherwise, go to Step 2.

Step 6 (Termination test). If X includes only one kernel, Stop. Otherwise, go to Step 7.

Step 7 (Backtrack). Remove from X the last two kernels included, say k_p and k_q. Set $j = p$ and go to Step 2.

Advantage is taken of ordering kernels so as to discard automatically all combinations whose populations are certainly larger than an already found $p > p_{MAX}$. For example, referring to Fig.2, let $\{k_1,k_2,k_3,k_4\}$ exceed p_{MAX}, then, it is useless to test $\{k_1,k_2,k_3,k_4,k_5\}$ and $\{k_1,k_2,k_3,k_5\}$. The first combination to be tested is $\{k_1,k_2,k_4\}$. The last feasible combination is $\{k_t\}$.

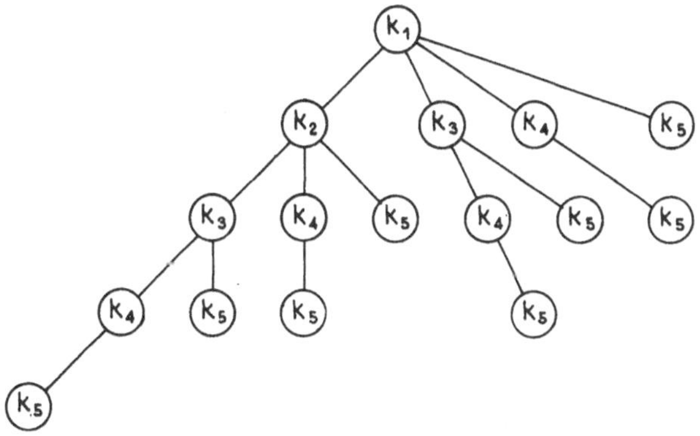

Fig. 2 - Generation of kernels

Whenever a new combination is generated, the corresponding subproblem is solved by following the procedure shown in the flow-chart of Fig. 3, which refers to the whole method, from the output of the T.P. to the search for feasible districts.

First of all, a new graph G is built whose nodes belong to two district classes. The first class (E, essential nodes) includes all components of each kernel of X and all n.u.a. communes assigned only to kernels of X. The second class (NE, non-

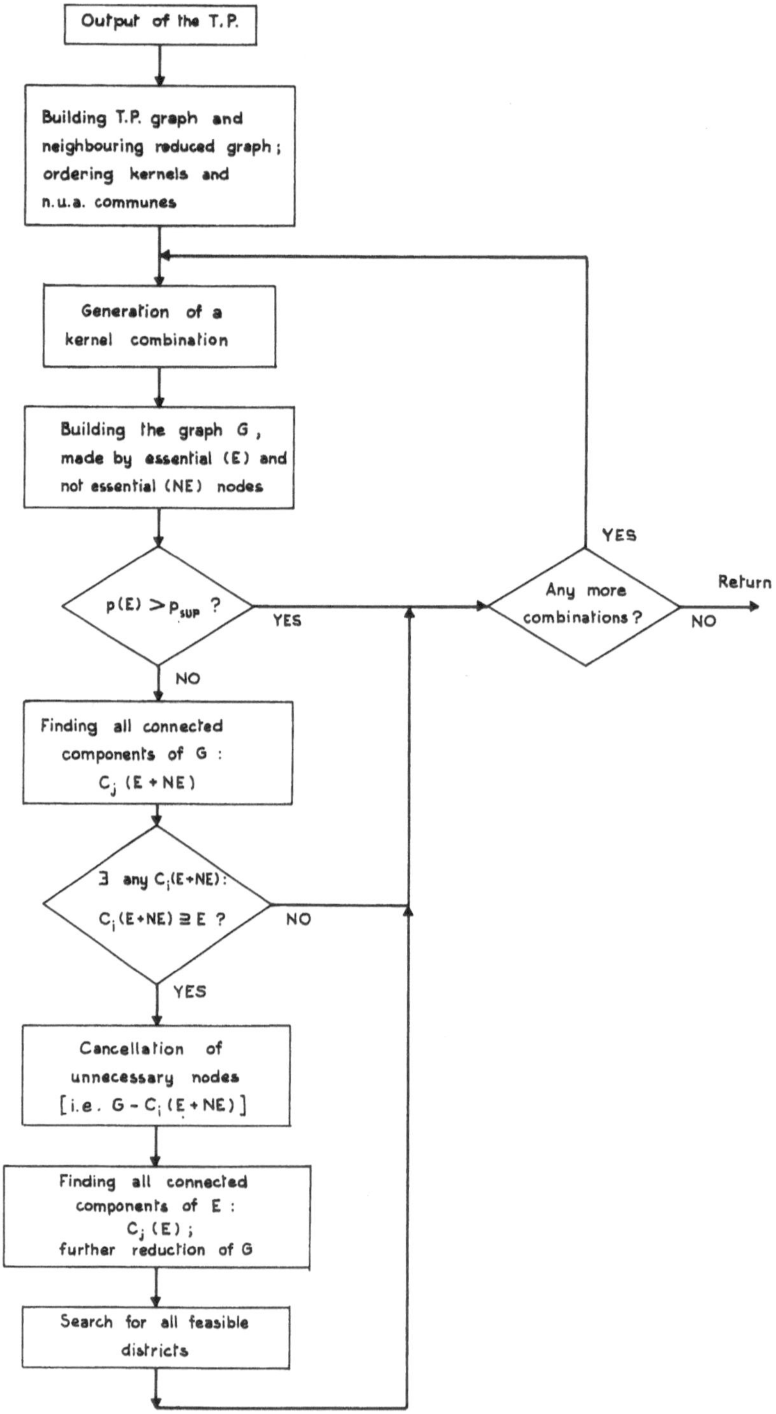

Fig. 3 - Flow-chart of the whole procedure to generate feasible districts

-essential nodes) includes all n.u.a. communes assigned, at the same time, to kernels of X and to kernels not belonging to X. Links of graph G correspond to neighbouring relationships. For instance, for the graph in Fig. 4, m_1, s_1^1 and s_1^2 are the three components of kernel 1, m_2 is the second kernel, e_1 is a n.u.a. commune assigned only to these two kernels and ne_1, ne_2, ne_3, ne_4 and ne_5 are non-essential n.u.a. communes.

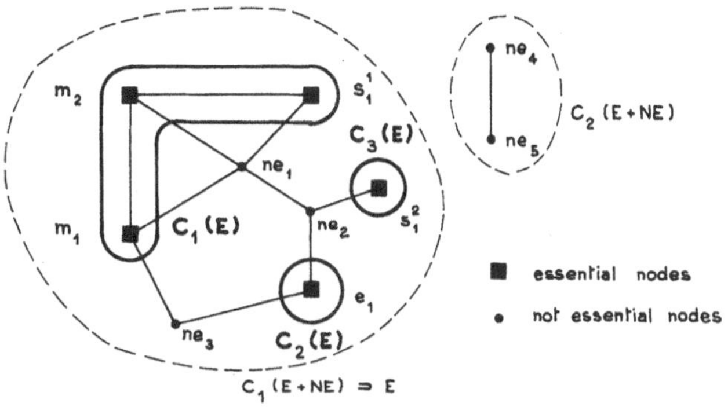

Fig. 4 - Components of graph G

As q is a fixed number (for a given combination X), the upper bound for the population will be $p_{SUP} \triangleq \min[p_{MAX}, \alpha_{MAX} \cdot q]$. If the population p(E) of all essential nodes exceeds p_{SUP}, then no feasible district with this combination X may be found.

If p(E) does not exceed p_{SUP}, the procedure goes on, and, by way of an algorithm described in [3], all connected components of G, $C_j(E+NE)$, j = 1,2,...,r, are found. Since no connection is possible between two of these components, then, if the set of essential nodes is shared among two or more of these components, no feasible district may be found. If, by contrast, a component, say $C_i(E+NE)$, includes all essential nodes, then all other components may be deleted, since none of their nodes may enter in a feasible district. Thus, in the graph of Fig. 4, ne_4 and ne_5 are deleted.

After deleting non-necessary nodes, all connected components of the essential-node graph are found, $C_j(E)$, j = 1,2,...,m, by using the same above mentioned algorithm. Then, these connected components become the new essential nodes, thus yielding a further reduction in graph G (see Fig. 5). Now we may begin the search

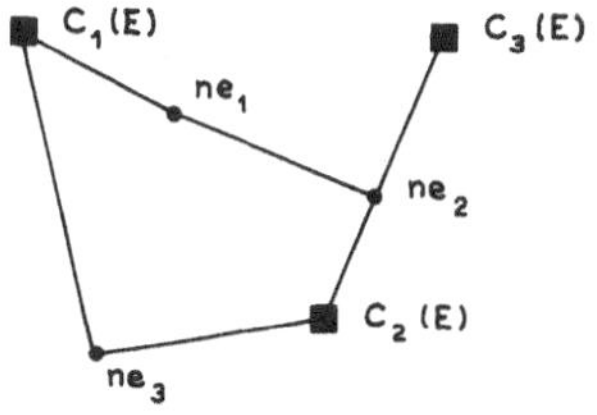

Fig. 5 - The reduced graph

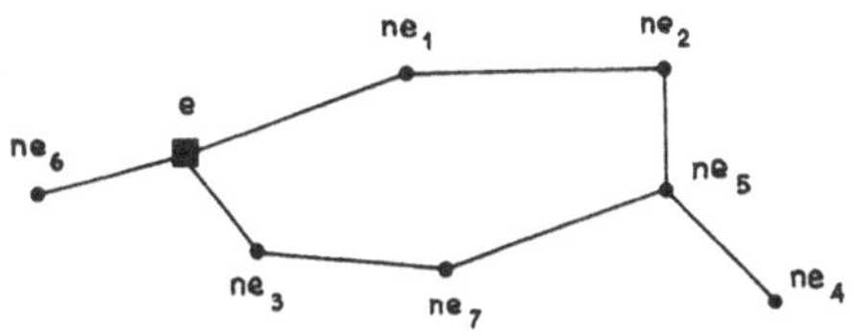

Fig. 6 - A possible situation with a single essential node

for all feasible districts, including all essential nodes and a combination of remaining non-essential nodes.

As regards this search, two different cases may occur: $m = 1$ (a single essential node) and $m > 1$. Let us consider first $m = 1$. In this case, supposing that the s non-essential n.u.a. communes are ordered from the smallest to the largest population, the following algorithm may be used.

Algorithm 2.

Step 1 (Starting). $W = \{e\}$ (e is the only essential node); $j = 1$. Store the district W (if $p(W) \geq p_{MIN}$).

Step 2 (Neighbouring test). If W is connected to ne_j, go to Step 3; otherwise, go to Step 13.

Step 3 (Population test). If $p(W)+p(ne_j)$ exceeds p_{SUP}, Return; otherwise, go to Step 4.

Step 4 (Fusion). Delete ne_j definitively. Fuse ne_j with W ($W = W \cup \{ne_j\}$) and store the district W^j (if $p(W) \geq p_{MIN}$).

Step 5 $k = 1$.

Step 6 If ne_k has been definitively or temporarily deleted or is fused in W, go to Step 10. Otherwise, go to Step 7.

Step 7 (Neighbouring test). If W is connected to ne_k, go to Step 8; otherwise, go to Step 10.

Step 8 (Population test). If $p(W)+p(ne_k)$ exceeds p_{SUP}, go to Step 11; otherwise, go to Step 9.

Step 9 (Fusion). Fuse ne_k with W ($W = W \cup \{ne_k\}$) and store the district W (if $p(W) \geq p_{MIN}$). Go to Step 5.

Step 10 If $k = s$, go to Step 11; otherwise, increment k and go to Step 6.

Step 11 If W includes only one non-essential node, set $W = \{e\}$, restore all temporarily deleted non-essential nodes and go to Step 13. Otherwise, go to Step 12.

Step 12 (Backtrack). Remove the last fused node from W and delete it temporarily. Let now i be the number of non-essential nodes fused in W: restore all non-essential nodes temporarily deleted when this number was (i+1). Go to Step 5.

Step 13 If $j = s$, Return; otherwise, increment j and go to Step 2.

When this algorithm is used, the connection tests always regard only two nodes. To test the connection of a combination, use is made of the results of testing previous combinations. For instance, referring to fig. 6, districts are found in this order: $\{e\}$, $\{e,ne_1\}$, $\{e,ne_1,ne_2\}$, $\{e,ne_1,ne_2,ne_3\}$, $\{e,ne_1,ne_2,ne_3,ne_5\}$, $\{e,ne_1,ne_2,ne_5\}$, $\{e,ne_1,ne_2,ne_5,ne_4\}$, $\{e,ne_1,ne_3\}$, $\{e,ne_1,ne_6\}$, $\{e,ne_3\}$, $\{e,ne_6\}$ (suppose that the other combinations exceed p_{SUP}).

If $m > 1$ (this is the case in Fig. 5), the problem of finding all feasible districts is more complicated, but we can try to reduce it some way to the previous case (m=1); so the above algorithm can still be used. If another algorithm quite similar to Algorithm 1 is used, all combinations of non-essential nodes are found. Suppose that one of these combinations, say Y, forms, with essential nodes, a connected district, with $p < p_{SUP}$. Then, all nodes of this district are fused together to form a new (temporary) essential node, and the search for all feasible districts that can be formed with combinations descending from Y (that is, combinations Y' such that $Y' \supset Y$) is carried out by means of Algorithm 2. Once this search has been concluded, a new non-descending combination is generated, and the procedure goes on.

A quite different strategy from the one stated above involves the search for connecting sets. A connecting set is defined as a non-redundant set of non-essential nodes that connects the set of essential nodes (for instance, $\{ne_1, ne_2\}$ in Fig. 5). This second strategy is more complicated than the previous one (since a connecting set cannot be definitively deleted, after all feasible districts including it have been found). However, a heuristic hybrid type strategy may be chosen: search for all connecting sets (if any) including only one node; delete each of these nodes (after finding all feasible districts including it) and then go on, with the remaining non-essential nodes, following the procedure which generates combinations.

3. Some comments about results

Let us now show and comment some results obtained for the province of Savona. For this province, $\bar{\alpha}$ (i.e., the ratio between its actual population and actual number of beds) is 185. As stated above, α must be $\geq \bar{\alpha}$, and increasing α results in a trend towards least-distance assignation. For linear cost functions of distances (and hence for all monotonically increasing cost functions of distances) least-distance assignation is found for $\alpha \geq 2.5\bar{\alpha}$. As an example, Fig. 7 shows the output of the T.P. for $\alpha = 1.1\bar{\alpha}$.

The main results obtained are summarized in Table 1. Here N is the number of T.P. links less the number of communes with no hospital; ND is the number of dependent components of kernels (this number equals zero where all kernels are connected sets of communes) and tg θ determines the cost functions (of distances) chosen. For our purpose, d (see Fig. 1) has been chosen equal to about 11 km. The lower bound p_{MIN} is 30.000 inhabitants; the value of p_{MAX} is 135.000; α_{MAX} is 290.

Observe that N is monotonically nonincreasing as α increases, for a fixed value of tg θ, and monotonically nonincreasing as tg θ increases, for a fixed value of α; it is not so for ND. We remark that, for $\alpha = 1.5\bar{\alpha}$, the three outputs of T.P. (for tg θ = 1,2,5) are identical. Note that neither the number of feasible districts nor the number of solutions is monotonically nonincreasing as α increases at a fixed value of tg θ (or as tg θ increases, at a fixed value of α).

If, as mentioned in Section 1, two different values of p_{MAX}, namely p^U_{MAX} and p^R_{MAX}, are considered (the former for urban, the latter for rural districts), differents results are obtained as regards the number of feasible districts and the number of solutions. These results are summarized in Table 2, for tg θ = 1 (linear cost functions), p_{MIN} and α_{MAX} being the same as above, p^U_{MAX} = 135.000 and p^R_{MAX} = 80.000.

To conclude these comments, let us point out that α cannot be chosen greater than α_{MAX}. In fact, an $\alpha > \alpha_{MAX}$ could yield kernels with an inhabitants/beds ratio greater than α_{MAX} (so each of these kernels, even if connected, could not be a feasible district). Finally, an $\alpha < \alpha_{MAX}$, but too close to α_{MAX}, can also lead to the impossibility of finding solutions. Consider, for instance, a kernel with an inhabitants/beds ratio very close to α_{MAX}, and suppose it cannot form any combination with neighbouring kernels owing to overcoming p_{MAX}. Let this kernel consist of two components separated by a n.u.a. commune that cannot be aggregated, owing to overcoming p_{MAX}. Then, this kernel cannot be included in any district, and so no solution can be found.

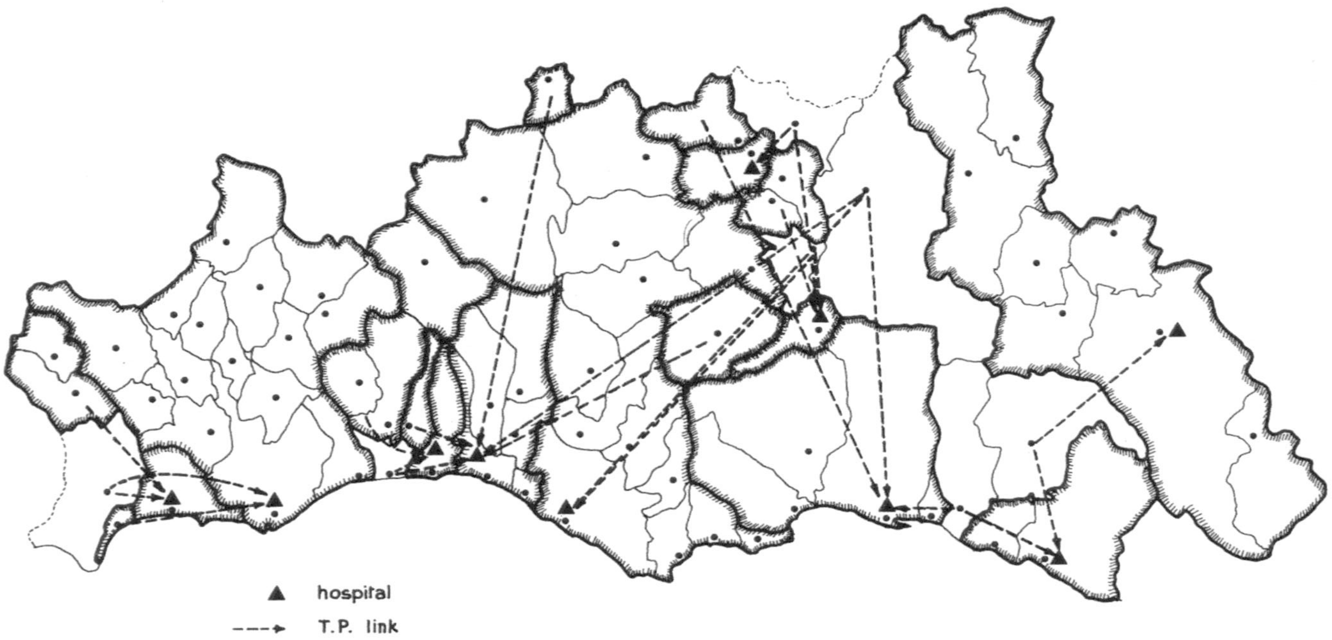

Fig. 7 - Output of T.P. for $\alpha = 1.1\overline{\alpha}$ and tg $\theta = 1$ (linear cost functions).
N = 8.ND = 8.

Table 1 - A summary of the results obtained with a single value of p_{MAX}

	$\alpha/\bar{\alpha}$	N	ND	number of feasible districts	number of solutions
tgθ = 1 (linear cost functions)	1	9	4	748	84
	1.05	8	5	197	16
	1.1	8	8	271	8
	1.2	5	1	167	13
	1.4	4	0	115	24
	1.5	2	0	67	28
tgθ = 2	1	9	5	637	>80
	1.05	8	6	383	21
	1.1	8	8	255	37
	1.2	5	3	232	32
	1.4	3	0	54	21
	1.5	2	0	67	28
tgθ = 5	1	8	5	360	37
	1.05	7	6	342	87
	1.1	7	9	252	27
	1.2	5	3	159	25
	1.4	3	0	54	21
	1.5	2	0	67	28

Table 2 - Results obtained with two different values of p_{MAX} (p^{U}_{MAX} and p^{R}_{MAX})

	$\alpha/\bar{\alpha}$	N	ND	number of feasible districts	number of solutions
tgθ = 1	1	9	4	150	14
	1.05	8	5	54	8
	1.1	8	8	70	4
	1.2	5	1	67	8
	1.4	4	0	50	11
	1.5	2	0	30	11

Acknowledgement

This work was supported by the National Council of Research of Italy (CNR).

References

[1] C.GHIGGI, P.P.PULIAFITO, R.ZOPPOLI: "A Combinatorial Method for Health-Care Districting", Optimization Techniques, Part 1, Proceedings 7th IFIP Conference, Nice, Sept. 1975, Springer Verlag 1976, pp. 116-130.

[2] R.E.BELLMAN, L.A.ZADEH: "Decision-Making in a Fuzzy Environment", Management Science, Vol. 17, Ser. B, No. 4, 1970, pp. 141-164.

[3] N.DEO: "Graph Theory with Applications to Engineering and Computer Science", Prentice Hall, Englewood Cliffs, N.J., 1974.

A MODEL OF HOUSING DEVELOPING COSTS RELATED TO LOCATION

by

M.L. Costa Lobo

L. Valadares Tavares

Rui Carvalho Oliveira

Technical University of Lisbon, IST, and Centre of Urban and Regional Systems of the Universities of Lisbon (INIC) - IST - Av. Rovisco Pais Lisbon, Portugal.

1. INTRODUCTION

The analysis of housing costs for new urban developments as a function of their sites is an essential step to evaluate alternative locational strategies and to optimize decisions in Urban and Regional Planning.

The purpose of this study is developing a methodology of data collection, information processing, cost analysis and iterative search in order to select the most economic sites for new urban developments within a given region.

Often, practical applications of Mathematical Programming are not produced in this field, either because the available data are unsufficient or because the number of alternative and feasible solutions are too big for an economic search of the optimum decision. Usually, the definition of an objective function is also a difficult task because trade-offs between costs of distinct nature have to be assumed.

Several suggestions on the procedure adopted for the spatial discretization of the studied region, the structure and definition of economic and social costs, the estimation of missing data and the iterative technique used to search the optimum solutions are presented in this paper.

With this objective, a set of special programs for small size computing facilities were also developed.

As a case-study, this methodology is applied to select the location of a new urban area ($4\ Km^2$) in Coimbra *concelho** (Fig. 1) and several conclusions are presented herein.

---

* *Concelho* is a portuguese administrative local division (in average, smaller than an english county. The area of Coimbra *concelho* is about $314\ Km^2$).

2. ECONOMIC AND SOCIAL COSTS

The formulation of housing costs is a debatable issue and in this work they are agregated into three major groups:

(a) Land cost

(b) Construction costs (including equipment and infrastructures)

(c) Social costs (running costs)

(a) and (b) are initial costs but the nature and the degree of controllability of the land cost and of the construction costs are quite distinct and therefore they were not included in the same group. Social costs include all the annual costs which result from the use of the new urban development and also from preventing other types of use for that land (e.g., the cost of not using that area for agricultural production). The cost of commuters daily trips and the cost of water and sewage pumping drainage, house heating, etc., are examples of group (c).

The estimation of all these costs for each elementary area i (i=1,...N) of the studied region implies a considerable amount of work which is partly unnecessary because only some terms are function of location. Therefore only these cost components are analysed:

(a) Land cost

This cost, C_{IL}, is a function of i and it is equal to the price by which each plot of land can be purchased if a land market exists and if it should be respected.

Real data about the land price can be only obtained for areas where some transactions have recently occurred and usually this happens in a few spots. Thus, for most sites, price has to be deduced from their features using an acceptable model. For Coimbra *concelho* some information on land prices was given (1) and a price model is developed and tested as it is presented in chapter 4.

(b) Construction costs

The following terms are the major construction costs which are dependent on location:

C_{I1} - Cost of required soil levelling works which are a function of i average steepest slope, S (i). C_{I1} includes the embankment and digging costs and if S (i) > 12% the cost of retaining walls is also considered.

C_{I2} - Cost of additional foundation required by the embankments in order to reach bedrock. (Obviously, $C_{I2} = 0$ if $S(i) = 0$).

C_{I3} - Additional cost of the sewage systems due to inadequate slope: a pumping system is required if $S(i) < 1\%$ and earthworks are needed to eliminate the slope excess above 4%.

C_{I4} - Cost of the surface water drainage systems (A network of open channels crossed by small viaducts is assumed and the trapezoid cross-section area is computed using Manning-Strickler formula). This cost is a function of the draining area.

The considered initial cost, C_I, is given by:

$$C_I = C_{IL} + C_{IS} \quad \text{with} \quad C_{IS} = C_{I1} + C_{I2} + C_{I3} + C_{I4}$$

A general land-use model for the new urban area has to be assumed in order to determine C_{IS} as a function of $S(i)$ and therefore a general pattern was adopted.

(c) Social costs

The new urban area is sufficiently large to include basic social equipment (schools. hospitals, etc.) and to offer employment for a fraction, LAP, of the new active population, AP, in self supporting and servicing activities (local commerce, local public services, etc.). This fraction is estimated assuming that the ratio, R, between the number of jobs in these activities and the total number of jobs is known, namely equal to R estimate obtained for the studied region before the new urban area is developed. (In Coimbra concelho, $R \simeq 0.4$). Thus, major annual costs are the average cost, C_{AE}, of work journeys for EAP population with EAP = AP - LAP, the average travelling cost to major urban centres a few times a year with non-working purposes for the new resident population, C_{AC}, and the annual pumping cost for water supply, surface water drainage and sewage transport, C_{AW}.

Other terms such as house heating were not included in this study because they are negligible when compared to the other annual terms.

The altitude and the approximate location of a source and a sink have to be known to determine C_{AW} which is then a function of the average alititude of unit i, $H(i)$. (In Coimbra case-study it is assumed that they can be located at Coimbra town). C_{AC} can be easily computed if a reasonable average number of non-working trips to major centers per person is assumed (10 trips/inhabitant is acceptable for Coimbra area) and doubtless C_{AE} is usually the most difficult component of annual costs to be estimated because it depends on the predicted spatial distribution of jobs for the new resident population. The discussion of an estimation method for C_{AE} is also important because C_{AW} and C_{AC} are often dominated by C_{AE}.

The probability that a generic EAP individual with residence at unit i will work at unit j, P_{ij}, is assumed to be proportional to the *employment attractiveness* of unit j to a unit i resident, A_{ij}, and this attractiveness is proportional to a function of the number of jobs in unit j, E_j, and inversely proportional to a function of the travelling distance (or cost) between i and j, d_{ij}:

$$A_{ij} = K.\ E_j^{\alpha}\ .\ d_{ij}^{-\beta} \quad \text{with } \alpha,\beta > 0.$$

Then, one has:

$$P_{ij} = \frac{E_j^{\alpha}\ d_{ij}^{-\beta}}{V_i} \qquad \text{with } V_i = \sum_{j=1}^{N} E_j^{\alpha}\ d_{ij}^{\beta}$$

With $d_{ii} = 0.5q$ for a square unit with area equal to q^2 and V_i can be interpretated as an *employment potential.*

The cumulative distribution function of home-work distance for any EAP individual with residence at unit i, F (li), can be easily computed after P_{ij} being determined for i,j = 1,...N. Therefore an average ℓ_i, $\overline{\ell}_i$, can be estimated and $C_{AE}(i) = \gamma.\ \overline{\ell}_i.\ (1 - R).$ AP where γ is a constant, if it is assumed that the cost due to trips is negligible.

For Coimbra *concelho* , α and β are made equal to 1.0 and 2.0, respectively, as a first step on an iterative development expected for the study.

3. SPATIAL DISCRETIZATION AND ESTIMATION OF INDICATORS FOR COST ANALYSIS

A similar shape and size for each discretized unit are required by the computing process and squared units are usually adopted. The most convenient size for the grid unit is a debatable issue because the estimation errors for some magnitudes (usually, geophysical ones such as the average altitude, the surface drainage area and the average steepest slope) will increase with q but for other indicators (usually, demographic and economic ones such as the number of jobs per unit) their errors can be significantly increased if q is lower than some threshold which is a function of the quality of the social and economic available data. On the other hand, the volume of information to be collected and processed and the number of alternative solutions to be evaluated are inversely proportional to q^2 and severe restrictions on q size can be obtained for large regions if scarce human resources and small computing facilities are available. Often, basic geophysical indicators can be computed for small grid units (e.g. q = 100m; 200m) within some small sub-regions and therefore the errors introduced by using a greater q (e.g. q = 100m) can be estimated. These results are useful to correct the estimates obtained for the whole region using larger grid units.

Estimation of the number of jobs per grid unit for Coimbra *concelho* is a difficult task even with q = 1Km because the number of jobs in Agriculture, Industry and Commercial Activities and Services given by official statistics for this *concelho* are not spatially distributed. Then. q < 1Km is unacceptable for this case and a grid unit of $1Km^2$ is adopted. (fig. 1).

For Industry, jobs are distributed proportionally to the number of industrial units located in each Km^2 as the location of major industries is given (1) and for agricultural activities jobs are allocated proportionally to culture areas classified in three groups according to their needs of human work.

The distribution of commercial areas is known (1) for a central area of Coimbra *concelho* and a strong correlation between this area per Km^2 and the corresponding resident population is obtained: $\rho = 0.908$. Then, this result is used to estimate the commercial areas outside this sub-area and again jobs are distributed proportionally to these areas. Finally, the distribution of jobs in Services is done according to the location of major entities in this class and most jobs are concentrated in Coimbra town as it could be expected.

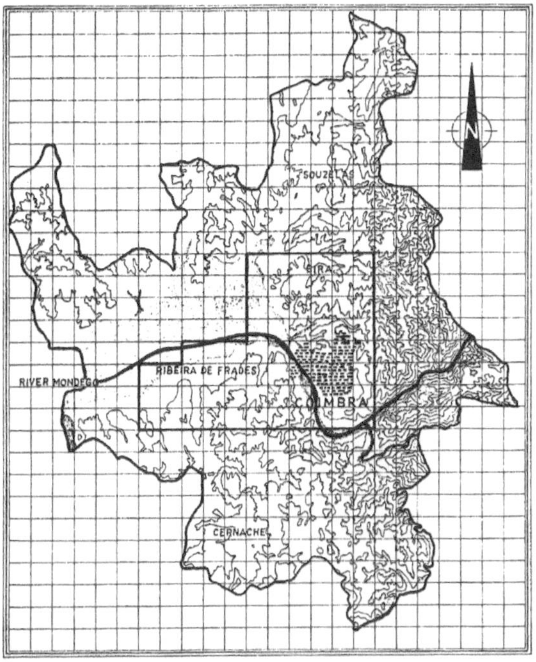

Fig. 1 Coimbra *concelho* Topographic and toponimic map.

Average altitude, steepest slope and surface drainage area for each spatial unit are the geophysical indicators required by costs analysis presented in chapter 2. Average altitude can be easily estimated from a topographic map with negligible errors and new algorithms were developed to compute the steepest slope and drainage area from the altitudes map using smaller computer facilities (2). Some of the errors introduced by usual procedures are eliminated by these algorithms but the adoption of a 1 Km^2 grid unit can be responsible for a significant under-estimation of the steepest slope. Thus, a sensivity analysis of S (i) as a function of q was carried out for a sub-region with q = 100 m and q = 1 Km, so that the S (i) expected error due to adopting q = 1 Km can be estimated and used to compensate S (i) for each Km^2 grid unit.

4. TOWARDS A MODEL OF LAND PRICE

As it was explained in chapter 2 a model relating the average land price with major features of each grid unit is particularly useful and therefore for Coimbra *concelho* several indicators were selected with this purpose (3):

X_1 - A geophysical indicator function of the average altitude, H (i), steepest slope and slope orientation.

X_2 - A roads indicator function of the number and class of roads by which a spatial unit is crossed.

X_3 - An indicator on the electricity, water and sewer systems of each unit (i).

X_4 - Urbanized area indicator defined by:

$$X_4\ (i) = n_i p_i$$

where p_i is the percentage of urbanized area in unit i and n_i is the corresponding estimated average number of floors.

X_5 - Urban centrality indicator:

$$X_5\ (j) = \sum_{i=1}^{N} \frac{P_i \cdot n_i}{d_{ij}^{\alpha}}$$

Tests were carried out with α equal to 2 and 3 and this last value was adopted because the correlation of X_5 with C_{IL} is increased and the similarity between the spread of X_5 and C_{IL} iso-lines is also improved.

The results obtained from a multiple correlation analysis between C_{IL} and $X_1,\ldots, X_5$ are presented in Table 1 (excluding the central units of Coimbra town for which the price is strongly dependent on special speculative factors) and two conclusions for this specific case-study can be drawn up:

a) X_5 (or also X_4) is a master indicator strongly correlated with the land price.

b) The additional capacity of other indicators to explain C_{IL} is quite negligible.

	C_{IL}	X_1	X_2	X_3	X_4	X_5
C_{IL}	1.00	0.06	0.38	0.54	0.79	0.87
X_1		1.00	-0.00	0.09	0.05	0.09
X_2			1.00	0.40	0.32	0.36
X_3				1.00	0.55	0.56
X_4					1.00	0.86
X_5						1.00

Table 1. Correlation analysis

C_{IL} related to X_5 is presented in Fig. 2 and it is shown that:

(a) C_L is almost constant for $X_5 \leq 2.5$ and it is linearly increased by X_5 for $X_5 > 2.5$.

(b) The dispersion of C_{IL} is also increased by X_5 with $X_5 > 2.5$ and particularly for $X_5 > 6.5$.

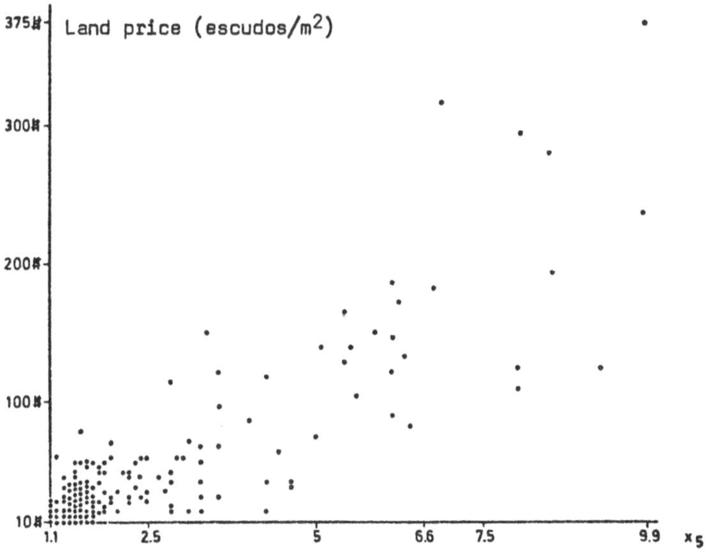

Fig. 2 Land price related to X_5

Thus, other factors have to be searched to explain C_{IL} when $X_5 \leqslant 2.5$ and research on statistical models with non-constant unexplained variance is recommended to relate C_{IL} with X_5 for $X_5 > 2.5$.

5. OPTIMIZATION

Initial costs (excluding land price), initial costs (including land price) and annual running costs were computed for Coimbra *concelho* using the methodology presented in previous chapters. The definition of a total cost, C_T, implies the adoption of a discount factor, f, and the definition of an economic horizon T for this project of urban development. 0.91 is reasonable value for f and the economic life of such a project is usually taken equal or greater than 25 years. As one has

$$C_T = C_I + b.\ C_A$$

with b = 9.2 and 10.2 for T = 25 and ∞, respectively, it was adopted b = 10 and the computed total costs are presented in Fig. 3.

Three distinct optimization problems are then studied:

(a) - Selecting the most economic spatial units without any other restrictions.

(b) - Selecting the most economic location of a 4 Km^2 area with a given shape (e.g. a square).

(c) - Selecting the most economic location of a 4 Km^2 continuous area with a free shape. An area is considered continuous if any unit is at least connected to the remaining set by a common grid unit side.

(a) is essential to evaluate the individual potentiality of each grid unit for urban development purposes and (b) & (c) are useful to study the sensivity of the optimum, C_T^*, and of its location to prescribed shapes. Optimization in (a) can be achieved by straightforward ordering but for (b) and (c) more convenient methods have to be used. Branch and bound or some other implicit enumeration technique is recommended to solve these problems but it has to be adapted in order to determine sub-optimum solutions which are required to evaluate alternative locational strategies. This can be achieved by giving the length, Δ, of the interval $|C_T^*, C_T^* + \Delta|$ in which the eventual sub-optimum solutions have to be calculated and by using the condition $C_T^o\ (B) \geqslant (C_T^*\ (t) + \Delta)$ in each fathoming step (instead of the usual restriction $C_T^o\ (B) \geqslant C_T^*\ (t)$) where C_T^o is a lower bound for branch B and where $C_T^*\ (t)$ is the best feasible solution obtained at iteration t.

The most economic grid units and the optimum solutions for a square and a free shape continuous area with 4 Km^2 are indicated in Fig. 3 and the cost components for units with $C_T \leqslant 500 \times 10^6$ escudos are presented in Fig. 4.

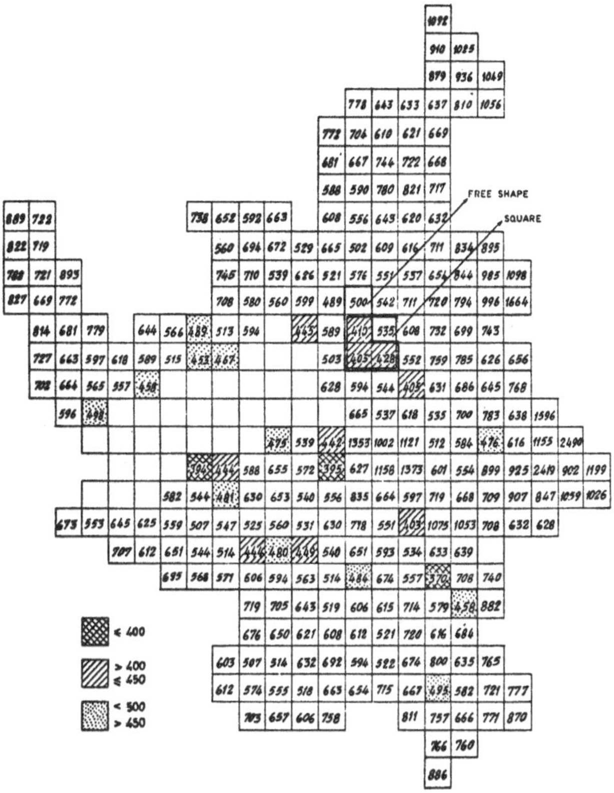

Fig. 3 Total costs (10^6escudos/km^2)

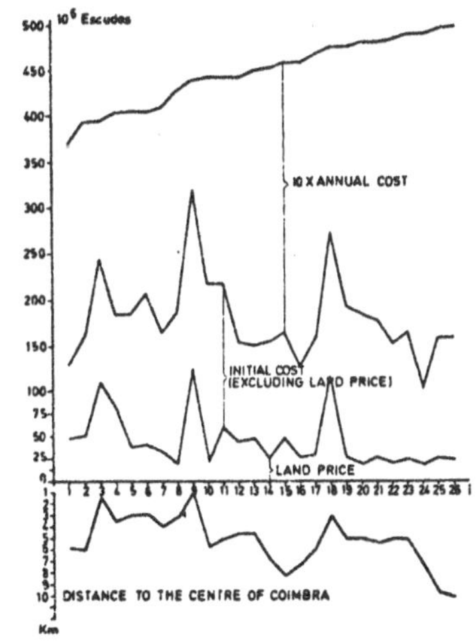

Fig. 4 Total cost components for grid units with $C_T \leq 500 \times 10^6$ escudos.

6. CONCLUSIONS

(a) A methodology was successfully developed and applied to Coimbra *concelho* in order to select most economic sites for the location of urban developments using scarce demographic, social and economic data.

(b) A set of programs applicable to the study of large regions was developed to process collected data and to compute some basic indicators (e.g. surface water drainage area and steepest slope) using small computing facilities. The compensation of systematic errors due to the grid unit size was also considered.

(c) A probabilistic model for work trip distribution and the identification of a master indicator to explain the spatial variation of land price was carried out.

(d) Major components of initial and annual costs which are a function of location are identified and their relative importance for Coimbra *concelho* is shown in Fig. 4.

(e) A modified version of branch and bound technique is used to optimize the location of a new urban development assuming a given (or a free) shape (Fig. 3.).

(f) Social and economic transformation induced over the studied area by the new urban centre were not considered and this second order analysis will object of further research because these effects can be responsible for other significant cost components not included in this study. The hypothesis of ignoring interactions between the analysed area and the "rest of the world" can also be improved in future studies.

(g) In this paper, the decision of location to a new urban area is considered a single-step process but it can be modelled as a sequential decision and the controllability of land price through a convenient use of the land market is high because this price is strongly related to the urban centrallity indicator.

This subject in connection with the financial analysis of sucessive expenses and revenues supported and received by the public authority in charge of the new urban development can be studied by O.R. techniques such as Dynamic Programming and Games Theory and they are a very promising reserach area particularly important for countries where public authorities have scarce financial resources and where the land market is hardly controlled.

References

(1) M.L. COSTA LOBO "Plano de Urbanização de Coimbra", author ed., 1970 & 1974.

(2) M.L. COSTA LOBO, L. VALADARES TAVARES, DINIS MAIA REBELO, P. BARBOSA VEIGA & RUI OLIVEIRA, "Computarização de Análises Regionais. Aplicação ao Concelho de Coimbra", CESUR, 1976.

(3) M.L. COSTA LOBO & L. TADEU ALMEIDA, "Elaboração de um Modelo de Preços de Terrenos", CESUR, 1976.

AN OPTIMUM SURVEILLANCE AND CONTROL SYSTEM FOR SYNCHRONIZED TRAFFIC SIGNALS

A. Cumani, R. Del Bello, A. Villa

ABSTRACT

This paper develops a traffic-responsive control system for a line of synchronized traffic lights. Traffic status information detected by suitable sensors is fed to an optimization procedure which computes the optimum settings by simulating future traffic behaviour. In order to evaluate the feasibility of such a control system, the simulation program is used to analyze the variations of the average travel time along an artery as a function of the control parameters (split and offset).

- The authors are with the Istituto Elettrotecnico Nazionale Galileo Ferraris, Corso Massimo d'Azeglio 42 - I-10125 Torino (Italy).

- This work was partially supported by Consiglio Nazionale delle Ricerche, Rome (Italy), in the "Progetto Finalizzato Energetica".

AN OPTIMUM SURVEILLANCE AND CONTROL SYSTEM FOR SYNCHRONIZED TRAFFIC SIGNALS

A. Cumani, R. Del Bello, A. Villa

INTRODUCTION

In recent years a large amount of work has been done to develop simulation models and mathematical techniques for the optimization of traffic control systems. The most widely used technique is the traffic-signal control. It includes the point control in which the green and red periods of a traffic signal are adjusted in accordance with incoming traffic volume to the intersection, the arterial control in which the phase of one-dimensionally distributed traffic signals along an artery are adjusted in accordance to the traffic volume in the up and down streams, and the area control in which the timings of two-dimensionally distributed traffic signals are adjusted in accordance with the two-dimensional traffic in the area.

A modern signal control system generally assumes a typical configuration of an on-line real-time computer system whose functions are the following:

1) Detection of traffic status information by means of suitable detectors (magnetic loop, ultrasonic, etc.) and transmission of these data over communication lines to the control computer.

2) Processing of status information for the determination of an optimum control strategy and optimum signal parameters.

3) Generation of control signals in accordance with the results of the optimization procedure, and transmission of such signals to the local signal controllers.

With regard to the data processing, present efforts are directed to the development of mathematical models which can simulate the urban traffic behaviour as fast as possible with a good degree of accuracy and to the design of suitable control algorithms, based on such models.

This problem has been studied by different points of view.

There is a strictly analytical approach applying traffic flow theory and network theory to develop a flow pattern that is mathematically optimum in terms of whatever criterion has been selected.

A second approach involves simulation of traffic for individual locations such as specific intersections, extended lengths of arterial streets and complete street networks. Such models use as inputs mathematical relationships describing the traffic process to be studied and data describing traffic and road conditions, such as typical vehicle arrival distributions and physical characteristics of streets and intersections.

This approach to the traffic control problem belongs to a research field widely studied in the literature [1]. Some authors have tried to apply the results of optimal control theory to simplified models of the traffic flow dynamics, generally of macroscopic type. Straightforward application of these results, however, generates some difficulties, as for example the necessity of solving metrix Riccati equations of large order [2][3].

On the other hand, some authors [4] use simulation models to design and evaluate off-line some sequences of traffic signals.

According to this approach, our work develops a feed-back control method to design an optimum progressive traffic signal siming for a line of synchronized lights. The control problem may be outlined as follows (see Fig. 1).

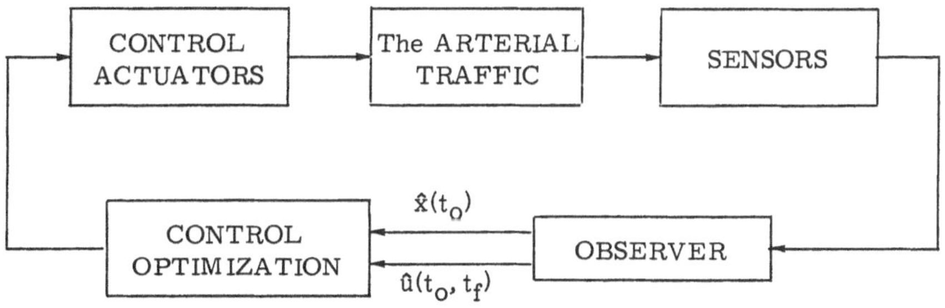

Fig. 1 - The on-line arterial control system

Sensors placed at the beginning and the end of each link continuously detect the passage of vehicles. These informations are transmitted to the "observer", which consists of a microsimulation program giving, at each instant of time, an estimate of the microscopic status ($\hat{x}$) of the network (position and speed of each vehicle), and of a predictor giving an estimate of future inputs ($\hat{u}$) on the basis of past data.

The optimization procedure consists of a copy of the microsimulation program used in the observer, and of an optimization algorithm. This procedure works as follows. The signal parameters (offset and split) are kept constant over periods of a prefixed length T, which may be one or two sycles. Let (t_1, t_2) be one of such intervals (see Fig. 2). At a prefixed instant $t_o < t_1$, the observer's outputs $\hat{x}(t_o)$ and $\hat{u}(t_o, t_2)$ are fed to the optimization program, which computes the optimum signal settings for the interval (t_1, t_2) by simulating the traffic behaviour over (t_o, t_2). This requires a fast simulator program in order to ensure that the optimization procedure may terminate before t_1. The optimized offsets and splits are then applied at time t_1 to the traffic lights.

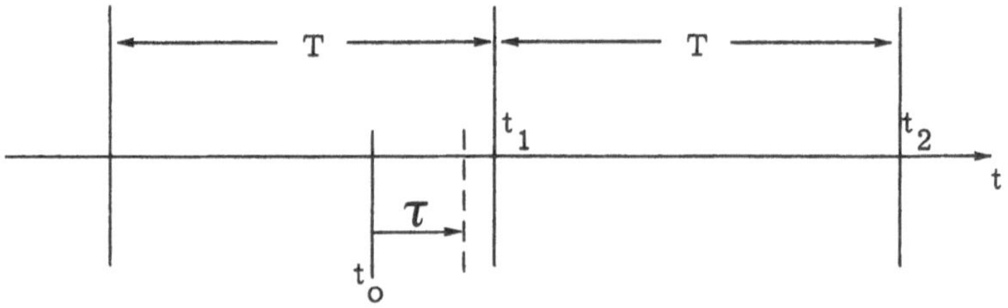

Fig. 2 - Intervals in which the optimization program works (τ) and the optimized control parameters are kept constant (T)

THE SIMULATION MODEL

The proposed traffic model requires the simulation of the trajectory of each individual vehicle. The vehicles move according to a simple car-following law of the type:

$$V_F(t+1) = a\, V_F(t) + b\, V_L(t) + c\; X_L(t) - X_F(t) - d_o \quad (1)$$

where: $X_L(t)$ = position of the "leader" car at time t

$V_L(t)$ = speed of the "leader" car at time t

$X_F(t)$ = position of the "follower" car at time t

$V_F(t)$ = speed of the "follower" car at time t

and a, b, c, d_o are parameters characterizing different types of vehicles and of driver behaviour. Average values of these parameters have been identified from data collected during a traffic survey carried out on roads of an Italian city.

As a vehicle approaches an intersection, it may turn right, left, proceed through or stop at a red light. Turning movements are in conformity with the turning percentages estimated by the observer.

The response of a driver to an amber signal depends on his speed and distance from the intersection and on his behavioral parameters.

The deceleration phase triggered by a red traffic signal or by the need to make a right or left turn at an intersection is simulated by the presence of a dummy leader car with zero speed and in a suitable position.

The speed change given by (1) is limited between a minimum and a maximum value, and for each car speed is limited below a "desired" value which is a characteristic of the link itself.

The output of the simulation model consists in a wide variety of traffic performance measures. Cumulative statistical data for each link as well as for the whole network are available, including the number of vehicles which have traversed each street, average speed, average delay per vehicle, mean queue length and travel time per vehicle.

The informations required by the simulation model are generally available to traffic engineers, that are typical geometric data such as dimension of each street or arterial section, number and widths of travel and parking lanes, parking restrictions and lane use (e.g. right turn only).

A flow chart of the simulation program is shown in Fig. 3.

THE OPTIMIZATION PROCEDURE

The method of optimization suited for a particular problem is largely dependent on the nature of the object function and the subsidiary conditions. In our case in which the object function has no analytical definition but may only be computed by simulation, a heuristic neighborhood search method is used.

As object function one may consider, for example, the average speed or, for a sufficiently long simulation interval, the average travel time over some predefined paths.

The control parameters used are the offset and the split of each traffic light, while the cycle length is fixed. Due to the hardware design of most signal controllers, the signal settings can be altered only in discrete steps (1 s.). As a consequence, the performance index need to be defined only over a discrete solution space.

The optimization is simplified if first only the offsets are varied, then the split optimization is carried out keeping constant the determined offset values.

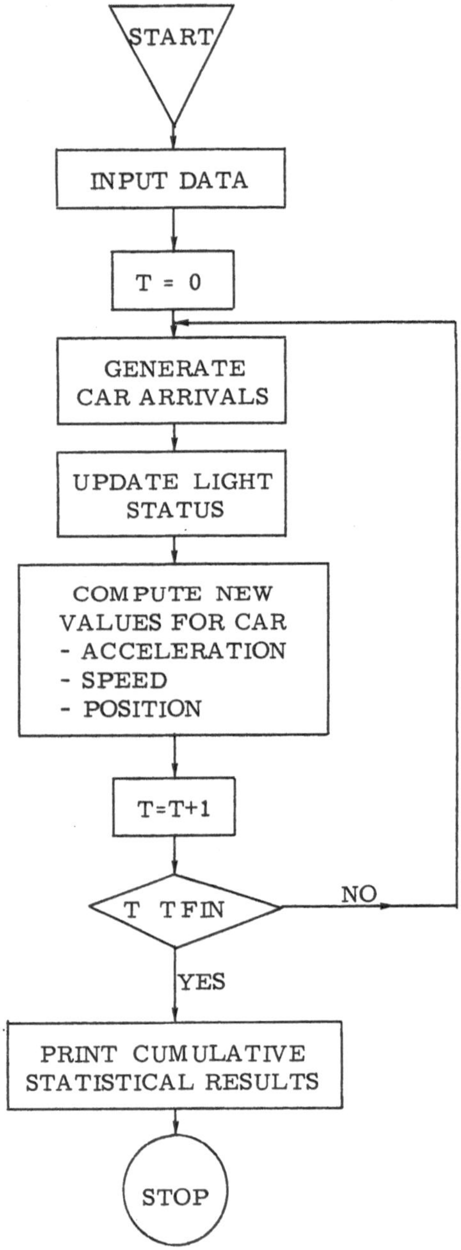

Fig. 3 - A simplified flow-chart of the simulation program

Green-red and offset optimization iteratively succeed each other until no further improvement upon the solution can be obtained.

EVALUATION OF THE SIMULATION MODEL

In order to evaluate the applicability of such a control system to actual traffic, the simulation program has been used to analyze the variations of the average travel time along an artery as a function of the control parameters.

A simple road network consisting of two junctions was considered, as shown in Fig. 4.

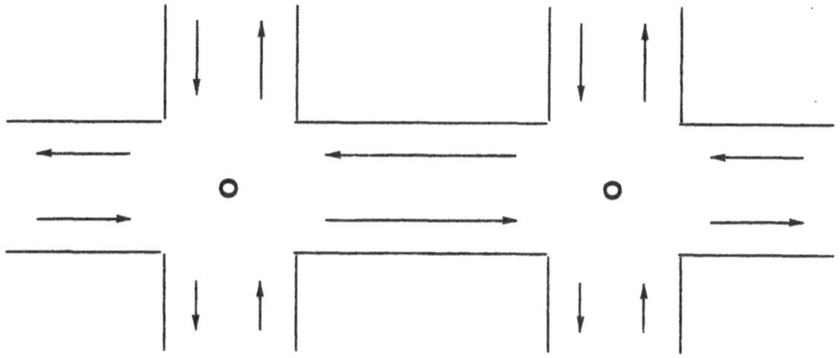

Fig. 4 - The road network under study

The values of the parameters of the car-following model are:

$a = 0.35$; $b = 0.547$; $c = 0.0679$; $d_o = 5.044$.

The inputs to the network are assumed to have a Poisson distribution with fixed arrival rates.

First, the relationship between the average travel time and the cycle length with constant split was studied on a single intersection. Fig. 5 shows the results of a simulation over ~ 4500 seconds of simulated time with various values of traffic flow. Such results are in accordance with theoretical results by many authors [5][6] which indicate that an optimum cycle length exists for a given traffic volume.

Fig. 6 shows the average travel time on the artery of Fig. 5 as a function of the offset and for two different cycle lengths. This diagram suggests that the optimum offset value is not too much influenced by the cycle length.

In Fig. 7 the cycle length is kept constant, while the traffic demand is varied. It may be observed that the optimum offset increases with the traffic demand, due to the lowering of the average speed.

The above simulated tests, implemented on a HP-21MX computer, have shown that the simulation program is sufficiently fast (1" of simulated time $\simeq$ 60" of actual time for an incoming traffic of 0.2 cars/s in each link), however such a program may be used in the optimization routine, for an on-line control, only for simple networks.

Future developments of the outlined traffic control program consist in the implementation of a microprogrammed car-following simulator, in order to obtain a significant reduction of simulation time. With this improvement, the proposed control system could be effectively used for the on-line control of an actual artery.

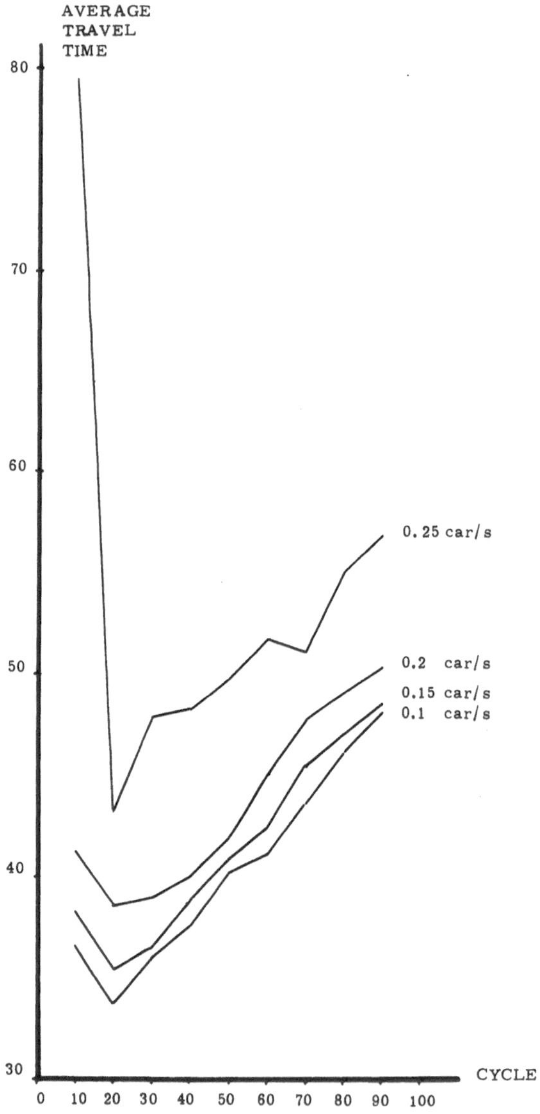

Fig. 5 - Mean travel time at an intersection as a function of cycle length and for various values of the arrival rate

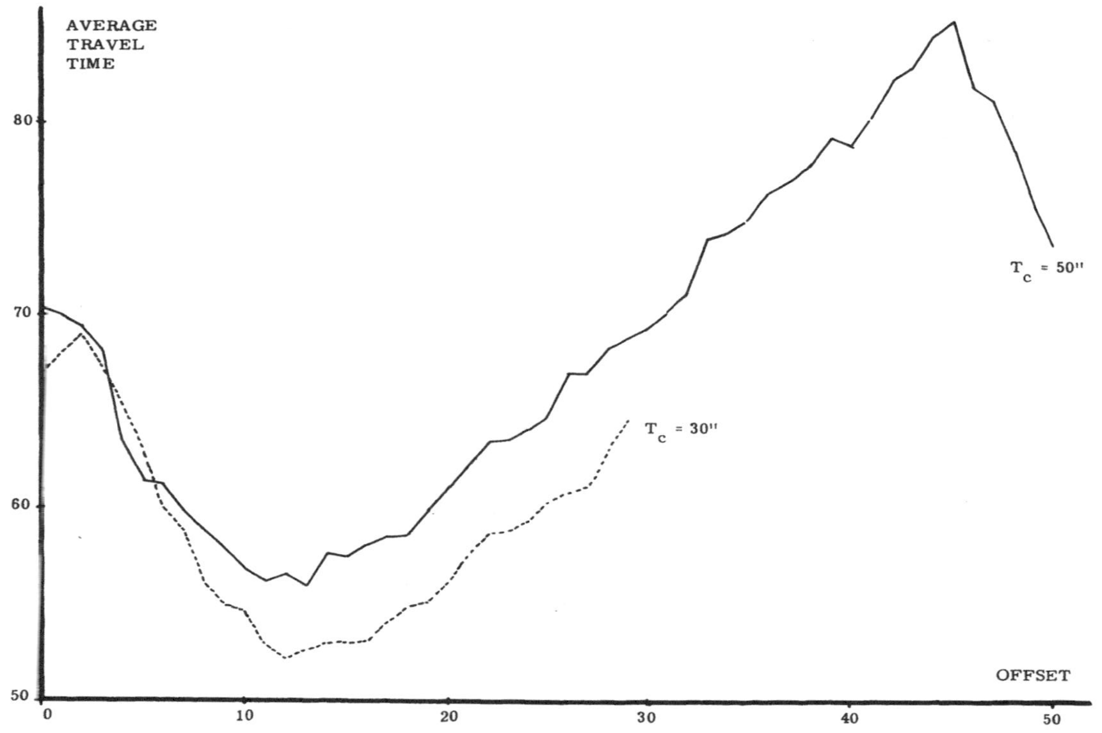

Fig. 6 - Mean travel time along the artery as a function of offset and for different cycle lengths

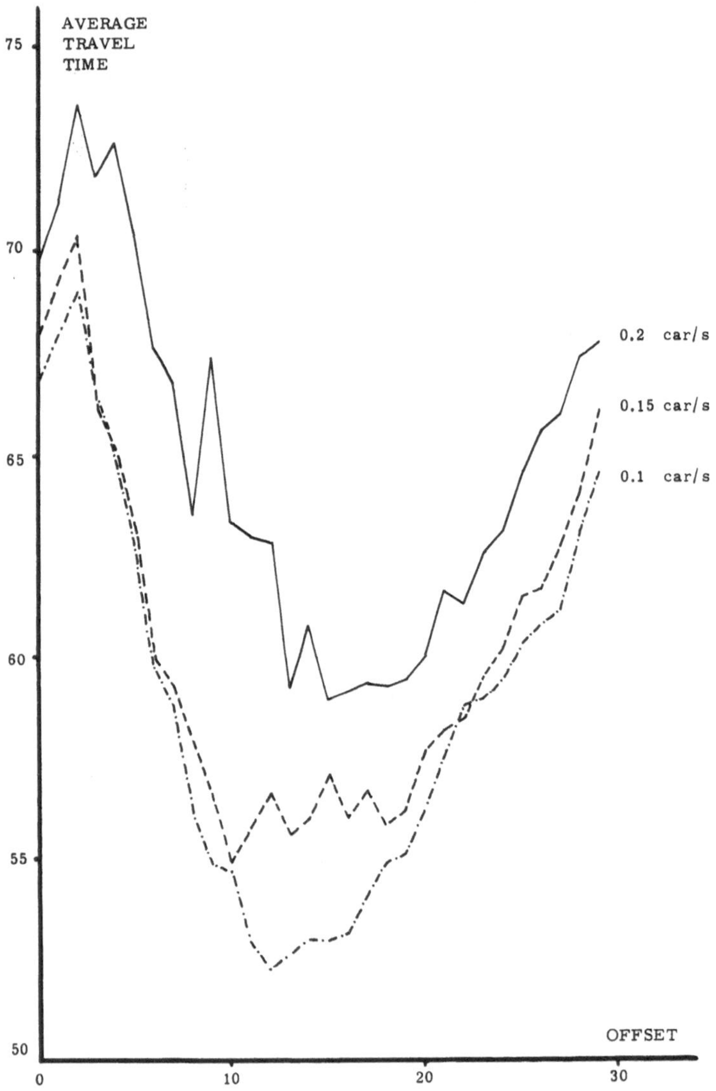

Fig. 7 - Mean travel time along the artery as a function of offset and for various values of traffic volume

REFERENCES

1. A.D. May, "Some fixed-time signal control computer programs", Proc. 2nd IFAC/IFIP/IFORS Symp. on Traffic Control and Transportation Systems, Monte Carlo, 1974, pp. 209-239.

2. H.J. Payne, W.A. Thompson, L. Isalsen, "Design of traffic-responsive control systems for a Los Angeles freeway", IEEE Trans. on Syste. Man and Cybern., vol. SMC-3, n° 3, pp. 213-224, May 1973.

3. H. Akashi, T. Yoshikaw, H. Imai, "Urban traffic control system based on optimal control theory", Proc. 5th IFAC World Congress, Part 2, paper n° 12.5, Paris, June 1972.

4. A. Muzyka, "Urban/freeway traffic control", Proc. IEEE Decision and Control Conf., paper n° FA3-4, New Orleans, December 1972.

5. T.A. Hillier, R. Rothery, "The synchronization of traffic signals for minimum delay", Transportation Science, vol. 1, pp. 81-94, 1967.

6. H. Inose, "Road-traffic control with the particular reference to Tokyo traffic control and surveillance system", Proc. IEEE, vol. 64, n° 7, pp. 1028--1039, July 1976.

REGIONAL CLASSIFICATION PROBLEM AND WEAVER'S METHOD

K. YAJIMA
INSTITUTE OF JUSE
TOKYO, JAPAN

The various kinds of clustering techniques have been applied to regional researches with a great diversity. Fig. 1 shows one example of such applications on demographic data. The analysis is based on data file comprising of four thousands 10-kilometer square cells over the Japanese islands. Three kinds of demographic figures are used and they are primary industry population, secondary and third population in each cell. In Fig. 1 there possibly exist 16 kinds of classes including non population cell. The classes are designated by character symbols, for instance character F indicates 1-3 as shown in the right hand side and it means that the first and third populations are prevailing in this district. This figure is drawn using the method proposed by J.C. Weaver and it will be explained later. In this method the location parameter is not utilized.

```
LJOIIIIIHIFFKFKO
MMKHHIIINIIFFF
MJKHIHH/ LJFK
GK/MHII AJAJK
IMOMHHI/GAAFF
NHIHHIIFKFFFF
KCI  /IFFFFFF
HC     FFFFFF
NM      FAF
NO/     GF
GN/     GF
FG GG
N  GG
N
   J
   /
   N
  N/
  K
     N
    NN
```

A	1	308
B	2	2
C	3	6
D	1-2	16
E	2-1	5
F	1-3	1155
G	3-1	68
H	2-3	108
I	3-2	141
J	1-2-3	303
K	1-3-2	842
L	2-1-3	87
M	2-3-1	123
N	3-1-2	151
O	3-2-1	98
total		3413
/	NONE	914

Fig. 1 Regional classification (Tokyo area)

All of the clustering procedures can be applied to the regional researches. From the point of view of location parameter there are special methods which are taking into consideration of adjacency of regions. These methods are originated from the experience that often we get the result in which with long distance two districts are classified into the same category.

J.C. Weaver proposed a method of regional classification in his study of American midwestern crop-combinations. He used the percentages of total harvested cropland of each county occupied by p individual crops as the indices of the land-occupancy strength of crops. Then as the theoretical standards following models are introduced : monoculture (1-crop) = 100 % in one crop; 2-crop combination = 50 % in each of two crops; ... ; p-crop combination = 100/p % in each p crops. With regard to these models the square sum of deviations, more precisely the variation is calculated for possible combinations of 1, 2, 3, ... , p-crop and the minimum variance assigns the type of regional structure. Let $u_1, u_2, \ldots, u_p$ be observed values in each region. Along with the Weaver's method let transform u_i to percentage, x_i. Suppose $x_{(1)}, x_{(2)}, \ldots, x_{(p)}$ be the non-ascending squence and hereafter simply express it as $x_1, x_2, \ldots, x_p$. We calculate the following variances:

$$\begin{aligned} \sigma^2(1) &= (x_1 - 1)^2 \\ \sigma^2(2) &= \sum_{i=1}^{2} (x_i - 1/2)^2/2 \\ \sigma^2(3) &= \sum_{i=1}^{3} (x_i - 1/3)^2/3 \qquad (1) \\ &\cdots \\ \sigma^2(p) &= \sum_{i=1}^{p} (x_i - 1/p)^2/p \end{aligned}$$

Let the minimum variance be $\sigma^2(k)$ then we adopt k indices to this district. In Fig. 1 a symbol F indicated that $\sigma^2(2)$ is the minimum variance and the first and the third populations are dominating.

There are 1155 districts to which labels F are sticked and these regions consist of the largest class of 34 percent. The second largest class is labelled as K and in this class three populations are all dominating with the non-ascending order as first, third and second. We do not utilize the location parameter but actively make use of similarity existing between adjacent districts. On the other hand it is recognized that there are 1604 cells in total to which three kinds of labels are attached. In other words, the total number of classes comprising J to O comes up to 47 percent. Generally this percentage depends on the data structure, slightly modified method in which reduction of this percentage will be realized was suggested by K. Doi (1954). His method is founded on the idea to omit the denominator k.

Let modify equations (1) to the following

$$\sigma^2(1) = (x_1 - 1)^2$$

$$\sigma^2(2) = \sum_{i=1}^{2} (x_i - 1/2)^2 \qquad (2)$$

$$\sigma^2(3) = \sum_{i=1}^{3} (x_i - 1/3)^2$$

$$\cdots$$

then find the minimum variance as we had before. The similar method also had been presented by E.N. Thomas (1960?). Amending the variances (1) to

$$\sigma^2(1) = ((x_1 - 1)^2 + \sum_{i=2}^{p} x_i^2)/p$$

$$\sigma^2(2) = (\sum_{i=1}^{2} (x_i - 1/2)^2 + \sum_{i=3}^{p} x_i^2)/p \qquad (3)$$

$$\sigma^2(3) = (\sum_{i=1}^{3} (x_i - 1/3)^2 + \sum_{i=4}^{p} x_i^2)/p$$

$$\cdots$$

then we proceed in similar way. The precending two methods do not use values x_{k+1}, x_{k+2}, ... , x_p to calculate $\sigma^2(k)$, it means that these two ptocedures accept k-component model (1/k, 1/k, ... , 1/k), while the later makes use of p-component model

$$(1/k, 1/k, \ldots , 1/k, 0, \ldots , 0). \qquad (4)$$

The researcher who engaged in the work illustrated in Fig. 1 evaluated various clustering methods and pointed out that a small number of three-label regions is preferrable. Keeping this in mind, we proceed to discuss the nature of these three deviation methods. To get geometric interpretation let p equal to 3, that is let x_1, x_2, x_3 be such that

$$x_1 + x_2 + x_3 = 1$$

$$0 \leqq x_1, x_2, x_3 \leqq 1. \qquad (5)$$

These three values can be designated by the triangle coordinates. Fig. 2 shows that triangle is decomposed to 15 small domains and these 15 areas compose three larger domains, namely one-label, two-label and three-label regions.

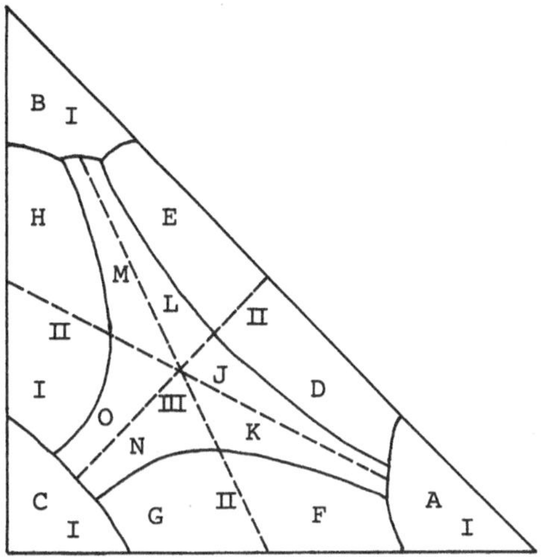

Fig. 2 Domain decomposition (Weaver)

Using this triangle representation, the decompositions of three methods are shown in Fig. 3. If we pay attention to the areas of three-label domains, the smallest domain is attained by Doi's methos then Weaver and Thomas follow.

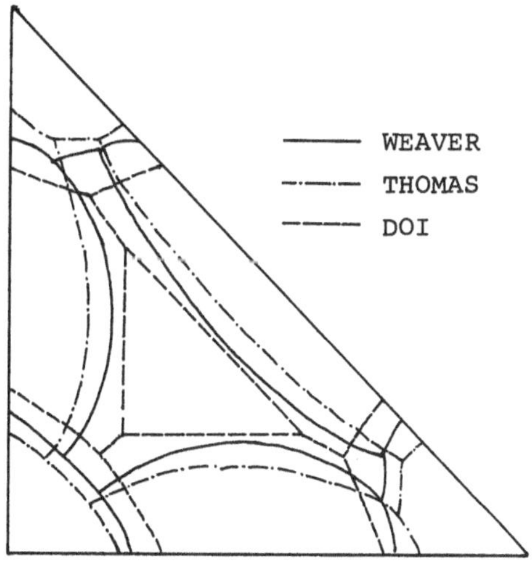

Fig. 3 Domain decomposition (three methods)

If we confine ourselves to evaluate those three deviation methods, it can be said to put Doi's method above is due to the minimum area of three-label domain. These three methods are dealing with percentage data, it means that the original values should be transformed into the same category or dimention.

Finally it should be pointed out that these method are based on the idea of using the affinity measure of two distributions. So we can choose the other definitions, for example

$$A(k) = \sum_{i=1}^{p} t_i \, \ell og(t_i/x_i) \qquad (6)$$

where we accept (t_i) $(i=1, \dots, p)$ as sequence (4). It is found that in this method we can not get the enough small three-label domain although it can be handled theoretically.

Reference:

(1) Weaver, J.C. (1954): Crop-Combination Regions in the Middle West, Geog. Rev., vol.44, pp.175-200.

(2) Doi, K. (1957): Industrial Structure of Japanese Prefecture, Preceedings I.G.U. Regional Conf. in Japan, pp.310-316.

A MATHEMATICAL MODEL FOR DECISION MAKING IN PUBLIC SERVICE PLANNING

P. Vicentini
B. Zanon
IRTU - Facoltà di Architettura
Università di Firenze - ITALIA

In Italy the development of major cities during the last thirty years has determined a situation characterised by: first a qualitative and quantitative gap between the social demand of social services and the public offer, second spaces highly saturated in historical downtowns as well as in peripheral areas.

Consequently there is a lack of land resources and it becomes very difficult locate public services.

Furthermore we still have the typical problems connected with the planning of an efficient net of services, for instance, the environmental characteristics, the limits derived from the dimension of users'area, the technical tresholds, the land cost evaluation, the structures flexibility concerning qualitative and quantitative variation of the social demand.

All these aspects are very complex and very interrelated with each other. The proposed approach relates all these aspects in their complexity and gives evaluation criteria, that is a series of parameters that clarify the users'interest and strategically orient the planning choice.

The proposed model

To meet these requirements we have used a combinatorial optimisation procedure; the programe name is DOT (Decision Optimising Technique) developed in 1975 by Dr. Stan Openshaw, Dept. of Town and Country Planning, University of Newcastle upon Tyne, G.B.

We have used the first release of the model, which is still beeing improved by the author.

The logical scheme of the DOT looks like the more famous AIDA (Analysis of Interconnected Decision Areas) by the Institute of Operational Research of London: a problem is handled by dividing it into many interconnected sub-problems, called Decision Areas. Therefore we can state: "A Decision Area is any area of choice within which there are mutually exclusive alternatives from which we have to choose one" (1).

The interconnection between the ares means that we cannot make a decision in one area without reference to the decision to be made in the others. The alternative decisions in each area are called options. The solution of the problem is given by a complete set of options, one from each decision area.

It is possible to define a set of constraints and a set of weights, which reflect either the preferences of the decision maker and/or other kinds of constraints, when present (that is physical, budget, and so on). To prevent solutions defined by incompatible or undesired options we can define option bars, which inhibit the coexistence of the two options barred. The term bar is drawn by the use of representing the problem with graphs.

The DOT implements this model through the use of a set of key-words, which permit to control the input-output and the data processing in a very simple way. This means that it is user-oriented and can be used by planners with no computer programming experience.

---

(1) A. Hickling (5)

An integer linear programming routine is used to achieve the solution.

A mathematical formulation of the model

Let

A_k be the k-th decision area, k=1,...n

x_{i_k} be the i_k decision variable, that is the i-th option in A_k
$i_k =1,\dots n_k$

x_{i_k} is equal 0 or 1;
$x_{i_k}=1$ means that the i-th option is chosen

Then we obtain the first constraint:

$$\sum_{i_k=1}^{n_k} x_{i_k}=1$$ (no more than one option may be chosen in each A_k)

A similar constraint holds for every k.

Incompatibility bars may be represented as:

$$x_{i_k} + x_{j_m} \leqslant 1$$

Such a constraint holds for all incompatible couple (i_k, j_m).

Let us denote by C the set of such incompatible couples.

Physical, budegct and other kinds of constraints may be represented as:

$$f(x_1 \dots . x_n) \leqslant 0$$

Assuming a linear form for function f, this inequality may be restated:

$$\sum_{k=1}^{n} \sum_{i_k=1}^{n_k} a_{i_k j} \; x_{i_k} \leqslant b_j$$

The objective function is a weighted sum of decision variables:

$$z=\sum_{k=1}^{n} \sum_{i_k=1}^{n_k} C_{i_k} x_{i_k}$$

which has to be maximised.

An application of the model

The experience we explain in this paper is one out of several possible ways to solve the problem by the use of DOT program.

The urban area considered in this analysis is about one square kilometer, located to the East of Florence and is a typical suburban area with a population of 20000 people (200 persons per hectare). This area is characterised by the lack of public services, as schools, playgrounds, and so on. The problem to be solved by applying the DOT was the location of these structures.

The Master Plan's forecast is a null demographic balance until 1982. For this reason the demographic aspect was excluded from the model. We have calculated the needs of each type of service according to the national standards, which prescribe 18 square metres per person of public services. The difference between this quantity and the quantity already satisfied is the need of public services.

We have defined a decision area for each type of service for which there is a need. Therefore we have defined 9 decision areas: 4 areas for 4 different school-levels, 1 for recreational structures, 1 for parks, 1 for large commercial structures, 1 for parkings, 1 for health-care centres.

For each decision area we have found the options that satisfy the demand with regard to the quantitative as well as the qualitative needs. This is defined by the dimension of the users'area (person per service), the home-to-service distance, the law prescriptions relatively to the physical aspects of the buildings.

Each option defined in a decision area can satisfy the need of the service corresponding to the area; these alternative options are differentiated by the quality and the quantity of the service supplied, by its costs, and by the type and extension of the land used. The options thus defined are 24 in all, from a minimum of two to a maximum of four in each decision area. Then we made a hypotesis on the offer of land resources and the areas for which it is possible to make a functional convertion.

Thus we have 11 alternative plans of land acquisition by the Local Government. At this point the model contains 10 decision areas and 35 options.

The next step is the definition of the constraints, which may be given in two ways: 1) incompatibility bars: the options which are in the same area are considered mutually exclusive automatically; so we only have to define the incompatibilities between options in different areas. 2) inequalities: we have introduced 10 constraint inequalities regarding the land balance and the size of the parcels of land: the negative values are considered demand and the positive values offer. These inequalities control that 1) the combinatorial model doesn't make the land offer exceed the demand; 2) the land residuals is minimised; 3) the demand of the more important services is satisfied. Other inequalities are relative to the economical balance.

Finally we have defined the coefficients of the objective function; these coefficients reflect the planner's preferences for the different options.

The DOT evaluates a set of feasible solutions picking up one option from each decision area and, according to the constraints set, gives the combinations which maximise the objective function value.

It is very easy to modify the constraints set and the coefficients of the objective function when a real time terminal can be used. In this way we can perform a sensitivity analysis, by considering the different view-points of the social classes involved in the planning process.

Conclusions

The model presented needs to be improved expecially as far as computing time is concerned. The problems dealt with DOT require usually a large number of decision variables to be realistic, but no efficient optimisation technique is available for 0-1 linear programming with a large number of variables.

Another drawback of the model concerns the handling of disomogeneus variables, for instance in the objective function.

Anyhow this approach may be fruitfully used by local planning authorities to test a set of alternative decisions.

References

(1) P. Baldeschi - P. Scattoni, Pianificazione come scelta strategica, in: "Urbanisticaipotesi", 1, LEF, Firenze, 1974

(2) CES - Center for Evironmental Studies, The Logimp Experiment, London, 1970

(3) J.K. Friend - Power - Yewlett, Public Planning - The Intercorporate Dimension, London 1974

(4) J.K. Friend - W.N. Jessop, Local government and Strategic Choice, Tavistock, London, 1969

(5) A. Hickling, Aids to Strategic Choice (ms for publication in UK), IOR - Institute for Operational Research, London, 1974

(6) S. Openshaw, An alternative approach to Structure Planning: the Structure Plan Decision Making Model(SPDM), in: "Planning Outlook", 17, 1975

(7) S. Openshaw - P. Whitehead, A Decision Optimising Technique for Planners, in: "Planning Outlook, 1975

(8) S. Openshaw, DOT1, (users manual), Newcastle upon Tyne, 1975

(9) Stanley - Openshaw - Withehead, Decision-making in Local Plans, in: "Planning Outlook", 1, 1977

OPTIMAL CONTROL OF REGIONAL ECONOMIC GROWTH.

Medard T. Hilhorst, Faculty of Theology, Free University, Amsterdam*)
Geert Jan Olsder, Rens C.W. Strijbos, Department of Mathematics, Twente University of Technology, Enschede, Netherlands.

Abstract

This paper is in the field of application of optimal control methods to economy. The basic question in this paper is what part of the production should be invested and what part should be consumed. The decisions are both time and location dependent and the underlying system is a distributed parameter model. The criterion to be maximised is prosperity. First-order optimality conditions together with Green's theorem yield a unique solution. The optimal decision may be either bang-bang or have a singular part, depending on initial and boundary conditions. The pieces of the state trajectory, corresponding to these singular parts, lie on the so-called turnpike. Some possible extensions and limitations of the model will be dealt with.

1. Introduction and statement of the problem.

One of the basic questions in any economy is what part of the production should be invested and what part should be consumed. Decisions in regional economics are both time and location dependent. Mathematically speaking such problems can be quite naturally formulated as control problems for distributed parameter systems. One of the first models on control of regional growth has been published by Isard and Liossatos [1], [2]. In the current paper such a model will be given by a nonlinear hyperbolic partial differential equation, in which time t and geographical location x are the independent variables. The state variable, K, stands for capital and satisfies the balance equation

$$\frac{\partial K(x,t)}{\partial t} = - v\,\frac{\partial K(x,t)}{\partial x} + R(x,t)K^a(x,t) - C(x,t), \qquad (1.1)$$

where $C(x,t) = u(x,t)R(x,t)K^a(x,t)$. The term $\partial K(x,t)/\partial t$ is interpreted as the

---

*) Formerly Mathematisch Centrum, Amsterdam.

investment (at x and t). The term $v\partial K(x,t)/\partial x$ denotes the capital flow with constant, positive velocity v, and reflects the concept of growth poles, see remark 3. The function $R(x,t)$, which defines the relation between capital K and production P according to $P(x,t) = R(x,t)K^a(x,t)$, where a is a positive constant, is assumed to be given and characteristic for the area under consideration:

$$R(x,t) = m \exp\{nt-kx\}, \tag{1.2}$$

m, k and n are positive constants. A fraction $u(x,t)$ of the production is used for consumption $C(x,t)$. The fraction of P not used for consumption is invested. The decision variable $u(x,t)$ must satisfy

$$0 \le u(x,t) \le 1. \tag{1.3}$$

The evolution of K will be considered for a prescribed one-dimensional area $x_0 \le x \le x_f$ and time interval $t_0 \le t \le t_f$. Initial and boundary conditions for K are assumed to be given. The control $u(x,t)$ must be chosen in such a way as to maximize prosperity W, which is defined as

$$W \triangleq \int_{x_0}^{x_f} \int_{t_0}^{t_f} u(x,t)R(x,t)K^a(x,t) \exp\{-bt + gx\}dt\, dx \tag{1.4}$$

where b,g are positive constants.

<u>Remark 1.</u>

The constants have economic interpretations. Constant b reflects, for example, population growth and inflation, g gives the possibility to weigh the same consumption at different locations differently, n reflects technological improvement, k reflects local differences in technological knowledge.

<u>Remark 2.</u>

By the law of deminishing returns we assume $a < 1$; $a = 0$ and $a = 1$ represent degenerate cases of the model.

Remark 3.

The capital flow reflects the concept of growth poles [3]. A capital source (or growth pole) is supposed to be located at $x = x_0$. It is represented by the boundary condition $K(x_0,t)$ and spreads its influence by means of a capital flow $- v \partial K / \partial x$ in the positive x-direction. Therefore we shall consider cases where $K(x,t)$ is a decreasing function of x.

2. Description along characteristic curves.

Eq.(1.1) is a first-order partial differential equation. From the theory of P.D.E. it follows that it can be considered as a O. D.E. along its characteristic curves $x - vt$ = constant. The distributed control problem (1.1) - (1.4) can be transformed into an ordinary (finite-dimensional) control problem as follows. Define

$$w = x + vt, \quad w^* = x - vt, \tag{2.1}$$

then (1.1) yields after the transformation (2.1)($K(x,t)$ is written as $K_{w^*}(w)$, etc.)

$$\frac{dK_{w^*}(w)}{dw} = \frac{R_{w^*}(w)K^a_{w^*}(w))}{2v}(1 - u_{w^*}(w)) \tag{2.2}$$

with

$$R_{w^*}(w) = m \exp\{(w - w^*)(n - vk)/2v\}, \tag{2.3}$$

$$0 \le u_{w^*}(w) \le 1 . \tag{2.4}$$

Because along each characteristic curve w^* is a constant, the criterion along each characteristic curve is:

$$W^* \triangleq \int_{w_0(w^*)}^{w_f(w^*)} u_{w^*}(w)R_{w^*}(w)K^a_{w^*}(w) \exp\{-w(b - vg)/2v\} dw, \tag{2.5}$$

which must be maximized w.r.t. the control $u_{w^*}(w)$. On each characteristic curve w_0 en w_f are known, and can be expressed in terms of x, t, initial and final conditions $K_{w^*}(w_0)$ and $K_{w^*}(w_f)$.

3. Applications of the maximum principle.

The maximum principle of Pontryagin [4] can be applied. We shall minimize $-W^*$ instead of maximize W^*. The Hamiltonian H is:

$$H = uRK^a \exp\{-w(b - vg)/2v\} + \lambda RK^a(1 - u)/2v$$

$$= uRK^a (\exp\{-w(b - vg)/2v\} + \lambda/2v) + \lambda RK^a/2v \qquad (3.1)$$

where the arguments have been omitted as well as the index w^*. The Lagrange multiplier λ satisfies

$$\frac{d\lambda}{dw} = -\frac{\partial H}{\partial K} = auRK^{a-1} (\exp\{-w(b - vg)/2v\} + \lambda/2v) - a\lambda RK^{a-1}/2v \qquad (3.2)$$

The Hamiltonian must be minimized with respect to u, which leads to three different cases.

Case 1. If $\lambda < -2v \exp\{-w(b - vg)/2v\}$ then H is minimized by choosing $u = 0$. Then we have from (2.2) and (3.2):

$$\frac{dK}{dw} = \frac{RK^a}{2v}, \qquad (3.3)$$

$$\frac{d\lambda}{dw} = -\frac{a\lambda RK^{a-1}}{2v}. \qquad (3.4)$$

The solution, provided with a subscript 1 because of case 1, is:

$$K_1(w) = \left(\frac{m(1-a)}{n-vk} \exp\{(w-w^*)(n-vk)/2v\} - (1-a)\phi_1(w^*)\right)^{1/(1-a)} \qquad (3.5)$$

$$\lambda_1(w) = (\exp\{(w-w^*)(n-vk)/2v\} - (n-vk)\phi_1(w^*)/m)/\psi_1(w^*) \qquad (3.6)$$

Notice that ϕ_1 and ψ_1 are independent of w: these functions are determined by $K(w_0)$ and $K(w_f)$.

Case 2. If $\lambda > -2v \exp\{-w(b-vg)/2v\}$ then H is minimized by choosing $u = 1$. The functions K and λ corresponding to this case will be denoted by K_2 and λ_2:

$$K_2(w) = \phi_2(w^*) \qquad (3.7)$$

$$\lambda_2(w) = \psi_2(w^*) + \frac{2avm\phi_2^{a-1}(w^*)}{n-vk - b+vg} \exp\{\frac{(n-vk)(w-w^*)}{2v} - \frac{(b-vg)w}{2v}\} \quad (3.8)$$

and ϕ_2 and ψ_2 are determined by $K(w_0)$ and $K(w_f)$.

Case 3. If $\lambda = -2v \exp\{-w(b-vg)/2v\}$ then H is independent of u. Substituting this λ, to be denoted by λ_3, in (3.2) and substituting (2.3) for R yields (K is denoted by K_3):

$$K_3(w) = (\frac{b-vg}{am})^{1/(a-1)} \exp\{\frac{(n-vk)}{2v(a-1)}(w-w^*)\} \quad (3.10)$$

From (3.2) and (3.10) it follows that

$$u = 1 + \frac{n-vk}{b-vg} \cdot \frac{a}{a-1} \quad (3.11)$$

which is a constant, being the ratio of consumption and production. This case is the so-called internal or singular situation ($0 < u < 1$). Notice that K_3 in (3.10) represents a given curve in the (K,w)-plane (fig.1). independent of initial and final condition. In fig.1 case 2 represents the set of horizontal trajectories, case 1 the set of steepest trajectories. For a given initial and final condition, there are still many policies possible for a path from initial to final point. E.g. a path form A to C via B, or via EF, etc. In the next section we will show how to find the optimal path. For every path AC the unknown $\phi_1, \psi_1, \phi_2, \psi_2$ can be solved. For path ADC for example, $K(w_0)$ yields ϕ_1, $K(w_f)$ yields ϕ_2 and the switching point w_s satisfies $K_1(w_s) = K_2(w_s)$. For λ to be a continuous function we must have $\lambda_1(w_s) = \lambda_2(w_s) = \lambda_3(w_s)$, which enables us to determine ψ_1 and ψ_2.

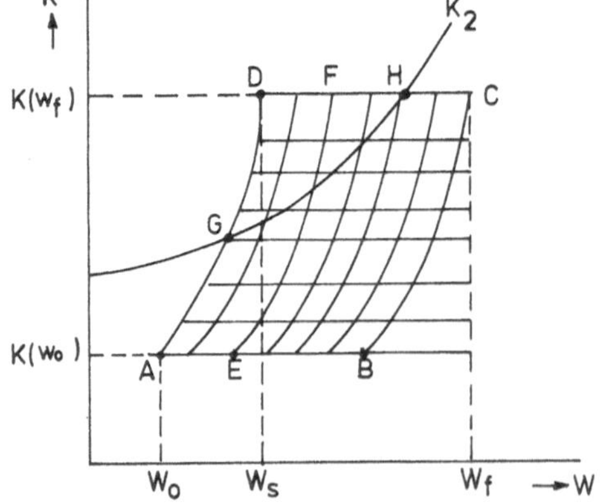

Fig.1. (w^* = constant)

Remark.

To guarentee the existence of a solution the given function $K(w_f)$ should satisfy certain restrictions with respect to $K(w_0)$. The function $K(w_f)$ must be reachable, which means that $K(w_f)$ must lie within the region formed by the trajectories corresponding to $u = 0$ and $u = 1$ and starting in $K(w_0)$.

4. Application of Green's theorem.

To find the optimal path we use Green's theorem [5]. This theorem expresses a line integral as a surface integral:

$$\int_{\partial\Omega} (Pdx + Qdy) = \iint_{\Omega} \left(\frac{\partial Q}{\partial x} - \frac{\partial P}{\partial y}\right) dx\, dy \qquad (4.1)$$

where $\partial\Omega$ is a closed curve (with counterclockwise orientation), Ω the area enclosed by $\partial\Omega$, P and Q functions of x and y.
To be able to apply this to our problem we eliminate $u(w)$ from (2.5) using (2.2), which yields

$$W^* = \int_{\sigma} \exp\{-w(b-vg)/2v\} \,.\, (RK^a dw - 2vdK) \qquad (4.2)$$

Here σ denotes an arbitrary path from $K(w_0)$ to $K(w_f)$.
Further we choose a closed integration path $\partial\Omega$. This can be every path formed by 2 trajectories. In fig. 2 the closed curve ABCD is formed by the trajectories ABC and ADC.

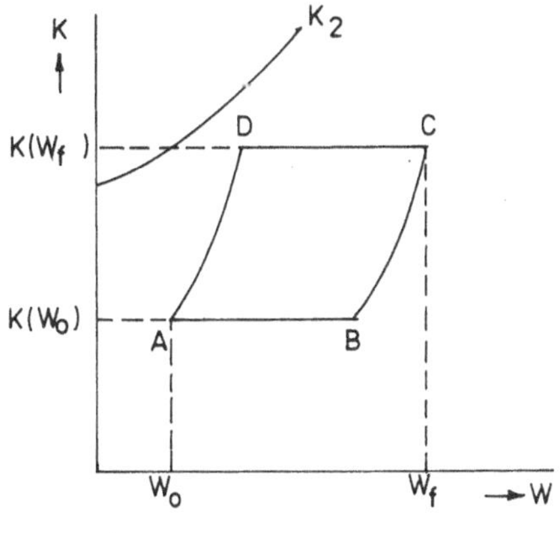

Fig.2. ($w^* = $ constant)

Now Green's theorem is applicable and we find

$$I_{\sigma\sigma'} \triangleq \int_{\sigma-\sigma'} \exp\{-w(b-vg)/2v\}\ (RK^a dw - 2vdK) \qquad (4.3)$$

$$= \iint_{\Omega_{\sigma\sigma'}} \exp\{-w(b-vg)/2v\}\ (b-vg - aRK^{a-1})dwdK \qquad (4.4)$$

where $\sigma - \sigma'$ denotes the closed curve formed by the two paths σ and σ', and $\Omega_{\sigma\sigma'}$ is the area bounded by $\sigma - \sigma'$.

Now it immediately follows that

a) $I = 0$ if $b - vg - aRK^{a-1} = 0$, or equivalently $K = (b-vg/aR)^{1/(a-1)}$

which is equal to the internal capital function K_3 in eq.(3.10), thus

$I = 0$ if $K = K_3$ (4.5)

b) $I < 0$ if $K < K_3$ (4.6)

c) $I > 0$ if $K > K_3$ (4.7)

This enables us to compare two paths. For instance, in fig.2, $I = W^*_{ABC} - W^*_{ADC}$ yields with eq.(4.6): $W^*_{ADC} > W^*_{ABC}$, and by the same reasoning W^*_{ADC} is larger than the prosperity along any other path. In fig.1 we can conclude that W^* along path AGHC is maximal. In general: the optimal path is the path which follows the singular curve as long as possible or is the path nearest to the singular curve. This curve is called the turnpike of the system. Notice that for $a > 1$ we get the opposite result: the optimal path is the path as far as possible from the sigular curve. All this holds for any w^*. For each w^* there are at most two switchpoints.

We now give a numerical example and the complete solution for the original x, t-plane.

Let be given: $x_0 = 0$, $x_f = 10$, $t_0 = 0$, $t_f = 10$, $b = 0.42$, $a = 0.5$, $m = 0.5$, $n = 0.08$, $g = 0.3$, $k = 0.06$, $v = 1$.

Further the conditions $K(x,t_0) = 5 \exp\{-0.1x\}$, $K(x,t_f) = 25 \exp\{-0.1x\}$ and

$K(x_0,t) = 5$ for $0 \le t \le 5$, $K(x_0,t) = 4t - 15$ for $5 \le t \le 10$,

$K(x_f,t) = (0.8 - e^{-1})\ t + 5e^{-1}$ for $0 \le t \le 5$,

$K(x_f,t) = (5.e^{-1} - 0.8)\ t + 8 - 25e^{-1}$ for $5 \le t \le 10$.

In fig. 3^a we draw all possible switch-points w_s as a function of w^*, denoted by 1 - 2, 2 - 3, etc. Using the turnpike-result we can extract the optimal path: along each characteristic curve we stay in the singular case as long as possible. The results are given in fig. 3b. Note that along a line x = constant more than 2 switch-points can exist.

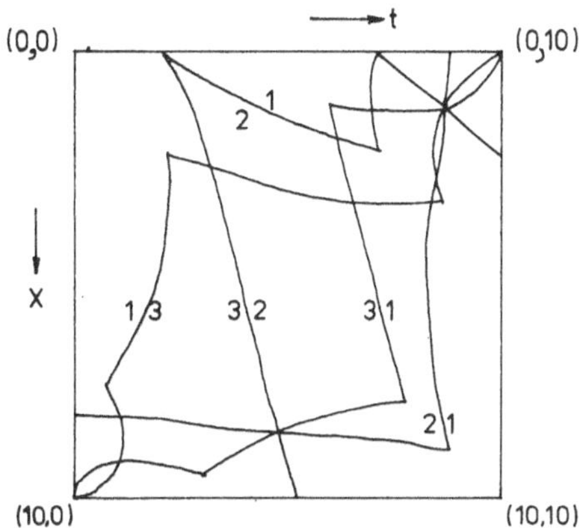

Fig.3a.

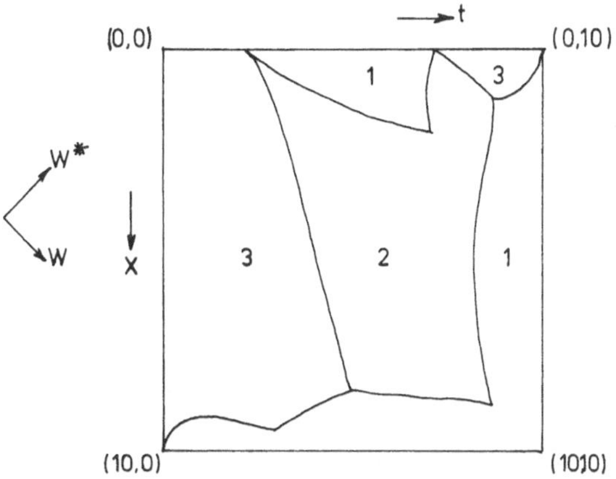

Fig.3b.

5. Free final condition.

Consider problem (1.1) - (1.4) but now with different boundary conditions: $K(x,t_0)$ and $K(x_0,t)$ are given, $K(x,t_f)$ and $K(x_f,t)$ not given. In terms of w and w^* : $K(w_0)$ given, $K(w_f)$ free. For this free-endpoint problem $\lambda(w_f) = 0$, which means for case 2 in section 4:

$$u(w_f) = 1.$$

In order to see whether Green's theorem can be applied we notice that (4.3) and (4.4) still hold when $\sigma - \sigma'$ is replaced by the closed curve $\partial\Omega$ and $\Omega_{\sigma\sigma'}$ by Ω. We consider the case given in fig.4. Path AF cannot be optimal (prosperity zero). From section 5: ABC is the best of all paths ending in C. Therefore we compare path ABC and AE, or BC and BE.

We have $I = W^*_{BE} + W^*_{EC} - W^*_{BC} > 0$ and from eq.(4.2) $W^*_{EC} < 0$.

So $W^*_{BE} > W^*_{BC}$.

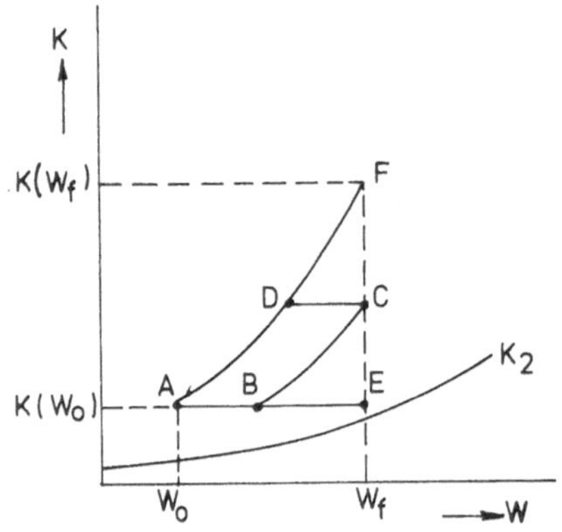

Fig.4.

However, when $K < K_3$ for some $K(w)$ no conclusion can be drawn. In general it can be concluded that for $K(w_0) \geq K_3(w_f)$ maximal consumption is optimal.

6. Second-order model.

When we express the capital flow in terms of a diffusion, we get instead of (1.1):

$$\frac{\partial K}{\partial t} = v \frac{\partial^2 K}{\partial x^2} + (1-u)RK^a \tag{6.1}$$

This model is more realistic because the capital now flows in both directions on account of the diffusion term. Here no simplification along characteristics is possible and the maximum principle for distributed parameter systems must be applied. However, analogous results such as in the previous sections can be obtained. There are again three cases, $u = 0$, $u = 1$ and u is internal and constant. Further K_1 and K_2 are dependent on, and K_3 is independent of the initial, final and boundary conditions. Green's theorem, now being a relation between a (closed) surface and a volume integral, can be applied only to some special cases concerning the boundary conditions.

7. Non-linear criterion.

In problem (1.1) - (1.4) we change the integrand of eq.(1.4) in $\sqrt{C(x,t)}$. This makes the criterion non-linear with respect to $u(x,t)$. For this problem the maximum principle again can be applied. Green's theorem, however, is not applicable. Take for simplicity $a = 1$ and R = constant [1],[2]. Then we get two cases:

$$\text{for } \lambda \geq \frac{-1}{2\sqrt{RK}} \quad : u = 1 \tag{7.1}$$

$$\text{for } \lambda < \frac{-1}{2\sqrt{RK}} \quad : u(w^*,w) = \frac{-1}{1-\phi(w^*) \exp\{R(w-w^*)/2v\}} - 1 \tag{7.2}$$

Eq. (7.2) is the internal solution and now depends on the initial $K(w_0)$ and final $K(w_f)$. This divides the K, w-plane into several regions.

Acknowledgement.

We are indebted to Mr. J. Spit who considered part of this problem in his M. Sc.-thesis.

References.

[1] W. Isard, P. Liossatos, Papers of the Regional Science Association, vols. 29, 30, 31 (1972, 1973).

[2] W. Isard, P. Liossatos, "Trading Behaviour (Transport), Marco and Micro, in an Optimal Space-Time Development Model", in London papers in Regional Science", E.L. Cripps, (ed.), Pion, London (1975).

[3] A. Kuklinski (ed.) "Growth Poles and Growth Centres in Regional Planning", Regional Planning series, vol. 5 (1972).

[4] A.P. Sage,"Optimal Systems Control",New York (1968).

[5] A. Miele, "Extremization of linear Integrals by Green's Theorem", in "Optimization Techniques", G. Leitmann (ed.), New York (1962).

SYSTEM MODELING FOR INTERCONNECTED DYNAMIC ECONOMY AND THE DECENTRALIZED OPTIMAL CONTROL

H. Myoken* and Y. Uchida**

* Faculty of Economics, Nagoya City University, Nagoya 467, Japan

** Research Institute for Economics and Business Administration, Kobe University, Kobe 657, Japan

Abstract

This paper is concerned with the decentralized optimal control of interconnected macroeconometric models composed of subsystems. In the paper we present the effective method of POEM(Policy Optimization using Econometric Models) introducing the concept of the control test that plays an important role in the choice and the evaluation of models. The simulation method of the decentralized POEM is developed, and then its empirical validity is investigated from the practical point of view.

1. Introduction

There have recently appeared various applications of control theory to national policy and planning implemented by the use of econometric models with a single controller[4]. On the other hand, the large-scale macroeconomic policy can be typically represented as the product of a decentralized control process, in which different agencies control different set of policy instruments. This paper presents the optimization method of macroeconomic policy decision under decentralized control, which is entirely unexplored field. Such study also implies a dynamic extension of the static decentralized macroeconomic policy that has been developed in the economic literature[8].

As is well-known, the national economy in question can be described by macroeconometric models in any of the structural form or the reduced form[2]. The application of optimal control theory to the analysis of policy decision employing econometric models has led to the introduction of the state-space form as an alternative model representation. The conversion of the linearized econometric model under centralized control into the state-space form is relevantly presented in Section 2. Secondly, we provide the control simulation method of the centralized model based on POEM(Policy Optimization using Econometric Models). In Section 3 we introduce the new criterion of the control test that plays an important role in the choice and the evaluation of models and discuss the POEM calculation for optimal control policy by the use of this criterion. The results obtained in Section 2 and Section 3 are applied to the decentralized optimal control problems in Section 4; it is shown that the approach proposed in the paper is effective in both the prediction and the control, and that it is expedient to deal with the problems under consideration from the practical point of view. Further research in this direction is suggested in Section 5.

2. The Optimal Control of the Centralized Econometric Model

2.1 The System Model and the Performance Criterion

Consider a general stochastic non-linear dynamic model described by the following structural form :

$$y(t) = \tilde{\alpha}(y(t), p(t)) + u(t) \tag{2.1}$$

where all the variables have the compatible dimensions ; y, p and u are the vectors of the endogenous, predetermined and random disturbance variables, respectively. It is assumed that the random disturbances are independent Gaussian vectors with statistics:

$$\varepsilon\{u(t)\} = 0, \quad \varepsilon\{u(t)u(\tau)'\} = \delta_{t\tau}\Omega \tag{2.2}$$

where $\delta_{t\tau}$ is the Kronecker delta. The predetermined variables classified as $p(t) \in \{x(t), z(t), \tilde{s}(t+1)\}$, where x, z, and $\tilde{s}$ are the vectors of the control, exogenous and lagged variables, respectively. The lagged variable vector is defined by $\tilde{s}(t) \in \{x(t-1), x(t-2), \ldots, y(t-1), y(t-2), \ldots, z(t-1), z(t-2), \ldots\}$. Then

$$\tilde{s}(t) = LS\tilde{s}(t-1) + LXx(t) + LYy(t) + LZz(t) \tag{2.3}$$

which is defined by the difference equation of the first-order, where the elements in the matrices LS, LX, LY and LZ consist of zero or unity. (2.1) becomes

$$y(t) = \tilde{\alpha}(\tilde{s}(t-1), x(t), y(t), z(t)) + u(t) \tag{2.4}$$

Thus it follows from (2.3) and (2.4) that the system of high-order economy is ended up with the first-order non-linear equations system as :

$$s(t) = \alpha(s(t), s(t-1), x(t), z(t)) + C'u(t) \tag{2.5}$$

$$y(t) = Cs(t) \tag{2.6}$$

where

$$s(t) = \begin{pmatrix} y(t) \\ \tilde{s}(t) \end{pmatrix}, \qquad C = (I, 0)$$

Suppose the performance measure of the control simulation is given by

$$\phi = \Sigma_{t=1}^{T} \phi(t)$$

$$\phi(t) = (y(t)-\hat{y}(t))'Q(t)(y(t)-\hat{y}(t)) + (x(t)-\hat{x}(t))'R(t)(x(t)-\hat{x}(t)) \tag{2.7}$$

where $\hat{y}$ and $\hat{x}$ are the desired values for the endogenous and control variables, respectively, and where $Q \geq 0$ and $R > 0$. The optimal control problem of the centralized econometric model is to find the control sequences $\{x(t)\}_{t=1}^{T}$ so as to minimize $\varepsilon\{\phi\}$ subject to the dynamic constraints (2.2), (2.5) and (2.6).

2.2 The Linearized Approximation

In general economic systems are non-linear in structure, but very often we must resort to some approximations. We now can approximate the system (2.5) about the nominal values $\{s(0),\ s(t)^n,\ x(t)^n,\ z(t)^n\}_{t=1}^{T}$ as :

$$s(t)^n = \alpha(s(t)^n,\ s(t-1)^n,\ x(t)^n,\ z(t)^n) \tag{2.8}$$

Namely

$$s(t) \cong \alpha(s(t)^n,\ s(t-1)^n,\ x(t)^n,\ z(t)^n) + \frac{\partial\alpha^n}{\partial s(t)}(s(t) - s(t)^n)$$
$$+ \frac{\partial\alpha^n}{\partial x(t)}(x(t) - x(t)^n) + \frac{\partial\alpha^n}{\partial z(t)}(z(t) - z(t)^n) + C'u(t) \tag{2.9}$$

Assuming that $z(t) = z(t)^n$, it follows from (2.8) and (2.9) that the linearized approximation system is given by

$$s(t) = A(t)s(t-1) + B(t)x(t) + b(t) + C(t)u(t) \tag{2.10}$$

where

$$(A(t),\ B(t),\ C(t)) = (I - \frac{\partial\alpha^n}{\partial s(t)})^{-1}(\frac{\partial\alpha^n}{\partial s(t-1)},\ \frac{\partial\alpha^n}{\partial x(t)},\ C')$$

$$b(t) = s(t)^n - A(t)s(t-1)^n - B(t)x(t)^n$$

2.3 The Optimal Control of Linear Model

The optimal control solutions $\{x(t)^*\}_{t=1}^{T}$ so as to minimize $\varepsilon\{\phi\}$ subject to (2.2), (2.6) and (2.10) are

$$x(t)^* = \hat{x}(t) - x(t)^m \tag{2.11}$$

$$x(t)^m = g(t) + G(t)\varepsilon\{s(t-1)\} \tag{2.12}$$

where

$$g(t) = \{R(t) + B(t)'(H(t) + C'Q(t)C)B(t)\}^{-1}B(t)'\{h(t) - C'Q(t)\hat{y}(t)$$
$$+ (H(t) + C'Q(t)C)(b(t) + B(t)\hat{x}(t)\} \tag{2.13}$$

$$G(t) = \{R(t) + B(t)'(H(t) + C'Q(t)C)B(t)\}^{-1}B(t)'(H(t) + C'Q(t)C)A(t) \tag{2.14}$$

$$H(t-1) = (A(t) - B(t)G(t))'(H(t) + C'Q(t)C)A(t) \tag{2.15}$$

$$h(t-1) = (A(t) - B(t)G(t))'\{h(t) - C'Q(t)\hat{y}(t) + (H(t) + C'Q(t)C)(b(t)$$
$$+ B(t)\hat{x}(t)\} \tag{2.16}$$

$(t=1,\ 2,\ \ldots,\ T)$

Also, the value for ϕ is

$$\phi^*_L = \phi(\{x(t)*\}_{t=1}^T)$$

$$= s(0)'H(0)s(0) + 2h(0)'s(0) + q(0) + \phi^*_s \qquad (2.18)$$

where

$$q(T) = 0$$

$$q(t) = q(t) + \hat{y}(t)'Q(t)\hat{y}(t) + g(t)'R(t)g(t) + (b(t) + B(t)\hat{x}(t) - B(t)g(t)'\{2(h(t) - C'Q(t)y(t)) + (H(t) + C'Q(t)C)(b(t) + B(t)\hat{x}(t) - B(t)g(t))\}, \quad (t = 1, 2, \ldots, T) \qquad (2.20)$$

$$\phi^* = \text{trace}\ \Omega\ \Sigma_{t=1}^T C(t)'H(t-1)C(t) \qquad (2.21)$$

2.4 The Control Simulation of Non-linear Model

By using the resluts obtained in 2.1 through 2.3, we can present the control simulation algorithm of non-linear model under centralized control process as follows:

[1] START: Determine the desired values $\{\hat{x}(t), y(t)\}_{t=1}^T$ and μ, where

$$Q(t)_{ij} = \delta_{ij}/\hat{y}(t)_i^2 \qquad (2.22)$$

$$R(t)_{ij} = \mu \frac{\delta_{ij}}{\hat{x}(t)_i^2}, \quad \mu > 0 \qquad (2.23)$$

[2] Obtain the solutions $\{s(t)^0\}_{t=1}^T$ of the system (2.5) for $\{s(0), x(t)^0 = \hat{x}(t), z(t)^n\}_{t=1}^T$, where $s(0)^0 = s(0)$; $l = 0$. GO TO [7].

[3] Perform the linearized approximation of the system (2.5) about $\{s(0), s(t)^n, x(t)^n, z(t)^n\}_{t=1}^T$ and obtain $\{A(t), B(t), b(t), C(t)\}_{t=1}^T$

[4] Obtain $\{g(t), G(t)\}_{t=1}^T$ from (2.13) through (2.17), where t=1.

[5] Find $x(t)^m$ from

$$x(t)^m = g(t) + G(t)s(t-1)^0 \qquad (2.24)$$

and also obtain the optimal value $x(t)^0$ from (2.11).

[6] Find the optimal value $s(t)^0$ as the solution to

$$s(t)^0 = \alpha(s(t)^0, s(t-1)^0, x(t)^0, z(t)^n) \qquad (2.25)$$

GO TO [9] when $t = T$. Otherwise, GO TO [5], where $t = t+1$.

[7] Obtain the optimal value ϕ_l^0 from

$$\phi_l^0 = \Sigma_{t=1}^T\{(y(t)^0 - \hat{y}(t)'Q(t)(y(t)^0 - \hat{y}(t)) + (x(t)^0 - \hat{x}(t))'R(t)(x(t)^0 - \hat{x}(t))\} \qquad (2.26)$$

GO TO [9] when $l = 0$.

[8] GO TO [10] when

$$\left|\frac{\phi_l^0 - \phi_{l-1}^0}{\phi_{l-1}^0}\right| < \text{EPS}$$

where $\phi^* = \phi_l^0$; $\{s(t)^* = s(t)^0,\ x(t)^* = x(t)^0\}$. Otherwise, GO TO [3].

[9] GO TO [3], where $\{s(t)^n = s(t)^0,\ x(t)^n = x(t)^0\}_{t=1}^T$; $l = l + 1$.

[10] Find ϕ_S^* and ϕ_L^* from (2.18) and (2.21), where

$$\phi_N^* = \phi^* - \phi_L^* \tag{2.27}$$

Also, it is assumed that the error ratio due to uncertainty is

$$\sigma_S = 100 \cdot \frac{\phi_S^*}{\phi^*} \tag{2.28}$$

and that the error ratio due to non-linearity is

$$\sigma_N = 100 \cdot \frac{|\phi_N^*|}{\phi^*} \tag{2.29}$$

STOP.

Notice that the weights Q(t) and R(t) are defined by (2.22) and (2.23) in order to minimize the error ratio for the desired value. It can be also tell by experience that μ is less than 0.1 in order to start from $\{x(t)^n = \hat{x}(t)\}_{t=1}^T$. Consequently, ϕ_0^0 is

$$\phi_0^0 = \Sigma_{t=1}^T (y(t)^0 - \hat{y}(t))'Q(t)(y(t)^0 - \hat{y}(t))$$

which is identical with the value of the cost performance given before implementing the control. Therefore we observe that the performance value is desirable to be $\phi_0^0 < \phi^*$ when the control is implemented the optimal control solution becomes $\{x(t)^* = \hat{x}(t)\}_{t=1}^T$ when $\phi_0^0 \geq \phi^*$. When σ_N in (2.29) is very large, the control calculation may be performed by means of another algorithm.

3. The Control Test

3.1 Testing the Model

The control simulation is performed by using the past statistical data (observed values) as the desired values. We call this simulation the control test. ϕ^* indicates the evaluation value for the model based on the control test. On the other hand, the value ϕ_0^0 in the control test corresponds to the evaluation value for the model based on the final test in the econometric literature [3]: Letting

$$\phi_F = \phi_0^0 \tag{3.1}$$

$$\phi_C = \phi \tag{3.2}$$

ϕ_F and ϕ_C indicate the evaluation values for the model based on the final test and the control test, respectively, where the desired values are the same as the statistical data. The usual simulation tests have been so far developrd based on the partial, total, and final tests. In addition, the new testing criterion proposed here plays an important role in both the prediction and the control.

3.2 The Choice and the Evaluation of Alternative Models

The value for the final test ϕ_F becomes small when the model is fitted to the statistical data. As is well-known, the general interdependence of economic phenomena may easily result in the appearence of approximate linear relationships in time series data. This phenomenon is known as multicollinearity. The value for the control test ϕ_C becomes small when the model is appropriate to economic theory, and when there do not exist the multicollinear relationships. Accordingly, when the model is fitted to the data only, then

$$\phi_F << \phi_C \tag{3.3}$$

The control test does not imply simply a matter of the fitness to the data. When the model is more appropriate to relationships of economic theory from given data and the multicollinearity is avoided, there are the relation such that

$$\phi_C < \phi_F \tag{3.4}$$

The satisfactory criteria of the control test are stronger than these of the final test.

Following the definition of the weight Q in (2.22), the performance criterion function includes all the endogenous variables as the evaluation variables. We now modify this point in some ways. Let the evaluation variable vector e be the subset of the control variable vector plus the endogenous variable vector, defined as $e(t) \in \{x(t), y(t)\}$, where e containes the M-components; all the control variables are supposed to be the evaluation variables. The weight Q is determined as follows:

$$Q(t)_{ij} = \begin{cases} 0, & y(t)_i \in e(t) \\ \dfrac{\delta_{ij}}{\hat{y}(t)_i^2}, & y(t)_i \in e(t) \end{cases} \tag{3.5}$$

In other words, the evaluation variables stand for the sum of the control (instrument) variables and the target variables.

Next we consider the criterion about the choice of alternative models. Recalling the definition of ϕ_C in (3.2), in general, ϕ_C varies with the size of the model and the controlled periods. Thus the criterion about the choice of the models may be defined as:

$$\pi = 100 \cdot \sqrt{\frac{\phi_C}{MT}} \tag{3.6}$$

We perform the control test for each model to obtain the evaluation value π, and then the best model is chosen so as to minimize π.

Remarks:

As seen in the control calculation, the model used has to pass more or less in either the control test or the final test. The evaluation value π is desirable to be at least less than 5%. It may be difficult to obtain desirable results, when the control calculation is performed based on the model such as π is large[6].

4. Optimal Control of Decentralized Systems

4.1 Modeling of Decentralized Systems

Results obtained in the previous sections are applied to the overall integrated dynamic optimization problem of decentralized systems composed of N subsystems. The n-th subsystem is described by

$$s_n(t) = \alpha_n(s_n(t),\ s_n(t-1),\ x_n(t),\ z_n(t)) + C_n' u(t) \tag{4.1}$$

$$y_n(t) = C_n s_n(t) \tag{4.2}$$

which correspond to the centralized systems (2.5) and (2.6). The exogenous variable vector z_n consists of

$$z_n(t) \in \{d_n(t),\ x_{\bar{n}}(t),\ y_{\bar{n}}(t)\ ;\ \bar{n}=1,\ 2,\ \ldots,\ n-1,\ n+1,\ \ldots,\ N\}$$

where d_n is the data variable vector ; $x_{\bar{n}}$ and $y_{\bar{n}}$ are the control variable vector and the endogenous variable vector contained in the subsystems except for the nth-subsystem, respectively. Thus $z_n(t)$ can be expressed as :

$$z_n(t) = \beta_n(d_n(t),\ v_n(t)) \tag{4.3}$$

where

$$v_n(t) \in \{x_{\bar{n}}(t),\ y_{\bar{n}}(t)\ ;\ \bar{n}=1,\ 2,\ \ldots,\ n-1,\ n+1,\ \ldots,\ N\}$$

where v_n indicates the interaction variable vector in the n-th subsystem. It is also assumed that the random disturbances are independent Gaussian vectors with statistics:

$$\varepsilon\{u_n(t)\} = 0, \quad \varepsilon\{u_n(t)u_n(\tau)'\} = \delta_{t\tau}\Omega_n \tag{4.4}$$

Suppose the performance measure is expressed by

$$\phi_n = \Sigma_{t=1}^{T}\ \phi_n(t) \tag{4.5}$$

$$\phi_n(t) = (y_n(t) - \hat{y}_n(t))' Q_n(t)(y_n(t) - \hat{y}_n(t)) + (x_n(t) - \hat{x}_n(t))' R_n(t)(x_n(t) - \hat{x}_n(t))$$

Then the performance measure for the whole subsystems is given by

$$\phi = \Sigma_{n=1}^{N}\ \phi_n \tag{4.6}$$

The optimal control problem of the decentralized systems is to find the control sequences $\{x_n(t)\}_{t=1,\ n=1}^{T\quad N}$ so as to minimize ϕ in (4.5) and (4.6) subject to the dynamic constraints (4.1) through (4.4) ; $n=1,\ 2,\ \ldots,\ N$.

4.2 Control Simulation Method of Decentralized Systems : Decentralized POEM

The method of centralized POEM is applied to the decentralized systems case. As mentioned in Section 4.1, for example, a part of exogenous variables (i.e. interaction variables) contained in the first subsystem are the control variables or the endogenous variables contained in other subsystems. Thus the control calculation of decentralized systems is performed by assuming that control agents exchange the optimal values for their control and endogenous variables.

In this paper we consider two level structure of decentralized POEM. (see Fig. 1.)

Fig. 1: Two Level Structure of Decentralized POEM

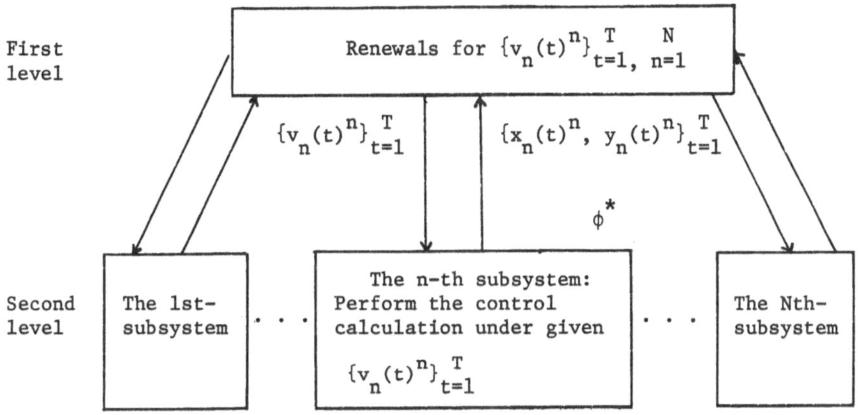

The agent at the first lavel calculates the nominal values for interaction variables $\{v_n(t)^n\}_{t=1,\ n=1}^{T\quad N}$ based on $\{\phi_n^*,\ x_n(t)^*,\ y_n(t)^*\}_{t=1,\ n=1}^{T\quad N}$ obtained in the calculations of centralized POEM for subsystems. The n-th subsystem at the second level calaulates the centralized POEM under given $\{v_n(t)^n\}_{t=1}^{T}$. The control simulation of decentralized POEM is implemented by considering the information exchanges between the first level and the second level. Thus the information exchanges between two levels are transmitted by the interaction variables, which differ from the Lagrange multipliers as developed in Mesarovic et. al. [5].

In the following we present the algorithm of the interacting prediction method at the second level by considering the information exchanges of the n-th subsystem.

[1] START : Obtain the desired values $\{\hat{x}_n(t),\ \hat{y}_n(t)\}_{t=1}^{T}$ from the n-th subsystem. Let the desired values be denoted by $\{x_n(t)^*,\ y_n(t)^*\}_{t=1}^{T}$. Then, register the data bank, where $r = 1$.

[2] Determine the interaction variables $\{v_n(t)^n\}_{t=1}^{T}$ from $\{x_{\bar{n}}(t)^*,\ y_{\bar{n}}(t)^*\}_{t=1,\ \bar{n}=1}^{T\quad N}$ in the data bank.

[3] The information exchanges of the n-th subsystem : Give $\{v_n(t)^n\}_{t=1}^{T}$ to obtain ϕ^* and $\{x_n(t)^*,\ y_n(t)^*\}_{t=1}^{T}$.

[4] Let $\phi_{n,\ r}^* = \phi_n^*$. Replace $\{x_n(t)^*,\ y_n(t)^*\}_{t=1}^{T}$ by the values in the data bank. GO TO [2] when $r = 1$. Otherwise, GO TO [5].

[5] STOP when

$$\left| \frac{\phi_{n,r}^* - \phi^*}{\phi_{n,r-1}^*} \right| \leqq EPS$$

Otherwise, GO TO [2], where $r = r+1$.

5. Applicability and Further Research

Recent years have witnessed increasing interest in large-scale systems, and in particular, the relationship of large-scale system theory to certain application areas. Main concerns to economists have been the problems of centralized and decentralized operation of large-scale economic systems. The effective applications to the decentralized POEM proposed in this paper are certainly illustrated in various macroeconomic models.

As pointed out by Pindyck[7], for example, macroeconomic control in the United States is obviously characterized by decentralization in decision making. In particular, monetary and fiscal policy are exercised by separate authorities that are largely independent of each other. Macroeconomic systems can be represented by interconnected decentralized systems composed of subsystems each defined in terms of government and the central bank. Two control agencies may have conflicting objectives. The decentralized POEM is available for such case. In Myoken et. al.[6], the national economy is decomposed into the public sector and the private sector based on some criteria. The policies between two sectors definitely have conflicting objectives. Even in such case, there exist the mutually independent determinants (control variables), and then the economic systems are implemented by certain coordination operations (the interacting prediction methods) developed in this paper.

The international linkage of national economic models has been developed by many econometricians from many countries [1][10]. Link model for each country can be expressed by each subsystem in the decentralized POEM, and a part of linkages in the model corresponds to that of interaction variables. Such study along these lines is also applied to policy and planning in a system of regions [9]. Of special significance to regional theoriest is the concept of hierarchy, and there has been still not explicitly quantitative treatment of this idea in existing regional econometric models.

References

[1] R. J. Ball (ed.): The International Linkage of National Economic Models, North-Holland Pub. Co., 1973.

[2] C. F. Christ: Econometric Models and Methods, John Wiley & Sons, 1966.

[3] A. S. Goldberger: Impact Multipliers and Dynamic Properties of the Klein-Goldberger Model, North-Holland Pub. Co., 1959.

[4] D. A. Kendrick: "Application of Control Theory to Macroeconomics", Annals of Economics and Social Measurement, 5(1976), 171-190.

[5] M. D. Mesarovic, D. Macho, and Y. Takahara: Theory of Hierarchical Multilevel Systems, Academic Press, 1970.

[6] H. Myoken, H. Sadamichi and Y. Uchida: "Decentralized Stabilization and Regulation in Large-scale Macroeconomic-Environmental Models, and Conflicting Objectives", in the Proceedings of IFAC Workshop on Urban, Regional & National Planning, Pergamon Press, 1977.

[7] R. S. Pindyck: "The Cost of Conflicting Objectives in Policy Formulation", Annals of Economic and Social Measurement, 5(1976), 239-248.

[8] J. Tinbergen: Economic Policy: Principles and Design, North-Holland Pub. Co., 1956.

[9] Y. Uchida: "A Hierarchical Multilevel Approach to Environmental Planning in a System of Regions", in the Proceedings of IFAC Environmental Systems Planning, Design and Control, Pergamon Press, 1977.

[10] J. L. Waelbroeck (ed.): The Models of Project LINK, North-Holland Pub. Co., 1976.

Economic Consequences of a Change in Demographic Patterns: A Linear Programming Model

Mikuláš Luptáčik, Technische Universität Wien

The subject of this contribution is the analysis of the economic consequences of population change in an economy described by - a little modified - open dynamic Leontief model in the form of a linear programming model.

The population creates the basis of an economy. People perform economic activities as consumers and / or producers as well as savers and / or investors. The economic consequences of a change in demographic patterns are therefore very different. Population growth has a number of desirable economic effects as well as a number of undesirable ones. The net result depends on a balance of effects - and probably no single effect is very large.

We consider an economy with n sectors of economic activity (agriculture, manufacturing, construction, energy etc.) connected by the input-output system and described by the matrix of input coefficients $\underline{A}$.

The product of each sector which is not used as an input for current production in another sector (or in this sector itself) may be consumed (private and government consumption) invested or exported. We consider only the government expenditures as given exogenously; the other components of the final demand are considered as endogenous variables.

The gross investment is determined by the matrix of investment coefficients $\underline{B}$ and by the utilisation of the productive capacities. The coefficients b_{ij} of the matrix $\underline{B}$ give us the investment in sector i for the increase of the product of sector j by one unit.

The private consumption in our model is considered as a function of income or wages created in the economy and of the structure of population described by the different types of households according to their number of children and to the number of income receivers.

The domestic production may be increased by the imports. The basis system of balance equations has then the following form:

$$\underline{A}.\underline{x} + \underline{B}.\Delta\underline{x} + \underline{v} + \underline{c} + \underline{g} = \underline{x} + \underline{d} \quad (1.1)$$

where $\underline{A}$ is the matrix of input coefficients; $\underline{B}$ the matrix of investment coefficients; $\underline{x} = (x_1, x_2, \dots x_n)$ the vector of gross production, $\Delta\underline{x} = \underline{x}(t+1) - \underline{x}(t) = (\Delta x_1, \Delta x_2, \dots, \Delta x_n)$ the vector of production increment over one period; $\underline{v} = (v_1, v_2, \dots, v_n)$ the vector of exports; $\underline{d} = (d_1, d_2, \dots, d_n)$ the vector of imports; $\underline{c} = (c_1, c_2, \dots, c_n)$ the vector of private consumption and $\underline{g} = (g_1, g_2, \dots, g_n)$ the exogenously given vector of government expenditures.

I denote, that the variables will be dated (e.g. $\underline{x}(t)$) only when they refer to periods preceding or following period t in question. Thus, while $\underline{x}(t-1)$, $\underline{x}(\tau)$, etc. will always be noted explicitly, $\underline{x}$ will always refer to $\underline{x}(t)$. We write the condition (1.1) in the form of inequalities:

$$\underline{A}.\underline{x} + \underline{B}.\Delta\underline{x} + \underline{v} + \underline{g} + \underline{c} \leq \underline{x} + \underline{d} \quad (1.2)$$

or

$$(\underline{E}-\underline{A})\underline{x} - \underline{B}.\Delta\underline{x} - \underline{v} + \underline{d} - \underline{c} \geq \underline{g} \quad (1.3)$$

where $\underline{E}$ is an n x n identity matrix.

This formulation is in accordance with algebraic convention where the endogenous variables (unknowns) appear on the left side and the exogenous variables on the right side of the equation. (1.2) is the classic inequality of sources and uses for each sector of the economy. It states that the sum of the demands for a product of each sector may not exceed the supply.

Total production in each sector is upper bounded by productive capacity:

$$\underline{x} \leqq \underline{z} \tag{2}$$

where $\underline{z} = (z_1, z_2, \dots z_n)$ is a vector of productive capacities. Capacity is measured in the same units as production. The productive capacity may be increased by investment. The capacity constraints in period t+1 have the following form:

$$\underline{x}(t+1) - \Delta\underline{x}(t) \leqq (1-\beta)\underline{z}(t)$$

where β is the depreciation rate of productive capacities.

$$\Delta x_j(t) = x_j(t+1) - x_j(t) \quad j = 1,2, \dots, n.$$

For the analysis of the foreign trade relationships we distinguish between competitive or complementary and non-competitive imports. The competitive import is a import of goods, which may be also domestically produced. The vector $\underline{d}$ in the conditions (1.1) - (1.3) is the competitive import. Non-competitive import is an import of goods, which cannot be produced domestically (e.g. some raw materials). We define the import coefficients m_j (j = 1,2, ..., n) which give us the proportion of non-competitive import to the production of the sector j. It would be possible to consider the non-competitive import in a more disaggregated form. In this case we have not a vector but a matrix of import coefficients.

Exports of each sector are assumed to be upper bounded in order to allow for frequent inelasticities in world demand:

$$v_j \leqq r_j \quad j = 1,2, \dots, n. \tag{3}$$

Deficit or surplus on current account of balance of trade is defined as the sum of imports minus exports in the following way (in domestic currency; e.g. in Austrian shillings):

$$- \sum_{j=1}^{n} m_j x_j - \sum_{j=1}^{n} d_j + \sum_{j=1}^{n} v_j - \xi + \phi = 0 \qquad (4)$$

where ξ is the surplus and ϕ the deficit on balance of trade. It would be possible to consider also factor payments to abroad e.g. as a constant proportion of GNP.

The variable $\phi-\xi$ represents the surplus of imports over exports. It may be positive or negative, in order to allow for the non-negativity condition imposed by the Simplex-method this variable is written as the difference between two non-negative variables. In the solution one of these variables will be zero, i.e. there will be either an export surplus or an import surplus (see MORGENSTERN-THOMPSON, 1976, p. 237).

The supply of the primary factor labour is mainly determined by demographic patterns and in our model is given exogenously. This is a one channel for the influence of population change on the economy (the population as producer). The demand for manpower (measured in man or man-hours) is an endogenous variable determined by the model under the assumption of a Leontief production function for each sector of the economy.

We define the labour coefficients a_{oj} ($j = 1,2, \ldots, n$) which express the input of labour per unit of production of sector j. The balance equation for labour has then the following form:

$$\sum_{j=1}^{n} a_{oj} x_j - f + u = L_p \qquad (5)$$

where $L_p = L - L_g$. L is the total supply of domestic labour force, L_g represents the people employed in the public sector and L_p is the supply of labour for the production sectors. These variables are exogenous and mainly determined by demographic development. f represents the foreign workers and u shows unemployment. In the solution one of these variables (or both) will be zero, i.e. we need either the foreign workers or we have unemployment. From equation (5) we can see the consequences of population change for the labour force market.

To the labour coefficients a_{oj} correspond the wage coefficients w_j, which express the wages per unit of production of sector j (j = 1, 2, ..., n). The manpower L_g employed in the public sector receive salaries w_g. The total income of the households denoted by I is:

$$\sum_{f=1}^{n} w_j x_j + w_g + w_t = I \tag{6.1}$$

or

$$\sum_{j=1}^{n} w_j x_j - I = - w_g - w_t$$

where w_t are the transfer payments (retirement payments). They are in our model exogenously given.

For the analysis of the consequences of population change on the private consumption we consider different types of households according to their number of children and to the number of income receivers. In accordance with Austrian statistics (KONSUMERHEBUNG 1974) we consider e.g. the following household types: one adult person household; one adult person and one or more children; a household with two adult persons and one income receiver; a household with two adult persons and two income receivers; a household with two adult persons and one child and with one income receiver; a household with two adult persons and one child and with two income receivers; a household with two adult persons and two children and with one income receiver; a household with two adult persons and two children and with two income receivers etc. We have according to the Austrian data 28 household types. The number of the households of type i (i = 1,2, ... h) is exogenously given and determined by demographic patterns, e.g. age-structure.

According to the total number of income receivers living in the particular household types the total income will be distributed to the particular household types in the following way:

$$\alpha_i I - l_i = 0, \quad i = 1,2, \ldots h \tag{6.2}$$

where $\sum_{i=1}^{h} \alpha_i = 1$ and α_i is the proportion of income receivers living in the household type i (i = 1,2, ... h) to the total number of income receivers. l_i (i = 1,2, ... h) express the income of the household type i (i = 1,2, ... h).

The structure of private consumption and the propensity to save in the particular household types is very different (see e.g. KONSUMERHEBUNG 1974). We denote by s_i (i = 1,2, ... h) the saving rate in the household type i. Then the consumption expenditures of the household type i denoted by e_i are:

$$(1-s_i)l_i = e_i \quad (i = 1,2, \ldots, h). \tag{7}$$

The savings in the period t increase the disposable income of the households in the period t+1:

$$s_i . l_i(t) + \alpha_i(t+1) . I(t+1) = l_i(t+1) \quad i = 1,2, \ldots, h.$$

The structure of private consumption in the particular household types is described by the matrix $\underline{C}$. The coefficients of this matrix c_{ki} give us the proportion of consumption of commodity group k (k = 1,2, ... q) to the total consumption of the household types i. The columns of matrix $\underline{C}$ express the consumption structure of the household types i. The total consumption of the particular commodity groups $\underline{y}$ is now:

$$\underline{y} = \underline{C} . \underline{e} \tag{8}$$

where $\underline{C}$ is a q x h matrix, $\underline{y} = (y_1, y_2, \ldots, y_q)$ a vector of private consumption by commodity and $\underline{e} = (e_1, e_2, \ldots e_h)$ a vector of households consumption.

In this way the private consumption is also a function of the population and its age composition. This is a further channel for the influence of population change on the economy (the population as consumer).

In order to transform the vector of private consumption by commodity group $\underline{y}$ into the vector of private consumption by sector $\underline{c}$ we write

$$\underline{c} - \underline{D}.\underline{y} = 0$$

and the condition (1.3) in the following form:

$$(\underline{E}-\underline{A})\underline{x} - \underline{B}.\Delta\underline{x} - \underline{v} + \underline{d} - \underline{D}.\underline{y} \geqq \underline{g} \qquad (1.4)$$

where $\underline{D}$ is an n x q matrix. The coefficients of this matrix d_{jk} give us the proportion of private consumption of the sector j to the total consumption of commodity k. It is obviously that:

$$\sum_{j=1}^{n} d_{jk} = 1 \qquad k = 1,2, \ldots q.$$

The government expenditures $\underline{g}$ are exogenous, but also a function of population development. They represent total government expenditures in the initial period modified by the rate of growth of population. In this way government expenditures per capita are not decreasing over time.

We denote the population living in the household type i by P_i and the consumption per capita by λ_i:

$$\lambda_i = \frac{e_i}{P_i} \qquad i = 1,2, \ldots, h.$$

In order to achieve more stable development of the consumption per capita in the particular household types, we consider only one - the minimal - consumption per capita :

$$e_i - \lambda P_i \geqq 0, \quad i = 1,2, \ldots, h \tag{9}$$

$$\lambda \leqq \frac{e_i}{P_i}, \quad i = 1,2, \ldots, h$$

and

$$\lambda = \min_i \frac{e_i}{P_i}$$

(See also KANTOROVIČ-GORSTKO 1976, p. 82 and LAŠČIAK 1968, p. 204.) This variable will be used as an indicator for the appreciation of the economic consequences of a change in demographic patterns.

In order to compare the consequences of two different goals of economic policy we consider two models. First, the maximization of consumption per capita λ as the second goal of modern economic policy by TINBERGEN, 1956 (model I). For the more stable development of consumption per capita over time we maximize λ only in the last period resp. at the end of the considered horizon with the additional constraints such that

$$\lambda(t) \leqq \lambda(t+1) \quad t = 1,2, \ldots, T-1$$

i.e. consumption per capita is not allowed to decrease over time (see e.g. LAŠČIAK 1968, p. 205).

Second, we consider the full employment as a goal of economic policy. The objective function is the minimization of deviations from equilibrium on the labour market (uneployment on the one side and the number of foreign workers on the other side) over the considered horizon as a problem of goal programming. The objective function has the form:

$$\text{minimize} \sum_{t=1}^{T} (-f(t) + u(t)) \quad \text{(model II).}$$

Since the concept of full employment is relative rather than absolute, the model does in fact aim at full employment as for as this is possible.

The finite time horizon T of the model would lead to consumption of everything available in the last period if no provisions were introduced to ensure sufficient investment in the last period that would allow for continuation of the development after the considered horizon. For this reason we impose additional constraint on the investment. A very simple condition is the following:

$$\sum_{t=1}^{T} \Delta \underline{x}(t) \geqq \beta \sum_{t=1}^{T} \underline{z}(t) \quad \text{where} \quad \underline{z}(t+1) = (1-\beta)\, \underline{z}(t)$$

i.e. the increment of the productive capacities over the considered horizon is not smaller than the depreciation of the productive capacities. Another procedure see e.g. BENARD-VERSLUIS 1974, pp. 39-40.

The aim of this model is not the determination of the optimal production plan, but the comparative-static analysis of the economic consequences of a change in demographic patterns. We change the exogenously given - or in a demographic model determined - demographic parameters (the population and its age structure P_1, P_2, ... P_h; the manpower supply, the government expenditures g_1, g_2, ..., g_n and the coefficients α_1, α_2, ... α_h) and analyse the implications for the production of goods, for the labour market, private consumption per capita, investment, export and import and the balance of trade. Very useful information give us the shadow prices with their well know economic interpretation as marginal values.

It would be easy to make a long list of deficiencies of this type of model. There are, however, some obvious improvements and extensions that could be made. But I believ, that the model offers both pedagogical and operational interest. Its pedagogival interest lies not so much in its schematic, but at the same time in the description of the <u>simultaneous</u> economic effects of a population change.

The operational use of the model lies in the possibility of the <u>quantitative</u> estimation of the influence of demographic changes on the economy.

<u>References</u>:

BENARD J. - VERSLUIS J. 1974: "Employment planning and optimal allocation of physical and human resources", International Labour Office, Geneva.

KANTOROVIČ L.V. - GORSTKO A.B. 1976: "Optimálne rozhodnutia v ekonomike" (Optimal decisions in the economy), Pravda, Bratislava.

KONSUMERHEBUNG 1974, Ergebnisse für Österreich, Beiträge zur Österreichischen Statistik. Herausgegeben vom Österreichischen Statistischen Zentralamt, Heft 420, Wien 1976.

LAŠČIAK A. 1968: "Optimálny lineárny model rastu národneho hospodárstva. Model Bratislava" (The linear growth model of the economy. Model Bratislava) in: LAŠČIAK A. (Ed.): Pokroky operačnej analýzy (Advances in Operations Research) School of Economics Bratislava.

MORGENSTERN O. - THOMPSON G.L. 1976: "Mathematical Theory of Expanding and Contracting Economies", Lexington Books, D.C. Heath and Company Lexington, Massachusetts Toronto, London.

TINBERGEN J. 1956: "Economic Policy: Principles and Design". Vol. XI of Contributions to Economic Analysis. North-Holland Publishing Company, Amsterdam.

THE MULTIPLE COVERING PROBLEM AND ITS APPLICATION TO THE DIMENSIONING OF A LARGE SCALE SEISMIC NETWORK

F.Archetti and B.Betrò
Istituto di Matematica - Università di Milano
Via Cicognara 7 - 20129 Milano

ABSTRACT

The dimensioning of a large scale network, for the seismic monitoring, is modelled as a continuos covering problem: a discrete approximation of it is given, whose optimal solutions can be computed by a new effective technique, particularly aimed at this kind of problems. Subsequently a general theoretical framework of continuos coverings and their discrete approximations is outlined and a theorem is finally proved about the convergence of the optimal discrete solution to the optimal continuos covering.

INTRODUCTION

Italy is a seismically active country: for large areas, an assesment of the seismic risk, should be a relevant factor in a long term urban and industrial planning. The increasing awareness of this fact led the National Research Council to set up, in 1976, the " Geodynamics Project " aimed at the seismic and volcanic monitoring and protection of the italian territory.

In order to provide reliable instrumental data about the seismic activity, a major task of the Project is to design a large scale seismic network, which will be operated by a national seismic service.

As a part of a complex allocation process, involved in the design of the network, a study has been carried out, in order to highlight the relation between the specifications of the network and the number of stations required.

§ 1. THE DISCRETE MODEL OF THE NETWORK.

Let's introduce the basic network specifications: the hypocentre of any shock of magnitude $M \geq M_0$, is to be identified. Thus, at least 4 independent records of the event are needed and any part of the territory must lie within the range of at least 4 stations, whose pairwise distances must be not smaller than δ.

The range of a station is assumed to be a circle whose radius depends on M_0 and the value of the noise-signal ratio in that site.

Some areas show too bad a value of this ratio to be candidate for the location of a station, but nevertheless are to be covered by the network; some locations are fixed in advance, where stations with good equipment are already operating.

Now we are going to show a simple discrete model of the network. A regular triangular net of the whole area to be covered, with mesh size $h=\delta$, is considered: the above network specifications are easily introduced in the model, with some errors depending on h.
Let T_ℓ, $\ell \in I_T \equiv \{1,2,\ldots,N_T\}$ be a collection of N_T triangles such that $\bigcup_{\ell \in I_T} T_\ell$ contains the whole area to be covered and let V_j, $j \in I_V \equiv \{1,2,\ldots,N_V\}$ be a grid-point of the net.

It is easy to point out "forbidden" grid-points, where a station cannot be located due too bad a value of the noise-signal ratio, and grid-points fixed a priori. Let V_j, $j \in I_R$ be the remaining grid-points where the stations not yet fixed can be located, while the area to be covered is $\bigcup_{\ell \in I_T} T_\ell$. For any V_j, $j \in I_R$, the noise-signal ratio will be expressed by an integer p(j), which means that the station located in V_j, can detect shocks of magnitude $M \geq M_0$, within a distance $p(j)\cdot h$.

Now we build the constraint system expressing the fact that any triangle T_ℓ must be"covered" by at least b_ℓ ($b_\ell \geq 4$) stations.

Let $C=[C_{\ell j}]$, $\ell \in I_T$, $j \in I_R$, where $C_{\ell j}=1$ if the triangle T_ℓ is covered by a station located in V_j, i.e. all its vertexes are less distant than $h\cdot p(j)$ from V_j and $C_{\ell j}= 0$ otherwise. For any T_ℓ, $\ell \in I_T$, we define $d_\ell = b_\ell - n_\ell$ where n_ℓ is the number of stations, fixed a priori, covering T_ℓ: let $d=(d_1,d_2,\ldots,d_{N_T})$; for any node V_j, $j \in I_R$, a boolean variable z_j is defined such that $z_j=1$ if a

station is located in V_j and $z_j=0$ otherwise.
Considering the whole network and the vector z, whose components are z_j, we obtain the constraints system:

$$Cz \geq d$$

Among all feasible solutions, i.e. station allocations which satisfy the geometric specifications of the network, the optimal one, which we look for, is that one with the minimum number of stations. Thus the combinatorial optimization problem can be formulated as:

$$(*) \min \sum_{j\varepsilon I_R} z_j \quad \text{with the constraints } Cz \geq d$$

Some areas could require a closer monitoring of the seismic activity and thereby an higher network density: this can be accomplished either by stiffening the covering conditions, i.e. increasing b_ℓ for the corresponding triangles, or defining a "seismic weight" w_j, inversely related to the seismic activity near V_j, and solving the problem

$$(**) \min \sum_{j\varepsilon I_R} w_j z_j \quad \text{with the constraints } Cz \geq d$$

Problems (*) and (**) can be solved either by a general purpose integer programming code or more effectively by some special covering technique particularly designed for the multiplicity requirement. One such technique has been developed by C. Vercellis ([1]): it seems quite effective and can handle easily the 100 variables-400 constraints problem arising, from the discrete model of the network, with h=50 km.

As the influence of observational errors in the accuracy of the hypocenter location depends on the distribution of observation points ([2],[3]), the algorithm in [1] is designed in order to give all optimal solutions: they can be subsequently ranked according to criteria of the kind given in [2] : the results of some actual computations are reported in [1] .

In the next section the influence of the mesh size h on the number of stations of an optimal solution is considered in a general framework.

§ 2. THE δ - MULTIPLE COVERING PROBLEM AND ITS DISCRETIZATION

In this section we set in a formal way a new class of continuos covering problems and outline a theoretical framework for their discrete approximation. Finally a theorem is proved about the convergence of the discrete solution to the optimal continuos covering. We introduce some notations which are required in order to state the δ - multiple covering problem.

1) Two sets are given: a compact $K \subset R^N$ and $C \subset R^N$ (K is the set to be covered while C is the set where the centres can be chosen);
2) $\rho(x,y)$ is a function satisfying the following conditions, for $x,y,z \in R^N$

   i) $\rho(x,y)=\rho(y,x)$ ii) $\rho(x,y) > 0$, $\rho(x,x)=0$

   iii) $\rho(x,y) \leq \rho(x,z)+\rho(z,y)$ iv) $\rho \in C(H)$;
3) two functions are given: an integer valued positive bounded function $b(a)$, $a \in K$ and a continuos function $r(c)$, $c \in C$ such that $\inf_{c \in C} r(c) > 0$.

Now we set the following definition:

Def.1. A δ-multiple covering (δ-MC) of K is any collection of circles $D(c,r(c)) \equiv \{ x \in R^N : \rho(c,x) \leq r(c) \}$, $c \in C$, such that any point a K belongs to at least b(a) different circles of the collection. The centres of the circles are required to be at a mutual distance not smaller than δ ($\delta \geq 0$).

The δ-MC, as it has been introduced by the above definition, generalizes the usual covering in that the sets C and K' do not necessarily coincide, a covering index b(a) is explicitely prescribed while b(a)=1 in the classical covering problem and some minimum distance is prescribed between the centres of the covering.

The dimensioning of the network, discussed in §1, which cannot be modelled in the classical framework of covering problems, can easily recognized as a particular case of this more general theory.

The main interest in applications lies in coverings with a finite number of circles. The problem arises quite naturally of finding the optimal one, i.e. that covering (generally not unique) which satisfies the prescribed covering conditions with the last possible number of circles.

Def. 2. A δ-MC of K is said to be optimal if there is no δ'-MC with $\delta' \geq \delta$, with a lower number of circles.

By δ-MC problem (δ-MCP) we mean the problem of computing a δ-MC optimal in the sense of the above definition.

Now we build a discrete model of the δ-MCP. For any $0<h<\inf_{c\in C} r(c)$ the sets A_i^h, $i=1,\dots,m(h)$ are defined such that $A_i^h \cap A_j^h = \emptyset$, $i \neq j$, $A_i^h \cap K \neq \emptyset$, $\bigcup_{i=1}^{m(h)} A_i^h = Q^h \supset K$, and $\mathrm{diam}(A_i^h) \leq h$, where $\mathrm{diam}(A_i^h) = \sup_{a,a' \in A_i^h} \rho(a,a')$. Let C^h be a set with a finite number $n(h)$ of points, such that $\bigcup_h C^h$ is dense in C. For any $c \in C^h$, we define a 'discretized' circle $D^h(c)$, with centre c, $D^h(c) = \bigcup_{i \in I_c^h} A_i^h$, where $I_c^h \equiv \{i : A_i^h \cap D(c,r(c)) = A_i^h\}$. Those sets A_i^h not fully contained in $D(c,r(c))$ must lie outside the circle $D(c,r(c)-h)$ and thus $D^h(c) \supset D(c,r(c)-h)$. For any point $a \in Q^h$ we can define the integer $b^h(a)$ such that $b^h(a) = \max\ \{b(a'),\ a' \in K\ A_i^h$, i such that $a \in A_i^h$ }. Hence, for any $a \in Q^h$, there exists $a' \in K$ such that $b^h(a) = b(a')$ and $\rho(a',a) \leq h$.

We are now in conditions of defining the discrete δ-MC and its optimality in the following way:

Def.3. A discrete δ-MC (Dδ-MC) of Q^h is any collection of 'discretized' circles $D^h(c)$, $c \in C^h$, such that any point $a \in Q^h$ is at least in $b^h(a)$ different circles D^h of the collection. The centres of the 'discretized' circles are required to be at a minimum distance not smaller than δ.

Def.4. A Dδ-MC of Q^h is said to be optimal if there is no Dδ'-MC,

$\delta' \geq \delta$, of Q^h, with a lower number of 'discretized' circles.

The discrete δ-MC problem (Dδ-MCP) is the problem of finding an optimal Dδ-MC, according to def. 4.

The Dδ-MCP can be solved by inspection of a finite number of cases, as the set C^h is, by definition, finite. The inspection can be effectively performed solving an integer linear programming problem, defined in the following way:

i) for any point c_i, $i=1,\ldots,n(h)$, define a boolean variable z_i, where $z_i=1$ if $D^h(c_i)$ belongs to the Dδ-MC, $z_i=0$ otherwise;

ii) for any set A_i^h, $i=1,\ldots,m(h)$ let $b_i^h=b^h(a)$, $a\varepsilon A_i^h$, and consider the constraint:

$$\sum_{j\varepsilon J_i} z_j \geq b^h(a_i) \qquad \text{where } J_i \equiv \{k:\ D^h(c_k) \supset A_i^h;\ c_k \varepsilon C^h\};$$

iii) for any point $c_i \varepsilon C^h$ let $L_i \equiv \{k:\ \rho(c_k, c_i) \leq \delta\ ,\ k \neq i,\ c_k \varepsilon C^h\}$ and consider the constraint:

$$\sum_{j\varepsilon L_i} z_j + n_i z_i \leq n_i \qquad \text{where } n_i \text{ denotes the number of elements of } L_i.$$

We remark that this last constraint is automatically satisfied in the model of the network outlined in §1, where h has been chosen equal to δ.

iv) minimize $\sum_{i=1}^{n(h)} z_i$ subject to the above constraints.

If the Dδ-MCP has a solution for some h then, as can be easily seen by the very construction of the discrete model, the centres of the optimal Dδ-MC, actually computed, are the centres of a δ-MC of K. The question naturally arises wether, for some small h, the centres of an optimal Dδ-MC of Q^h are the centres of an optimal δ-MC of K. Under some conditions, the answer is positive, as stated by the following theorem.

<u>Theorem</u>. Let the δ-MCP of K admit a solution $\{D(\bar{c}_i, r(\bar{c}_i))\}$, $\bar{c}_i \varepsilon C$, $i=1,\ldots,n^*$, such that the circles $D(\bar{c}_i, r'(\bar{c}_i))$ are still a δ-MC of K, where $r'(\bar{c}_i)=r(\bar{c}_i)-\varepsilon$. Moreover, assume that the n^*-uple $(\bar{c}_1, \bar{c}_2, \ldots, \bar{c}_{n^*})$

is the limit of a sequence of n*-uples $(c_1^h, c_2^h, \ldots, c_{n*}^h)$, $\bar{c}_i^h \varepsilon C^h$, such that $\rho(\bar{c}_i^h, \bar{c}_i^h) \geq \delta$ for $i,j=1,\ldots,n*$, for any positive h.

Then, for sufficiently small h, the D δ-MCP has a solution with n* discrete circles.

<u>Proof</u>. It is easily seen that n points $c_i \varepsilon C$, $c_i \neq c_j$ for $i \neq j$, are the centres of a δ-MC of K if and only if the following condition holds, for any $a \varepsilon K$:

$$(1) \qquad \min_{\substack{1 \leq i_1, \ldots, i_{b(a)} \leq n \\ i_k \neq i_\ell \text{ for } k \neq \ell}} \; \max_{j=1,\ldots,b(a)} \{\rho(c_{i_j}, a) - r(c_{i_j})\} \leq 0.$$

An equivalent condition could be stated introducing the function $\phi(a, c_1, \ldots, c_n)$ defined by the left side of (1):

$$(2) \qquad \operatorname*{Sup}_{a \varepsilon K} \phi(a, c_1, c_2, \ldots, c_n) \leq 0 .$$

By the assumptions of the theorem, (1) holds for $\bar{c}_1, \bar{c}_2, \ldots, \bar{c}_{n*}$, with $r'(\bar{c}_j)$ replacing $r(\bar{c}_j)$. Hence

$$(3) \qquad \operatorname*{Sup}_{a \varepsilon K} \phi(a, \bar{c}_1, \bar{c}_2, \ldots, \bar{c}_{n*}) \leq -\varepsilon .$$

It is useful to introduce the discrete analogous of ϕ, for $c_1, c_2, \ldots, c_n \varepsilon C^h$

$$\phi^h(a, c_1, c_2, \ldots, c_n) = \min_{\substack{1 \leq i_1, \ldots, i_{b^h(a)} \leq n \\ i_k \neq i_\ell \text{ for } k \neq \ell}} \; \max_{j=1,\ldots,b^h(a)} \{\rho(a, c_{i_j}) - r(c_{i_j})\} .$$

As $D^h(c)$, $c \varepsilon C^h$ contains $D(c, r(c)-h)$, if Q^h cannot be covered by a Dδ-MC, with n discrete circles $D^h(c, r(c))$, a fortiori the covering requirements $b^h(a)$, $a \varepsilon Q^h$, cannot be satisfied by n circles $D(c_i, r(c_i)-h)$, $c_i \varepsilon C^h$, $i=1,\ldots,n$, $\rho(c_i, c_j) \geq \delta$ for $i \neq j$. This implies

$$(4) \qquad \operatorname{Sup} \phi^h(a, c_1, c_2, \ldots, c_n) \geq -h.$$

In order to prove the theorem, we assume, by way of contradiction, that for any $h<\inf_{c\in C} r(c)$ n^* circles $D^h(c_i)$, $c_i \in C^h$, are not sufficient for a $D\delta$-MC of Q^h. Therefore from (4) it follows

$$\operatorname{Sup}_{a\in Q^h} \phi^h(a,c_1,\ldots,c_{n^*}) > -h$$

for any n^*-uple of points in C^h such that $\rho(c_i,c_j)\geq\delta$, $i\neq j$.

By the definition of $b^h(a)$, $a\in Q^h$, there exists $a'\in K$ such that $b^h(a)=b(a')$ and $\rho(a',a)\leq h$; therefore

$$\rho(c_i,a')\geq\rho(c_i,a)-\rho(a',a)\geq\rho(c_i,a)\ -h$$

and

$$\begin{aligned}&\phi(a',c_1,c_2,\ldots,c_{n^*})\geq\\ &\geq\min_{1\ i_1,\ldots,i_{b^h(a)}}\ \max_{j=1,\ldots,b^h(a)}\ \{\rho(c_{i_j},a)-h-r(c_{i_j})\} =\\ &=\phi^h(a,c_1,c_2,\ldots,c_{n^*})-h.\end{aligned}$$

Thus

$$(5)\quad \operatorname{Sup}_{a\in K}\ \phi(a,c_1,c_2,\ldots,c_{n^*})\ \geq \operatorname{Sup}_{a\in Q^h}\phi^h(a,c_1,c_2,\ldots,c_{n^*})-h\geq\ -2h$$

for any n^*-uple of points c_i, $i=1,\ldots,n^*$, $c_i\in C^h$, $\rho(c_i,c_j)\geq\delta$ for $i\neq j$ and for any $h<\inf_{c\in C} r(c)$.

Now we show the continuity of $\operatorname{Sup}_{a\in K}\ \phi(a,c_1,c_2,\ldots,c_n)$ in the variables c_i. Indeed, if $c_i^k\to c_i$, $i=1,\ldots,n$, as $k\to\infty$, by the continuity of r and ρ we have, for any $\eta>0$, that, for large k, $|r(c_i^k)-r(c_i)|$ + $\rho(c_i,c_i^k)<\eta$ for $i=1,\ldots,n$. Hence

$$|\rho(c_i,a)-r(c_i)-\ [(c_i^k,a)-r(c_i^k)]|\leq\rho(c_i,c_i^k)+|r(c_i^k)-r(c_i)|<\eta,$$

$i=1,\ldots,n$. Thus

$$-\eta\ +\rho(c_i^k,a)-r(c_i^k)<\rho(c_i,a)-r(c_i)<\eta+\rho(c_i^k,a)-r(c_i^k).$$

Performing for all the sides of the above relation the operations required by the definition of $\phi(a,c_1,c_2,\ldots,c_n)$ we get

$$-\eta+\phi(a,c_1^k,c_2^k,\dots,c_n^k)\le \phi(a,c_1,c_2,\dots,c_n)\le \phi(a,c_1^k,c_2^k,\dots,c_n^k)+\eta.$$

Taking the supremum with respect to a we obtain, for sufficiently large k,

$$-\eta+\operatorname*{Sup}_{a\varepsilon K}\phi(a,c_1^k,c_2^k,\dots,c_n^k)\le\operatorname*{Sup}_{a\varepsilon K}(a,c_1,c_2,\dots,c_n)\le$$
$$\le\operatorname*{Sup}_{a\varepsilon K}\phi(a,c_1^k,c_2^k,\dots,c_n^k)+\eta.$$

This proves, by the arbitrariety of η, the continuity of $\operatorname*{Sup}_{a\varepsilon K}\phi(a,c_1,c_2,\dots,c_n)$.

The continuity of $\operatorname*{Sup}_{a\varepsilon K}(a,c_1,c_2,\dots,c_n)$ and relation (5) imply, for the sequence $(\bar{c}_1^h,\bar{c}_2^h,\dots,\bar{c}_{n*}^h)$, that

$$\operatorname*{Sup}_{a\varepsilon K}\phi(a,\bar{c}_1,\bar{c}_2,\dots,\bar{c}_{n*}) = \lim_{h\to 0}\ \operatorname*{Sup}_{a\varepsilon K}\phi(a,\bar{c}_1^h,\bar{c}_2^h,\dots,\bar{c}_{n*}^h)\ge 0$$

which is against (3).

Thus the Dδ-MCP has a solution of n* circles, for a sufficiently small h, and the proof is complete.

Once the theorem can be applied, an optimal δ-MC of K can be obtained through an optimal Dδ-MC of Q^h, for sufficiently small h.

ACKNOWLEDGEMENTS

The authors are very grateful to Prof. M. Cugiani for fruitful suggestions and valuable help in the writing of this paper.

REFERENCES

[1] VERCELLIS,C., An Algorithm for the Multiple Covering Problem, to appear in Quaderni del Dipartimento di Ricerca Operativa e Scienze Statistiche- Università di Pisa.

[2] SATO,Y. and SKOKO,D., Optimum Distribution of Seismic Observation Points,II, Bull. Earthq. Res. Inst., vol. 45 (1965), pp. 451-457.

[3] BULAND,R., The Mechanics of Locating Earthquakes, Bull. Seism. Soc. Am., vol. 66 (1976), n.1, pp. 173-187.

A REMARK ON ECONOMETRIC MODELLING, OPTIMIZATION AND DECISION MAKING °

M. Cirinà

Istituto Matematico del Politecnico

Torino, Italy

This paper is concerned with an operational issue about a common type of econometric model viewed as a decision making tool. It is generally recognized that the predictive performance of many presently available econometric models may be considered somewhat below a desirable standard for decision making; indeed facts such as large variances of parameter estimates, fast changes in the system being modelled, presence of random disturbances, and flattening of nonlinearities into linear approximants may be responsible for reducing the usefulness of the model as an operational tool in decision making. This paper reports the initial part of an investigation on whether it is possible to improve the predictive performance of a given econometric model without modifying the form of the model equations. As a first step towards such an end, we introduce here a calibration problem - that is a certain mathematical program -, apply it to a specific econometric model, and report the preliminary results obtained.

1. INTRODUCTION

Econometric data can be made into econometric models in many ways. One general procedure - based mainly on statistics - assumes that the underlying (economic) process can be identified from the data describing (a sufficient part of) its evolution, and generates models by making use of time series analysis to "reduce the data to white noise", i.e. by removing the random pertubations and extracting the meaningful information from

---

° This research has been supported in part by the Ministero della Pubblica Istruzione (Art. 286 T.U.).

the observed data, see for instance Box and Jenkins [2] . The typical econometric model produced in this way is characterized in general by the fact that it interrelates few variables, that it has good forecasting properties, and that it is of modest value in decision making because it does not allow a compared evaluation of different policy options. Another procedure, of general purpose, has at its bottom an understanding of the process generating the data, i.e. an economic theory, makes use of such understanding to specify the form of a set of equations interrelating the relevant variables and containing in general numerous parameters or coefficients whose values (not known a priori) are to be estimated; this is often accomplished by fitting the model to the data (the time series of the dependent and independent variables) with a least square procedure, see for instance the models reported in Banca d'Italia [1] and Hickman [5] .
The typical econometric model obtained in this way is characterized in general by the fact that it interrelates many variables (several hundreds in the larger models), that it is structurally suited for evaluating policy alternatives - hence to be an operational tool in decision making - and that it may predict reasonably the evolution of the modelled system if it is not mis-specified. Mis-specifications are however rather common. This paper is concerned with models of this latter type, from an operational point of view. In section 2 a specific class of such models is singled out and the occurrence of mis-specifications considered. In section 3 a calibration problem is introduced as a mean for possibly improving the operational performance of such models without altering the form of the model's equations and thus without disturbing the underlying economic theory. Such calibration problem is then applied to a specific model concerning the Italian economy and from the preliminary results obtained it seems that in some cases it may be useful.

2. MIS-SPECIFICATIONS

Many econometric models for a given year (quarter, semester, ...), say for the year N, - see for instance Banca d'Italia [1] , the papers in

Hickman [5] and their references, and Sylos Labini [7]- are of the form

(2.1) $$Ax + By = 0$$

where $x \in R^m$ is a vector whose components $x_1, \ldots, x_m$ are the endogenous variables; $y \in R^{\bar{m}}$ is a (known) vector whose components $y_1, \ldots, y_{\bar{m}}$ are called pre-determined variables , i.e. y_i is either a lagged endogenous variable, or an exogenous variable, or a control variable; A is a m×m real matrix - briefly $A \in R_{m\times m}$ - invertible, and $B \in R_{m\times\bar{m}}$; also for each i,j the element $(A)_{ij}$ of A and $(B)_{ij}$ of B will be in general

(2.2) either 0, or +1, or a dynamical term, or an estimated coefficient,

where a dynamical term is typically a simple function of a lagged variable (i.e. of an observed value of some variable in some past year), for instance a term such as $-100/(\tilde{x}_k)_{-1}$, where $(\tilde{x}_k)_{-1}$ is the observed value of the variable x_k 1 year earlier; for a specific instance, see the matrices A and B (Tab. 1 and 2) corresponding to the model dealt with in section 4. When it will be convenient to emphasize the year N implicitly understood in (2.1), we shall write A_N, x_N, B_N, y_N instead of A, x, B, y. Since in this paper we are mainly concerned with the elements of A and B of the last type said in (2.2), it is convenient to introduce the set of indices E_A defined by

$$E_A = \{(i,j) \mid 1 \leq i,j \leq m, \text{ and } (A)_{ij} \text{ is an estimated coefficient}\}$$

and let E_B be defined analogously. Of course the case of (2.1) with non zero right hand side b, is included in the writing (2.1), for it suffices to include -b in B as last column and to give the value 1 to the corresponding last component of y. Solving the model (2.1) simply means to find x satisfying (2.1) for given A, B and y; since A is an invertible matrix, for any policy option - i.e. any choice of the control variables in y - the model (2.1) gives the decision maker (an indication of) the resulting effect on the state x of the system. Let us call $\tilde{x}$ the observed value - i.e. the actual value in the system - of the endogenous vector in the year N; an important fact is that in general $\tilde{x}$ does not solve (2.1), i.e. that in general the residual vector

(2.3) $$A\tilde{x} - By$$

is not zero and therefore that there are discrepancies between the value of the state variable x given by the model at any year N, and the value $\tilde{x}$ actually observed in the same year (for a specific instance see fig. 1); thus in general

(2.4) $$\tilde{x} - A^{-1}By$$

is not zero; of course the closeness of the value $\tilde{x}$ that is observed when a certain policy has been enforced, to the value x given by the model in correspondence of the same policy, is the crucial fact that has to be true if the model (2.1) is to be valuable as an operational tool in decision making. It may be useful to mention that from the smallness of the residual vector (2.3) it does not follow in general that of the vector (2.4).

There are many reasons that may contribute to large discrepancies (2.4); for instance, a variable may be unduly included among the endogenous variables x_i's or unduly excluded from them; a similar situation may occur for some exogenous variables in y; or perhaps there is some strongly nonlinear relationship among some variables that in (2.1) is flattened into a linear one; also the variances of some estimated coefficients appearing in A and B may happen to be rather large.

Improving a specific model (2.1) by removing or attenuating the misspecifications in it, such as those just mentioned, may be considered to need a medium to long term econometric research investment, involving a careful scrutiny and refinement of the model's equations.

Our objective here is of different nature because it is aimed at establishing whether it is possible or not to improve the operational performances of an econometric model (2.1) as a decision making tool, without altering the form of the model's equations. To motivate the procedure that we shall introduce in section 3 as a first step towards such objective, it may be useful to make some observations.

REMARKS 2.1 Loosely speaking, a large variance in an estimated coefficient $(A)_{ij}$ indicates that the form of some equation in (2.1) and the data (i.e. the time series) used in the estimation process, fail to determine $(A)_{ij}$; also, in the estimation procedure leading to the model

(2.1) for the year N, the observed data of the year N-1 have just the same weight as those of say the year N-10; and moreover the estimated coefficients in A_{N-10} and in A_{N-1} are the same while the system being modelled has often changed considerably over a 10 year period.

3. A CALIBRATION PROBLEM

Motivations such as those mentioned above induce us to introduce the following mathematical programming problem (CP).

CALIBRATION PROBLEM By taking into account the range of incertitude of the estimated coefficients in A and B - as indicated for instance by their standard deviations - choose the matrices A_+, A_-, B_+, B_- satisfying

(3.1) $A_- \leq A \leq A_+$, elementwise, with equality for all $(i,j) \notin E_A$,

(3.2) $B_- \leq B \leq B_+$, elementwise, with equality for all $(i,j) \notin E_B$,

and such that the first of the sets

(3.3) $\mathbb{A} = \{ A \in R_{m\times m} \ : \ A_- \leq A \leq A_+ \}$, $\mathbb{B} = \{ B \in R_{m\times \bar{m}} \ : \ B_- \leq B \leq B_+ \}$,

is made up by invertible matrices A, let N be a year for which y_N and the observed value $\tilde{x}_N$ are available, and define

(CP) $(A_{N+1}, B_{N+1}) = \arg\min \{ | A\tilde{x}_N - By_N | \ : \ (A,B) \in \mathbb{A} \times \mathbb{B} \}$.

A property that it would be useful having, i.e.

(3.4) $(A_{N+1}, B_{N+1}) = \arg\min \{ | \tilde{x}_N - A^{-1}By_N | \ : \ (A,B) \in \mathbb{A} \times \mathbb{B} \}$,

cannot be enforced directly, because in general the description of A^{-1}, $A \in \mathbb{A}$,is not available. Let us notice that if in particular N is the last year for which the data are available, property (3.4) would say that - within their range of incertitude - (A_{N+1}, B_{N+1}) are as close as pos sible to give reproduction of the latest observed values $\tilde{x}_N$. As however it will be seen below, under hypotheses that are often met in practice, the calibration problem (CP) makes property (3.4) true, or approximately so. To see this - that actually holds in settings more general than the one used so far - let X, Y be Banach spaces, $L(Y,X)$ be the B-space of bounded linear operators from Y to X - with the operator norm $\|B\| = \sup\{|By|_X/|y|_Y \ : \ 0 \neq y \in Y\}$ -, fix $\mathcal{A} \subset L(X,X)$, $\mathcal{B} \subset L(Y,X)$, $x \in X$, $y \in Y$, and consider the following problems

(P) $\qquad \inf \{ |Ax - By| \ : \ (A,B) \in \mathcal{A} \times \mathcal{B} \} =: v(P)$,

(Q) $\qquad \inf \{ |x - A^{-1}By| \ : \ (A,B) \in \mathcal{A} \times \mathcal{B} \} =: v(Q)$.

The following proposition is about what it can be said on (Q) when a minimizer (A_p, B_p) of (P) is known (of course there are sets of conditions that guarantee the existence of such minimizer, as for instance compactness of $\mathcal{A} \times \mathcal{B}$).

PROPOSITION 3.1 With the definitions above, suppose $\mathcal{A}$ is a subset of invertible operators, and let $(A_p, B_p) \in \mathcal{A} \times \mathcal{B}$ satisfy $|A_p x - B_p y| = v(P)$. Then

(i) $\quad v(P) = 0 \ \Rightarrow \ A_p, B_p$ are also a solution of (Q).

(ii) If $v(P) \neq 0$ then A_p, B_p need not solve (Q).

(iii) Moreover if $(A_q, B_q) \in \mathcal{A} \times \mathcal{B}$ satisfy $|x - A_q^{-1} B_q y| = v(Q)$, then

$$v(Q) \leq |x - A_p^{-1} B_p y| \leq \mu \, v(Q) \ , \text{ with } \mu = \|A_p^{-1}\| \, \|A_q\| \ .$$

Proof

(i) Since A_p is invertible, from $|A_p x + B_p y| = 0$ it follows that

$$|x + A_p^{-1} B_p y| \leq |x - A^{-1} By| \quad \forall \ (A,B) \in \mathcal{A} \times \mathcal{B}.$$

(ii) Let

$$A(s) = A(s_1, s_2) = \begin{bmatrix} s_1 & 1 \\ -1 & s_2 \end{bmatrix} \ ; \ x = \begin{bmatrix} 1 \\ 1 \end{bmatrix} \ ; \ y = \begin{bmatrix} -2 \\ 2 \end{bmatrix} \ ;$$

$\mathcal{A} = \{ A(s) \mid -2 \leq s_1 \leq 0 \ , \ 2 \leq s_2 \leq 4 \}$ and $\mathcal{B} = \{I\}$, where I is the identity in $R_{2 \times 2}$. Then it is easily seen that the hypotheses of the proposition are satisfied with $A_p = A(-2,3)$, $B_p = I$, and that $v(P) = 1$; but for $A = A(-2,4) \in \mathcal{A}$ and $B \in \mathcal{B}$ one has $|x - A^{-1}By| = \sqrt{10}/7 < |x - A_p^{-1} B_p y| = \sqrt{10}/5$.

(iii) It follows from $|x - A_p^{-1} B_p y| \leq \|A_p^{-1}\| \, |A_q x - B_q y|$, from $|A_q x - B_q y| \leq \|A_q\| \, |x - A_q^{-1} B_q y|$, and from the fact that the definition of $v(Q)$ implies

$$|x - A_q^{-1} B_q y| \leq |x - A_p^{-1} B_p y| \quad .$$

Let us notice that the assertion (iii) in the proposition above says that if the operators A_p, A_q are "well conditioned" in the sense that $\mu = \|A_p^{-1}\| \, \|A_q\|$ is not a large number, then A_p, B_p may be considered as approximate solution of problem (Q), the more so the closer μ is to 1.

4. PRELIMINARY RESULTS

A preliminary testing of the concept of calibration introduced above is being conducted on what is possibly the simplest meaningful model of the Italian economy introduced originally by Sylos Labini [7]. Actually we have been using the version of this model given by Del Monte [4, p. 92] and named MOSYL DF/70. Let us first put such model in matrix form. Each variable - say LIQTOT (=*liq*uidità *tot*ale) - appears in MOSYL or as such, or as first difference DLIQTOT, or as rate of change LIQTOTP, defined by

$$(4.1) \quad \text{DLIQTOT} = \text{LIQTOT} - \widetilde{\text{LIQTOT}}_{-1} \quad , \quad \text{LIQTOTP} = 100\ \text{DLIQTOT}/\widetilde{\text{LIQTOT}}_{-1}$$

where $\widetilde{\text{LIQTOT}}_{-1}$ is the observed value of LIQTOT 1 year earlier; therefore by adding to the 16 (affine) equations of MOSYL DF/70 enough defining identities (4.1) - and by ordering them properly - one can write it in the form (2.1) with A - having 1's along the diagonal - as given in Table 1 and B as in Table 2; A and B correspond indeed to the ordering of the variables given on such tables; that is, the endogenous variables CAPINU, CORTOT, ..., SALINDP are respectively, the components x_1, $x_2, \ldots, x_{33}$ of $x \in R^{33}$, and the predetermined variables CORTOT_{-1}, IMPORT_{-1}, ..., PREMONP are respectively the first 22 components of $y \in R^{23}$, y_{23} being always fixed to the value 1 to take care of the constant terms. Also let us notice that the matrix form given here contains precisely the 16 equations in [4, p. 92] and - among the added definitions (4.1) - includes all 9 identities in [4, p. 92]; moreover for graphical reasons, the dynamical terms in matrix A (Table 1) have been written in a simpler notation, i.e. $\overline{XK}$ and $\underline{XK}$ stand respectively for $-(\tilde{x}_k)_{-1}/100$ and $-100/(\tilde{x}_k)_{-1}$, where it is understood that $(x_k)_{-1}$ is the observed value of the variable x_k 1 year earlier.

$$
\begin{bmatrix}
1 &&&&&&&&&& .128 &&&&&&&&&&&&&&& .577 &&&&&&& \\
& 1 && \overline{X2} &&&&&&&&&&&&&&&&&&&&&&&&&&&&& \\
& -1 & 1 &&&&&&&&&&&&&&&&&&&&&&&&&&&&&& \\
&&& 1 &&&&&&&&&&&&&&&&&&&&&& -.608 &&&&&&& \\
&&&& 1 &&&&&&&&&&&&&&& -.958 &&&&&&&&&&&&& \\
&&&&& 1 &&& -.00051 &&&&&&&&&&&&&&&&&&&&&&&& \\
&&&&&& 1 & -1 &&&&&&&&&&&&&&&&&&&&&&&&& \\
&& -.0265 &&&&& 1 && -.662 &&&&&&&&&&&&&&&&&&&&&&& \\
&&&&&&&& 1 && \overline{X9} &&&&&&&&&&&&&&&&&&&&&& \\
&&&&&&&& -1 & 1 &&&&&&&&&&&&&&&&&&&&&&& \\
2.69 &&&&&&&&&& 1 &&& -.3959 &&&&&&& -.898 &&&&&&&&&&&& \\
&&&&&&&&&&& 1 & -1 &&&&&&&&&&&&&&&&&&&& \\
&&&&&&&&&&&& 1 &&&&&&&&&&&& -1.434 &&&&& 1.729 &&& \\
&&&&&&&&&&&& \underline{X12} & 1 &&&&&&&&&&&&&&&&&&& \\
& .0025 &&&&&&&&&&&&& 1 &&&&&&&&&&&&&&&&&& \\
&&&&&&&&&&&&&& -1 & 1 &&&&&&&&&&&&&&&&& \\
&&&&&&&&&&&&&&& \underline{X19} & 1 &&&&&&&&&&&&&&&& \\
&&&&&&&&&&&&&&&&& 1 &&&&&&&&&&&&&&& \\
&&&&&&&&&&&&&&&& -.169 & -.791 & 1 &&&&&&&&&&&&&& \\
&&&&&&&&&&&&&&&&&& -.416 & 1 &&&&&&&&& -.33 &&&& \\
&&&&&&&&&&&&&&&&&&&& 1 && \overline{X21} &&&&&&&&&& \\
&&&&&&&&&&&&&&&&&&&& -1 & 1 &&&&&&&&&&& \\
&&&&&&&&&&&&&&&&& -.67 &&&&& 1 &&&&&&&&&& \\
&&&&&&&&&&&&&&&&&&&&&&& 1 & -1 &&&&&&&& \\
&&&&& 3352.8 &&&&&&&&&&&&&&&&&&& 1 &&&&&&& -41.55 & \\
&&&&&&&&&&&&&&&&&&&&&&&& \underline{X24} & 1 &&&&&&& \\
&&&&& 61.03 &&&&&&&&&&&&&&&&&&&&& 1 &&&&&& \\
&&&&&&&&&&&&&&&&&&&&&&&&&& -1 & 1 &&&&& \\
&&&&&&&&&&&&&&&&&&&&&&&&&&& \underline{X27} & 1 &&&& \\
&&&&&& 12.01 &&&&&&&&&&&&&&&&&&&&&&& 1 &&& \\
&&&&&&&&&&&&&&&&&&&&&&&&&&&&&& 1 && \overline{X31} \\
&&&&&&&&&&&&&&&&&&&&&&&&&&&&&& -1 & 1 & \\
&&& -.72 & -17.68 &&&&&&&&&&&&&&&&&&&&&&&&&&&& 1
\end{bmatrix}
$$

CAPINU, CORTOT, DCORTOT, CORTOTP, COSVITP, DISOCC, IMPORT, DIMPORT, INVIND, DINVIND, INVINDP, LIQTOT, DLIQTOT, LIQTOTP, PREAGR, DPREAGR, PREAGRP, PREINDP, PREINGP, PREMINP, QUOPRO, DQUOPRO, QUOPROP, RELADI, DRELADI, RELADIP, SALCOM, DSALCOM, SALCOMP, SALDOM, SALIND, DSALIND, SALINDP.

Table 1 - Matrix A and the ordering of the endogenous variables

			.016													-.192						15.414
1																						
-1																						
																						2.429
										.091												
		.00094																				-.085
	1																					
														28.674								-18.061
		1																				
		-1																				
													18.34									-51.99
				1																		
												830.286						.586				886.398
																				-.257		109.66
					-1																	
											.297								.316		.459	-.167
																						.053
						1																
						-1																
											-.967											-1.355
							1															
																						-78.48
									.685													39.594
								-1														
																	10.27					
									1													
									-1													
															6.58							-.76

$CORTOT_{-1}$, $IMPORT_{-1}$, $INVIND_{-1}$, $INVIND_{-2}$, $LIQTOT_{-1}$, $PREAGR_{-1}$, $QUOPRO_{-1}$, $RELADI_{-1}$, $SALCOM_{-1}$, $SALIND_{-1}$, AFFITTP, COSLAVP, DUMMYB, DUMMYS, DUMMIK, DUMSAL, ESPINDP, ESPORT, INDPUB, MATINDP, OFFDIS, PREMONP.

Table 2 - Matrix B and the ordering of the predetermined variables

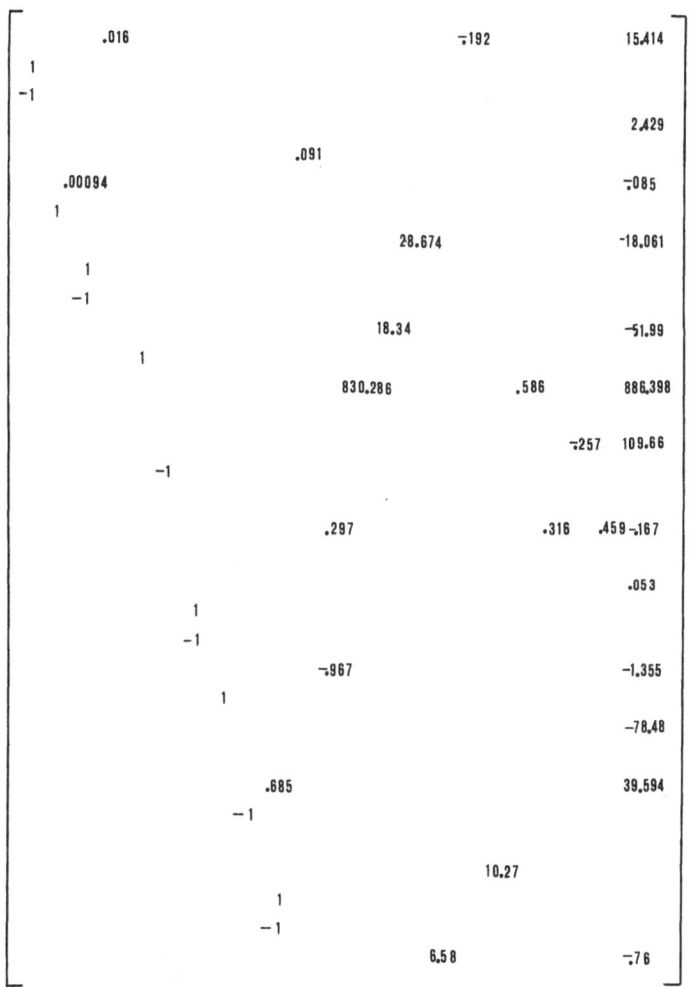

Now one has to solve the calibration problem (CP) for N = 1955,...,1969, i.e. 15 (mildly) nonlinear problems; this introduces no special difficulty and can be done by standard nonlinear programming methods, see for instance [6] for a general reference, and [3] for a specific one. In the process of solving such 15 problems (CP) it was realized that some of the data in [4, pp. 105-132] do not entirely correspond to the data used to estimate the coefficients of MOSYL DF/70 [4, p. 92] . To get nevertheless a preliminary indication as to whether (CP) may be useful or not, the problems (CP) were solved with the data in [4, pp.105-132] , except for IMPORT (p. 131) and INVIND (p. 128) whose values have been divided by 10 for a comparison purpose, i.e. in order to make the results we were able to obtain in absence of calibration closer to the corresponding ones in [4, pp. 114-131, columns marked M.T.].
For an indication on the effects of (CP), one has to compare

(4.2) the observed values $\tilde{x}_N$ and the model values x_N (with and without calibration), for N=1956,...,1970;

a synthetic indication is provided by the coefficient

$$r_k = 1 - \frac{|(x_k) - (\tilde{x}_k)|^2}{|(\tilde{x}_k)|^2}, \quad k = 1,2,\ldots,33$$

where (x_k) -and analogously $(\tilde{x}_k)$- is the vector whose components are the values of x_k for N=1956,...,1970, and $|\cdot|$ is the Euclidean norm.
For the situation described above - where part of the mis-specification is accidental, being due to misprints in the data - whe have obtained that the average value of r_k goes up from about .82 without calibration, to about .87 with a calibration (CP) corresponding to a certain choice of the sets (3.3) ; thus an average improvement of 5 percentage points.
In Fig. 1 it is shown the full comparison (4.2) for one single variable, i.e., for (x_{20}) that corresponds to PREMINP. Of course what matters here is not really the amount of improvement -which depends in particular on the amount of mis-specification in the model- but the fact that there are choices of (3.3) for which (CP) does give some improvement. In conclusion, it seems likely that (CP) as defined here may be useful in presence of larger mis-specification; smaller ones are the object of a further development not reported here.

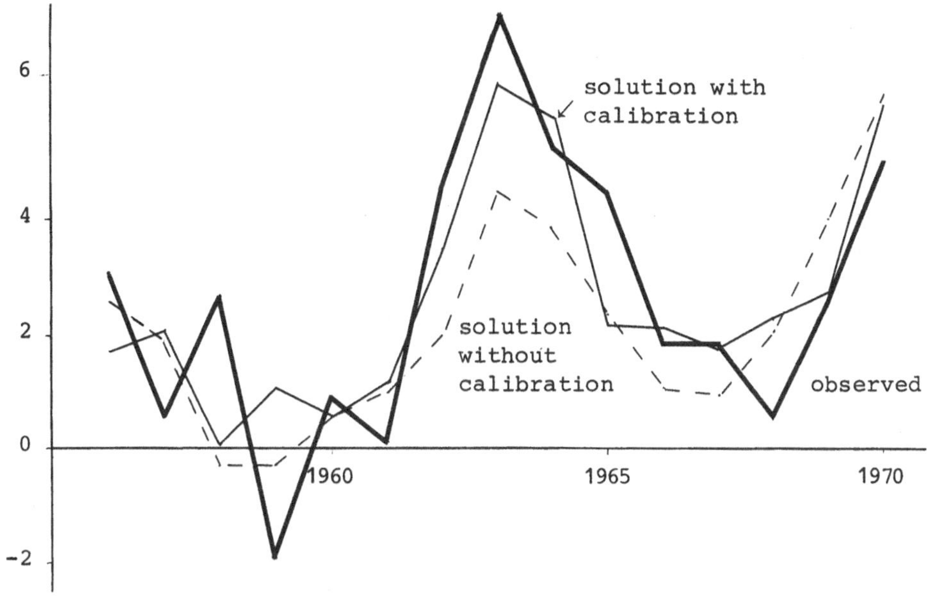

Figure 1 - PREMINP (rate of change of retail prices)

REFERENCES

[1] Banca d'Italia, "Un modello econometrico dell'economia italiana (M1BI)", Reports n. 1 to 7, Banca d'Italia, Roma, 1970.

[2] Box G.E.P. and Jenkins G.M., "Time series analysis, forecasting and control", Holden Day, 1970.

[3] Cirinà M., "A scaled multiplier method", presented at the II European Congress on Operations Research (Stockholm, 1976), in "Recent Advances in Operations Research", M. Roubens (Ed.), North-Holland, 1977, 111.

[4] Del Monte C., "Un modello econometrico per l'economia italiana utilizzato a fini previsivi", Rassegna Economica, XXXVII n.1, 1973, 69.

[5] Hickman B.G. (Ed.), "Econometric models of cyclical behavior", Columbia University Press, 1972.

[6] Stoer J. and Witzgall C., "Convexity and optimization in finite dimensions", Springer-Verlag, 1970.

[7] Sylos Labini P., "Prezzi, distribuzione e investimenti dal 1951 al 1966: uno schema interpretativo", Moneta e Credito, Sett.1967,264.

Józef Grabowski
Technical University of Wrocław
Institute of Technical Cybernetics
Wrocław, Poland

Formulation and solution of the sequencing problem with parallel machines

1. Introduction

In many branches of industry (metallurgy, machine construction) the production process is characterized by a flow of elements in the technological sequence. These elements are processed on successive machines. The processing times of individual operations are different for different elements and machines. Therefore, the problem arises to determine such a sequence of operations that the total time of performing all operations is minimal.

In this paper this problem with parallel machines is presented and solved. We assume that the duration of operations may be different for different machines from the set of parallel machines. Therefore, the problem considered is more general than those examined in the papers [4], [5], [8] and [9]. The problem in which the set of parallel machines contains exactly one machine was presented in the papers [1]-[3] and [5].

The mathematical model of the sequencing problem is formulated by using disjunctive graphs. The notions from [6] are generalized and some new ones are introduced. A number of properties are proved which allow us to construct a new, relatively effective algorithm. These problems may be modelled and solved by using zero-one linear programming [8]. The efficiency of algorithms which solve the zero-one programming problems is very low, so one can solve only sequencing problems which are of small size. Besides, to construct a zero-one programming model, the considered period time must be divided into a certain number of time periods. The greater the number of these periods, the better the accuracy of the solution. On the other hand, a greater number of periods gives a greater size of problems, so it is difficult to obtain a solution. For the models which are presented in the papers [4] and [5], in each iteration of the algorithm the coefficient of internal stability has to be computed, so it increases the computation time. The mathematical models of sequencing problems which are constructed by using disjunctive graphs have not these weaknesses.

2. Mathematical formulation of the problem

Let

$$N = \{1,2,\dots,n\}$$

be the set of operations number of operations which should be carried out by using the set of various type machines

$$B = \{1,2,\dots,b\}.$$

Each an individual operation $j \in N$ should be carried out by using exactly one of the subset from family

$$R_j = \{B_1,B_2,\dots,B_{r_j}\}, \quad B_p \subset B, \quad j \in N.$$

Therefore, the problem arises to assign the subset of machines to the individual operations and to determine such sequence of operations that the total time of all operations is minimal and the following assumptions are satisfied:

1. The operations should be carried out according to a required technological order.
2. The duration of individual operations are fixed and different for different subset of machines.
3. The sequence of operations on each machine can be arbitrary.
4. Every operation can not be interrupted.
5. The set-up times are equal to zero.

This problem is more general than those which have been considered so for [1] - [9].

Let

$$RT \subset N \times N$$

be the set of relations expressing the technological requirements of the operation order. Let $N_x \subset N$ be the set of operations which have no predecessors,

$$N_x = \{j \in N \mid \forall i \in N \quad \langle i,j \rangle \notin RT\},$$

and let $N_y \subset N$ be the set of operations which have no successors,

$$N_y = \{j \in N \mid \forall i \in N \quad \langle j,i \rangle \notin RT\}.$$

Let us introduce the following notation:

t_j^x - the starting time of the operation j;

t_j^y - the finshing time of the operation j;

c_{jp} - the duration of the operation j which is carried out by using the subset of machines $B_p (c_{jp} > 0,\ B_p \in R_j,\ j \in N)$;

$t_o = 0$ and t_z - the starting and the finishing times of all operations, respectively.

We formulate the sequencing problem as follows: Find t_z, t_j^x, t_j^y, (≥ 0), $j \in N$:

/1/ $$t_z = \min,$$

subject to:

/2/ $$(t_j^y - t_j^x \geq c_{j_1}) \vee (t_j^y - t_j^x \geq c_{j_2}) \vee \ldots \vee (t_j^y - t_j^x \geq c_{jr_j}), \quad j \in N,$$

/3/ $$t_j^x - t_j^y \geq 0, \qquad \langle i, j \rangle \in RT,$$

/4/ $$t_j^x - t_o \geq 0, \qquad j \in N_x,$$

/5/ $$t_z - t_j^y \geq 0, \qquad j \in N_y,$$

/6/ $$(t_j^y - t_j^x \geq c_{jp}) \wedge (t_i^y - t_i^x \geq c_{iw}) \Longrightarrow \qquad i, j \in N,\ i \neq j,$$
$$\Longrightarrow (t_j^x - t_i^y \geq 0) \vee (t_i^x - t_j^y \geq 0), \qquad B_p \cap B_w \neq \emptyset.$$

Constraint /2/ is disjunction and requires that the difference between the starting and finishing times of a given subset of machines is not less than the duration of this operation. Condition /3/ gives the required technological order of operations. Constraint /4/ and /5/ state the condition that t_o and t_z are the starting and finishing times, respectively, of all operations. Two different operations $j, i \in N$ which are to be carried out by using the two inseparable subset of machines can not be carried out at the same time, but one of these operations must be finished before the second one is started. This is expressed by constraint /6/ which is called the implicit condition.

For problem /1/ - /6/ we can construct a disjunctive graph

$$\bar{D} = \langle A, U; V, V^o \rangle$$

where

/7/ $$D = \langle A, U \rangle$$

is the graph.

The set of nodes equals

$$A = X \cup Y \cup \{o\} \cup \{z\},$$

where

$$X = \bigcup_{j \in N} \{x_j\}, \quad Y = \bigcup_{j \in N} \{y_j\},$$

and the nodes x_j, y_j, o, z are associated with the variables t_j^x, t_j^y, t_o, t_z, respectively.

The set of arcs equals

$$U = U_1 \cup U_2 \cup U_3$$

where

$$U_1 = \bigcup_{\langle i, j \rangle \in RT} \{\langle y_i, x_j \rangle\}, \quad U_2 = \bigcup_{j \in N_x} \{\langle o, x_j \rangle\}, \quad U_3 = \bigcup_{j \in N_y} \{\langle y_j, z \rangle\},$$

and the ares $\langle y_i,x_j\rangle$, $\langle o,x_j\rangle$, $\langle y_j,z\rangle$ are associated with constraints /3/, /4/, /5/, respectively.

To each constraint of the disjunctive form /2/ we can assign the set of disjunctive arcs in $\bar{D}$,

/8/ $$[\langle x_j,y_j\rangle_1 \vee \langle x_j,y_j\rangle_2 \vee \ldots \vee \langle x_j,y_j\rangle_{r_j}], \quad j \in N,$$

where $\langle x_j,y_j\rangle_p$ has the length c_{jp}, and let

/9/ $$V^o = \bigcup_{j \in N} V^{oj},$$

be the set of all these arcs, where

$$V^{oj} = \bigcup_{B_w \in R_j} \{\langle x_j,y_j\rangle_w\}, \quad j \in N.$$

To the right-hand-side of each constraint of the form /6/ we can assign the pair of disjunctive arcs in D, which have the length equals zero,

/10/ $$[\langle y_i,x_j\rangle \vee \langle y_j,x_i\rangle], \quad i,j \in N, \quad i \neq j,$$

and let

/11/ $$V = \{\langle y_i,x_j\rangle \in Y \times X \mid (i,j \in N) \wedge (\text{there does not exist a path from the node } y_i \text{ to } x_j \text{ in the graph } D\}.$$

The subset $S_p^o \subset V^o$ containing exactly one arc from each set of disjunctive set of disjunctive set /8/ is called a selection of sets, and let

$$R_s^o = \{S_1^o, S_2^o, \ldots, S_k^o\}$$

be the family of all selection of sets.

The subset $S_r \subset V$ containing at most one arc from each disjunctive pair /10/ is called a selection, and let

$$R_s = \{S_1, S_2, \ldots, S_l\}$$

be the family of all selection of pair.

Each selection $S_r \in R_s$ and each selection sets $S_p^o \in R_s^o$ generates a conjuctive graph

/12/ $$D_{rp} = \langle A,\ U \cup S_r \cup S_p^o\rangle.$$

Let

$$R_D = \{D_{rp} = \langle A,\ U \cup S_r \cup S_p^o\rangle\}$$

be the family of graphs of form /12/, and let

$$R'_D = \{D_{rp} \in R_D \mid D_{rp} \text{ has no circuits}\}$$

be the family of graphs without circuits.

In each graph $D_{rp} \in R'_D$, to each implicit condition /6/ we can assign the implicit condition of disjunctive arcs of the form:

/13/ $$(\langle x_j,y_j\rangle_k, \langle x_j,y_j\rangle_w \in S_p^o) \Longrightarrow (\langle y_i,x_j\rangle \in S_r \vee \langle y_j,x_i\rangle \in S_r),$$
$$i,j \in N,\ i \neq j,\ B_k \cap B_w \neq \emptyset.$$

Let

$$R'_{DJ} = \{ D_{rp} \in R'_D \mid D_{rp} \text{ satisfies the implicit condition /13/} \}$$

be the family of graphs for which the implicit condition is satisfied.

The critical path of the graph $D_{oo} \in R_{DJ}$ is called minimaximal path with the implicit condition /13/ in disjunctive graph $\bar{D}$, and the associated selection S_o and the selection of sets S_o^o are called optimal, if

/14/ $$L_{oo} = \min_{D_{rp} \in R'_{DJ}} L_{rp},$$

where L_{rp} is the lenght of critical path in D_{rp}.

Similarly as in [3],[7], we can prove that problem /1/ - /6/ is equivalent to finding the minimaximal path L_{oo} /14/ with the implicit condition /13/ and optimal selection S_o and S_o^o in disjunctive graph $\bar{D}$. The lenght L_{oo} of the critical path in D_{oo} is the optimal value of t_z, while the optimal values t_j^x, t_j^y, $j \in N$, can be obtained by applying the critical path method to the graph D_{oo}.

3. Algorithm

The minimal path with the implicit condition /13/ of the disjunctive graph $\bar{D}$ is obtained by generating a sequence of circuit graphs $D_{rp} \in R'_{DJ}$, finding the critical path for each D_{rp}, and testing its feasibility.

Let S_1^o be a complete initial selection of sets such that $S_1^o \in R_s^o$. It can easily be seen that the graph $D_1 = \langle A, U \cup S_1^o \rangle$ has no circuits. Further, let S_1 be a complete initial selection such that $S_1 \in R_s$ and the graph $D_{11} = \langle A, U \cup S_1 \cup S_1^o \rangle$ has no circuits. The disjunctive arcs $\langle a,b \rangle \in V$ and $\langle c,d \rangle \in V^o$ are called normal if $\langle a,b \rangle \in S_1$ and $\langle c,d \rangle \in S_1^o$. The disjunctive arcs $\langle a,b \rangle \in V - S_1$ and $\langle c,d \rangle \in V^o - S_1^o$ are called reverse. Each arc $\langle y,x \rangle \in S_1^o$ has $r_j - 1$ reverse arcs in the set $V^o - S_1^o$. Moreover, we assume that

/15/ $$\forall [i \in N \wedge \langle x_i, y_i \rangle \in S_1^o] \{ \forall \langle x,y \rangle \in V^{oi} [c(x,y) \geqslant c(x_i,y_i)] \} .$$

Starting with the graph $D_{11} = \langle A, U \cup S_1 \cup S_1^o \rangle$, we generate a sequence of graphs

$$D_{rp} = \langle A, U \cup S_r \cup S_p^o \rangle \in R'_{DJ}, \ S_r \in R_s, \ S_p^o \in R_s^o .$$

For the graph D_{rp}, the implicit condition /13/ must be satisfied. Hence the selection S_r is not necessarily complete. Thus, for each pair of disjunctive arcs of the selection S_r one of the following possibilities holds:

(a) it is a normal arc,

(b) it is a reverse arc,

(c) it is neither a normal nor a reverse arc from a disjunctive pair.

If (c) holds, we have an empty arc which is denoted by $\overline{\langle y,x\rangle}$ or $\overline{\langle u,v\rangle}$. Replacement of an arc by an empty arc is called eliminating.

Each graph D_{sq} is obtained from a certain graph D_{rq} or D_{sp} of the sequence by complementing or eliminating one normal arc from the selection S_r or by complementing one normal arc from the selection of sets S_p^o. Each arc from S_p^o is complemented on r_j-1 reverses. In case where the graph D_{sq} is obtained by eliminating arcs from the selection S_r, the implicit condition /13/ must be satisfied for this graph. The process of generating is presented in the form of a solution tree H. Each node in H corresponds to a graph D_{rp}. Each arc D_{sp} or D_{rq} is obtained from D_{rp} by complementing (or eliminating) one disjunctive arc from $\langle y,x\rangle \in S_r$ or $\langle y,x\rangle \in S_p^o$. Then the arc $\langle D_{rp}, D_{sp}\rangle$ in H represents the complement $\langle u,v\rangle \in S_s$ (or an empty arc $\overline{\langle y,x\rangle}$) of $\langle y,x\rangle \in S_r$, and the arc $\langle D_{rp}, D_{rq}\rangle$ in H represents the complement $\langle u,v\rangle \in S_q^o$ of $\langle y,x\rangle \in S_p^o$. We say that D_{rp} is the predecessor of D_{sp} or D_{rq} and D_{sp} is the successor of D_{rp} if there is a path in H between D_{rp} and D_{sp} or D_{rq}. The initial graph $D_{11} = \langle A, U \cup S_1 \cup S_1^o\rangle$ is the root in the solution tree H. The generation of a new branch in H is connected with the choice of a certain normal arc for complementing or eliminating from S_r or for complementing from S_p^o. This choice is called the operation of choice. For each graph D_{rp} from the sequence we perform:

(a) an operation of testing to check the critical path and the possibility of generation of a graph $D_{sq} \in R'_{DJ}$ with a critical path smaller than that already found;

(b) an operation of testing to check the implicit condition /13/.

If the result of testing (a) is negative, we abandon the considered graph D_{sp} or D_{rq} and backtrack the tree H to the predecessor D_{rp} from which the graph D_{sp} or D_{rq} was generated. Since the process of generating the graph D_{rp} is started from the complete selection S_1, a negative result of checking the implicit condition /13/ may appear if we perform the operation of eliminating an arc from S_r. If a new graph D_{sp} is obtained from the graph D_{rp} by complementing (or eliminating) a normal arc $\langle y,x\rangle \in S_r$, we temporarily fix a reverse arc $\langle u,v\rangle \in S_s$ (or an empty arc $\overline{\langle u,v\rangle}$) in D_{sp}.

This arc cannot be complemented (or eliminated) in any succesor D_{sp} in H. Howewer, if we backtrack the first time a reverse arc (or an empty arc) of a certain normal arc to D_{rp}, we momentarily fix this reverse arc (or an empty arc). If we backtrack the second time, we constantly fix this normal arc. The normal arc such that its reverse arc (or an empty arc) is momentarily fixed can be complemented (or eliminated) if we need to perform the second operation of eliminating (or complementing) which has not been performed yet for this normal arc. So, for each graph D_{rp}, we temporarily or constantly fix a subset $F_r \subset S_r$ and momentarily fix a set F_r^t of disjunctive arcs. The reverse arcs in F_r are temporarily fixed and represent the path from the root to D_{rp} in H. Each normal arc in F_r is constantly fixed and represents two arcs (reverse and empty) which have been abandoned during the backtracking process. Each momentarily fixed arc in F_r^t is a reverse (or empty) arc which has been abandoned during the backtracking process. The set F_r^t contains all arcs belonging to the path from the root to D_{rp} in H. No arc from the set F_r can be complemented and eliminated in any successors D_{tu} of D_{rp}. If a new graph D_{rq} is obtained from the graph D_{tp} by complementing a normal arc $\langle y,x\rangle \in S_p^o$, we temporarily fix a reverse arc $\langle u,v\rangle \in S_q^o$ in D_{rq}. This arc cannot be complemented in any successors of D_{rq} in H. However, if we backtrack the n-th time (where $n < r_j' = r_j - 1$) a reverse arc of a certain normal arc to D_{rp}, we momentarily fix this reverse arc. If we backtrack the r_j-th time, we constantly fix this normal arc. A normal arc such that its reverse arc is momentarily fixed can be complemented if we need to perform another operation of complementing which has not been performed yet for this normal arc. For each graph D_{rp}^o we temporarily or constantly fix a subset $F_p^o \subset S_p^o$ and momentarily fix a set F_r^{to} of disjunctive arcs. The comments relative to F_p^o and F_p^{to} are analogous to those relative to F_r and F_r^t, respectively.

3.1. Operation of testing the critical path.

The basic task of the operation of testing is the computation of the lower bound of the critical path for every possible successor $D_{sq} \in R_{DJ}'$ generated from the graph D_{rp}. The arcs from the sets F_r are fixed in any successors of D_{rp}. It follows from /15/ that the arcs from the selection of sets S_p^o do not decrease the length. Let

$$D(F_r \cup S_p^o) = \langle A, U \cup F_r \cup S_p^o \rangle \qquad /16/$$

be the graph from the sets F_r and S_p^o, and let $L(F_r \cup S_p^o)$ be the length of a critical path in this graph. Let L^* be the length of the shortest critical path found so far. Then if

$$L(F_r \cup S_p^o) \geqslant L^*,$$

we can reject the graph and all its successors. The value L^* is the upper bound of the length of the minimaximal path with the implicit condition /13/ in $\bar{D}$.

3.2. Operation of choice

The purpose of the operation of choice is to point out the normal arc for complementing or eliminating and to generate the successor in H. The arcs

$$E_r = S_r - F_r \text{ and } E_p^o = S_p^o - F_p^o$$

are free. We complement or eliminate only arcs of these sets which belong to the current critical path Crp, i.e.

/17/ $$K_r = E_r \cap C_{rp}$$

and

/18/ $$K_p^o = E_p^o \cap C_{rp}.$$

Let K_r' be the set of reverse and empty arcs with normal arcs belonging to the set K_r. The set K_r' is called the set of candidates. We want to choose a normal arc the complementing of which generates a successor with the possibly shortest critical path. This is especially important for the operation of testing. The choice criterionsfor an reverse arc of K_r' is the expression $\Delta_{rp}[(y,x),(u,v)]$ defined in [6], and for an empty arc of K_r' is the expression $\Delta_{rp}[(y,x),(\overline{u,v})]$ defined in [7].

Let $K_p^{o\prime}$ be the set of reverse arcs with normal arcs belonging to the set K_p^o. The set $K_p^{o\prime}$ is also called the set of candidates. We want to choose a normal arc the complementing of which generates a successor with the possibly shortest critical path. To choose a normal arc for complementing, we introduce the formula

/19/ $$\Delta_{rp}^o[(x,y),(u,v)] = c(u,v) - c(x,y).$$

3.3. Operation of testing the implicit condition

The implicit condition cannot be satisfied when we perform the operation of eliminating an arc from K_r. Each arc $\langle y_i,x_j\rangle \in K_r$ has two adjacent arcs $\langle x_j,y_j\rangle$ and $\langle x_i,y_i\rangle$ which belong to the set S_p^o. In the process of generating the sequence of graphs D_{rp}, the adjacent arcs may be complemented, so the logical result of the implicit condition may be changed. Therefore, we can check this condition and eliminate the arc $\langle y_j,x_i\rangle \in K_r$ the adjacent arcs of which are temporarily or constantly fixed. Let

$$R_r = \{\langle y_j,x_i\rangle \in K_r' \mid \langle y_j,x_i\rangle \text{ is an empty arc}\}$$

and

$$R_r' = K_r' - R_r$$

be the sets of empty and reverse arcs, respectively, on which we can replace arcs from the set K_r. Futher, let

$$R_{rp} = \{\langle y_j,x_i\rangle \in R'_r \mid (\langle x_j,y_j\rangle \in F_p^o) \wedge (\langle x_i,y_i\rangle \in F_p^o)\}$$

be the set of empty candidates the adjacent arcs of which are temporarily or constantly fixed. The normal arc from K_r can be eliminated if its empty arc belongs to the set R_{rp}.

3.4. Algorithm

We start with

$$D_{11}^o = \langle A, U \cup S_1 \cup S_1^o\rangle,\ F_1 = \emptyset,\ F_1^o = \emptyset,\ F_1^t = \emptyset,$$
$$F_1^{to} = \emptyset,\ L^* = \infty .$$

The graph D_{11} represents the root of the solution tree H.

Let $D_{rp} = \langle A, U \cup S_p \cup S_p^o\rangle$ be the current graph and let F_r, F_p^o, F_r^t, F_p^{to} be the current sets of temporarily, constantly and momentarily fixed disjunctive arcs in the $r+p$ -th iteration of the algorithm.

Step 1 (test step). Compute the lower bound $L(F_r \cup S_p^o)$ of the graph defined by /16/. If $L(F_r \cup S_p^o) \geqslant L^*$, then go to Step 4. Otherwise, go to Step 2.

Step 2 (evaluation step). Compute L_{rp}. If $L_{rp} < L^*$, then set $L^* = L_{rp}$.

Identify the sets K_r and K_p^o defined by /17/ and /18/, respectively. If $K_r = \emptyset$ and $K_p^o = \emptyset$, then go to Step 4. Otherwise, identify the sets of candidates K'_r and $K_p^{o\prime}$. Next perform

$$K'_r = K'_r - F_r^t \text{ and } K_p^{o\prime} = K_p^{o\prime} - F_p^{to}.$$

If $K'_r = \emptyset$ and $K_p^{o\prime} = \emptyset$, then go to Step 4. Otherwise, for each arc $\langle u,v\rangle \in K'_r$ compute $\Delta_{rp}[(y,x),(u,v)]$ and $\Delta_{rp}[(y,x),(\overline{u,v})]$. For each arc $\langle x,y\rangle \in K_p^{o\prime}$ compute $\Delta_{rp}^o[(x,y),(u,v)]$ defined by /19/. Then go to Step 3.

Step 3 (forward step). If $K'_r = \emptyset$ and $K_p^{o\prime} = \emptyset$, then go to Step 4. Otherwise, identify the sets R_r, R_{rp}, and R'_r. Next, choose $\langle u,v\rangle \in R'_r \cup R_{rp}$ or $\langle u,v\rangle \in K_p^{o\prime}$ (perhaps an empty arc) such that

$$\delta_{rp}[(y,x),(u,v)] = \min\Big\{ \min_{\langle c,d\rangle \in R'_r \cup R_{rp}} \Delta_{rp}[(a,b),(c,d)],$$
$$\min_{\langle c,d\rangle \in K_p^{o\prime}} \Delta_{rp}^o[(a,b),(c,d)]\Big\}.$$

If an empty arc $\langle \overline{u,v}\rangle$ is chosen and $\langle \overline{u,v}\rangle \in R_{rp}$, then check the implicit condition /13/ as follows. Let $\langle y,x\rangle = \langle y_j,x_i\rangle$ be the normal arc of the arc $\langle \overline{u,v}\rangle$ and let $\langle y_j,x_i\rangle$ have adjacent arcs of the forms $\langle x_j,y_j\rangle_k$ and $\langle x_i,y_i\rangle_w$, $i,j \in N$, $B_k \in R_j$, $B_w \in R_i$. If $B_k \cap B_w \neq \emptyset$, do not eliminate these normal arcs, perform

$$K_r = K_r - \{\langle \overline{u,v}\rangle\},\quad F_r^t = F_r^t \cup \{\langle \overline{u,v}\rangle\},$$

and go to the beginning of Step 3. If $B_k \cap B_w = \emptyset$, then generate a new graph D_{sp} by eliminating the normal arc $\langle y,x\rangle$ and perform

$$U_{sp} = U_{rp} - \{\langle y,x\rangle\} \;,\; F_s^t = F_r^t \,.$$

If the reverse arc $\langle u,v\rangle$ is chosen and $\langle u,v\rangle \in R_r'$, then generate a new graph D_{sp} by complementing the normal arc $\langle y,x\rangle$ and perform

$$U_{sp} = [U_{rp} - \{\langle y,x\rangle\}] \bigcup [\{\langle u,v\rangle\}] \,,$$
$$F_s = F_r \bigcup \{\langle u,v\rangle\} \,,\; F_s^t = F_r^t.$$

Simultaneously, add to the solution tree H a new node D_{sp} and a new arc $\langle D_{rp}, D_{sp}\rangle$ associated with an empty arc $\langle\overline{u,v}\rangle$ or a reverse arc $\langle u,v\rangle$ of the disjunctive graph $\bar{D}$. Go to Step 1.

If the reverse arc $\langle u,v\rangle$ is chosen and $\langle u,v\rangle \in K_p^{o\prime}$, then generate a new graph D_{rq} by complementing the normal arc and perform

$$U_{rq} = U_{rp} - [\{\langle y,x\rangle\}] \bigcup [\{\langle u,v\rangle\}] \,,$$
$$F_q^o = F_p^o \bigcup \{\langle u,v\rangle\} \,,\; F_q^{to} = F_p^{to}.$$

Simultaneously, add to the solution tree H a new node D_{rq} and a new arc $\langle D_{rp}, D_{sq}\rangle$ associated with the reverse arc $\langle u,v\rangle$. Then go to Step 1.

<u>Step 4 (backtracking step).</u> Backtrack to the predecessor D_{kp} or D_{rl} of D_{rp} in H. If D_{rp} has no predecessor, then the algorithm terminates, the selection S_* and the selection of sets S_*^o associated with the current L^* are optimal, and the longest path in D_{**} is minimaximal in $\bar{D}$. Otherwise, drop the data for D_{rp} and update the data for D_{kp} or D_{rl} as follows.

If D_{rp} is generated by an empty arc $\langle\overline{u,v}\rangle \in K_k'$, then perform

$$K_k' = K_k' - \{\langle\overline{u,v}\rangle\} \;,\; F_k^t = F_k^t \bigcup \{\langle\overline{u,v}\rangle\} \;.$$

If D_{rp} is generated by a reverse arc $\langle u,v\rangle \in K_k'$, then perform

$$K_k' = K_k' - \{\langle u,v\rangle\} \;,\; F_k^t = F_k^t - \{\langle u,v\rangle\} \;.$$

Besides, if the backtracking is performed the second time while considering the normal arc $\langle y,x\rangle$, then perform

$$F_k = F_r \bigcup \{\langle y,x\rangle\} \;,\; F_k^t = F_k^t - [\{\langle\overline{u,v}\rangle\} \bigcup \{\langle u,v\rangle\}] \,.$$

If the graph D_{rp} is generated by the reverse arc $\langle u,v\rangle \in K_1^{o\prime}$, then perform

$$K_1' = K_1' - \{\langle u,v\rangle\} \;,\; F_1^{to} = F_1^{to} \bigcup \{\langle u,v\rangle\} \;.$$

If the backtracking is performed the r_j-th time while considering the normal arc $\langle y,x\rangle$, then perform

$$F_1^o = F_1^o \quad \{\langle y,x\rangle\} \;,\; F_1^{to} = F_1^{to} - \bigcup_{w=1}^{r_j} \{\langle u,v\rangle_w\} \,,$$

where $\langle u,v\rangle_w$ is the reverse arc of $\langle y,x\rangle$. Go to Step 3■

References

[1] E.Balas, An additive algorithm for solving linear programs with zero-one variables, Opns. Res. 13 (1965), p. 517-546.

[2]-Finding a minimaximal path in a disjunctive PERT network,International Symposium on the Theory of Graphs, Rome 1966.

[3]-Discrete programming by the filter method, Opns. Res. 15 (1967), p.915-967.

[4]-Project scheduling with resource constraints, Report No. 185, Carnegie - Mellon University, Pittsburgh.

[5] S.Gorenstein, An algorithm for project (job) sequencing with resource constraints, Opns. Res. 20 (1972), p. 835-850.

[6] J.Grabowski, A new formulation and solution of the sequencing problem: mathematical model, Zastosow. Matem. 15 (1976), p.325--343.

[7] J.Grabowski, Zagadnienie sterowania produkcją w dyskretnym procesie przy ograniczonych środkach, Podstawy Sterowania, t.7 (1977), z.2, pp. 121-138.

[8] Z.Jankowska-Zorychta, Modele sekwencyjne i ich zastosowanie w planowaniu optymalnej organizacji w dyskretnych procesach produkcyjnych, Prace CO PAN, Warszawa 1973.

[9] L.Schrage, Solving resource-constrained network problem by implicit enumeration, Nonpreemptive case, Opns. Res. 18 (1970), p. 263-278.

STOCHASTIC MODEL OF RESOURCE ALLOCATION TO R&D ACTIVITIES UNDER COST VALUE UNCERTAINTY

Andrzej Jakubowski

Systems Research Institute, Polish Academy of Sciences
Newelska 6, 01-447 Warsaw, Poland.

Abstract - The paper is concerned with the development of a stochastic mathematical model for management of large-scale Research and Development program. The problem of optimal funding of an R&D complex program, consisting of several projects, their components and possible technical approaches is considered. It is assumed that the values of costs of technical approaches as well as the probabilities of technical succes are not known with certainty. So it is advantageous to perform a limited number of diagnostic experiments in order to reduce this uncertainty. The problem is to develop a policy for performing experiments and allocating resources on the basis of the results of the experiments. This policy is such that a chosen performance index is optimized. A computionally practical algorithm for obtaining an approximate solution using the basic stochastic dynamic programming approach is developed. This algorithm preserves the "closed loop" feature of the dynamic programming solution in that the resulting decision policy depends both on the results of past experiments and on the statistics of the outcomes of future experiments. In other words, the present decision takes into account the value of future information.

I. INTRODUCTION

Complex problems of Research and Development (R&D) planning and decision making require the use of quantitative methods. The principal purpose of applying these methods relays in determining the optimal courses of R&D funding under constrained resource conditions. However, two significant difficulties are encountered when scientific methods and models are used to transform the resource allocation R&D decision problems into computer programs and analyses.

The first one results from interactions occuring in structures of R&D complex programs which consist of projects, their components, and possible technical approaches. Due to this interactions the mathematical formulation of the optimal resource allocation problems as well as the analysis of the existence of optimal solutions become very complicated. Several models taking into account these interactions have been worked out [1].

The second difficulty results from the fact that the formulation of the model depends upon the knowledge of the estimates of the probabilities of technical succes as well as the costs of potential technical approaches. The accuracy of this estimation and the knowledge of relations between these two quantities determines the accuracy of R&D models. Due to the uncertainty in the costs and estimations of the probabilities of succes results of a given allocation of resources are not known with certainty. The paper is concerned with problems of this type. In other words, the presented stochastic model takes into account cost and payoff value uncertainties. It is assumed that a limited number of diagnostic experiments is performed to learn more about the values of costs and the probabilities of technical success.

The results obtained extend those achieved by B.V.Dean, T.S.Chidambaram, R.R.Palanki [2], Y.Bar-Shalom, R.E.Larson and M.A.Grossberg [3].

II. FORMULATION OF THE OPTIMAL RESOURCE ALLOCATION PROBLEM

Structure of R&D Complex Program

We assume that an R&D complex program to be considered consists of projects $k(k=1,\ldots,W)$, project components j $(j=1,\ldots,M)$ and potential technical approaches $i(i=1,\ldots,N)$.

Define

$I_j = \{i$: the technical approach i is capable of developing of the projects component $j\}$,

$J_k = \{j$: the projects component j is an element of the project $k\}$,

c_i - the estimated cost of the funding the i-th technical approach, if it is selected for funding

P_i - subjective likelihood or probability that the i-th technical approach is successful, if it is selected for funding

z_i - the binary decision variable; $z_i=1$ when the i-th technical approach is chosen to develop a given project, otherwise $z_i=0$

V_k - the value of the k-th project when developed

B - total amount of resources available.

The following structure of an R&D complex program is to be considered (Fig.1)

(i) Technical approaches are alternative means to develop project components. Any one technical approach i in the set I_j is capable of successfuly developing the project component j. The only way the project component j will not be successfuly developed is if all the technical approaches in I_j fail simultaneously.

(ii) All the project components in the set J_k must be developed for the project k to be achieved.

To simplify consideration it is assumed that R&D model to be discussed consists of unshared project components and unshared technical approaches. In other words, an R&D complex program with no interactions is analyzed.

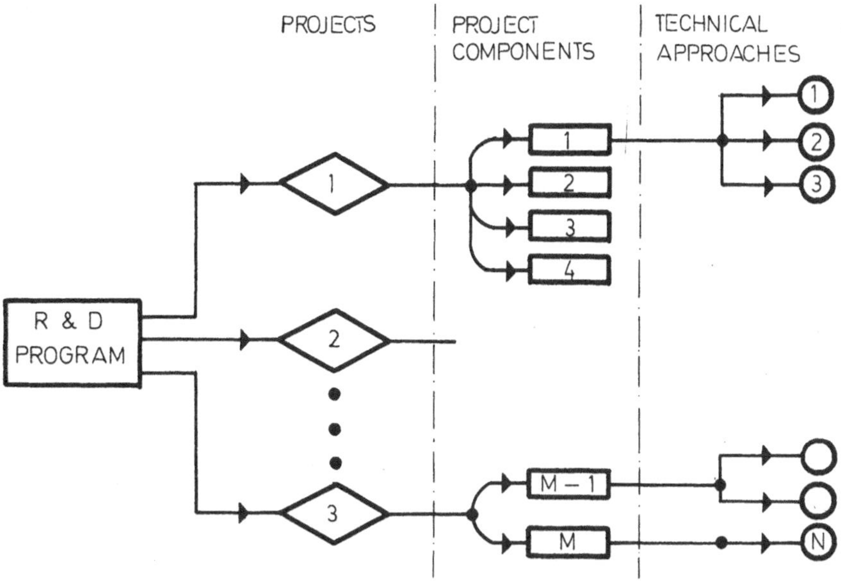

Fig.1. The structure of an R&D complex program

R&D Decision Problem

We consider the following decision problem:

(i) How should technical approaches be selected so as to maximize the probability of achieving a project component?

(ii) How should project components be funded so as to maximize the probability of achieving of a project?

(iii) How should total budget B be allocated accross projects so as to maximize the global performance index?

It is easy to show that the probability of the event that all the technical approaches in I_j fail simultaneously is $\prod_{i\in I_j}(1-z_iP_i)$, i.e. the probability that the project component j will be successfully developed is $\{1-\prod_{i\in I_j}(1-z_iP_i)\}$. Technical approaches are not shared; therefore the probability that the project k will be successfully achieved is $\prod_{j\in J_k}[1-\prod_{i\in I_j}(1-z_iP_i)]$. Multiplying this probability by the value V_k results in the expected value of the project k. The sum of this products is the expected value of the R&D complex program.

Hence, for the discussed R&D decision problem we can define the following optimal resource allocation problem:

Problem P: Find an optimal set $\{z_i\}$ of the decision variables z_i $(i=1,\dots,N)$ assuring the maximum of the expected value of the R&D complex program, i.e.

$$\hat{Q}_o = \max_{\{z_i\}} \sum_{k=1}^{W} V_k \left\{ \prod_{j \in J_k} \left[1 - \prod_{i \in I_j} (1 - z_i P_i) \right] \right\} \tag{1}$$

subject to

$$\sum_{i=1}^{N} z_i c_i \leq B \qquad \text{and} \quad z_i = \{0,1\} \tag{2}$$

The problem P is a zero-one non-linear programming problem. To solve this problem well known discrete optimization methods can be applied.

Data Requirements

R&D model discussed require the following data:

(i) the complete network structure of R&D complex program,

(ii) the values V_k of the all projects ($k=1,\ldots,W$),

(iii) estimates of the expected cost and estimates of the probability of success for each potential technical approach i ($i=1,\ldots,N$).

It is assumed that this information is provided by experts and the total budget B, is given. Especially, a diagnostic process of a prespecified nature will be carried out in order to determine estimates of the cost and probability of success for each technical approach.

III. MODELLING OF THE DIAGNOSTIC PROCESS

Determining of the Probability Distribution of Cost and its Optimal Estimator for a Potential Technical Approach

Let us assume that the cost c of a technical approach is a priori not known with certainty. To be more precise we assume that the cost c is a random variable ranging over the interval $[\underline{c}, \bar{c}]$ (subscript i has been suppreses for notational simplicity), where the lower and upper bounds $\underline{c}$ and $\bar{c}$ are estimated by experts. There is no other a priori information on a random cost c. Therefore it is supposed that the a priori probability density $p(c\,|\,\mathcal{Y}_o)$ of the cost c is of uniform type, i.e.

$$p(c\,|\,\mathcal{Y}_o) = \begin{cases} \dfrac{1}{\bar{c}-\underline{c}} \;; & \text{for } c \in [\underline{c}, \bar{c}], \\[2ex] 0 \;; & \text{for } c \notin [\underline{c}, \bar{c}], \end{cases} \tag{3}$$

where $\mathcal{Y}_o$ denotes the initial information on the cost c, i.e. the values $\underline{c}$ and $\bar{c}$.

More information on the cost c is provided by a multistage diagnostic process consisting in independent expert estimations of the real cost. A step of this process is called the diagnostic experiment.

Denote

y_ν - the estimate of cost c provided by ν-th expert, i.e. the result of the ν-th diagnostic experiment ($\nu=1,\ldots,k$),

$\underline{y}_k = [y_1,\dots,y\ ,\dots,y_k]'$ - the vector of k estimates of cost c,
n_ν - the error of estimation of cost c during the ν-th diagnostic experiment,
$\underline{n}_k = [n_1,\dots,n\ ,\dots,n_k]'$ - the vector of k errors of estimations of cost c,
$p(c|\underline{y}_k)$ - a posteriori probability density of cost c conditioned upon the estimates $\underline{y}_k$,
$\hat{c}_k$ - the optimal estimator of cost c resulting from a k-step diagnostic process.

We can assume that the estimate y_ν of cost c obtained as a result of the ν-th diagnostic experiment is the sum of the real cost c and the error of estimation n_ν, i.e.

$$y_\nu = c + n_\nu \tag{4}$$

for $\nu = 1,\dots,k$.

Let the quantities n_ν form an uncorrelated random sequence with the expected values $E\, n_\nu = 0$ and the variances $E\, n_\nu^2 = \sigma_\nu^2$ which are assumed to be given (for a method of evaluation see [6]) and estimate errors n_ν are of normal type. Hence $\underline{n}_k$ is the uncorrelated multidimensional Gaussian random process with the probability density

$$p(\underline{n}_k) = \frac{1}{2\pi^{\frac{k}{2}} \prod_{\nu=1}^{k} \sigma_\nu} \exp\left\{-\frac{1}{2}\sum_{\nu=1}^{k}\frac{n_\nu^2}{\sigma_\nu^2}\right\} \tag{5}$$

Due to the assumptions adopted we can treat the sequential process of estimating the real cost c as a process of reception of useful signals in the presence of random additive noise. Such problems are basic for the communication theory [4]. It can be shown that in the case under consideration the a posteriori probability density of the cost c is [5]:

$$p(c|\underline{y}_k) = \frac{\Lambda(\underline{y}_k,c)}{\Lambda(\underline{y}_k)}\ , \tag{6}$$

where

$$\Lambda(\underline{y}_k,c) = p(c|\mathcal{Y}_o) \cdot \frac{\exp\left\{-\frac{1}{2}\sum_{\nu=1}^{k}\frac{(y_\nu-c)^2}{\sigma_\nu^2}\right\}}{\exp\left\{-\frac{1}{2}\sum_{\nu=1}^{k}\frac{y_\nu^2}{\sigma_\nu^2}\right\}} \tag{7}$$

$$\Lambda(\underline{y}_k) = \int_{-\infty}^{+\infty}\Lambda(\underline{y}_k,c)\,dc\ , \tag{8}$$

and $p(c|\mathcal{Y}_o)$ is given by [3].
In statistical decision theory the quantities $\Lambda(\underline{y}_k)$ and $\Lambda(y_k,c)$ are called the likelihood ratios.

<u>The optimal estimator of cost c.</u> It is natural to define the optimal estimator

$\hat{c}_k$ of cost c as the most probable value of this cost, i.e. the value of c for which the probability density $p(c|\underline{y}_k)$ and the likelihood ratio $\Lambda(\underline{y}_k, c)$ take their maximum values [5].

It follows from the formulae (3), (6), (7) and (8) that the problem of determining the most probable value of cost c is equivalent to the application of the method of least squares, i.e. the maximum of $p(c|\underline{y}_k)$ or $\Lambda(c,\underline{y}_k)$ is achieved, when the expression

$$\varepsilon^2 = \sum_{\nu=1}^{k} \frac{1}{\sigma_\nu^2} (y_\nu - c)^2 \tag{9}$$

is a minimum.

Hence it is easy to show that the optimal estimator $\hat{c}_k$ is equal to

$$\hat{c}_k = \sum_{\nu=1}^{k} \frac{\vartheta_k}{\sigma_\nu^2} y_\nu, \quad \text{where} \quad \vartheta_k = \frac{1}{\sum_{\nu=1}^{k} 1/\sigma_\nu^2} \tag{10}$$

The a posteriori probability density of cost c. Using formulae (3), (6), (7), (8) and (10) the a posteriori probability density of cost c can be written in the form [6]

$$p(c|\underline{y}_k) = \begin{cases} \mathcal{H}(\hat{c}_k)^{-1} \dfrac{1}{\sqrt{2\pi}\sqrt{\vartheta_k}} \exp\left\{-\dfrac{1}{2} \dfrac{(c-\hat{c}_k)^2}{\vartheta_k}\right\} & \text{for } c \in [\underline{c}, \bar{c}] \\ 0 & \text{for } c \notin [\underline{c}, \bar{c}] \end{cases} \tag{11}$$

where

$$\mathcal{H}(\hat{c}_k) = \frac{1}{2}\left[\Phi\left(\frac{\bar{c}-\hat{c}_k}{\sqrt{\vartheta_k}}\right) - \Phi\left(\frac{\underline{c}-\hat{c}_k}{\sqrt{\vartheta_k}}\right)\right] \tag{12}$$

and $\Phi(\cdot)$ is a Laplace function well known in statistics, i.e.

$$\Phi(x) = \frac{2}{\sqrt{2\pi}} \int_0^x e^{-\frac{t^2}{2}} dt; \quad \Phi(-x) = \Phi(x) \tag{13}$$

Using the formula (11) it can be shown that if the number k of diagnostic experiments increases indefinitely, then the expected value $E(c|\underline{y}_k)$ of a posteriori distribution of real cost c converges to the optimal estimator $\hat{c}_k$ and the variance $\text{var}(c|\underline{y}_k) = E(c^2|\underline{y}_k) - E(c|\underline{y}_k)^2$ converges to zero. It is in a agreement with intuition that if the value of k increases, the optimal estimator $\hat{c}_k$ becomes more accurate.

Evaluating of the a posteriori Probability of Success of a Potential Technical Approach

Let us assume that the optimal estimator $\hat{c}_k$ of real cost c is considered as a planned cost of a potential technical approach. Hence, the a posteriori overall probability of technical success can be defined as the product of the conditional probability of success estimated under the assumption that real cost c will not exceed the planned cost $\hat{c}_k$ and the a posteriori probability that this assumption will take place.

Denote

t - the zero-one random variable; t=1 if the potential technical approach is successful, otherwise t=0,

$\Pi_k = P\{c \leq \hat{c}_k | \underline{y}_k\}$ - the a posteriori probability that the real cost c will not exceed the planned cost $\hat{c}_k$ conditioned upon the estimates $\underline{y}_k$.

$P_{|c \leq \hat{c}_k} = P\{t=1 | c \leq \hat{c}_k\}$ - the conditional probability of technical success estimated under the assumption that the real cost will not exceed the planned cost $\hat{c}_k$.

$P_k = P\{t=1 | \underline{y}_k\}$ - the a posteriori overall probability of technical success.

The probability Π_k is

$$\Pi_k = P\{c \leq \hat{c}_k | \underline{y}_k\} = \int_{-\infty}^{c_k} p(c|\underline{y}_k)dc \tag{14}$$

From formulae (11) and (14) it follows that

$$\Pi_k = \begin{cases} 0 & \text{for } \hat{c}_k < \underline{c} \\ \frac{1}{2}\, \mathcal{H}(c_k)^{-1}\, \Phi\left(\frac{\hat{c}_k - \underline{c}}{\vartheta_k}\right), & \text{for } \hat{c}_k \in [\underline{c}, \bar{c}] \\ 1 & \text{for } \hat{c}_k > \bar{c} \end{cases} \tag{15}$$

Hence, we have

$$P_k = P\{t=1 | \underline{y}_k\} = P_{|c \leq \hat{c}_k} \cdot \Pi_k \tag{16}$$

where Π_k is given by (15) and $P_{|c \leq \hat{c}_k}$ is provided by experts for each potential technical approach.

IV. THE STOCHASTIC CONTROL FORMULATION OF THE SEQUENTIAL DECISION PROCEDURE

In Section III the diagnostic process has been performed in order to determine the estimates of cost, $\hat{c}_k$, and the probability of success, $\hat{P}_k$, for each potential technical approach.

It is assumed that the diagnostic process performed for a given R&D complex program consists of n step. If the number n is very large, then uncertainties in the value of cost and probability of success can be reduced by allocating a sufficient number of diagnostic experiments to each potential technical approach.

Let us set

$$c_i = \hat{c}_k^i \quad \text{and} \quad P_i = P_k^i \tag{17}$$

where c_k^i and P_k^i are the cost and probability estimates given for each technical approach $i(i=1,\ldots,N)$ by formulae (10) and (16). In such a case the optimal resource allocation problem is reduced to the problem (1), (2), which, alone, is relatively simple.

However, in many practical situations the number n of diagnostic experiments is limited. So, it is advantageous to perform such a reasonable allocation of a limited number of the diagnostic experiments to the technical approaches, which allow to determine the most efficient allocation of the financial resources across R&D projects, project components and potential technical approaches. The development of such an allocation procedure via a stochastic control approach will be discussed in this section.

The Optimal Planning of the Diagnostic Process and the Optimal Resource Allocation Problem

Denote

$\underline{\mathcal{X}} = [\mathcal{X}_1,\dots,\mathcal{X}_k,\dots,\mathcal{X}_n]'$ - a vector whose k-th component $\mathcal{X}_k$ specifies on which technical approach the k-th diagnostic experiment is carried out.

Since the estimates of the costs, $c_i = \hat{c}^i_n$, and the probabilities of technical success $P_i = P^i_n$ are random variables at the start of the diagnostic process, the performance index to be considered is the following

$$Q = E\{Q_o;\underline{\mathcal{X}}\} = E\left\{\sum_{k=1}^{W} V_k \left\{ \prod_{j\in J_k} \left[1 - \prod_{i\in I_j} (1-z_i\cdot P_i)\right]\right\};\underline{\mathcal{X}}\right\} \qquad (18)$$

subject to

$$\sum_{i=1}^{N} z_i E\{c_i;\underline{\mathcal{X}}\} \le B;\ z_i = \{0,1\} \qquad (19)$$

where the expectations are to be taken over the unknown results $\underline{y}_n = [y_1,\dots,y_k,\dots,y_n]'$ of the experiments to be performed.

It follows from Section III that the values of the estimates $\hat{c}^i_k$ and P^i_k (k==1,...,n) will change as diagnostic experiments proceed. So, the actual allocation of the experiments and of the financial resources over time is done as follows:

(i) The diagnostic experiments are sequentially allocated in order to reduce the uncertainty in c_i and P_i. After the k-th experiment, the result y_k is obtained and then one proceeds to the next experiment unless k=n. The accumulated set of data is $\underline{y}_k = [y_1,\dots,y_k]'$; k=1,...,n.

(ii) Finally, when k=n, the allocation of financial resources is based upon the data obtained from all the n experiments.

These decisions are made sequentially and the information state $\underline{y}_k$ evolves in time. The allocation of the experiments at every step has to account for the value of the information to be obtained from the subsequent results. This value is measured in terms of the improvement of the allocation of the financial resources: the less uncertainty about the c_i, P_i, the more efficiently the financial resources can be allocated.

It is well known that the absolute maximum of the performance index (18) can be obtained via the stochastic dynamic programming equation [7]:

$$\hat{Q} = \max_{\mathcal{X}_1} E\{\ldots \max_{\mathcal{X}_n} E\{\max_{\{z_i\}} E[Q_o | \underline{y}_n] | \underline{y}_{n-1}\} | \ldots | \mathcal{Y}_o\} \qquad (20)$$

where $\mathcal{Y}_o$ is the initial a priori information (see Section III).

The allocation $\hat{\underline{\mathcal{X}}}$ resulting from (20) has the following two properties:

(i) uses the results of the previous experiments, i.e.

$$\hat{\mathcal{X}}_k = \hat{\mathcal{X}}_k(\underline{y}_{k-1}); \qquad k=1,\ldots,n$$

(ii) takes into account its effects on the future allocation via all the observations to be made after the present experiment. In this manner the decision $\hat{\mathcal{X}}_k$ at every stage is a function of the future observation program and the associated statistics $\mathcal{Y}_{k,n}$ as well, i.e.

$$\hat{\mathcal{X}}_k = \hat{\mathcal{X}}_k(\underline{y}_{k-1}, \mathcal{Y}_{k,n}) \ ; \qquad k=1,\ldots,n.$$

In other words, the allocation policy $\hat{\underline{\mathcal{X}}}$ is of the closed-loop type [7].

Since the implementation of the closed-loop policy is not feasible in our case, the following simplification is made here. Rather than attempting to find a decision $\hat{\mathcal{X}}_k$ that will account for all the future measurements, it will be limited to be a function of the statistics of only the next measurement. This is called <u>one-measurement feedback (1-MF) policy</u> [7].
Thus, the allocation of the k-th diagnostic experiment according to the 1-MF policy is obtained from the following:

$$Q_k = \max_{\mathcal{X}_k} E\{\max_{\{z_i\}} E(Q_o | \underline{y}_k) | \underline{y}_{k-1}\} \qquad (21)$$

subject to

$$\sum_{i=1}^{N} z_i \ E(\hat{c}_k^i | \underline{y}_{k-1}) \leq B; \quad z_i = \{0,1\}; \qquad k=1,\ldots,n \qquad (22)$$

As can be seen in the above formulae, the decision $\hat{\mathcal{X}}_k$ takes into account only the next measurement, but, since it is not yet available, it "averages it out". Equation (21), even though simpler that (20), is still an infinite-dimensional optimization problem, because the inner maximization depends on a continuous random variable y_k. Next, a slightly modified version of (21), (22) is developed such as to reduce the problem to a finite-dimensional search.

<u>The Stochastic Allocation Algorithm</u>

The optimization problem (21), (22) requires evaluation for all i=1,...,N of

$$E\{\max_{\{z_i\}} E(Q_o | \underline{y}_k) | \underline{y}_{k-1}; \quad \mathcal{X}_k = i\} \qquad (23)$$

and

$$E(\hat{c}_k^i | \underline{y}_{k-1} \; ; \quad \mathfrak{X}_k = i) \tag{24}$$

Specifically, given the result y_k of the experiment on the i-th technical approach, one has to compute the optimal performance and then average it over all the possible observations y_k resulting from this experiment. Therefore, from formulae (15) - (19) it follows that the resulting financial resource allocation is a function of

$$P_i = P\{t_i = 1 | \underline{y}_k\} \tag{25}$$

When evaluating the outside expectation in (23), conditioned upon $\underline{y}_{k-1}$, the term (25) is a random variable.

Therefore, to evaluate (23) one needs

$$p(P_i | \underline{y}_{k-1}) = p[P\{t_i=1 | \underline{y}_k\} | \underline{y}_{k-1} \; ; \quad \mathfrak{X}_k = i] \tag{26}$$

the probability density /conditioned on $\underline{y}_{k-1}$/of the /future/ probability that the i-th technical approach will be successful after the k-th experiment will be performed on it.

Also, in order to evaluate the expectation given by (24) one needs

$$p(\hat{c}_k^i | \underline{y}_{k-1} \; ; \quad \mathfrak{X}_k = i) \tag{27}$$

the probability density /conditioned upon $\underline{y}_{k-1}$/ of the /future/ optimal estimator of the cost of the i-th technical approach after the k-th experiment will be performed on it.

The procedure to obtain this probability densities which utilizes the results given in Section III and method of moments well known in statistics, is described in [6]. It is via these probability densities (26) and (27) that the present decision $\hat{\mathfrak{X}}_k$ /on which technical approach the k-th experiment should be carried out/ is obtained by taking into account its effect on the subsequent decision /the financial resource allocation/.

Simplification. Since computing of the optimum $\hat{\underline{z}} = \{\hat{z}_1, \dots, \hat{z}_N\}$ for every y_k when the experiment is performed on every technical approache, as required in (23), might be too tedious, the following modification is introduced. Assume that after the (k-1)-st experiment one has the allocation

$$\underline{z}_{k-1} = [z_{k-1,1}, \dots, z_{k-1,N}]'$$

The payoff for a diagnostic experiment is obtained by perturbing this allocation, i.e. by evaluating the benefit of "reallocating" one technical approach after the result of the experiment will be available. This reallocation has to be done in such a way that the resulting decision vector $\underline{z}_k$ satisfies the constraint (22).

So, the simplified algorithm is the following:

Let the performance index corresponding to the perturbed allocation be as follows

$$Q(\underline{z}_k) = Q_{k-1} + \Delta Q_{k-1}, \qquad \text{where} \quad Q_{k-1} \triangleq Q(\underline{z}_{k-1})$$

and ΔQ_{k-1} is the change in Q_{k-1} when only one technical approach can be "reallocated". Then (23) is approximated by a maximization over this perturbation $\Delta \underline{z}$, i.e.

$$E\left\{\max_{\{z_k\}} E\left[Q(\underline{z}_k)|\underline{y}_k\right]|\underline{y}_{k-1} ; \quad \mathcal{X}_k = i\right\} \approx$$

$$\approx E\{Q_{k-1}|\underline{y}_{k-1}\} + E\left\{\max_{\Delta \underline{z}} E\left(\Delta Q_{k-1}|\underline{y}_k\right)|\underline{y}_{k-1} ; \quad \mathcal{X}_k = i\right\}$$

The last term in the expression above is the marginal return of allocating one experiment in conjunction with a possible subsequent "reallocation" of an technical approach.

Therefore, this procedure makes it possible to reduce the infinite-dimensional problem of stochastic optimization (21), (22) to a finite-dimensional search at every stage.

At every stage the diagnostic experiment is allocated such as to maximize the expected increase in performance.

For more details, see [6].

V. CONCLUSION

In the paper the problem of the optimal resource allocation to R&D activities in a case in which the values of costs of a potential technical approaches and the probabilities of technical success are not known with certainty has been considered. It is shown that for this class of problems, a significant improvement in the performance or saving in resources can be achieved when the allocation is made using stochastic control methods that have the closed-loop property. In these methods the present decision takes into account the value of future information. A computationally practical suboptimal algorithm having the above property has been derived.

REFERENCES

[1] Jakubowski A.: "Mathematical Models of Resource Allocation to Interrelated R&D Activies". Proceedings of IFAC Workshop on Systems Analysis Application to Complex Programs, Bielsko-Biała, Poland, June 1-6, 1977, Pergamon Press Ltd.

[2] Dean B.V., Chidambaram T.S. and Palanki R.R.: "Project selection under cost and payoff value uncertainties", Technical Memorandum No 79, Operation Research Group, Case Institute of Technology, 1967.

[3] Bar-Shalom Y., Larson R.E. and Grossberg M.A.: "Application of Stochastic Control Theory to Resource Allocation under Uncertainty", IEEE Trans.on Automatic Control, Vol.AC-19, No 1, February 1974.

[4] Middleton D.: "An Introduction to Statistical Communication Theory", McGraw--Hill Book Company, Inc., 1960.

[5] Wainstein L.A., Zubakov V.D.: "Extraction of Signals from Noise", Prentice--Hall, Inc. 1962.

[6] Jakubowski A.: "Stochastic Algorithms of Resource Allocation to R&D Activity", Internal Report of Systems Research Institute /in Polish/, Warsaw, Poland,1977.

[7] Meier L., Larson R.E., and Tether A.J.: "Dynamic Programming for Stochastic Control of Discrete Systems", IEEE Trans. on Automatic Control, Vol. AC-16, December 1971.

SCHEDULING OF TRAINS - AN OPTIMIZATION APPROACH

C. Colding-Jørgensen

O. Holst Jensen
The Technical
University of
Denmark

P. Stig-Nielsen
The Danish State
Railways

INTRODUCTION

A wide range of problems in transportation can succesfully be approached by OR-methods. This applies especially to many scheduling problems in public transportation e.g. routing, frequencing, time-tabelling, crew and mode rostering.

The common attribute of all these real life problems is that they are enormous as regards the number of variables and a feasible solution has to fulfill a lot of practical conditions. Therefore it is a very troublesome task merely to find a feasible solution and in fact impossible to find an optimal solution. Especially if the elaboration of a feasible solution is done purely manually this is very manpower demanding and no optimization is tried.

What seems to be a succesful approach is to use rather simpel optimization-models which ensure the great majority of restrictions to be fulfilled. A manual verification of the solution must then ensure all restrictions to be fulfilled. Thus when passing from manual methods into OR-based methods the improvements appear primarily as more cheaply found solutions and secondarily as better solutions.

The problem considered in this paper is the time-tabelling of freight-trains. The demand is measured in the number of cart-loads per day and is indicated by an O-D (Origin-Destination) matrix. The demand is assumed not to be affected by the time-tabelling. Before the actual time tabelling is done the demand is transformed into a list of trains. A train is defined by the nodes in the railway network to be served by the train and by the number of the cart-loads carried by the train at each section.

The paper is based on a case study from the Danish State Railways.

THE PROBLEM

Before presenting the mathematical models we shall point out the assumptions and the limitations of the models.

We consider a certain railway network and a certain list of trains to be scheduled as defined in the introduction. A train is defined by indicating only those nodes in which the train has to exchange cart-loads (C-nodes), while the intermediate nodes (I-nodes) are found as the nodes of the shortest path between the C-nodes. Thus each train is defined as a sequence of C- and I-nodes. In order to avoid too much marshalling the number of C-nodes per train is upper bounded.

We introduce

O_i = the origin node for trains i

D_i = the final destination node for train i .

The problem is to schedule the trains. This must be done with consideration to the following:

i) The total travel time of a train is the travel time at the links in the network, the waiting time in the nodes and the time in C-nodes used for marshalling, brakingtests etc.

ii) Each link in the network is restricted either by a maximum bumber of trains per hour or the link is accessible only at specified hours during the day. The latter are denoted channels. The introduction of channels is due to the following facts:

a) Because of other trains, mainly passenger trains, the tracks are open for freight trains only at some discrete hours.

b) Some of the links in the Danish railway network are railway ferries. These links can be passed only according to the timetable of the ferries. The capacity for ferrying freight trains is the excess capacity after having scheduled the passenger trains. Thus, a channel of this type can be passed only by a limited number of cartloads equal to the capacity of the ferry.

iii) Train i starting at node O_i must depart within a predescribed interval of time $[S_i, S_i']$ and must arrive at node D_i within the time interval $[T_i, T_i']$.

Finally the following constraint is optionally included:

iv) In order to avoid marshalling at the ferry berths it is required that a train is not divided and shipped by more than one ferry.

In the introductory chapter the question of finding an optimal solution against a feasible solution has been mentioned. In fact, the optimal time-table is the time-table requiring the minimum number of engines. However using this criterion would require a simultaneous elaboration of time-tables and engine rotation-plans which is possible only in some special cases not met very often in practise (Colding-Jørgensen

1977). Therefore, a sequential approach is used as shown in figure 1.

Step 1	Transformation of the demand (O-D matrix) into freight trains.
Step 2	Time-tabelling of the freight-trains generated in step 1.
Step 3	Optimal rotation plans for the engines.

Figure 1 The total stepwise model used to obtain an optimal time-table for freight trains. The solution of the time-tabelling problem in step 2 is evaluated in step 3 by the number of engines required.

Only the problem in step 2 is dealt with in this paper.

Now consider train i which can be used in the interval $[S_i, T_i']$. By a late dispatch the engine will be available in node O_i from S_i until the dispatch, while an early dispatch from O_i implies that the engine will be available in D_i that much earlier. Ignoring changes in the total trip time, if any, the question is how you want to divide a certain amount of available engine time between the nodes O_i and D_i.

The criterion used is that engine time must be allocated to the node with the greatest demand of engines.

Thus to each node in the railway network a weight is assigned expressing the node's demand for engines. The value of the weight is found heuristicly as an over-all measure of the total demand for freight of the node and the accessibility of the node. i.e. the number of and the distance to neighbouring nodes. The weight of node ℓ is denoted w_ℓ .

The criterion and the constraints will be presented mathematically in the following section.

When considering the Danish railway network especially one section is restricted. This is the railway ferries at the Great Belt where the number of channels is small as well as every channel has a limited capacity.

Therefore, in the following we shall consider a simplified case of the general time-tabelling problem considered so far. We consider one link in the network - say the Great Belt - and the problem is to schedule the trains passing this link.

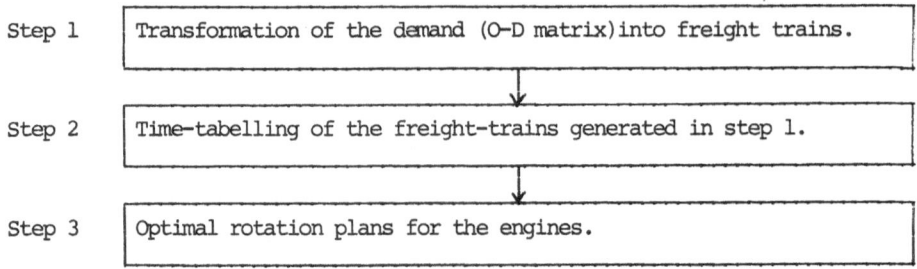

THE MODELS

The models all consider one directed link in the railway network in this particular case the Great Belt. For this section a number of channels is given. A channel indicates a period of time at which the section is accessible for a freight train.

Let j denote the channel number $j=1,2,\ldots,J$.

To each link in the network a figure is associated indicating the travel time. Thus, the total travel time (including stops at C-nodes, if any) from the origin O_i to the Great Belt and from the Great Belt to the destination D_i is calculated by finding the shortest path in the network.

The total travel time of train i combined with the two time intervals $[S_i,S_i']$ and $[T_i,T_i']$ define a set J_i of feasible channels as shown in figure 2.

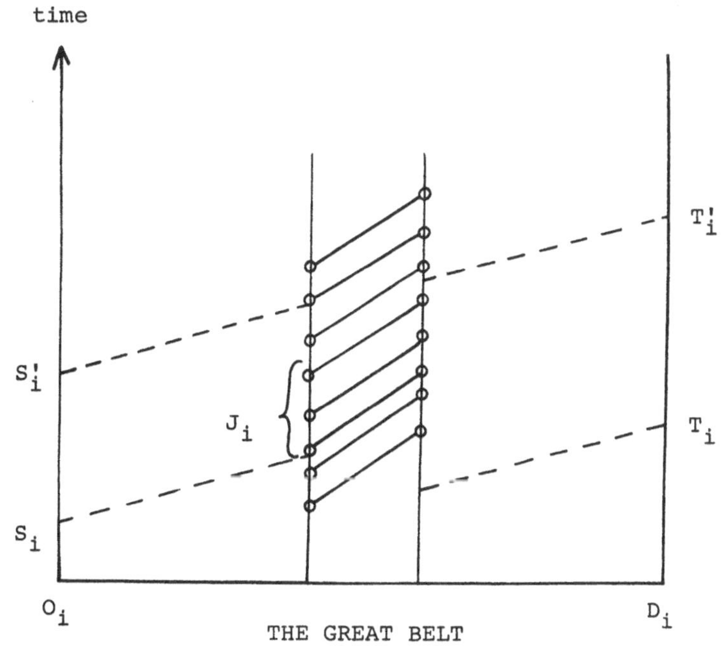

Figure 2. Time-location diagram for train i which has to depart from O_i in the time-interval $[S_i,S_i']$ and arrive at D_i between T_i and T_i'.
o-o indicates a ferry crossing the Great Belt. The feasible departures (channels) are denoted J_i. N.B. $K_j \geq L_i$ for all $j \in J_i$.

The following parameters and variables are introduced:

I = the number of trains.

J = the number of channels.

a_i = the trip time from O_i to the Great Belt.

b_i = the trip time from the Great Belt to D_i.

w_ℓ = the weight associated with node ℓ (see the previous section).

x_{ij} = the fraction of the cart-loads in train i ferried by ferry departure j.

y_{ij} = the number of the cart-loads in train i ferried by ferry departure j.

L_i = the number of cart-loads in train i.

K_j = the capacity of channel j measured in cart-loads.

τ_j = the departure time of the ferry at The Great Belt using channel j.

τ'_j = the arrival time of the ferry at The Great Belt using channel j.

As mentioned in the previous section we want to examine the following two cases:

1. The case where a train must not be divided and shipped by more than one ferry (restriction iv). This case is denoted the integer case and the associated model the integer model.

2. The case where a train is allowed to be divided and shipped by more than one ferry. This case is denoted the continous case and the associated model the continous model.

The integer case is represented by the following model:

$$\max \sum_i \sum_{j\in J_i} [w_{O_i}(\tau_j - a_i - S_i) + w_{D_i}(T'_i - (\tau'_j + b_i))]x_{ij} \qquad (1)$$

$$\sum_{j\in J_i} x_{ij} = 1 \qquad \forall_i \qquad (2)$$

$$\sum_i x_{ij} \leq 1 \qquad \forall_j \qquad (3)$$

$$x_{ij} \in \{o,1\} \qquad \forall(i,j) \qquad (4)$$

By entering some dummy 'trains' in the model the inequality (3) is converted into an equality and the model (1-4) is a classic assignment model.

The restriction (4) which is added in order to avoid marshalling may imply a poor utilization of the capacities of the ferries. Therefore case 2 is examined, too, and this is done by the following model:

$$\max \sum_i \sum_{j \in J_i} w_{O_i} (\tau_{j'} - a_i - S_i) + w_{D_i} (T_i' - (\tau_{j''}' + b_i)) \tag{5}$$

$$j' = \min\left\{ j \mid x_{ij} > 0 \right\} \tag{6}$$

$$j'' = \max\left\{ j \mid x_{ij} > 0 \right\} \tag{7}$$

$$\sum_{j \in J_i} x_{ij} = 1 \tag{8}$$

$$\sum_i L_i \, x_{ij} \leq K_j \tag{9}$$

$$0 \leq x_{ij} \leq 1 \tag{10}$$

If a train is shipped by more than one ferry the train must arrive at the Great Belt in time for the first departure involved, and cannot depart from the other side of the Great Belt before the last freight cars have been shipped. This is assured by (6)-(7) and it is apparent that (4) in the integer case cannot merely be substituted by $0 \leq x_{ij} \leq 1$.

The model (5-10) is a non-linear model in the continous variables x_{ij} .

However, keeping in mind that the criterion used in the two models (1-4) and (5-10) is a heuristic approximation to the real objective, it seems not worth-while to apply (5-10), the solution of which requires non-linear programming algorithms. Therefore (5-10) is replaced by the following model:

$$\max \sum_i \sum_{j \in J_i} [w_{O_i} (\tau_j - a_i - S_i) + w_{D_i} (T_i' - (\tau_j' + b_i))] y_{ij} \tag{11}$$

$$\sum_{j \in J_i} y_{ij} = L_i \tag{12}$$

$$\sum_i y_{ij} \leq K_j \tag{13}$$

$$y_{ij} \geq 0 \quad \text{and integer} \tag{14}$$

Note that (11-14) results from (5-10) by relaxing (6) and (7).

The model (11-14) is the classic transportation model.

As well (1-5) and (11-14) can be regarded as a linear integer programming model which may be solved by e.g. Branch & Bound or in the case of (1-5) by Implicit Enumeration

In real life problems the number of variables, x_{ij} or y_{ij}, will be very great and presumably Implicit Enumeration or Branch & Bound will turn out to be inefficient.

However, the two problems can be transformed into network flow problems which makes possible the application of more efficient algorithms as e.g. out-of-kilter or the simplex based network methods.

The integer model (1-4) can be represented as a graph as sketched in figure 3.

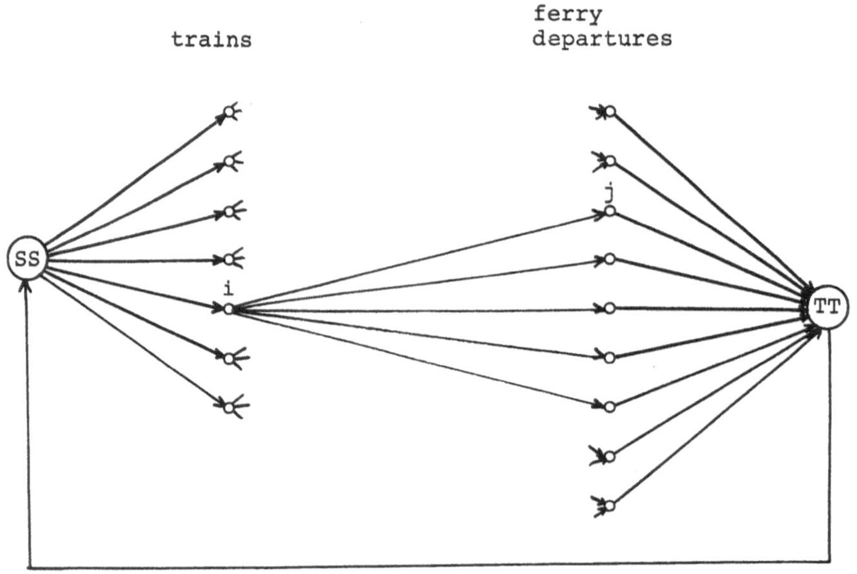

Figure 3. The model (1-4) represented by a graph. The train i is represented by node i and the ferry departure j is represented by node j.

In general there are three parameters and one variable associated with each arc in the graph.

c is the unit cost of the arc

q is the capacity of the arc

r is the minimum required flow in the arc

f is the flow in the arc.

The graph contains a node i for each train to be scheduled and a node j for each ferry departure. From a 'train node' i arcs are leading to those 'ferry nodes' J_i which are feasible for the train in question (J_i is defined in figure 2).

Other constraints could easily be added. If e.g. a certain maximum total dispatching time from O_i to D_i are decisive this is used to define J_i by only making feasible those ferries which keep this constraint.

For all arcs leading from a 'train node' i to a 'ferry node' j the following values of the parameters apply:

$$f_{ij} = x_{ij} \qquad r_{ij} = 0 \qquad q_{ij} = 1 \tag{15}$$

$$c_{ij} = w_{O_i}(\tau_j - a_i - S_i) + w_{D_i}(T_i' - (\tau_j' + b_i)) \tag{16}$$

The problem can now very easily be converted into a general assignment problem by adding some dummy nodes and arcs, but for computional reasons we want the problem converted into a circulation problem. Therefore a supersource SS and a superterminal TT are introduced.

The supersource is connected to each 'trainnode' i by an arc with the following parameters:

$$r_{SS,i} = 1 \qquad q_{SS,i} = 1 \qquad c_{SS,i} = 0 \tag{17}$$

Each 'ferry node' j is connected to the superterminal TT by an arc with the parameters:

$$r_{j,TT} = 0 \qquad q_{j,TT} = 1 \qquad c_{j,TT} = 0 \tag{18}$$

Finally the arc (TT,SS) enters with the following parameters:

$$r_{TT,SS} = I \qquad q_{TT,SS} = I \qquad c_{TT,SS} = 0 \tag{19}$$

Now let k indicate an arbitrary node in the graph while $\Gamma(k)$ and $\Gamma^{-1}(k)$ indicate the set of successors and the set of predecessors of node k in the graph.

Our scheduling problem (1-4) is now

$$\max \sum_{k} \sum_{\ell\in\Gamma(k)} c_{j\ell} \cdot f_{k\ell} \tag{20}$$

$$\sum_{m\in\Gamma^{-1}(k)} f_{mk} - \sum_{\ell\in\Gamma(k)} f_{k\ell} = 0 \qquad \forall k \tag{21}$$

$$r_{k\ell} \leq f_{k\ell} \leq q_{k\ell} \qquad \forall k \qquad \forall \ell\in\Gamma(k) \tag{22}$$

The model (20-22) is a standard minimum circulation problem. This is easily seen by substituting $c_{k\ell}$ by $- c_{k\ell}$ and max by min in (20).

The model (20-22) can be solved by some standard method as 'out-of kilter' or a simplex based algorithm. The latter was preferred in the case study. For results see the following chapter.

Note that the requirement for $f_{k\ell}$ to be integer is not included in the model (20-22). This is not necessary as it can be proved (e.g. Garfinkel and Nemhauser (1972) p. 66-74) that with integer parameters in the network, which is the case in our problem, there is an integer optimal solution to (2o-22).

In a quite similar way the model (11-14) can be modelled on a graph. The structure of the graph is the same as on figure 3 ; only the parameters are changed. The parameters are now:

$$f_{ij} = y_{ij} \qquad r_{ij} = 0 \qquad q_{ij} = L_i \tag{23}$$

$$c_{ij} = w_{O_i}(\tau_j - a_i - S_i) + w_{D_i}(T_i' - (\tau_j' + b_i)) \tag{24}$$

$$r_{SS,i} = L_i \qquad q_{SS,i} = L_i \qquad c_{SS,i} = 0 \tag{25}$$

$$r_{j,TT} = 0 \qquad q_{j,TT} = K_j \qquad c_{j,TT} = 0 \tag{26}$$

$$r_{TT,SS} = \sum_i L_i \qquad q_{TT,SS} = \sum_i L_i \qquad c_{TT,SS} = 0 \tag{27}$$

With these parameter values the model (20-22) still applies. As in the former case the existence of an integer optimal solution to (20-22), if any feasible solution at all, can be proved.

So far we have only considered the scheduling of trains at one section in the railway network. In fact the whole network should be considered simultaneously as any link in the network is accessible only at some discrete hours. In principle this overall problem could also be presented as a network flow problem. This appears to be a multicommodity problem that can be transformed to a single commodity problem which however implies a graph with an anourmous amount of nodes and arcs.

Some approximations can be done which reduce the multicommodity problem to a single commodity problem with dimensions comparable to the models presented in this paper. For details, see Jensen (1977).

SOME RESULTS

In the case study (Colding-Jørgensen 1977) - the transformation of the O-D matrix into a transportation demand expressed in trains to be scheduled, resulted in 96 freight trains. The size of the trains range from 10 to 66 cart-loads (average 55.3 cart-loads). 50 of these freight-trains pass the Great Belt using some of the 55 ferry-departures during the period of 24 hours. The total ferry-capacity measured in the number of cart-loads is about 20% greater than the demand.

Two problems are considered:

1. The direction from east to west (Sealand-Funan/Jutland)
2. The direction from west to east (Funen/Jutland-Sealand).

The two time-tabelling problems, each including 25 trains - are solved by the minimum cost circulation model (20-22) and (23-27) as only the model (11-14) is considered in the following i.e. the case where a train is allowed to be divided and shipped by more than one ferry.

The associate graph contains 82 nodes and about 1200 (1188 and 1222) arcs. The algorithm applied is the PNET-1 (Glover, Karney and Klingman 1973) taking advantage of an especially adapted version of the simplex method using threaded index for the tree-indexing (Glover, Klingman and Stutz 1974). An 'out-of-kilter' algorithm could have been used instead but is inferior compared to the PNET-1 in great network flow problems ((Karnye & Klingman 1973) and (Colding-Jørgensen 1977)).

The CPU-time (IBM 370/90) for solving each of the time-tabelling problems was less than 4 sec. with a storage demand of 82 K bytes.

As the engines are not ferried across the Great Belt the rotation of the engines separates in two problems,one for Sealand and one for Funen/Jutland. For algorithms for simultaneous rotation and time-tabelling (only for trains not crossing the Great Belt) see Colding-Jørgensen (1977).

Now considering Sealand the time-tabelled transports and the transports within Sealand (12 freight-trains) imply the demand of 11 engines with an average utilization of 63.2% during a period of 24 hours. Of this 17.2% is for waiting, 40.5% for driving with freight and 5.5% for driving without freight.

For the territory Funen/Jutland - with 34 freight-trains within Funen/Jutland and 25 freight-trains crossing the Great Belt - the demand of engines was 15 with an average utilization of 84.8% (12% waiting, 64% driving with freight and 8.8% driving without freight).

It is obvious that the relative size of the node-weights have significant influence upon the timing of freight-trains. The optimal set of weights is defined to be the set resulting in the minimum number of engines required. This optimal set is depending on the number of freight-trains, the ferry-time-table etc. and might be found by simulation.

In the case described above the weights applied range from 1 to 10 implying a demand of 26 engines.

However,using a set of weights all equal to 1 , $w_\ell = 1 \; \forall \ell$; the demand of engines was 29.

CONCLUSION

A model for scheduling freight-trains, considering one section in the railway network, has been developed. The model is a network flow model with ensuring efficient algorithms. The model has been tested on data from the Danish State Railways.

The model may as well be used for longterm planning e.g. finding the optimal capacity of the section considered as for short term control e.g. the bi-annual changes of the schedule.

The model is set up with reference to real life applications and has appeared to be able to include the major part of the technical and operational constraints.

In real life the scheduling of trains is a very complicated and troublesome task to do manually. A mathematical model can not entirely make up for the experience and the ability of an experienced scheduler. What the model can do is to help him get a survey of the problem and to do some of the boring computional work. Therefore, in this paper it has been the general aim to set up rather simple models which partly optimize, partly ensure most constraints from the application in question to be fulfilled. Thus, the model is an addendum to the scheduler who has to verify the solution given by the model, who can preditermine parts of the solution, or in other ways combine his general knowledge of the railway system with the flexibility and the rapidity of the model. This might in general be the road ahead for using optimization models in real life applications.

REFERENCES

Colding-Jørgensen, C. (1977): Omløbsplanlægning for lokomotiver (in Danish). Master thesis. The Institute of Mathematical Statistics and Operations Research (IMSOR), The Technical University of Denmark, DK 2800 Lyngby, Denmark.

Garfinkel, R.S. and Nemhauser, G.L. (1972): Integer Programming. John Wiley & Sons.

Glover, F., Karney, D. & Klingman, D. (1973): Implementation and computational comparisons of primal, dual and and primal-dual computer codes for minimum cost network flow problems, Research Report CS 136, Center for Cybernetic Studies, University of Texas, Austin, TEX.

Glover, F., Klingman, D. & Stutz, J. (1974): Augmented treaded index method for network optimization, INFOR, vol. 12, no. 3, pp. 293-298.

Jensen, O.H. (1977): Scheduling of trains, a network approach. Working paper. The Institute of Mathematical Statistics and Operations Research (IMSOR). The Technical University of Denmark, DK-2800 Lyngby, Denmark.

Karney, D. & Klingman, D. (1973): Implementation and computational study on an in-core out-of-core primal network code, Research Report CS 158, Center for Cybernetic Studies, University of Texas, Austin, TEX.

OPTIMAL PORTFOLIOS WHERE PROCEEDS ARE A FUNCTION OF THE CURRENT ASSET PRICE

by

BRUCE L. MILLER
DEPARTMENT OF SYSTEM SCIENCE
UNIVERSITY OF CALIFORNIA, LOS ANGELES 90024

Abstract

We consider a portfolio model with one riskless and one risky asset where the distribution function of random proceeds is a monotone function of the current price of the risky asset. Several examples of models in the literature with this property are briefly discussed. The main objective is to compare for the monotone price dependent case and the usual price independent case the amount (suitably interpreted) invested in the risky asset. The result is that if proceeds are monotone decreasing in price, then for $U(x) = \frac{1}{\gamma}(x+A)$, one invests less in the price dependent case, but more if $U(x) = \frac{1}{\gamma}(x+A)^{\gamma}, \gamma < 0$, or $U(x) = -e^{-\alpha x}$. For $U(x) = \ln(x+A)$ the amounts are the same. The reverse inequalities hold when random proceeds are monotone increasing in price. The last section which is not included in this abridged version established the fact that if random proceeds are based on a conditional Markov chain of a particular but plausible form then those proceeds are a monotone function of their price.

This research was supported by the National Science Foundation under contract ENG 76-12230. I thank Mark Rubinstein for several useful comments on an earlier draft.

Introduction

The assumption that the random proceeds per unit of capital invested in an asset is a function of the current price of that asset has been made occasionally in the literature. By the proceeds or return in a period i, we mean the price of the asset in period $i+1$ divided by the price in period i. We will assume that this price dependence has a monotone property, that the distribution function describing proceeds is either less favorable with increasing price or more favorable. When the price dependent assumption has been made in the literature it usually has had the monotone property. A brief survey of some rationales which lead to monotone price dependent random proceeds is given at the end of the introduction.

The standard assumption in multiperiod portfolio problems (Hakansson [4], Mossin [8], and Neave [9]) is that the probability distribution of random proceeds at any period i, is known in period one, and thus is independent of events that will occur between periods 1 and i. In particular it is not a function of asset price changes. However, the important work of Rosenberg and Ohlson [10] shows that this kind of assumption is not consistent with the assumption of portfolio separation, which says that an investor can separate his investment decision and first decide what proportions to purchase of available risky assets then how to allocate his wealth between risky and safe investments. Precisely, they prove that the assumption that the probability distribution of random proceeds at any period i, is independent of i, and the assumption of portfolio separation where the proportions of available risky assets are constant over time are inconsistent. In the second paragraph of their conclusion they state, "It is clear one assumption that should be relaxed in the interest of greater realism is the one of stationary and serially independent return distributions, independent of the market process which generates equilibrium prices. Since price changes (capital gains) is the most important component of returns, the ex ante return distribution should be a function of current prices, with any change in the observed current equilibrium prices inducing a change in the return-generating process, conditional upon this new set of prices." After having undermined the standard assumptions in multiperiod portfolio problems, they point to the Hakansson model [5] as a promising way to revise the assumptions. The Hakansson model allows random proceeds to be a function of the state of the economy and allows for a finite number of different states. The state of economy changes from period to period according to a Markov chain. Their results add credence to our assumption that random proceeds be a function of asset prices. However a predetermined monotone property is an admittedly strong additional assumption. A quite different approach (Rubinstein [11]) is a multiperiod financial model where the prices of securities are endogenously determined. Then the solution of the model would say whether or not proceeds are a monotone function of price.

The model considered is in discrete time with a finite horizon, one riskless asset, and one risky asset with monotone price dependent random proceeds. The investor is assumed to have either an isoelastic or an exponential utility function of consumption and terminal wealth. When this model is formulated as a dynamic programming problem there are two state variables, wealth and the price of the risky asset. If the set of prices is finite then the problem is computationally equivalent to the model of Hakansson [5] where price in our model plays the role of his states of the economy. Hakansson shows how his problem can be solved computationally (see also Ziemba [13] for some mathematical programming details), and one observes that the computational requirements in the price dependent situation are approximately P times those of the usual model where proceeds are independent of the current asset price, where P is the number of possible prices. These computational requirements as well as the additional data requirements (the distribution function of proceeds must be specified for each possible price) may make it burdensome to compute the optimal policy in the price dependent case. Therefore one may have to resort to solving a price independent model with the same current period investment opportunity set as the price dependent model. The main effort of this paper is to compare for the price dependent and price independent cases the ratio of the amount invested in the risky asset to the sum of the amount invested in the risky and the riskless asset. This is the comparison that is made for the isoelastic utility functions. For the exponential utility functions we we compare the amounts invested in the risky asset. These comparisons are possible because in the price independent case the above ratio or amount does not depend on investment opportunities in future periods as shown by Mossin [8].

Our survey of some rationales which lead to monotone price-dependence begins with the "asymtotic normal price level" hypothesis described by Merton [7]. In this case prices are moving toward some "normal" or long-run price level and hence when prices are low they have more of a tendency to move up than when they are high. A second example is that of Winkler [12]. He assumes that the proceeds are identically and normally distributed each period. However, this normal distribution has a known variance but an unknown mean whose Bayesian prior is described by a normal distribution. For this situation Winkler comes up with a closed form solution for proceeds in period i [12, p. 391], and the updated mean depends only on the prior estimate of the mean and the current price of the asset. This price dependent case is monotone increasing as high past proceeds necessarily imply both that the current price is high and that we are more likely to enjoy higher proceeds in the future by the Bayesian hypothesis. In his empirical work on commodity prices Houthakker [6] finds evidence that proceeds are monotone increasing in price. His (cautious) explanation is that the wealth effect is likely to be of importance to many speculators, and that this would reinforce downward and upward price movements.

An interesting hypothesis which we showed in the expanded paper leads to monotone price dependent returns is that of a conditional Markov chain. Boyce [1] considered a bond selling problem where the bond's price was described by the usual random walk assumption, but a predicted terminal price distribution was superimposed on the random walk resulting in a conditional Markov chain. This paper motivated two papers containing probability theory results, one by Föllmer [2] working in continuous time and the other by Griffeath and Snell [3] in discrete time. Using some results of Griffeath and Snell [3] we show in the unabridged version that the conditional Markov chain resulting from a random walk is always either (1) strictly monotone increasing (2) constant or (3) strictly monotone decreasing in the state value when the predicted terminal state distribution is of the discrete normal form. Thus when proceeds are based on this type of conditional Markov chain the results of Section 2 will apply.

2. The Model and the Main Result

We make the usual assumptions of no taxes, transaction costs, or indivisibilities. The risky asset has no dividend payouts. We need the following terms to describe the n-period model where the subscript i refers to the period.

r_i : the riskless return received at the end of period i from the investment of one unit at the beginning of period i .

p_i : the price of the risky investment at the beginning of period i . p_1 is given and $p_{i+1} = p_i\beta_i(p_i)$ where $\beta_i(p_i)$ is described below. The p_i may take on values over any given subset of $(0,\infty)$.

$\beta_i(p_i)$: the random proceeds received at the end of period i from the investment of one unit at the beginning of period i when the price of the risky asset at the beginning of the period is p_i . The $\beta_i(p_i)$ are independent over time but not necessarily identically distributed over time. We assume that there is a β_i with $E(\beta_i) < \infty$ such that $\beta_i(\omega) \geq \beta_i(p_i,\omega)$, and for some $\varepsilon > 0$ $P\{(\beta_i - r_i) < \varepsilon\} > \varepsilon$. Here and throughout the text E stands for expected value. This is Hakansson's "no easy money" condition. Other conditions on $\beta_i(p_i)$ are that $E[\beta_i(p_i)] \geq r_i$, that $\beta_i(p_i,\omega) > 0$, that $p_i\beta_i(p_i,\omega)$ is an increasing function of p_i , and that $\beta_i(p_i,\omega)$ is a monotone function of p_i . We need either that $E[\beta_i(p_i) \geq r_i]$ or no short selling. Otherwise an unfavorable $\beta_i(p_i)$ could lead to a favorable investment opportunity through short selling and the results to follow would not hold.

c_i : the consumption in period i .

s_i : the amount of capital at the beginning of period i.

z_{1i} : the amount invested in the riskless investment in period i .

z_{2i} : the amount invested in the risky investment in period i . The assumption that $E[\beta_i(p_i)] \geq r_i$ implies that $z_{2i} \geq 0$ since our utility functions are concave.

The performance criterion is $U(c_1,\ldots,c_n,s_{n+1}) = \sum_{i=1}^{n} U_i(c_i) + U_{n+1}(s_{n+1})$. The U_i come from one of three classes:

A-1 $\quad U_i(x) = K_i \frac{1}{\gamma} x^\gamma \quad 0 < \gamma < 1 , x > 0$

A-2 $\quad U_i(x) = K_i \frac{1}{\gamma} x^\gamma \qquad \gamma < 0 , x > 0$

B $\quad U_i(x) = -K_i e^{-\alpha x} \qquad \alpha > 0$

C $\quad U_i(x) = K_i \ln x , \quad x > 0$

where $K_i > 0$. The flexibility of the K_i obviously permits a discount factor. For utility functions A and C we define $U_i(x) = -\infty$ for $x \leq 0$. A useful generalization of the utility functions A and C is $U_i(x) = K_i \frac{1}{\gamma} (x+A_i)^\gamma$ and $U_i(x) = K_i \ln(x+A_i)$. In the appendix of the unabridged version we showed how the results for our model include these generalized utility functions. The idea is that the A_i play the role of a deterministic income or obligation (A_i negative), and that they can be discounted into a revised state variable. This result is similar to Merton [7] equations 47-49, and Hakansson's idea of revising the state variable by discounted future labor income [4] p. 600.

The state equation of our model is

$$(1) \quad s_{i+1} , p_{i+1} = r_i(s_i - c_i - z_{2i}) + \beta_i(p_i) z_{2i} , p_i \beta_i(p_i)$$

$$= r_i(s_i - c_i) + (\beta_i(p_i) - r_i) z_{2i} , p_i \beta_i(p_i) .$$

If we let $V_i(s_i, p_i)$ be the optimal return function of dynamic programming, the equation of optimality is

$$(2) \quad V_{n+1}(s_{n+1}, p_{n+1}) = U_{n+1}(s_{n+1}) , \text{ and}$$

$$V_i(s_i, p_i) = \sup_{c_i, z_{2i} \geq 0} E\{U_i(c_i) + V_{i+1}(s_{i+1}, p_{i+1})\}$$

where s_{i+1} , p_{i+1} are given by (1).

We will state our results for the case where $\beta_i(p_i)$ is monotone decreasing distribution function in p_i only. The cases where $\beta_i(p_i)$ is monotone increasing distribution function in p_i , and where the model is changed by not considering consumption but only terminal reward will be stated in two propositions at the end of this section.

Theorem 1 For the model of this section

$$V_i(s_i, p_i) = f_i(p_i, \gamma) s_i^\gamma \qquad \text{(case A)}$$

$$= f_i(p_i, \alpha) e^{-\alpha D_i s_i} \qquad \text{(case B) where}$$

$$D_i = \prod_{j=i}^{n} r_j / \sum_{k=i}^{n} \prod_{j=k}^{n} r_j , \text{ and } D_{n+1} = 1 .$$

$$= L_{1i} \ln x + L_{2i}(p_i) . \qquad \text{(case C)}$$

where $f_i(p_i,\gamma)$ is positive for A-1 and negative for A-2, $f_i(p_i,\alpha)$ is negative and L_{1i} is positive. The function $L_{2i}(p_i)$ is bounded. The functions $f_i(p_i,\gamma)$ and $f_i(p_i,\alpha)$ satisfy $0 < m_{\gamma,i} < |f_i(p_i,\gamma)| < M_{\gamma,i}$ | and $0 < m_{\alpha,i} < | f_i(p_i,\alpha)| < M_{\alpha,i}$. Furthermore $f_i(p_i,\gamma)$ and $f_i(p_i,\alpha)$ are monotone decreasing functions of p_i .

Comment. For those curious about D_i we note that when $r_i = r$, $D_i = r^n/1 + r^2 + \ldots + r^n$, and as $n \to \infty$, $D_1 \to (r-1)/r$, the constant which appears in eq. (58) of Hakansson [4]. The upper bounds on $f_i(p_i,\gamma)$ and $f_i(p_i,\alpha)$ are needed to ensure that $g_i(p_i,\gamma)$ and $g_i(p,\alpha)$ in (3) and (4) are finite. The boundeness of $L_{2i}(p_i)$ is needed to ensure that $E(L_{2i}(p_{i-1}\beta_{i-1}(p_{i-1})))$ is finite. Our proof is general enough to be valid when $\beta_i(p_i)$ is monotone increasing in p_i . The functional forms Case A and Case C of Theorem 1 are the same as those in Hakansson [5] and as those in the complete securities model of Rubinstein [11], pages 27 and 28. In our development the p_i are not restricted to a finite set.

The proof of Theorem 1 is omitted in this abridged version. In the proof it is shown that the amount to invest in the risky asset is $\theta_p^*(s_i-c_i)$ for utility functions (A) and (C) and $\theta_p^*/\alpha D_{i+1}$ for utility function B where the value of θ^* is the maximizing θ in

(3) $g_i(p_i,\gamma) = \max_{\theta\geq 0} E[f_{i+1}(p_i\beta_i(p_i),\gamma)(r_i + (\beta_i(p_i)-r_i)\theta)^\gamma]$ (case A),

(4) $g_i(p_i,\alpha) = \max_{\theta\geq 0} E[f_{i+1}(p_i\beta(p_i),\alpha)e^{-(\beta_i(p_i)-r_i)\theta}]$ (case B), and

(5) $g_i(p_i) = \max_{\theta\geq 0} E[\ln(r_i + (\beta_i(p_i)-r_i)\theta)]$ (Case C) .

Now consider the price independent random proceeds model where the state equation is given by

(6) $s_{i+1} = r_i(s_i-c_i) + (\beta_i-r_i)z_{2i}$.

It is known (Mossin [8] Hakansson [5] that a myopic policy is optimal. The optimal amount to invest in the risky asset is $\theta^*(s_i-c_i)$ for utility functions (A) and (C) and $\theta^*/\alpha D_{i+1}$ for utility functions B where the value of θ^* is the maximizing value of θ in

(7) $h_i(\gamma) = \max_{\theta\geq 0} E[\frac{1}{\gamma}(r_i + (\beta_i-r_i)\theta)^\gamma]$

(8) $h_i = \max_{\theta\geq 0} E[-e^{-(\beta_i-r_i)\theta}]$, and

(9) $h_i = \max_{\theta\geq 0} E[\ln[r_i + (\beta_i-r_i)\theta]]$.

Recall that we may restrict our attention to $\theta \geq 0$ since we assume $E(\beta_i) \geq r_i$. We now compare θ^* and θ^*_p .

Theorem 2. For (7), (8) and (9) let r_i equal the r_i in (3), (4) and (5) respectively and $\beta_i = \beta_i(p_i)$. Then $\theta^*_p \leq \theta^*$ for utility functions A-1 and $\theta^*_p \geq \theta^*$ for utility functions A-2 and B . $\theta^*_p = \theta^*$ for case C.

Proof. It is evident that $\theta^*_p = \theta^*$ for case C. For utility function A-1 $\theta = \theta^*_p$ satisfies

$$(10)\quad \int_{\omega:\ \beta_i(p_i) \geq r_i} f_{i+1}(p_i\beta_i(p_i),\gamma)\gamma(r_i + (\beta_i(p_i)-r_i)\theta)^{\gamma-1}(\beta_i(p_i)-r_i)$$

$$= \int_{\omega:\ \beta_i(p_i) < r_i} f_{i+1}(p_i\beta_i(p_i),\gamma)\gamma(r_i + (\beta_i(p_i)-r_i)\theta)^{\gamma-1}(r_i-\beta_i(p_i)) .$$

Since f_{i+1} is monotone decreasing in p_{i+1} by Theorem 1, the righthand side of (10) is smaller than the lefthand side when the $f_{i+1}\gamma$ terms are dropped and $\theta = \theta^*_p$. However, (10) without the $f_{i+1}\gamma$ terms is the equation θ^* must satisfy. The righthand side is increasing in θ and the lefthand side is decreasing in θ . Therefore $\theta^* \geq \theta^*_p$.

For utility function A-2 $f_{i+1}\gamma$ is positive but monotone increasing in p_{i+1} since $\gamma < 0$ and f_{i+1} is monotone decreasing in p_{i+1} . Therefore the righthand side of (10) is greater than the lefthand side when the $f_{i+1}\gamma$ terms are dropped and $\theta = \theta^*_p$. As before (10) without the $f_{i+1}\gamma$ terms is the equation θ^* must satisfy. Therefore $\theta^* \leq \theta^*_p$.

For utility function B, $\theta = \theta^*_p$ satisfies

$$(11)\quad \int_{\omega:\ \beta_i(p_i) \geq r_i} -f_{i+1}(p_i\beta_i(p_i),\alpha)e^{-(\beta_i(p_i)-r_i)\theta}(\beta_i(p_i)-r_i) .$$

$$= \int_{\omega:\ \beta_i(p_i) < r} -f_{i+1}(p_i\beta_i(p_i),\alpha)e^{-(\beta_i(p_i)-r_i)\theta}(r_i-\beta_i(p_i)) .$$

Since $-f_{i+1}$ is positive and monotone increasing, the righthand side of (11) is greater than the lefthand side when the $-f_{i+1}$ terms are dropped and $\theta = \theta^*_p$. However, (11) without the $-f_{i+1}$ terms is the equation θ^* must satisfy. The righthand side is increasing in θ and the lefthand side is decreasing in θ . Therefore $\theta^* \leq \theta^*_p$ which completes the proof.

The model where $\beta_i(p_i)$ is monotone increasing in p_i can be analyzed in the same way. The change we get is that f_{i+1} is monotone increasing in p_{i+1} . Therefore the inequalities of Theorem 2 are reversed and we state this result as

Proposition 1. Consider the original model except that $\beta_i(p_i)$ is a monotone increasing distribution function in p_i . For (7), (8) and (9) let r_i equal the r_i in (3), (4) and

(5) respectively and $\beta_i = \beta_i(p_i)$. Then $\theta^*_p \geq \theta^*$ for case A-1 and $\theta^*_p \leq \theta^*$ for A-2 and B. $\theta^*_p = \theta^*$ for case C.

Finally let us consider the case of maximizing terminal wealth only. Then (1) becomes

$$(12) \quad s_{i+1}, p_{i+1} = r_i s_i + (\beta_i(p_i) - r_i) z_{2i} \;,\; p_i \beta_i(p_i)$$

and (2) becomes

$$(13) \quad V_i(s_i, p_i) = \max_{z_{2i} \geq 0} E\{V_{i+1}(s_{i+1}, p_{i+1})\}$$

where s_{i+1}, p_{i+1} are given by (12). Lemma 1 of the unabridged version which is used in the proof of Theorem holds with no changes (for $\beta_i(p_i)$ monotone increasing or monotone decreasing), and Lemma 2 of the unabridged version is no longer needed. Theorem 1 holds with an easier proof with D_i changed to $\sum_{j=i}^{n} r_j$. Theorem 2 and Proposition 1 hold without changes, and we state this last result as Proposition 2.

Proposition 2. Consider the original model except that $c_i \equiv 0$, $i = 1, \ldots, n$, and $U(0, \ldots, 0, s_{n+1}) = U_{n+1}(s_{n+1})$. This is the maximization of expected terminal wealth criterion. Then Theorem 2 and Proposition 1 remain valid.

References

1. Boyce, W., "Stopping Rules for Selling Bonds," Bell J. of Economics, 1, (1970), pp. 27-53.
2. Föllmer, H., "Optimal Stopping of Constrained Brownian Motion," J. Appl. Probability, 9, (1972), pp. 557-571.
3. Griffeath, D. and L. Snell, "Optimal Stopping in the Stock Market," The Annals of Probability, 2, (1974), pp. 1-13.
4. Hakansson, N., "Optimal Investment and Consumption Strategies Under Risk for a Class of Utility Functions," Econometrica, 38, (1970), pp. 587-607.
5. Hakansson, N., "Optimal Entrepreneurial Decisions in a Completely Stochastic Environment," Management Science, 17, (1971), pp. 427-449.
6. Houthakker, H., "Systematic and Random Elements in Short-Term Price Movements," American Economic Review, 51, (1971), pp. 164-172.
7. Merton, R., "Optimal Consumption and Portfolio Rules in a Continuous Time Model," J. of Economic Theory, 3, (1971), pp. 373-413.
8. Mossin, J., "Optimal Multiperiod Portfolio Policies," J. of Business, 41, (1968), pp. 215-229.
9. Neave, E., "Multiperiod Consumption Investment Decisions and Risk Preference," J. of Economic Theory, 3, (1971), pp. 40-53.
10. Rosenberg, B. and J. Ohlson, "The Stationary Distribution of Returns and Portfolio Separation in Capital Markets: A Fundamental Contradiction," J. of Financial and Quantitative Analysis, 11, (1976), pp. 393-402.
11. Rubinstein, M., "A Discrete-Time Synthesis of Financial Theory: Part 1. Optimal Decision and Sharing Rules," Working Paper No. 20, Institute of Business and economic Research, U. of California, Berkeley, 1974.

12. Winkler, R., "Bayesian Models for Forecasting Future Security Prices," J. of Financial and Quantitative Analysis, 8, (1973), pp. 387-405.

13. Ziemba, W., "Note on Optimal Portfolio Growth When Yields are Serially Correlated," J. of Financial and Quantitative Analysis, 7, (1972), pp. 1195-1199.

AN ALGORITHM FOR OPTIMAL SEQUENCING OF AIRCRAFT IN THE NEAR TERMINAL AREA

L. Bianco, B. Nicoletti, S. Ricciardelli
Centro di Studio dei Sistemi di Controllo e Calcolo
Automatici - Consiglio Nazionale delle Ricerche
Via Eudossiana 18 - ROMA - ITALY

ABSTRACT

In this paper, the optimal sequencing of aircraft in the near terminal area with a single runway for both landings and take-offs, is considered. This problem is imbedded in the class of problems related to sequencing n jobs (either landings or take-offs) on one machine (the runway). A difference immediately arises, since the jobs are not simultaneously available and once begun must be fully completed. A branch and bound algorithm is developed, by making use of some peculiar properties which hold in this particular case.
The approach proposed looks promising, since it could be used to study more complex problems such as those which arise in the case of more runways or taking into account several classes of aircraft.

1. INTRODUCTION

For several reasons, automated air-traffic management systems have recently been proposed and investigated as a solution of what is usually referred to as the air traffic control problem. As a consequence, a need exists for a mathematically tractable model and a comprehensive analysis of the components of a workable semi/fully automated air traffic control system. In this paper, only a small portion of such a model is analyzed. Particularly, optimal sequencing of aircraft in the near-terminal area is considered. This is, approximately, an area within a 50-mile radius surrounding an airport (see fig.1). Within it, the approach (or take off) paths are prestructured and "fixed" in space. The landings conflict with the take-offs, because they compete for the runways available. In this paper only one runway is considered.

Under the preceding hypothesis, given a certain number of aircraft in the near-terminal area, every time that a new aircraft enters the system, a conflict situation could arise, if the estimated arrival times at the "merging gate" of two successive aircraft do not satisfy the constraint relative to the separation standard. In such a case, it is necessary to modify these times of a certain quantity δt, whose minimum and maximum values are constrained by the performance of the aircraft considered. Therefore, a constrained sequencing problem must be solved in order to avoid all the potential conflicts at the merging gate. Without making any distinction between incoming and departing aircraft, let define the nominal time of arrival at the merging gate as the sum of the actual time of arrival at the feeder fix (or at the taking off) point and of the time necessary to fly from this point to the merging gate thru an optimal path. The sequencing problem can be stated as follows: "Given a set of nominal times at the merging gate and corresponding maximum and minimum delay capabilities, determine that sequence of arrival times in the merging point which insures proper separation and minimizes a measure of the total system delay".

In a paper, presented by Stephen K.Park et al. in 1972 [1], three approaches have been proposed to solve this problem. In the first approach, the conflict resolution problem is stated as a linear programming problem in which, once fixed "a priori" a certain merging sequence, a weighted sum of the resulting delays is minimized.

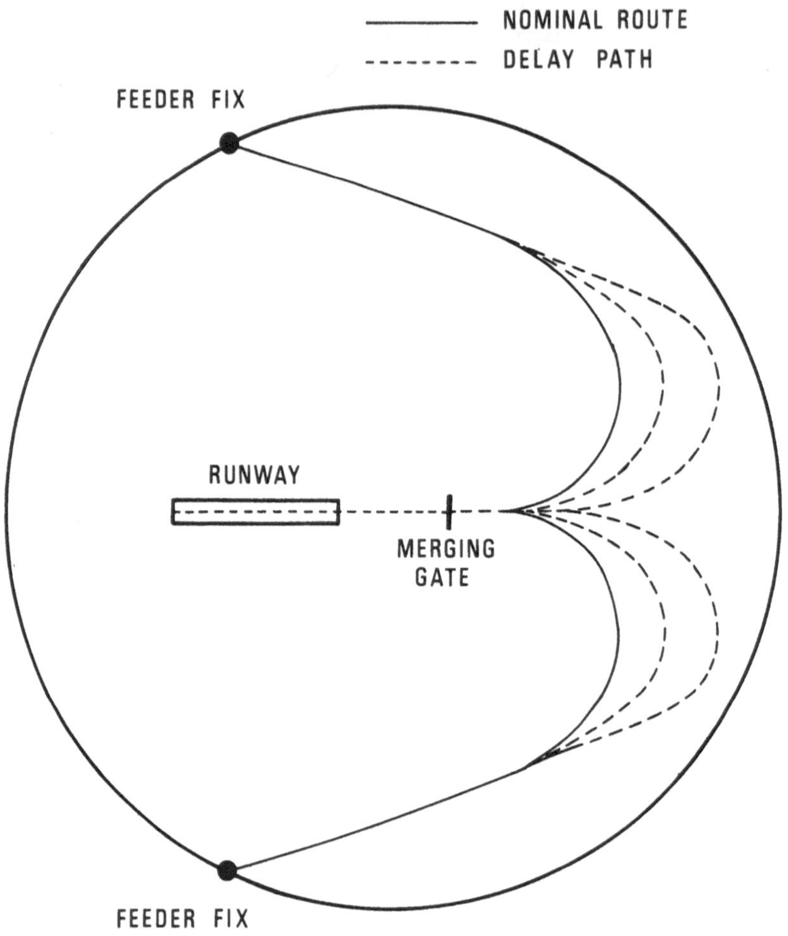

FIG.1 - STRUCTURE OF THE TERMINAL AREA

In the second approach, the following assignment problem is solved: "Given n fixed equally spaced merging opportunities (time slots), assign merging times to the n planes, one to a merging time slot, in such a way that a weighted sum of the resulting delays is minimized".

Both these approaches present heavy limitations, since, in the first one, the merging order is fixed "a priori", while, in the second one, the sequencing time interval is fixed, disregarding of the type of the aircraft.
In the latter approach there might occur, for some aircraft, deviations from the nominal trajectories which are not necessary.

The third approach proposed is a combination of the preceding ones and the problem is solved via enumeration. This leads to a solution in an acceptable time for low values of the n aircraft in a conflict situation. However, the larger the n the larger the number of solutions to be considered, since it increases with n!. This severely reduces the applicability of the method.

In this paper, an algorithm is proposed to deal, in an optimal way, with heavy traffic conditions in the near terminal area. In such a case, a pure enumerative approach does not look feasible.

LIST OF THE SYMBOLS

A = set of the aircraft to be sequenced;

a_i = an aircraft of the set A;

$\bar{t}_i$ = nominal time of arrival at the merging gate of a_i ;

d_i = time interval between the instant a_i passes thru the merging gate and that one in which the runway is cleared (or viceversa); it is also called "service time";

t_i = actual arrival time of a_i at the merging gate;

n = number of aircraft in the set A;

s_k = a sequence of k aircraft of the set A;

L = a level of the tree at which a sequence s_L is examined;

N_L = node at level L associated at a certain sequence s_L ;

τ_i^t = time at which the servicing of a_i can start, t being the effective time;

$F_{s_k}^t$ = final time of servicing of s_k, t being the time at which s_k can start;

$D_{s_k}^t$ = total delay of s_k, t being the time at which s_k may start; (if t=0 $D_{s_k}^o = D_{s_k}$);

A_{s_k} = set of all the possible successors of s_k at level k+1;

$\{s_k\}$ = set of all the aircraft which belong to s_k;

s_{n-k} = complement of s_k on the set A;

s_k^* = sequence of k aircraft of the set A which minimizes their total delay;

B_{s_k} = Lower bound at node N_k associated to s_k ;

$\delta_{-1}(t)$ = unit step function

s_n^o = initial solution

$\left|A_{N_L}\right|$ = dimensions of the set A_{N_L}

2. THE SEQUENCING PROBLEM AS A "n JOBS-ONE MACHINE" PROBLEM

Given a set $A = \{a_1, a_2, \dots, a_n\}$ of aircraft in the near-terminal area at time t, each one being characterized by a nominal time $\bar{t}_i$ and a service time d_i, which depends in general on the required operation (either landing or take-off) and the aircraft class, they conflict with each other when

$$\bar{t}_i - \bar{t}_1 < \sum_{h=1}^{i-1} d_h \quad , \qquad \forall i \in \{1,2,\dots,n\}$$

(the general assumption has been made that the aircraft are ordered according to the nominal sequence $\bar{t}_1 \leq \bar{t}_2 \leq \bar{t}_3 \leq \dots \leq \bar{t}_n$). If no constraint exists on the total service time for the set A, the conflict resolution problem can be stated in the following way: find the sequence which minimizes

$$z = \sum_{i=1}^{n} (t_{[i]} - \bar{t}_{[i]}) \tag{1}$$

with

$$t_{[i+1]} \geq t_{[i]} + d_{[i]} \tag{2}$$

where the subscript [i] denotes the aircraft which, in the optimal sequence, is in the ith position and $t_{[i]}$ is the corresponding actual arrival time at the merging gate.

The runway can be seen as a facility which must process the activities (either landings or take-offs). With reference to fig.1, the sequencing problem may be stated as a job-shop scheduling problem: in particular, in the case considered, a n jobs-one machine problem . One can then apply the results already available for this class of problems [2],[3],[4]. Particularly, if $\bar{t}_i = \bar{t}$ = constant $\forall i$, the optimal sequence can be immediately found by applying the "shortest service time" rule : that is, sequence the aircraft according to the order

$$d_{[1]} \leq d_{[2]} \leq d_{[3]} \leq \dots \leq d_{[n]} \tag{3}$$

However, in general the nominal times are not equal, so that it is necessary to find in real time the optimal solution.

3. THE ALGORITHM

The proposed algorithm is based on a branch and bound approach, whose initial solution can be obtained by an heuristic approach.

Let L be the current level at which all partial sequences s_L, with L aircraft fixed, are examined; the search tree starts at a dummy root node at level L = 0. All the possible successors of N_o (that is s_o), at level L = 1, are determined and for each node N_L , thus obtained, one finds all possible successors at level L = 2 and so forth. The level at which the search should be stopped for every branch of the tree is determined by making use of some peculiar properties which hold in this particular case. Let begin, however, with some general definitions.

DEFINITION 1. If a_i is an element in a sequence and t is the actual time, the instant at which the servicing of a_i can begin, depends on t and it is defined as

$$\tau_i^t = \max(t, \bar{t}_i) \tag{4}$$

DEFINITION 2. Given a sequence $s_k = s_{k-1}\, a_k$, the time of completion of service for s_k depends on the initial time t and may be defined as

$$F_{s_k}^t = \max(F_{s_{k-1}}^t, \tau_k^t) + d_k \tag{5}$$

For simplicity, if $t = 0$, $F_{s_k}^o = F_{s_k}$.

DEFINITION 3. Given a sequence s_n, if t is its initial time, the total delay may be expressed as

$$D_{s_n}^t = \sum_{k=1}^{n} \delta_{-1}\left[F_{s_{k-1}}^t - \bar{t}_k\right] * \left[F_{s_{k-1}}^t - \bar{t}_k\right]$$

or equivalently

$$= \sum_{k=1}^{n} \sum_{i=1}^{n} \delta_{-1}\left[F_{s_{k-1}}^t - \tau_i^{F_{s_{k-2}}^t}\right] * \left[F_{s_{k-1}}^t - \tau^{F_{s_{k-2}}^t}\right] \tag{6}$$

Let analyse in some details the branch and bound phases of the algorithm at a certain step k, in which the partial sequence s_{k-1} is considered to be fixed.

a) *Branching phase*

In this phase, starting from a node $N_{L=k-1}$, one determines all possible successors, at level L = k, of the associated sequence s_{k-1}.

The following property has been rigorously demonstrated [5]: Given a partial sequence s_{k-1}, its possible successors at level k are all and only the elements of the set

$$A_{s_{k-1}} = \left\{a_i \in A-\{s_{k-1}\} : \tau_i^{F_{s_{k-1}}} < F_{s_{k-1}\tilde{a}_i}\right\} \tag{7}$$

where

$$F_{s_{k-1}\tilde{a}_i} = \min_{i:a_i \in A-\{s_{k-1}\}} \left\{\tau_i^{F_{s_{k-1}}} + d_i\right\}$$

However, intuitive consideration also lead to this result. Let consider an aircraft a_i in $A-\{s_{k-1}\}$, with time $\bar{t}_i > F_{s_{k-1}}$.

If an aircraft a_j exists such that $\tau_j^{F_{s_{k-1}}} + d_j \leq \bar{t}_i$, then, certainly, the sequence $s_{k-1}\, a_i\, s_{n-k}^*$ is worse than $s_{k-1}\, a_j\, a_i\, s_{n-k-1}^*$. As a matter of a fact, in such a case, the time window $(\bar{t}_i - F_{s_{k-1}})$ is sufficient for the processing of a_j without implying larger delays of a_i. This implies that $F_{s_{k-1}}\, a_i \geq F_{s_{k-1}}\, a_j\, a_i$.

b) *Bounding phase*

This phase consists of two steps. In the first one, on the basis of certain elimination criteria, among the nodes at level k, one identifies those that can be not considered, since they generate feasible solutions certainly worse than the other ones. In the second step, for each remaining node, one can compute a lower bound to be compared with the current best solution.

b_1) *Elimination conditions*

Let examine all pairs $(a_i,a_j) \in A_{s_{k-1}}$. Three cases are possible:

a) $\bar{t}_i \leq F_{s_{k-1}} \leq \bar{t}_j$

b) $F_{s_{k-1}} \leq \bar{t}_i \leq \bar{t}_j$

c) $\bar{t}_i \leq \bar{t}_j \leq F_{s_{k-1}}$

With reference to each case, one needs to consider the following sequences:

I) $s_{k-1}\ a_i\ a_j\ s_{n-k-1}$

II) $s_{k-1}\ a_j\ a_i\ s^{ji}_{n-k-1}$

where s^{ij}_{n-k-1} and s^{ji}_{n-k-1} are the optimal complements of $s_{k-1}\ a_i\ a_j$ and $s_{k-1}a_ja_i$ respectively. The corresponding delay for any pair (a_i,a_j) is given by:

$$D_I = D_{s_{k-1}} + \delta_{-1}\left[F_{s_{k-1}} - \bar{t}_i\right]*\left[F_{s_{k-1}} - \bar{t}_i\right]+\left[\max(F_{s_{k-1}},\bar{t}_i)+d_i - \bar{t}_j\right]+D^{F_{s_{k-1}a_ia_j}}_{s^{ij}_{n-k-1}}$$

$$D_{II} = D_{s_{k-1}} + \delta_{-1}\left[F_{s_{k-1}} - \bar{t}_j\right]*\left[F_{s_{k-1}} - \bar{t}_j\right]+\left[\max(F_{s_{k-1}},\bar{t}_j)+d_j-\bar{t}_i\right]+D^{F_{s_{k-1}a_ja_i}}_{s^{ji}_{n-k-1}}$$

By comparison, one can conclude that

$$D_I < D_{II} \quad \text{iff}$$

$$(d_i - d_j) < 2f+\left(D^{F_{s_{k-1}a_ia_j}}_{s^{ji}_{n-k-1}} - D^{F_{s_{k-1}a_ia_j}}_{s^{ij}_{n-k-1}}\right)$$

where $f = \max\left\{0,\ \bar{t}_j - \left[\max(\bar{t}_i,\ F_{s_{k-1}})\right]\right\}$. "f" has the physical meaning of the time window, in which the runway is not utilized, even if some aircraft are waiting to land or to take off.
Since $F_{s_{k-1}a_ja_i} = F_{s_{k-1}a_ia_j} + f$, for lemma (1) of the Appendix one has that

$\left(D^{F_{s_{k-1}a_ja_i}}_{s^{ji}_{n-k-1}} - D^{F_{s_{k-1}a_ia_j}}_{s^{ij}_{n-k-1}}\right)$ is greater than zero. So a sufficient condition for

$D_I < D_{II}$ is that

$$(d_i - d_j) < 2f \tag{8}$$

In such a case, all complete sequences which derive from $s_{k-1}a_j a_i$ may be discarded, since $s_{k-1}a_i a_j s^{ij}_{n-k-1}$ is certainly the best one among them. Let suppose now that $s^{ji}_{n-k-1} \equiv s^{ij}_{n-k-1}$, under the hypothesis that the starting time is $F_{s_{k-1}a_j a_i}$ and $F_{s_{k-1}a_i a_j}$ and that, starting from these times, there is no time interval in which the runway is not utilized. From the preceding definitions one concludes immediately that

$$D^{F_{s_{k-1}a_j a_i}}_{s^{ji}_{n-k-1}} = D^{F_{s_{k-1}a_i a_j}}_{s^{ij}_{n-k-1}} + (n-k-1)f$$

If $\quad s^{ji}_{n-k-1} \neq s^{ij}_{n-k-1}$, one gets

$$D^{F_{s_{k-1}a_j a_i}}_{s^{ji}_{n-k-1}} < D^{F_{s_{k-1}a_i a_j}}_{s^{ij}_{n-k-1}} + (n-k-1)f$$

so that certainly

$$D_{II} \leq D_{s_{k-1}} + \delta_{-1}\left[F_{s_{k-1}} - \bar{t}_j\right] * \left[F_{s_{k-1}} - \bar{t}_j\right] + \left[\max(F_{s_{k-1}}, \bar{t}_j) + d_j - \bar{t}_i\right] +$$

$$+ D^{F_{s_{k-1}a_i a_j}}_{s^{ij}_{n-k-1}} + (n-k-1)f.$$

It is possible to conclude that $D_I > D_{II}$ if

$$(d_i - d_j) > (n-k+1)f \tag{9}$$

So that if (a_i, a_j) satisfies the preceding condition, the sequence $s_{k-1}a_j a_i s^{ji}_{n-k-1}$ is better than any sequence which derives from $s_{k-1}a_i a_j$ and the corresponding node can be not further searched.
Conditions (8) and (9) hold independently of the value k-1 and, as a consequence, may be used from the start of the algorithm for decreasing the set of nodes and branches of the tree to be searched.

b_2) *Lower Bound*

THEOREM:

Let s_k be a partial sequence on a set A of n elements and s_{n-k} be any complement of s_k. It is possible to compute a lower bound $B_{s_k} \leq D_{s_k s_{n-k}}$, where:

$$B_{s_k} = D_{s_k} + \sum_{h=1}^{k+1} \sum_{i=1}^{\ell_1+\ell_2+\ldots+\ell_p} \delta_{-1}\left[F_{s_{h-1}} - \tau^{F_{s_{h-2}}}_{k+i}\right] * \left[F_{s_{h-1}} - \tau^{F_{s_{h-2}}}_{k+i}\right] +$$

$$+ \sum_{j=1}^{p}\left[(\ell_j - 1) \min_{a_{k+1} \in \{s_{\ell_j}\}} F_{s_k a_{k+1}} - \sum_{i=1}^{\ell_j} \tau^{F_{s_k}}_{k+i} + \min_{i: a_i \in \{s_{\ell_j}\}} \tau^{F_{s_k}}_{k+i} + \sum_{i=1}^{\ell_j - 2} (\ell_j - 1 - i) d_{k+i}\right]$$

Where d_{k+i} is non-decreasing, when i = {1,2,...n} increases .

PROOF. Any sequence s_n , with fixed s_k, can be written as follows $s_n=s_k s_{\ell_1} s_{\ell_2} \dots s_{\ell_p}$ with $(\ell_1+\ell_2+\dots+\ell_p)=(n-k)$ and s_{ℓ_j} defined by the set

$$\left\{a_i\right\}: \min_{i:a_i\in\{s_{\ell_{j-1}}\}} (\tau_i^{F_{s_k}} + d_i) \leq \tau_i^{F_{s_k}} < \min_{i:a_i\in A-\{s_k s_{\ell_1} \dots s_{\ell_{j-1}}\}} (\tau_i^{F_{s_k}} + d_i)$$

ordered according to increasing nominal times. For lemma (2) in Appendix it is possible to write that

$$D_{s_k s^*_{n-k}} = D_{s_k} + D^{F_{s_k}}_{s^*_{n-k}} \geq D_{s_k} + D^{F_{s_k}}_{s^*_{\ell_1}} + D^{F_{s_k}}_{s^*_{\ell_2}} +\dots+ D^{F_{s_k}}_{s^*_{\ell_p}}$$

Let consider a sequence $s_k s_{\ell_j}$ and the partial sequence $s_{k+1} = s_k a_{k+1}$ where $a_{k+1} \in \{s_{\ell_j}\}$.

Let s_{ℓ_j-1} be a complement of s_{k+1}, then, for the given definition of s_{ℓ_j}, it is possible to conclude immediately that the minimum delay, cumulated in the system after time F_{s_k}, is given by

$$D^{F_{s_{k+1}}}_{s^*_{\ell_j-1}} = \sum_{i=1}^{\ell_j-1} (F_{s_{k+1}} - \tau_{k+1+i}^{F_{s_k}}) + \sum_{i=1}^{\ell_j-2} (\ell_j-1-i)d_{k+1+i}$$

where d_{k+1+i} is non-decreasing, when i increases. Of course, if $a_{k+1} \in \{s_{\ell_j}\}$ varies, $D_{s^*_{\ell_j-1}}$ varies.

The cumulated delay of the system up to time F_{s_k} may be expressed, by taking into account definition 3, as

$$\sum_{h=1}^{k+1} \sum_{i=h}^{k+\ell_j} \delta_{-1}\left[F_{s_{h-1}} - \tau_i^{F_{s_{h-2}}}\right] * \left[F_{s_{h-1}} - \tau_i^{F_{s_{h-2}}}\right] =$$

$$= D_{s_k} + \sum_{h=1}^{k+1} \sum_{i=1}^{\ell_j} \delta_{-1}\left[F_{s_{h-1}} - \tau_{k+i}^{F_{s_{h-2}}}\right] * \left[F_{s_{h-1}} - \tau_{k+i}^{F_{s_{h-2}}}\right]$$

where the second term represents the delay cumulated up to F_{s_k} by the elements of $\{s_{\ell_j}\}$. As a consequence

$$D^{F_{s_k}}_{s^*_{\ell_j}} = \sum_{h=1}^{k+1} \sum_{i=1}^{\ell_j} \delta_{-1}\left[F_{s_{h-1}} - \tau_{k+i}^{F_{s_{h-2}}}\right] * \left[F_{s_{h-1}} - \tau_{k+i}^{F_{s_{h-2}}}\right] + \min_{a_{k+1}\in\{s_{\ell_j}\}} D^{F_{s_{k+1}}}_{s^*_{\ell_j-1}}$$

It is also

$$\min_{a_{k+1}\in\{s_{\ell_j}\}} D^{F_{s_{k+1}}}_{s^*_{\ell_j-1}} \geq (\ell_j-1)\min_{a_{k+1}\in\{s_{\ell_j}\}} F_{s_{k+1}} - \sum_{i=1}^{\ell_j} \tau^{F_{s_k}}_{k+i} +$$

$$+ \min_{i:a_i\in\{s_{\ell_j}\}} \tau^{F_{s_k}}_{k+i} + \sum_{i=1}^{\ell_j-2} (\ell_j-1-i)d_{k+i} \ .$$

From the preceding expression, one can obtain a lower bound for $D^{F_{s_k}}_{s^*_{\ell_j}}$. If any $D^{F_{s_k}}_{s^*_{\ell_j}}$ is substituted by its lower bound, one obtains

$$D_{s_k s^*_{n-k}} \geq D_{s_k} + \sum_{h=1}^{k+1} \sum^{\ell_1+\ell_2+\ldots\ell_p} \delta_{-1}\left[F_{s_{h-1}} - \tau^{F_{s_{h-2}}}_{k+1}\right] * \left[F_{s_{h-1}} - \tau^{F_{s_{h-2}}}_{k+i}\right] +$$

$$+ \sum_{j=1}^{p}\left[(\ell_j-1)\min_{a_{k+1}\in\{s_{\ell_j}\}} F_{s_k a_{k+1}} - \sum_{i=1}^{\ell_j} \tau^{F_{s_k}}_{k+i} + \right.$$

$$\left. + \min_{i:a_i\in\{s_{\ell_j}\}} \tau^{F_{s_k}}_{k+i} + \sum_{i=1}^{\ell_j-2} (\ell_j-1-i)d_{k+i}\right]$$

This completes the proof.

c) *Optimality test*

When the following conditions hold true: $A_{s_k} = A-\{s_k\}$ and $\tau_i^{F_{s_k}} = \tau_{i+1}^{F_{s_k}}$ $\forall i: a_i \in A_{s_k}$, the optimal completion of s_k can be found immediately by applying the shortest service time rule. Besides, as it can be easily checked, B_{s_k} coincides with the delay of the optimal solution deriving from s_k; this one is also the new best solution.

A synthetic flow chart of the proposed solution is given in fig. 2.

4. COMPUTATIONAL RESULTS AND CONCLUSIONS

The algorithm has been implemented and some preliminary tests have been performed . The computation time ranged from few tens of milliseconds, for sequences of 7 conflicting aircraft, to few seconds for sequences of 15 conflicting aircraft. Therefore, the proposed approach seems to be promising in order to develop an efficient, real-time working, sequencing tool. Besides, concrete algorithm improvements, such as taking into account more runways, giving different weights to the various aircraft and an upper bound to their delays, have already been considered and will be the subject of a future work.

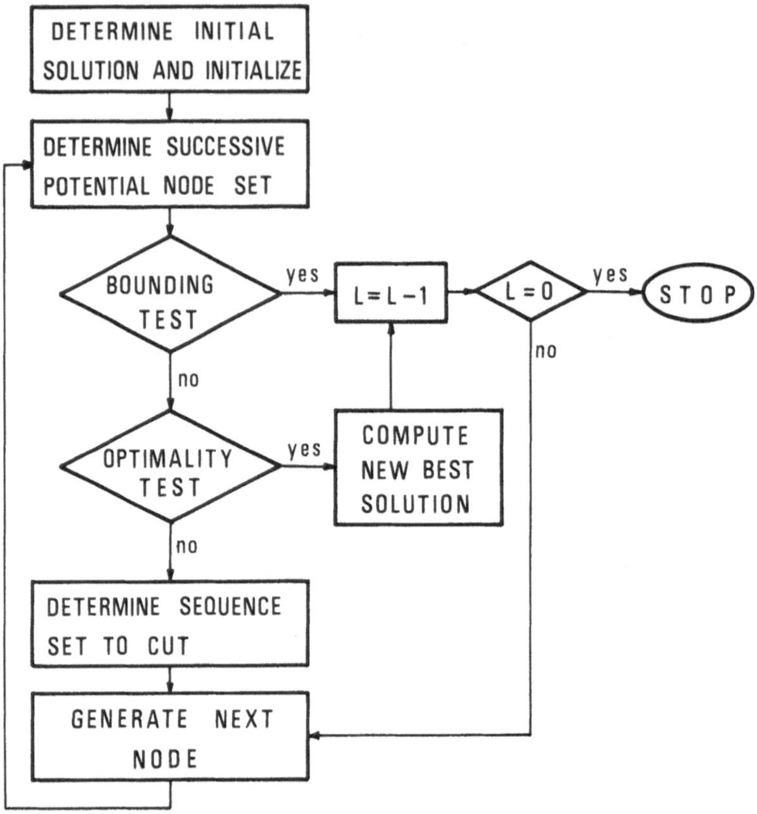

FIG. 2 - SINTHETYC FLOW-CHART OF THE ALGORITHM

APPENDIX

LEMMA 1. Given a sequence s_n, if s_n^* and $s_n'^*$ are the optimal sequences, with initial times t and t', respectively,
it is

$$D^{t}_{s_n^*} \leq D^{t'}_{s_n'^*} \qquad \text{for } t' > t$$

PROOF. Under the stated hypothesis on s_n^* and $s_n'^*$
it is

$$D^{t}_{s_n^*} \leq D^{t}_{s_n'^*}$$

Besides, it must also be (see [5]) $D^{t}_{s_n'^*} \leq D^{t'}_{s_n'^*}$

so that necessarily it is $D^{t}_{s_n^*} \leq D^{t'}_{s_n'^*}$.

This completes the proof.

LEMMA 2. Given a sequence $s_n = s_{k_1} s_{k_2} \dots s_{k_p}$, where $k_1+k_2+\dots k_p=n$, if $s_n^*=a_{[1]}a_{[2]}\dots a_{[i]}\dots a_{[n]}$ is the optimal sequence starting from time zero, then

$$D_{s_n^*} \geq D_{s_{k_1}^*} + D_{s_{k_2}^*} + \dots D_{s_{k_\ell}^*} + \dots D_{s_{k_p}^*}$$

PROOF. By definition, it is

$$D_{s_n^*} = \sum_{i=1}^{n} D_{a_{[i]}}^{F_{a_{[i-1]}}} = \sum_{\forall i: a_{[i]} \in \{s_{k_1}\}} D_{a_{[i]}}^{F_{a_{[i-1]}}} + \dots \sum_{\forall i: a_{[i]} \in \{s_{k_\ell}\}} D_{a_{[i]}}^{F_{a_{[i-1]}}} + \dots$$

$$\dots \sum_{\forall i: a_{[i]} \in \{s_{k_p}\}} D_{a_{[i]}}^{F_{a_{[i-1]}}}$$

Since $s_{k_\ell}^*$ is the optimal sequence of s_{k_ℓ}, starting from t equal to zero, one gets

$$\sum_{\forall i: a_{[i]} \in \{s_{k_\ell}\}} D_{a_{[i]}}^{F_{a_{[i-1]}}} \geq D_{s_{k_\ell}^*} , \quad \forall \ell=\{1,2,\dots p\}.$$ This completes the proof.

REFERENCES

[1] S.K.PARK, T.A.STRAETER, J.E.HOGGE, *An analytic study of near terminal area optimal sequencing and flow control techniques*. Agard CP 105.

[2] R.W.CONWAY, W.L.MAXWELL, L.W.MILLER, *Theory of Scheduling*. Addison-Wesley, Reading, Mass., USA, 1967.

[3] K.R.BAKER, *Procedures for sequencing tasks with one resource type*. Int.J.Prod. Res., Vol.11, n.2, 125-138.

[4] J.SHWIMER, *On the N-job, one-machine, sequence independent scheduling problem with tardiness penalties: a branch-bound solution*. Mgmt.Sci., vol.18, n.6, Feb. 1972.

[5] G.RINALDI, A.SASSANO, *On a job scheduling problem with different ready times: some properties and a new algorithm to determine the optimal solution*. Centro di Studio dei Sistemi di Controllo e Calcolo Autom. C.N.R. Roma. R 77-24. To appear.

A MATHEMATICAL MODEL OF TRAFFIC IN COMMUNICATION NETWORKS

B. CAMOIN

Centre National d'Etudes des Télécommunications

92131 - Issy + les - Moulineaux , FRANCE

1. THE OBJECTIVES OF THE STUDY

Communication networks take place among the more complex man-made systems. In order to control them, the various system analysis and optimization techniques must be used.

The size and structure of a telephone network are planned so as to satisfy the forecasted traffic demands with a fair grade of service at the lowest possible cost. However the actual working of the System is not always as satisfactory as it should be. It can be explained either by the fact that the real traffic demand may differ from what was expected or by the various disturbances which can affect the network such as :

- traffic overloads (increases of the level of offered traffic)
- casual failures of equipments, such as transmission links between the nodes of the networks or control units in the exchanges, (reduction of link capacities or node capacities).

Thus, a further stage in the study of the telecommunication networks is to take into account the concept of "reliability" ; this idea appears also in other fields such as energy networks. The final aim of this study, which has just been initiated, is to find how to build more reliable networks at a reasonnable cost, by selecting sturdiest structures and adequate real time control systems. Practically, to achieve this end, the first step is to be able to predict the efficiency or the "response" when submitted to any possible disturbance, of any network ; involving possibly a

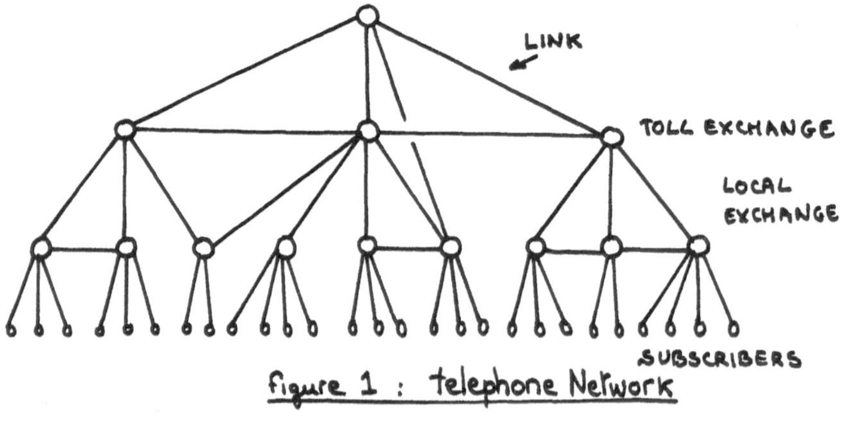

figure 1 : telephone Network

control command or a regulating system ; as the subscribers and the present public network cannot be considered as guinea pigs to test the effects or disturbances or tentative control commands, the only way to do it is to simulate the system by the means of models which nevertheless must be validated and identified. Once such models of a network are available, they can be included in an optimisation loop in order to compute a good control command for a given distrubance ; then having estimated the probability distributions of these perturbations considered as random events, one can estimate the probability distribution of the response of the network when properly controlled; reliability criteria can then be derived from this distribution, for example, if a network is built to provide a point to point call loss probability lower than n% between any two users, when no trouble occurs, a reliability parameter is the probability that this call loss rate exceed n% ; and the problem will be to find the best network and command structure with respect to these reliability criteria.

The purpose of this paper is to describe a basic model for these studies.

2. THE MAIN FEATURES OF THE "TELEPHONE NETWORK" SYSTEM

A telephone network can be seen a graph (figure 1) the terminal nodes are the subscribers, the non terminal nodes are the telephone exchanges. A "link" between two exchanges is composed of several "circuits" ; one circuit can carry one call at the same time (the number n of circuits built on link AB depends on the probability function of the number of calls likely to be simultaneouly in progress between A and B). A call originating from A is established step by step towards its destination. In each exchange a control unit receives the dial numbers and selects the next link to be used ; these control units are based on a relay logic or on programmed computers. The real time management of the trafic flows consists in modifying the "routing" of the calls and sharing the flows between several "routes" (paths) of the network.

Then, we can see that :

- Beside, this is a large scale problem, since the network studied (maybe a of a larger one) may have hundred of nodes.

- The system is highly interactive, the various parameters interacting all the stronger as the flows are more disturbed.

- Random processes are present in the system : on one hand he subscribers behaviour (rate of arrivals of call, delay before a new attempt of an unsuccessfull call, ...) is random ; on the other hand, the working of very intricated systems such as switching gears or control units can be regarded more simply as probalistic queuing systems.

- This is a multicriteria study ; two kinds of criteria are mainly used : the grade of service, which is the ratio of successfull calls to total call attempts, and the income resulting from paid traffic ; these parameters have to be carried out separately for each traffic flows, to ensure a minimum grade of service on each flow.

All these features of the system cannot be taken simultaneously into account within a single model ; several models have been developped, each one being more relevant to a specific use. These models are either simulations using pseudorandom generators (Monte-Carlo simulation) or mathematical models.

3. A MATHEMATICAL MODEL OF THE TRAFFIC IN NETWORKS

This model is used to simulate the dynamical behaviour of a network quite precisely, the random processes being described rather roughly by their expectations. The model consists in a state representation :

$$\frac{dY}{dt} = F\ (Y,E)$$

$$S = G\ (Y,E)$$

where E is the input vector ; it is composed of :

- the matrix of traffic demands on the different origine-destination pairs of the network.

- the parameters of the subscribers'behaviour (mean communication time, ...)

- the parameters of the network (link capacities, holding times of control units, ...)

S is an output vector, it includes :

- various criteria (grade of service, waiting time before the conversation begins, taxed traffic, ...)

- observation parameters (number of busy circuits, ...)

Y is the state vector.

The progress of a call in a network can be seen as a sequence of time-consuming phases (figure 2). We describe the process in an exchange in the following way :

- A queuing period before one of the control processors is available,

- A processing period during which the call is routed throughout the exchange,

- In each phase the call can be interrupted by the subscriber or blocked and given the busy tone.

We now assume that the calls from a given flow being at the same time in the same phase, cannot be distinguished. We can then build for each flow of the network a diagram of transitions and phases as shown on figure 2. The state variables are the numbers of calls of each flow in each phase at time t. In fact these numbers being random processes, we consider only their expectations. Considering the average number of links and of paths in the networks which can be used by the calls of any given flow, the average number of state variables is about 20 . Let N be the number of local exchanges of the studied network (origins and destinations of calls) then the dimension of the state vector Y is 20 N^2.

figure 3 : elementary state

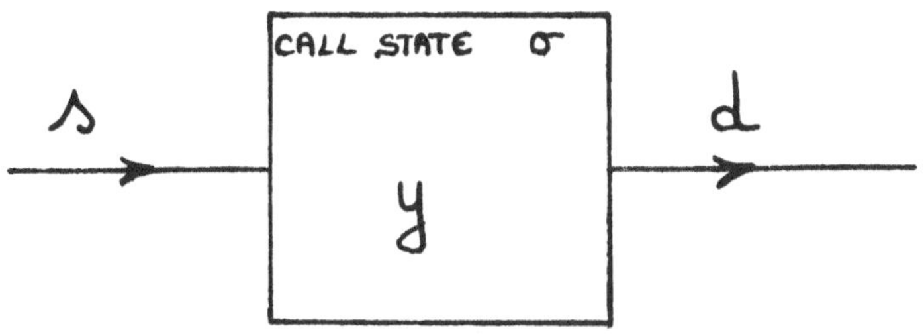

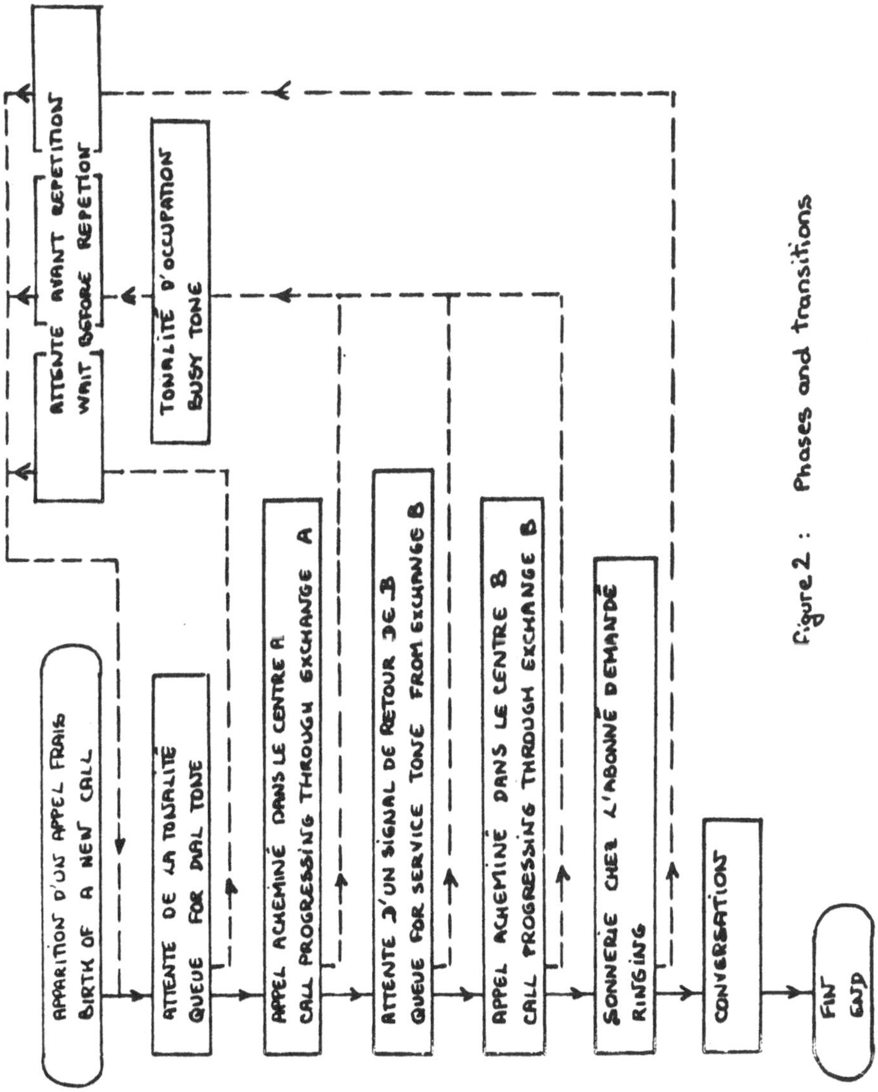

figure 2 : Phases and transitions

If we consider a state σ (figure 3) and its associated state variable y, we define s and d as the transitions rates of incoming and outgoing calls ; the basic relation is then the following :

$$\frac{dy}{dt} = s - d$$

We must now evaluate these transition rates. Some are very easy to compute :

- Birth process of calls in the network : s is the parameter of a poisson process and is given in vector E.

- if σ represents an exponential service time, the mean holding time being T, we have :

$$d = \frac{Y}{T}$$

- if σ represents an independant random service delay which is not exponential ; r (t) being the density probability, we can approximate it with a multiexponential density

$$r(t) = Be^{At}C$$

and then use a multidimensionnal model for the service phase σ :

$$\frac{d\eta}{dt} = A\eta + Cs$$

$$d = B\eta$$

were η is a K-dimension state vector ; A, B, C are respectively KxK , Kx1 , 1xK matrices.

Other processes are more complex and non linear, for instance :

- Random choice in a queue : let us consider a queue associated to a group of servers, let d(t) be the rate of transitions of calls ending their service period. When a server becomes available it chooses immediatly a new call in the queue, if the queue is composed of n kinds of calls, the number of waiting calls of kind i being $y_i(t)$, the rate of kind i calls being taken by servers is :

$$d_i(t) = \frac{y_i(t)}{y_1(t) + \ldots\ldots + y_n(t)} d(t)$$

- Call congestion of n circuits of a link : if z(t) is a Poisson birth process, and if any call which arrives when all the circuits are busy, is lost, then the transition rate s(t) of calls being taken by circuits is given by :

$$s(t) = (1-P(X)) \cdot z(t)$$

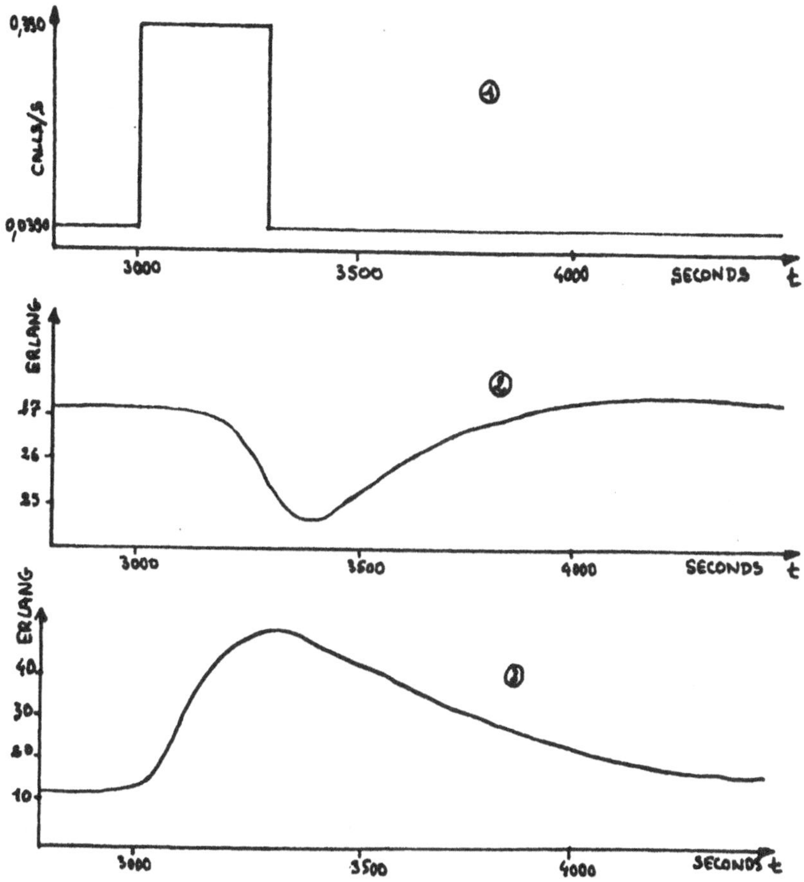

Figure 4: Example of time evolution of a Network.

① INPUT : traffic peak on a flow φ (offered traffic)

② Output : Decrease of the carried traffic of a flow φ'.

③ Output : Increase of the carried traffic of the flow φ.

where $P(x)$ is the blocking probability of n circuits,

and X is the expectation of carried traffic ; $P(X)$ is computed, X being known, by solving the system :

$$X = A\ (1 - P)$$

$$P = \frac{A^n/n!}{1 + \ldots + A^n/n!} \qquad \text{(Erlang's forumla)}$$

Eventually, for each flow i, we have exhibited a subsystem of equations :

$$\frac{dy}{dt} = fi\ (Y_i\ ,\ P_{j_1}\ ,\ \ldots\ ,\ P_{j_k}\ ) \qquad i = 1, \ldots, \text{number of flows)}$$

were : - Y_i is the state vector of flow i

- Pj are the variables related to the blocking probabilities of the servers used by the flow i.

This variable P_j has been shown as depending on the total traffic carried by the servers j. This total traffic can be computed as the sum of different variables yi, which are the numbers of calls of phases i for which a server of group j can be used :

$$P_j = G_j(Y_{i_1}, \ldots, Y_{i_p}) \qquad (j = 1, \ldots, \text{number of groups of servers)}$$

The model is then decomposed into interactive subsystems :

- Flows interact because they share the same servers

- Servers interact because they are used by the same flows

If there is no gongestion, the blocking probabilities are very law and the flows can be considered as independant.

A discretization method provides the solution of the differential equations ; we have noticed that the numerical behaviour is generally surprisingly good for a system of that size ; this can be explained by the decomposition which underlies it. For a 13 nodes test network involving 1602 state variables, the simulation of one hour of operation in given within 20 minutes CPU and uses 30K memory words of an Honeywell-Bull 6080 computer. Examples of evolution of different parameters are given on figure 4.

4. VALIDATION OF THE MODEL

Many simplifications and many chancy approximations have been done in the description of the physical processes in order to reduce the complexity of the model and for specific uses (optimization) simpler models are even needed. We have to check their validity. On the other hand many input parameters are included which are unknow and have to be measured and even identified for example :

- Conversation times, holding times of servers.

- Traffic demands : the problem is that when we observe a birth process of calls in a local exchange, we cannot distinguish the first attempts (the demand) and the calls generated by the subscribers renewing previous unsuccessful attempts.

The observation of the whole network is not possible, so that we can only identify separately subsets of the system, assuming that the validity of the description of the subsets provides the validity of the complete model : this may be disputed, but no other way can be imagined. It is not even easy to decide wether a complete model is valid or not ; in fact, as in many other fields, the validity depends on what use will be made of the model, and intuition and practical experience are needed to confirm or deny its predictions. We have developped specific tools to collect experimental data, mainly at the level of a telephone exchange : among them, ARMORE (Adapted for Research and MOdelling Recording Equipement) is a minicomputer driven device which can convert electrical informations collected from an exchange through 1000 probes, into logical informations about traffic processes and store them on a magnetic tape, the data on the magnetic tapes are subsequently processed on an universal computer, using statistical analysis, or identification techniques (although proper methods in the field of jump processes are still rather mysterious).

We have also compared the results of this mathematical model with a monte-carlo simulation model which has been also developped at CNET, in order to verify whether considering only mean values was not too harsh an approximation. The fitting was good. The comparison was made only on stationnary states with a test network for a set of various values of the input parameters ; the mean stationnary state of the Monte-Carlo model is obtained by simulating one or two hours of the working of the network, after a stabilisation period starting from an initial state

5. APPLICATIONS

This model is used mainly as a simulation tool, and to test different control policies ; it is, nevertheless, one of the first attempts to quantify the interactivity of the various elements at a network level. A sensitivy analysis, asing a differentiated model, can be made, it consists in the calculation of criteria variations caused by a set of input-variations around any working rate. It allows to locate bottelnecks, to point out vulnerable flows and the best control parameters and also to neglect the model parameters which have the lesser weights so as to simplify the description. Using this model we have been able to find characteristics of traffic at a large scale network level where the usual traffic theory and queuing theory are no more relevant.

- Congestions can spread along the networks, the major cause being the increase of waiting times in the exchanges inducing longer holding time of control units (C.U.) of other exchanges (a C.U. usually waits an answer from a C.U. of the next exchange reached by the call) ; renewal of unsuccessful attempts within a short delay by the subscribers makes things even worse.

- The Interactivity between exchanges A and B is the smaller as the traffic carried between A and B is a less important part of the total traffic flowing through A or B : this is a "diffusion effect".

- When a traffic overload occurs, traffic flows of small size are generally more vulnerable (than larger)flows ; this can be explained by the fact that small flows , usually between distant exchanges, are often routed on several consecutive links through tandem exchanges, shared with other traffic flows ;on the contrary a direct link may be built between two local exchanges when the offered traffic is sufficient by high.

- Generally, when a transmission equipment (a cable, ...) fails, several links which are physically carried by the same cable, are simultaneously interrupted ; the effect of such correlated disturbances may be very severe.

Various methods to prevent the consequences of these disturbances are now being studied. Yet, many notions have been pointed out, among them :

- Sharing a traffic flow between several equivalent paths of the network instead of having one "normal" route occasionnaly overflowing (when saturated) on a "2nd choice" route.

- Notion of closed loop control of the traffic : when several routes are available from an exchange A to a destination B, each call destinated to B may be given the greatest odds by choosing the best route at this moment according to observation results (occupancy of distant links, grade of service on each route)

- Decomposition into several decentralized centers and hierarchical levels of the real time control structure.

However a lot of work has still to be done before the public telephone network is adpated according to these ideas.

TABLE OF CONTENTS

Bibliography and contributions to conferences

[1] B. CAMOIN : "Modèle analytique de trafic dans un réseau téléphonique", Annales des télécommunications, T-31, n° 7-8, juillet-août 1976, Pages 239-267.

[2] J. AGASSE, B. CAMOIN, J.M. CUVIER, M. PEYRADE : "La sécurité du réseau Téléphonique national", Echo des recherches n° 88, juillet 1976, pages 1-15

[3] A. TITLI : "Commande hiérarchisée et optimalisation des processus complexes", DUNOD éditeur, "série automatique", Paris (1975)

[4] P. LE GALL : "Les systèmes avec ou sans attente et les processus stochastiques" Dunod éditeur, Paris (1962), 482 pages.

[5] KLEINROCK : Queuing systems" : Volume 1 : theory" Wiley, collection "Wiley interscience", New-York (1975), 417 pages.

[6] B. HENNION, J.C. HENNET : "La commande en temps réel d'un réseau de telecommunications" presented at the AFCET congress on "Modelisation et la maitrise des Systèmes", November 1977, Versailles, FRANCE.

[7] L. CHAROY, P. MALVAUX, A. SPIZZICHINO : "Measurements for modelling of the telephone Network : the ARMORE project" presented at the URSI "international symposium on measurements in telecommunications" october 1977, Lannion, FRANCE.

[8] M. PEYRADE, A. SCHLATTER, J.F. CUVIER, M. LETERRIER : "Le superviseur du réseau téléphonique interurbain"
presented at the F.N.I./S.E.E. "international conference on automation sound and video broadcasting and transmission networks".
october 1976, Paris, FRANCE

[9] A. SCHLATTER, M. PEYRADE, A. SPIZZICHINO : "Study of the telephone traffic, using the national network supervision "presented at the URSI" international symposium on measurements in telecommunications", october 1977, Lannion, FRANCE.

PARTITIONING A COMPUTER NETWORK INTO SUBNETWORKS AND ALLOCATION OF DISTRIBUTED DATA BASES

Janusz Kacprzyk
Systems Research Institute
Polish Academy of Sciences
ul.Newelska 6
01-447 Warszawa, POLAND

Wiesław Stańczak
Institute of Computer Science
Polish Academy of Sciences
PKiN P.O.Box 22
00-901 Warszawa, POLAND

SUMMARY

A problem in the design of computer networks is considered. It consists in the decomposition of the network into subnetworks and the allocation, i.e. the determination of geographical positions, of local data bases to particular subnetworks. The partitioning into subnetworks is performed with respect to similarities of information used between particular computers. For solving it, the method of minimally interconnected subnetworks [1,2,4] is applied. In the second problem, the data bases in subnetworks are to be placed, so as to attain minimal data exchange costs. The partitioning of areas accupied by subnetworks into a number of squares transforms the problem to a discrete one and makes it possible to formulate and prove some properties of the cost function, which simplify to a great extent its minimization.

1. INTRODUCTION

A wide use of computers in various fields makes the design of computer networks be more and more relevant. The access to various data stored in data bases is there of a great importance. Not all the computers, however, need the same data and one central data base may be, therefore, unreasonable [3]. Smaller data bases servicing groups of computers -subnetworks- may be a better solution, in particular for computers using mostly some specific information, e.g. regional, branch etc. Evidently, it does not mean that subnetworks cannot get data from data bases other than their own. This cases are, however, assumed to be relatively rare and are excluded from considerations.

For solving the problem of partitioning a given computer network into subnetworks, the method of minimally interconnected subnetworks, as given in [1,2,4], is applied. The network is here represented by a weighted graph, in which the vertices represent the computers and the edges - the strength of similarity of information needed between the computers at the end vertices. The problem is to partition the network into disjoint subnetworks, such that the sum of strengths of similarity between computers in particular subnetworks is greater than that between them and those not in that subnetwork. This problem is solved by the determination of minimal groups for the given network.

Then, the data bases are to be allocated in particular subnetworks. It is meant here as the determination of their geographical positions, such that the data exchange cost is minimal. It is assumed that the area occupied by subnetworks is highly urbanized. For solving the problem, which is originally a difficult combinatorial one, the area is transformed into a discrete one. The properties of the cost function, which are stated and proved, facilitate now to a considerable extent the solution of the problem.

2. PARTITIONING A COMPUTER NETWORK INTO SUBNETWORKS

Let us denote the set of computers by $V=\{1,\ldots,N\}$. The computer system is represented by a complete, nonoriented and loopless graph $G(V,E)$, where $E=\{e_1,\ldots$ $\ldots,e_{1/2N(N-1)}\}$ is the set of edges.

Now, let a function be given

$$w : E \to R^{+} \cup \{0\} \tag{1}$$

whose values $w(e)$, denoted by w_{ij}, represent the strength of similarity of information needed between the i-th and j-th computer connected by $e=(i,j)$.

Let us denote

$$f(R,S) = \sum_{i \in R,\ j \in S} w_{ij} \tag{2}$$

for $R,S \subset V$, $R \cap S = \emptyset$. If S is the complement of R to V, then $f(R,S)$ will be denoted by r. Let us remark that (2) indicates the sum of strengths of information similarities between computers in R and S, while r - between computers in R and the rest of the network.

The problem of partitioning a given network into subnetworks consists in finding a family of subsets $\{V_1,\ldots,V_k\}$ of V, such that

$$V_i \cap V_j = \emptyset \qquad \text{for all } i,j \in \{1,\ldots,k\}, \quad i \neq j \tag{3}$$

$$\bigcup_{i=1}^{k} V_i = V \tag{4}$$

$$f(W, V_i \setminus W) > f(W, V \setminus V_i) \qquad \text{for all } i=1,\ldots,k \text{ and all } W \subset V_i,\ \emptyset \neq W \neq V_i \tag{5}$$

Thus, (5) has a simple interpretation: the computers in a subnetwork should be interconnected stronger than with the ones not in that subnetwork. Only the relation ">" must hold, the value is not given in advance.

As it will be seen, this problem may be solved by the method of minimally interconnected subnetworks consisting in the determination of minimal groups. The method was introduced in [4] and then generalized and extended by the authors in [1, 2]. Let us now present some basic notions, which are here of use.

Any subset $S \subset V$, taken with all its edges, is said to be the <u>group</u> S.

A nonempty group S, such that for every $R \subset S$, $\emptyset \neq R \neq S$

$$r > s \tag{6}$$

is said to be the <u>minimal group</u>. Moreover, a single vertex is the minimal group, by definition.

<u>Lemma 2.1</u>. The necessary and sufficient condition for a group S to be minimal is that for every $R \subset S$, $\emptyset \neq R \neq S$

$$f(R, S\setminus R) > f(S\setminus R, V\setminus S) \tag{7}$$

<u>Lemma 2.2</u>. Two minimal groups are either disjoint, or one of them is contained in the other.

Thus, the determination of minimal groups may lead to the solution of the problem considered.

Further properties of minimal groups and algorithms for determining them may be found in [1,2]. The algorithm given in [2] is particularly efficient.

<u>Example</u>. For the network of 15 computers 1,2,...,15, with the w_{ij}'s given by Tab.1 the minimal groups (subnetworks) determined are: {1,5,6,15}, {2,10,11,12,14} and {3,4,7,8,9,13}.

	1	2	3	4	5	6	7	8	9	10	11	12	13	14	15
1	10	0.1	0.2	0.9	2.8	1.6	0.2	0.1	0	0	0	0.1	0.1	1.2	0.8
2	0.1	10	0.1	0	0.1	2.8	0	0.1	0.1	7.5	2.2	0.1	0	5.7	0
3	0.1	0	10	0.8	0	0.1	1.3	4.3	5.3	0.2	0.1	0	0.7	0.1	0.2
4	1.1	0	1.2	10	0.1	0	0	1.4	1.7	0.4	0.2	0.3	0	0	0
5	3.2	0.2	0.1	0.1	10	5.9	0	0	0.3	0.6	0	0	0	0.2	3.1
6	5.4	0.2	0.2	0.1	3.1	10	1	0	0	0	0	0	0	0.2	2.4
7	0.2	0.1	1.7	0	0	1	10	2.7	0.2	0.1	0	0	0	0	0
8	0	0.1	4.7	1.6	0	0	1.3	10	8.7	0	0.3	0	0	0.1	0
9	0	0.2	3.7	1.3	0.2	0	0.1	0.3	10	0	0	0.1	1.8	0	0
10	0	0.5	0.2	0.2	2.4	0	0	0	0	10	6.6	4.1	0.1	0	0.1
11	0.1	4.8	0.1	0.1	0	0.1	0.1	0.1	0	1.4	10	8.2	1	7.6	1.1
12	0.1	0.3	0	0.2	0	0.1	0.1	0	0.1	5.9	1.8	10	0.5	3.2	0
13	0.2	0	2.3	0	0	0	0	0	1.2	2	1	1.5	10	0	0
14	0.8	0.3	0.1	0	0.1	0	0	0.5	0	0	1.4	1.8	0	10	0
15	4.2	0.1	0.2	0	4.9	1.6	0	0	0	0	0.9	0	0	0	10

Tab.1

3. ALLOCATION OF DISTRIBUTED DATA BASES IN SUBNETWORKS

After the partitioning of a given computer network into subnetworks, the data bases for each subnetwork are now to be allocated.

First, let us assume that: (1) each computer in a subnetwork is connected with its proper local data base and the cost of length unit is the same and equal to c; (2) the connection structure is of star-type with the data base in the middle, which gives: a simple algorithm for controlling data flows, an easy access to data, a short connection time, a low fault rate etc.; (3) the subnetworks are in an urbanized area.

Let us denote by: $W = \{1,\dots,n\}$ - the set of computers in the subnetwork, u_d - the position of data base, u_i - the position of i-th computer, U - the set of possible u_d's, $d(\cdot,\cdot)$ - the distance between two points specified, z_i - the number of necessary junctions between the data base and the i-th computer, K_d - a fixed cost of locating the data base in u_d, K - the cost function.

Now, the problem of allocating the data base is to find $u_d^* \in U_d$, such that

$$K(u_d^*) = \min_{u_d \in U} \left[c \sum_{r=1}^{n} d(u_i, u_d) z_i + K_d \right] \tag{8}$$

subject to:

$$z_i \geqslant 0 \qquad \text{for all } \; i \in W \tag{9}$$

$$\sum_{i=1}^{n} z_i = Z = \text{const} \tag{10}$$

The constraint (10) means that the number of links resulting from the existing amount of data and transmission is fixed.

Since the number of possible u_d's, i.e. dim U, is usually very high, the problem is computationally difficult. In order to simplify it, we proceed as follows. The map of the subnetwork's area is assumed to be a rectangle. It is partitioned by the lines parallel to its sides into m vertical and q horizontal stripes of equal dimensions, i.e. into mq squares. Let us now denote: $Q = \{1,\dots,q\}$, $M = \{1,\dots,m\}$, $U = \{1,\dots,mq\}$, u_{rs} - the position of i-th computer in the square at the crossing of r and s, u_{xy} - as u_{rs}, but for the data base, $a_{rs} = z_{rs}/Z$. Moreover, K_d = const.

Now, assuming that particular computers and the data base are placed in the middles of squares, the distance between u_{xy} and u_{rs} in an urbanized area may be approximated by [5]

$$d(u_{xy}, u_{rs}) = |r-x| + |s-y| \tag{11}$$

which means, roughly speeking, that links cannot go arbitrarily, but e.g. along the streets, roads etc.

Hence, the problem is to find (x^*, y^*), such that

$$k(x^*, y^*) = \min_{x \in Q, y \in M} \left[\sum_{r=1}^{q} \sum_{s=1}^{m} (|r-x| + |s-y|) a_{rs} \tag{12}$$

subject to:

$$a_{rs} \geqslant 0 \tag{13}$$

$$\sum_{r=1}^{q} \sum_{s=1}^{m} a_{rs} = 1 \tag{14}$$

Let us now remark that

$$\min_{x,y} k(x,y) = \min_{x} k_1(x) + \min_{y} k_2(y) \tag{15}$$

where:
$$k_1(x) = \sum_{r=1}^{q} \sum_{s=1}^{m} |r-x|\, a_{rs} \tag{16}$$

$$k_2(y) = \sum_{r=1}^{q} \sum_{s=1}^{m} |s-y|\, a_{rs} \tag{17}$$

Now we proceed to the presentation of some specific properties of $k(x,y)$, which simplify to a great extent the minimization problem (12) ÷ (14). Due to (15) they can be formulated for k_1 or k_2 - they hold analogously for the other one.

First, let us denote

$$k_M = \sum_{r=1}^{q} \sum_{s=1}^{m} s a_{rs} \tag{18}$$

$$k_Q = \sum_{r=1}^{q} \sum_{s=1}^{m} r a_{rs} \tag{19}$$

$$a_r = \sum_{s=1}^{m} a_{rs} \tag{20}$$

$$a_s = \sum_{r=1}^{q} a_{rs} \tag{21}$$

Evidently: $1 \leqslant k_M \leqslant m$, $1 \leqslant k_Q \leqslant q$.

<u>Lemma 3.1</u>. The function $k(x,y)$ on $Q \times M$ is - for a fixed $x \in M$ - decreasing, increasing and constant for all y's, such that

$$k_M - 1/2 > y \tag{22}$$

$$k_M - 1/2 < y \tag{23}$$

$$k_M - 1/2 = y \tag{24}$$

respectively.

<u>Proof</u>. Let $y_1, y_2 \in M$ and denote

$$\Delta k = k(x,y_1) - k(x,y_2) = \sum_{r=1}^{q} \sum_{s=1}^{m} (|s-y_1| - |s-y_2|) a_{rs} =$$

$$= \sum_{r=1}^{q} \sum_{s=1}^{m} [2s-(y_1+y_2)] a_{rs}$$

Substituting $y_2 = y_1 + 1$, we obtain

$$k = \sum_{r=1}^{q} \sum_{s=1}^{m} [2s - (2y_1 + 1)] a_{rs} = 2k_M - 2y_1 - 1$$

Hence, for k to be -respectively- decreasing, increasing and constant: $\Delta k > 0$ - - which implies (22), $\Delta k < 0$ - which implies (23) and $\Delta k = 0$ - which implies(24). Q.E.D.

The analogous property holds for a fixed $y \in Q(k_M \rightarrow k_Q)$. Evidently, k is convex and there is a solution in Q * M.

Let us denote: $y^* = \text{entier}(k_M - 1/2)$ $x^* = \text{entier}(k_Q - 1/2)$

Proposition 3.1. The minimal value of k subject to (13) and (14) is given by

$$k_{min} = \min_{x \in X^*, y \in Y^*} k(x,y) \tag{25}$$

where: $X^* = \{x^*, x^* + 1\}$, $Y^* = \{y^*, y^* + 1\}$.

Proof. Due to Lemma 3.1., k attains its minimum, for a fixed x, when y is near $k_M - 1/2$. Since y is an integer, then

$$k_M - 1/2 = y^* \tag{26}$$

implies $y = y^*$. If, otherwise, (26) does not hold, then $y = y^* + 1$. Analogously for x. Q.E.D.

Let us now remark that

$$k_1(x) = \sum_{r=1}^{q} \sum_{s=1}^{m} |r-x| a_{rs} = x + \sum_{r=1}^{q} \sum_{s=1}^{m} r a_{rs} +$$

$$- 2x \sum_{r=x+1}^{q} a_r - 2 \sum_{r=1}^{x} r a_r = a(x) + k_Q \tag{27}$$

In the sequel, the following difference will be often used (for $x_1 \geqslant x_2$)

$$a(x_1) - a(x_2) = (x_1 - x_2) - 2(x_1 - x_2) \sum_{r=x_1+1}^{q} a_r +$$

$$+ 2x_2 \sum_{r=x_2+1}^{x_1} a_r - 2 \sum_{r=x_2+1}^{x_1} r a_r \tag{28}$$

Theorem 3.1. If for $x,y > 1$

$$\sum_{r=1}^{x-1} a_r < 1/2 < \sum_{r=1}^{x} a_r \tag{29}$$

$$\sum_{s=1}^{y-1} a_s < 1/2 < \sum_{s=1}^{y} a_s \tag{30}$$

then (x,y) is the only optimal solution to the problem (12) ÷ (14). It is also true for a special case, when $1/2 < a_{r=1}$ and/or $1/2 < a_{s=1}$.

Proof. First, let $x,y > 1$. From (28) it follows that

$$a(x + 1) - a(x) = 1 + 2 \sum_{r=1}^{x} a_r \tag{31}$$

Moreover

$$a(x) - a(x - 1) = -1 + 2 \sum_{r=1}^{x-1} a_r \tag{32}$$

For (29) to hold, (31) and (32) imply that

$$a(x) - a(x - 1) < 0 < a(x + 1) - a(x) \tag{33}$$

Hence, a is decreasing for every $x_1 < x$ and increasing for every $x_2 > x$, i.e. $x \in X^*$. The reasoning for y is analogous. The point (x,y) is, therefore, optimal and, due to (33), it is unique.

Now, for x=1 we have

$$a(2) - a(1) = -1 + 2a_{r=1} \tag{34}$$

Thus, a is increasing in Q. Analogously for y. Q.E.D.

Proposition 3.2. Let there exist a $x \in Q \setminus \{1\}$ and a $t' \in T = \{0,1,\ldots,q - x - 1\}$, such that for every $t \leqslant t'$

$$\sum_{r=1}^{x-1} a_r < 1/2 = \sum_{r=1}^{x+t'} a_r \tag{35}$$

Then, there exists a $t_m \in T$, which is the greatest t', such that the problem $(12) \div (14)$ is solved by any $p \in \{x+t : t \in T, t \leqslant t_m\}$.

Proof. Due to Theorem 3.1., the left side of (35) is equivalent to $a(x)-a(x-1)<0$. Hence, the solution may only be $p \geqslant x$. The equality

$$1/2 = \sum_{r=1}^{x} a_r \tag{36}$$

is equivalent to $a(x+1) - a(x) = 0$.

Due to (35), $a(x+t) - a(x) = 0$, because $a_r=0$ for all $x+1 \leqslant r \leqslant x+t$.

It is now to notice that, if (35) would hold for $t = q-x$, then it would contradict (14). Thus a maximal t,t^*, must exist, satisfying (35) and $t^*<q-x$. Then, $a_{r=t^*+1} \neq 0$ and $1/2 < \sum_{r=1}^{r+t\ +1} a_r$, which - due to Theorem 3.1. - is equivalent to $a(x+t^*+2) - a(x+t^*+1) > 0$, if $x+t^*+2 \in Q$. Hence $t_m \leqslant t^*$. Moreover, since $a(x+t^*+1) - a(x) = 0$, then $t_m = t^*$. Q.E.D.

Corollary 3.1. Let there exist a $x \in Q \setminus \{1,2\}$, such that

$$\sum_{r=1}^{x-t-1} a_r = 1/2 < \sum_{r=1}^{x} a_r \tag{37}$$

Then, there exists a $t_m \in T$, $t_m < x-2$, which is the greatest possible t, such that the problem $(12) \div (14)$ is solved by any $p \in \{x-t : t \in T, t \leqslant t_m\}$.

The analogous properties hold also for y.

From Theorem 3.1., Proposition 3.2. and Corollary 3.1. there follows the following fundamental theorem.

Theorem 3.2. The solution of the problem $(12) \div (14)$ is any element of the set $X \times Y$, where

$$X = \{x : x \in Q, x_{min} \leqslant x \leqslant x_{max}\}$$

$$Y = \{y : y \in M, y_{min} \leqslant y \leqslant y_{max}\}$$

and x_{min}, x_{max}, y_{min} and y_{max} are determined form

$$\sum_{r=1}^{x_{min}-1} a_r < 1/2 \leqslant \sum_{x=1}^{x_{min}} a_r \tag{38}$$

$$\sum_{r=1}^{x_{max}-1} a_r \leqslant 1/2 < \sum_{r=1}^{x_{max}} a_r \tag{39}$$

$$\sum_{s=1}^{y_{min}-1} a_s < 1/2 \leqslant \sum_{s=1}^{y_{min}} a_s \qquad (40)$$

$$\sum_{s=1}^{y_{max}-1} a_s \leqslant 1/2 < \sum_{s=1}^{y_{max}} a_s \qquad (41)$$

respectively.

This basic result obtained makes it possible to devise a simple minimization algorithm. One computes particular a_r's and a_s's, sums them up and tests (38) ÷ ÷ (41). The appropriate indices give X × Y.

Example. Let the area of subnetwork be partitioned into 10 × 10 stripes as in Tab.2

r=	s=1	2	3	4	5	6	7	8	9	10
1	0.0125	0.015		0.01			0.005	0.0325		0.0275
2			0.0275		0.0225		0.0175			
3	0.02			0.0075	0.015		0.0025		0.0125	0.005
4			0.05			0.015		0.0225	0.0425	
5	0.0375	0.025		0.0225	0.0325				0.04	
6		0.0075					0.0175	0.015		
7	0.0225		0.0325	0.0125		0.01	0.0225		0.0025	0.015
8				0.02		0.0075			0.0125	
9	0.03	0.035	0.0025		0.015	0.0325		0.045		
10		0.0175		0.01		0.025		0.025	0.0475	

Tab.2

To save the space, the vertical and horizontal stripes are not equal. The numbers in squares indicate the computers and denote the values of a_{rs}'s.

Due to (20), we get the following subsequent a_r's for r = 1,...,10 from left to right :

0.1025 0.0675 0.0625 0.13 0.1575 0.04 0.1175 0.04 0.16 0.1225

Summing them up, i.e. $\sum_i a_{r=i}$, we get for subsequent i's

0.1025 0.17 0.2325 0.3625 0.52 0.56 0.7775 0.7815 0.8775 1

Hence, due to (38) and (39), $x_{min} = x_{max} = 5$. Analogously, due to (21) we get the following subsequent a_s's

0.1225 0.1 0.1125 0.0825 0.085 0.0875 0.065 0.14 0.1575 0.0475

Summing them up, i.e. $\sum_i a_{s=i}$, we get for subsequent i's

0.1225 0.2225 0.335 0.4175 0.5025 0.59 0.655 0.795 0.9525 1

Hence, due to (40) and (41), $y_{min} = y_{min} = 5$. Thus, the data base should be placed in the square r=5, s=5.

It should be noted that some results presented are similar to those in [6].

However, they are obtained here in a different, discrete way, which is particularly adequate to computer networks, where many a_{rs}'s are equal to zero.

4. CONCLUDING REMARKS

The design of a computer network is a very difficult problem. Not all of the aspects and factors may be analytically and quantitatively expressed. The paper presents an approximating approach, based on many simplifying assumptions. Its main feature, in particular in the allocation step, is the computational simplicity. Thus, in a short time one can get a good starting point to further, more precise investigations. Viewing from this perspective, the method proposed seems to be of use in practice.

REFERENCES

1. Kacprzyk J., Stańczak W.: On an Extension of the Method of Minimally Interconnected Subnetworks. Control and Cybernetics, No 4, 1976
2. Kacprzyk J., Stańczak W.: On a Further Extension of the Method of Minimally Interconnected Subnetworks. Control and Cybernetics /in press/.
3. Kulikowski J.L.: Program for the Development of Informatics and the Requirements for the Development of a National Data Transmision Network /in Polish/. Proc. Conference "Teleinformation Networks", Gdańsk /Poland/, 1975.
4. Luccio F., Sami M.: On the Decomposition of Network in Minimally Interconnected Subnetworks. IEEE Trans. on Circuit Theory, Vol.CT-16, No 2, 1969.
5. Nowicki T., Stańczak W.: Some aspects of Optimizing the Configuration of Zone Telephone Network /in Polish/. Rozprawy Elektrotechniczne, Vol.23, No 1, 1977.
6. Rapp Y.: Planning of Exchange Networks. Ericsson Technics, No 2, 1962.

EFFECTIVE FILE ALLOCATION METHOD ONTO DISC DEVICES

HIDEFUMI KONDO , IKUZO YOSHIDA
Systems Development Laboratory, Hitachi Ltd.,
5030 TOTSUKA-MACHI, TOTSUKA-KU, YOKOHAMA, 244 JAPAN
TAKASHI KATO
Software Works, Hitachi Ltd.,
15, 23-BAN, 6-CHOME, MINAMIOI, SHINAGAWA-KU, TOKYO,
140 JAPAN

ABSTRACT

A new file allocation method which effectively reduces file acces time is considered. Reductions in the total seek-time of magnetic disc heads are realized by predetermining the storage area of files on disc devices. Multiple files are effectively allocated on multiple disc packs by the approach outlined below. When file characteristics such as access frequency, access transition between files and distribution imbalance of access cylinder addressing are incorporated in a system, effective file allocation for reducing the total seek-time can be determined using a new hcuristic algolithm based on these characteritics. This method can be used effectively in the case of serial access file form on disc packs. In addition, this new method permits the calculation of the total seek-time for a specified file allocation and enables the effective allocation of files through the result of these calculations.

1. INTRODUCTION

The processing ability of a computer system, for example, feasible quality of data processing per hour, depends on the performance of the hardware units themselves. However, it also depends on the manner in which each unit is used to exhibit its full performance capabilities.

In computer systems, disc devices are usually composed of multiple disc packs. To date the optimal file allocation method for deciding the positions of multiple files on a single disc pack [1] , and the control method relative to the file access order a single disc pack [2] [3] have been studied. However, a file allocation method for multiple files on multiple disc packs has not been reported.

As a result, an effective method for allocating multiple files on multiple disc packs has been developed. A file application system can generally be characterized by some combination of the following : 1) Access frequency of each file, 2) Access transition between files, 3) Distribution imbalance of access cylinder addressing position in each file, 4) Movement, measured in cylinders, and relative movement frequencies of magnetic heads in each file, and 5) Volume of each file. The total movements of magnetic heads, namely total seek-time can be reduced by employing these characteristics. Accordingly an effective file allocation method using a heuristic algorithm has been proposed.

In computer systems, there are three kinds of file access processes for disc packs : a) Serial access, b) Parallel access, and c) Mixed access. In this study, the serial access form is considered for a file having fixed cylinder capacity. In addition, each file is assumed to be stored as one unit, is not broken, on one disc pack.

2. EFFECTIVE FILE ALLOCATION METHOD

The effective file allocation method consists of (a) a seek-time calculation method, and (b) a procedure for effective file allocation, as illustrated in Fig.1.

Input

1) Access frequency of each file
2) Access transition probability
3) Distribution imbalance of accessed cylinder addressing position in each file
4) Movement, measured in cylinders, and relative movement frequencies of magnetic heads in each file
5) Volume of each file
6) Type and number of disc pack
7) Initial allocation

(a) Seek-time calculation method

(b) Effective file allocation procedure

Output

(1) Effective file allocation
(2) Reduced ratio of seek-time

Fig. 1 Effective file allocation method using automatic synthesis

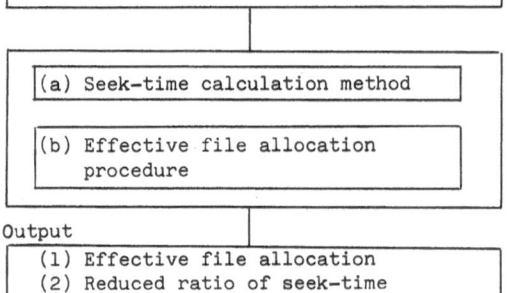

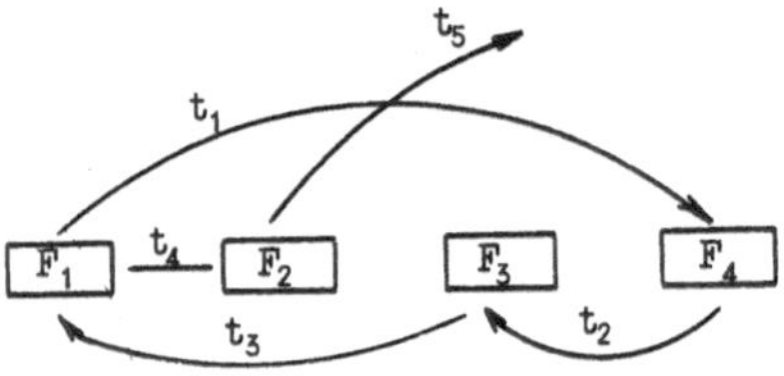

Fig. 2 Logical transition of historical file access

$$P=\begin{pmatrix} P_{11}\,P_{12}\cdots P_{1n} \\ P_{21}\,P_{22}\cdots P_{2n} \\ \vdots \quad \vdots \qquad \vdots \\ P_{n1}\,P_{n2}\cdots P_{nn} \end{pmatrix} \qquad N=\begin{pmatrix} N_1 & & & \\ & N_2 & & 0 \\ & 0 & \ddots & \\ & & & N_n \end{pmatrix}$$

(a) Logical access transition probability matrix

(b) Access frequency matrix

Fig. 3 Logical access transition probability between files and access frequency for each file

If the seven kinds of Input data, specified in Fig. 1, are given, this method generates effective file allocation and shows a reduced ratio of total seek-time.

2.1 Calculation Method of Seek-time

It is possible to calculate the total seek-time for any file allocation, which is an evaluation measure for deciding effective file allocation. In this section, the seek-time calculation method is explained.

2.1.1 Logical access transition frequency and distance between files

Files are accessed historically according to their applications, as exemplified in Fig. 2, t_1, t_2, ... These transitions are called logical access transitions, and a logical access transition probability matrix, P, can be constructed as shown in Fig. 3(a). P_{ij} indicates the probability of an access transition from file F_i to F_j. A logical access frequency matrix, N, which describes, the access frequency for each file is illustrated in Fig. 3(b).

The product of N and P is a logical access transition frequency matrix, Q,as specified in Fig.4(a). q_{ij} indicates the logical access frequency of a transition from F_i to F_j. Distance matrix L is defined in Fig.4(b). ℓ_{ij} is the mean movement distance of the magnetic head from F_i to F_j. Although the establishment of ℓ_{ij} is explained in detail in section 2.1.3, this distance is decided when the position of files on a disc pack is decided. The total seek-time can be calculated using q_{ij} and ℓ_{ij}, basically. However it is necessary to obtain the physical access transition frequency for each disc pack, in order to calculate the seek time corresponding to magnetic head movement.

$$Q = N \cdot P = \begin{bmatrix} q_{11} & q_{12} & \cdots & q_{1n} \\ q_{21} & q_{22} & \cdots & q_{2n} \\ \vdots & \vdots & & \vdots \\ q_{n1} & q_{n2} & \cdots & q_{nn} \end{bmatrix} \qquad L = \begin{bmatrix} \ell_{11} & \ell_{12} & \cdots & \ell_{1n} \\ \ell_{21} & \ell_{22} & \cdots & \ell_{2n} \\ \vdots & \vdots & & \vdots \\ \ell_{n1} & \ell_{n2} & \cdots & \ell_{nn} \end{bmatrix}$$

(a) Logical access transition frequency matrix

(b) Distance matrix

Fig. 4 Logical transition frequency and distance between files

2.1.2 Physical access transition matrix

Physical access transition frequencies can be calculated from two pieces of data, the logical access transition frequencies and probabilities stated in 2.1.1. These data can be collected from an actual system.

When multiple files are allocated on multiple disc packs, all the distances between files can not always be established. For example, in Fig.5, the distances between files F_i and F_k, and F_j and F_k can not been established. Then, logical access frequency matrix Q is modified in order to represent the access transition frequencies between files on only the same disc pack. This is accomplished in the following manner:

In Fig.5, ▲ is a position of magnetic head. When a logical access transition occurs from F_i to F_k, a magnetic head moves from F_k' to F_k. In addition when a logical access transition occurs from F_k to F_j, a magnetic head moves from F_i to F_j. $F_i \rightarrow F_j$ and $F_k' \rightarrow F_k$ represent physical access transitions of magnetic heads. The related physical transition frequencies are described by q_{ij}' and $q_{k'k}'$, respectively. It is assumed that the probability of accessing file F_k after accessing file F_i is decided by logical access transition probability P_{kj} when calculating q_{ij}'. Namely, the Markov chain is assumed, considering that the state is an accessed file. As mentioned above, additional physical transition frequency q_{ij}' can be calculated from the product of q_{ik} and P_{kj}.

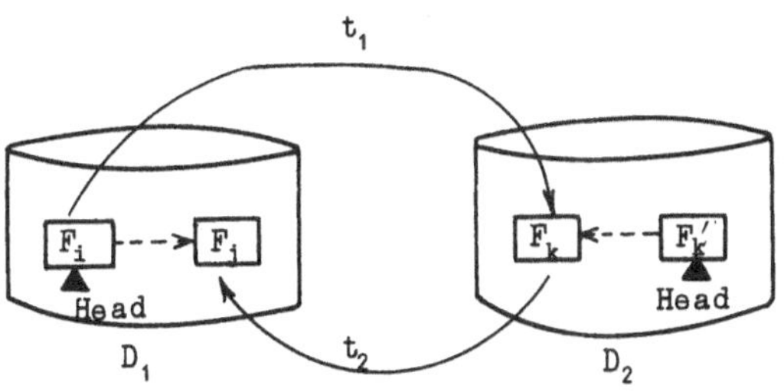

Fig. 5 Logical access transition and physical access transition

Physical transition $F_i \rightarrow F_j$ can also occur according to logical transition $F_i \rightarrow F_k \rightarrow F_k' \rightarrow F_j$, etc. Therefore, additional physical transition frequencies q_{ij}' can be expressed in the form:

$$\left.\begin{aligned} & \underset{i,j \in X}{q_{ij}'} = \sum_{k \notin X} q_{ik} p_{kj} + \sum_{k \notin X} \sum_{k' \notin X} q_{ik} p_{kk}' p_{k'j} + \cdots \\ & \underset{i \in X \;\; k \notin X}{q_{ik}'} = 0 \end{aligned}\right\} \qquad (2.1)$$

q_{ij}' the magnetic head frequency transformed from a logical access transition to a physical access transition.

X a set of files on the specified disc pack.

q_{ik} the logical access transition frequency.

p_{kj} the probability of a logical access transition.

If logical access transition $F_i \rightarrow F_j$ occurs, of course, physical access transition $F_i \rightarrow F_j$ also occurs. All physical transitions frequency can be expressed as follows:

$$\underset{i,j \in X}{q_{ij}^{(+)}} = q_{ij} + q_{ij}' \qquad (2.2)$$

As a result of this replacing of a logical access transition by a physical access transition, physical transitions between files on different disc packs are given by expression (2.3):

$$\underset{i \in X \;\; k \notin X}{q_{ik}^{(+)}} = 0 \qquad (2.3)$$

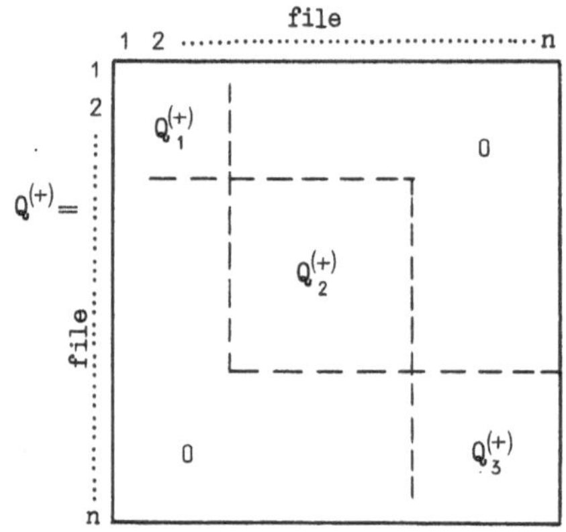

Fig. 6 Physical access transition frequency matrix $Q^{(+)}$

For example, a physical access transition matrix is modified as shown in Fig.6. Each part, $Q_1^{(+)}$, $Q_2^{(+)}$, $Q_3^{(+)}$, represents the physical access transition frequencies for their respective disc packs.

2.1.3 Establishment of mean movement distance between files

The mean movement distance of magnetic heads can be classified in two groups as follows:

- the mean movement distance between files F_i and F_j.
- the mean movement distance in one file F_i.

(1) The mean movement distance between files F_i and F_j.

A mean movement distance between files F_i and F_j, ℓ_{ij}, is calculated as as follows:

$$\ell_{ij} = \ell_{ix} + \sum_{k} \ell_k + \ell_{xj} \tag{2.4}$$

Where,

$$\ell_{ix} = \ell_i - \sum_{t=1}^{\ell_i} \rho_{it} \cdot t \qquad \rho_{it} = f_{it} / \sum_{s=1}^{\ell_i} f_{is}$$

$$\ell_{xj} = \sum_{t=1}^{\ell_j} \rho_{jt} \cdot t \qquad \rho_{jt} = f_{jt} / \sum_{s=1}^{\ell_j} f_{js}$$

ℓ_{ij} mean movement distance between files F_i and F_j.

ℓ_k file capacity in cylinders of file F_k, which exists between files F_i and F_j.

ℓ_i, ℓ_j file capacity in cylinders of file F_i or F_j, respectively.

f_{it} access frequency of relative cylinder address in file F_i; $f_{it} = 1, 2, \cdots, \ell_i$.

f_{jt} access frequency of relative cylinder address in file F_j; $f_{jt} = 1, 2, \cdots, \ell_j$.

The data for f_{it} and f_{jt} can be collected from an actual system.

(2) mean movement distance in one file.

Firstly, the mean seek-time is calculated using the following expression:

$$t_{ii} = \sum_{k=0}^{\ell_i - 1} p_k^i \, f(k) \tag{2.5}$$

Where,

$$p_k^i = n_k^i / \sum_{k'=0}^{\ell_i - 1} n_{k'}^i$$

t_{ii} mean seek-time of movement in file F_i.

n_k^i frequency of movement for k cylinders in file F_i.

p_k^i probability of movement for k cylinders in file F_i.

$f(k)$ function to transform the movement distance to seek-time.

The data for n_k^i can be collected from an actual system.

It is possible to change the seek-time into mean movement distance by the following expression which is applied to the calculation of total seek-time hereafter:

$$\ell_{ii} = f^{-1}(t_{ii}) \tag{2.6}$$

2.1.4 Calculation of total seek-time

Total seek-time T for any file allocation can be calculated with $q_{ij}^{(+)}$, ℓ_{ij}, ℓ_{ii} and $f(x)$ as mentioned above, in the following form:

$$T = \sum_{i,j} q_{ij}^{(+)} f(\ell_{ij}) \tag{2.7}$$

2.2 Effective File Allocation Procedure

Since total seek time can be calculated in case of any allocation of files, the optimal file allocation can be determined logically by calculating the total seek-time for all file allocation cases. However, the number of cases extremely large. Therefore a heuristic algorithm is employed to determine the file allocation with the lowest total seek-time by investigating each allocation individually. This procedure is described briefly as follows:

1) Let T be the total seek-time of any file allocation.
2) Move any file on one disc pack to another disc pack, or transpose two files on two different disc packs, respectively.
3) Rearrange the position of files on disc packs where procedure 2) was carried out.
4) Repeat procedures 2) and 3) for various allocations until the total seek-time is not reduced.

Details of this procedure are explained below: Firstly, seek time T_α of disc pack D_α is defined. When multiple files are on one disc pack D_α, let T_α be the seek-time of the file allocation with the lowest seek-time for D_α. T_α is calculated according to the flow-chart in Fig.7. T_α of D_α is calculated using expression (2.7). When the number of files is greater than k, $(r-k)$ files are allocated on both sides as indicated in box a. This is the application of the optimal file allocation on a single disc pack [1] . This method [1] was developed assuming

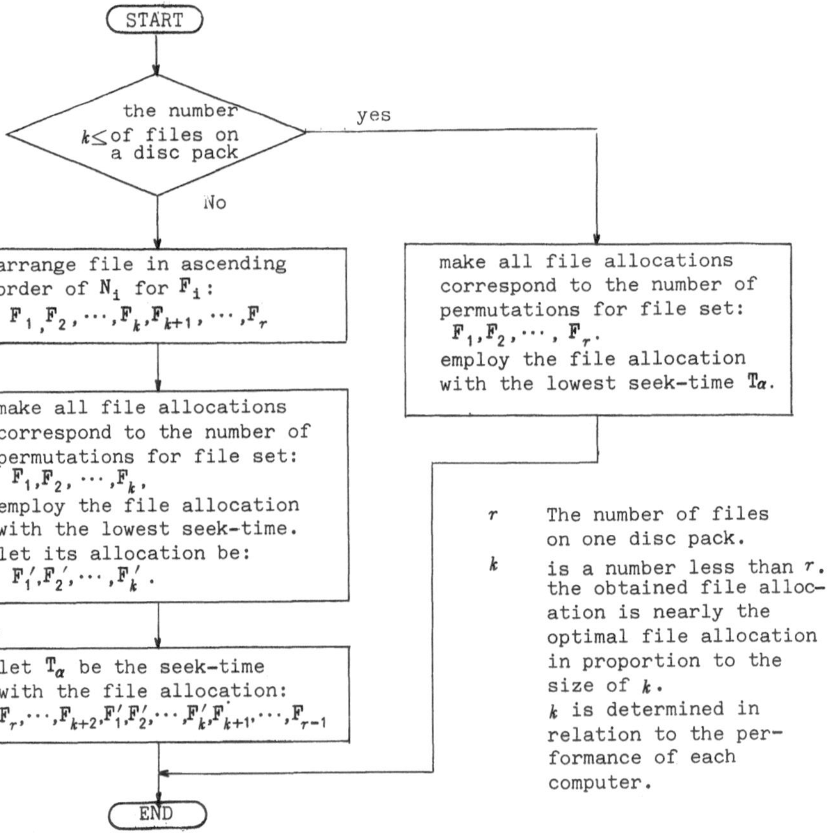

Fig. 7 Procedures to decide the effective file allocation on each disc pack.

that every file has the same capacity, and the access transition probability is determined by the access frequency of each file. In the present method, the method for single disc pack allocation [1] has been adopted approximately because the number of permutations becomes excessively large for a large number of files. k is established in relation to the performance of each computer.

Therefore the procedures for effective file allocation consist of the following four steps:

(1) INITIALIZATION

A total of n files are assumed. Furthermore, a total of m disc packs, which are empty, are assumed. Then it is assumed that total seek-time T is maximum at this time. Let τ_i be the seek-time for file F_i. Then total seek-time is established as follows:

$$T = \sum_{i=1}^{n} \tau_i = \sum_{i=1}^{n} N_i t \qquad (2.8)$$

t is assumed to be the seek-time relative to one file access, for example 250ms.

(2) ENTER

Any file, F_i, is placed on any disc pack, D_α, and the seek-time is calculated. Then, the file with the maximum reduced seek-time, $\triangle T_{i\alpha}$, is selected. Next, related file F_i is stored on the disc pack, D_α. $\triangle T_{i\alpha}$ is expressed in the following form:

$$\triangle T_{i\alpha} = (\tau_i + T_\alpha') - T_\alpha \qquad (2.9)$$

T_α' Seek-time for D_α before file F_i is stored in D_α.
T_α Seek-time for D_α after file F_i is stored in D_α.

Then, is established as:

$$T = T - \triangle T_{i\alpha} \qquad (2.10)$$

All files are entered on the disc packs by repeating this process.

(3) MOVEMENT

Any file, F_i, on one disc pack, D_α, is moved to another disc pack, D_β, and the reduction in total seek-time T, $\triangle T_{i\alpha\beta}$, is calculated. The combination of F_i and D_β which yields the maximum $\triangle T_{i\alpha\beta}$ is selected, and F_i is moved to D_β. If the capacity of disc pack is exceeded by the movement of F_i, the calculation is not carried out. When every $\triangle T_{i\alpha\beta} \leq 0$, this step ends.

$\triangle T_{i\alpha\beta}$ is calculated with the following expression:

$$\triangle T_{i\alpha\beta} = (T_\alpha' + T_\beta') - (T_\alpha + T_\beta) \qquad (2.11)$$

T_α', T_α Seek-time for D_α before and after F_i is moved from D_α, respectively.
T_β', T_β Seek-time for D_β before and after F_i is moved to D_β, respectively.

(4) EXCHANGE

Any file, F_i, on disc pack D_α, and any file, F_j, on disc pack D_β are exchanged. Then reduced seek-time $\triangle T_{ij\alpha\beta}$ is calculated. If the capacity of the disc pack is exceeded by this exchange, its calculation is not carried out. The combination of F_i and F_j, for which $\triangle T_{ij\alpha\beta}$ is maximum, is selected, and F_i on D_α is exchanged with F_j on D_β. $\triangle T_{ij\alpha\beta}$ is calculated as follows:

$$\triangle T_{ij\alpha\beta} = (T_\alpha' + T_\beta') - (T_\alpha + T_\beta) \qquad (2.12)$$

T_α', T_α Seek-time for D_α before and after F_i on D_α is exchanged with F_j on D_β, respectively.

T_β', T_β Seek-time for D_β before and after F_i on D_α is exchanged with F_j on D_β, respectively.

If every $\triangle T_{ij\alpha\beta} \leq 0$,this step ends. At this time, the file allocation determined here is the most effective file allocation.

3. EXPERIMENTAL RESULTS

In order to verify the effectiveness of this method, on-line system of a factory was selected. This system had 6 disc packs and 38 files.
Over 99 percent of the access frequency was concentrated on 11 files, then the effective file allocation method was applied to these 11 files. Data for file access frequency using this system is shown in Table 1 and Fig. 8. Table 1 denotes the access frequency of each file and Fig. 8 indicates the logical access transition probability matrix. The initial file allocation in this system is shown in Fig. 9 and the effective file allocation after effective file allocation method application is shown in Fig. 10. Further, the reduced seek-time was calculated for the case of only shifting the files as depicted in Fig. 11. Table 2 shows the total seek-time calculation results of these three allocations.
Then, the ratio of reduced seek-time was 32.8 and 14.5 percent, respectively, in relation to the initial allocation total seek-time.

This method provides an effective file allocation according to the characteristics of its file access frequency on the strength of file access data for a specific time period (e.g. a day). Therefore, the specified effective file allocation has less effect, if the file access characteristics are not continued for long periods of time. The data for 6 days (a, b, c, d, e, and f represent day 1, ··· , 6, respectively) was collected. This data was used to decide the effective file allocation on the strength of the file access data of a, and confirmed on the effect of each day $a \sim f$ for that allocation. In Table 3 the relative reduced ratios are indicated where the reduced ratio of a is 100.
This table indicates that the effect of the reduced ratio is stabilized in this system. Whenever this method is employed, it is necessary to check stability in this manner.

4. SEEK-TIME CALCULATION ACCURACY AND PROGRAM RUNNING TIME

The seek-time is always obtained using this method of calculation.
Then, it was necessary to verify the accuracy of the seek-time calculation. This was done in the following order:

(1) Make a file model on disc packs like an actual system.

(2) Access the files in the same order as an actual system according to file access data collected from an actual system, and measure the seek-time with a hardware monitor.

The 7 file allocation classifications were measured. And a comparison between the calculation and measurement results is listed in Table 4. As a resust of this experiment, it is recognized that the error rate is under 10 percent, Table 5 describes the results of the same experiment for the initial and effective file allocations. As shown in Table 5, the results of both the calculation and measurement indicate a 33-34 percent reduction in the ratio by the change to the effective file allocation.

A program has been implemented for the effective file allocation method. However, it is difficult to predict the running time of the program. Consequently, 5 different experiments were conducted as described in Table 6. The parameters used are the number of files and disc packs. The running time for the experiments

Table 1 Example of access frequency

File	F_1	F_2	F_3	F_4	F_5	F_6
Frequency	5252	4609	1956	1956	7480	1508
File	F_7	F_8	F_9	F_{10}	F_{11}	
Frequency	5119	5971	4846	3221	2362	

	F_1	F_2	F_3	F_4	F_5	F_6	F_7
F_1	.040	.169	.106	.025	.102	.025	.013
F_2	.842	.002	.030	.007	.034	.006	.010
F_3	.067	.015	.000	.836	.020	.005	.012
F_4	.020	.856	.009	.000	.027	.008	.013
F_5	.041	.092	.061	.005	.044	.015	.000
F_6	.031	.082	.050	.006	.048	.015	.000

Fig. 8 Example of logical access transition probability

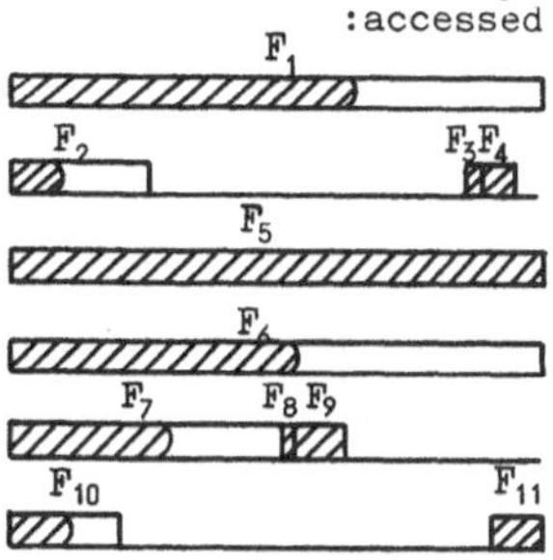

Fig. 9 Allocation before applying the effective file allocation method

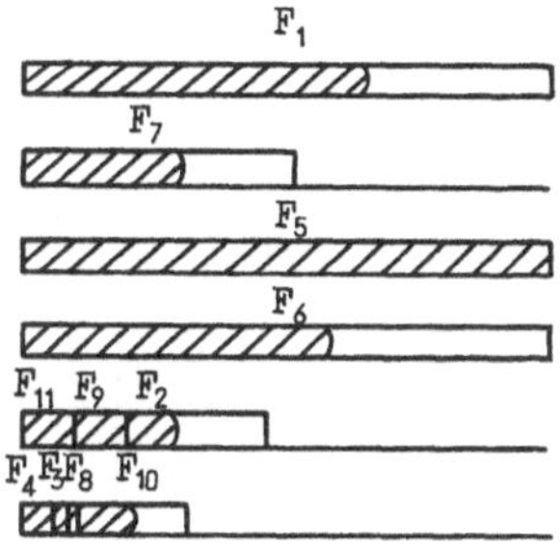

Fig. 10 Effective file allocation

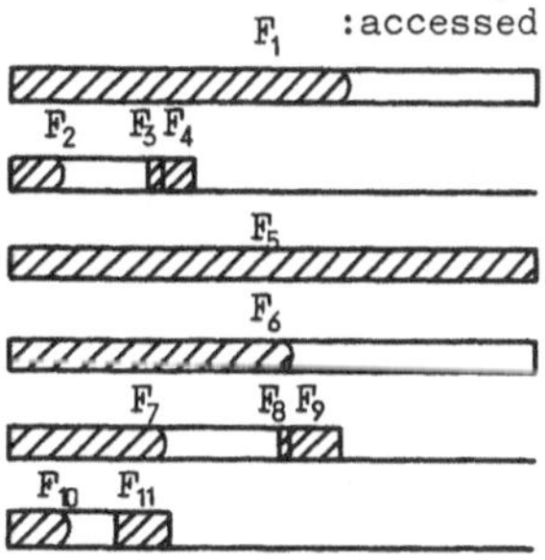

Fig. 11 File allocation after shifting

Table 2 Comparison of total Seek-times

File Allocation	Total Seek-time(sec)	Ratio	Reduced ratio
Initial allocation	2329.529	100	-
Shifting Only	1992.565	85.5	14.5
Effective allocation	1565.857	67.2	32.8

Table 3 Reduced ratio of total Seek-time value : relative value as *a* is 100

Day	*a*	*b*	*c*	*d*	*e*	*f*
Reduced ratio of total seek-time	100	122	97	127	116	120

Table 4 Accuracy of seek-time calculation

Case	Calculation (sec)	Measurement (sec)	Error Rate(%)
1	519.128	512.127	1.4
2	738.269	755.014	-2.2
3	337.726	311.395	8.5
4	329.663	334.935	-1.6
5	190.227	184.574	3.1
6	367.514	376.774	2.4
7	314.474	304.841	3.2

Table 5 Comparison between calculation and measurement results for total seek-time

	Calculation	Measurement
a) total seek-time of initial alloc. (sec)	2329.529	2257.038
b) total seek-time of effec. alloc. (sec)	1565.857	1497.411
c) ratio (b/a)x 100(%)	67.2	66.3
d) reduced ratio 1-c (%)	32.8	33.7

Table 6 Experimental parameter

No	1	2	3	4	5
Number of files	10	10	20	20	30
Number of disc packs	3	5	5	8	10

Number of Disc Packs \ Number of File	10	20	30
10	—	—	20.05
8	—	4.29	—
5	0.12	20.24	—
3	0.46	—	—

Fig. 12 Running time (min. sec)

are listed in Fig. 12. The computer used for the experiment is a H-8700(Similar to the IBM 370/158). Although the running time increases as the number of files increases, it is considered that this method is very satisfactory for practical use on the condition that the number of files is limited to about 20 or less.

5. CONCLUSION

An effective file allocation method has been developed to reduce the total seek-time of magnetic heads by utilizing the characteristics of file access frequency in the case that files on disc packs are accessed serially. File access data was collected from an actual system in order to verify the effectiveness of this method. The effective file allocation was then determined by this method, and the total seek-time was calculated. As a result, it was shown that the total seek-time can be reduced by about 30 percent, in this case.

In order to verify the seek-time calculation accuracy, the file access data was collected from an actual system, and the seek-time measured with a hardware monitor accessing model files according to the data collected. From these results of experiment, the error rate of the seek-time calculation was found to be under 10 percent. Therefore this method is considered highly satisfactory for practical application and now enables the useful allocation of multiple files on multiple disc packs.

ACKNOWLEDGMENT

The authors wish to thank Hajime Enomoto of the Tokyo Institute of Technology for participating in helpful discussions and offering valuable suggestions in connection with this paper.

REFERENCES

1. Grossman, D.D. and Silverman, H.F. Placement of Records on a Secondary Storage Device to Minimize Access Time, Journal of ACM, July 1973

2. Denning, P.J. Effect of Scheduling on File Memory Operations, SJCC 1976

3. Teory, T.J. and Pinkerton, T.B. A Comparative Analysis of Disc Scheduling Policies, Comm. of ACM, March 1972

OPTIMAL FILE ALLOCATION PROBLEM AND RELATIONAL DISTRIBUTED DATA BASES

G.Mazzarol IBM Venice Scientific Center

E.Tomasin Venice University - Centro di Calcolo

INTRODUCTION

The aim of this report is to discuss an optimization problem which should be as near as possible to real situations.
A certain general design of a relational distributed data base (RDDB) is developed and a particular view for the file allocation problem is obtained.
A mathematical model is proposed which takes into account some peculiarities of the relational data base nature.

1- OUTLINE OF AN ARCHITECTURE FOR THE DATA BASE

A hierarchically distributed system composed by two different kinds of computers is considered. There is a central computing facility C and n local nodes Ln. Each node has its own data base and the capability of a local data base management system .
A high level data base language (e.g. sequel (1))is available in every node. Each user is attached to a single node (C or Ln) and accesses the whole network resources pool through that node .
Let now the functional characteristics of the central node and of the local nodes, be described.

FUNCTIONS OF Ln. A given transaction (query or update) is originated at a node (C or Ln) and is captured by the data base management system (DBMS) in a processor of the node. The transaction will be subsequently parsed and possibly optimized.
Subqueries (3) are detected and sequenced in logical order. They

will be sent immediately through the network to the nodes where the corresponding files are located. Then there could be a further optimization or a direct execution (of course some subqueries can be executed locally if the data involved are present). The remote nodes that process the subqueries send back to the originating node data files containing temporary results to be used for further processing.
When the answers begin to return they will be frozen until the final result can be obtained. So some tasks of the processor could be:

- 1 a parser: it checks the syntactical correctness and semantics of the query using the catalog
- 2 a decomposer : it breaks the query into sequenced subqueries and fixes the overall processing strategy (appropriate routing etc.)
- 3 a local subquery optimizer: it can receive an input of subqueries also from remote nodes and optimize the execution in the node
- 4 an optimizer of "not only local" subqueries e.g. for a join when the relations involved are distributed into different nodes
- 5 a local subquery executor which uses the access method facilities and extracts the stored data
- 6 a sender: which assigns and changes the priority to the various kinds of traffic: system messages, commands, transactions and data. It can also act as switching node manager if some traffic is only transit. The overall traffic is routed properly.

The files managed by Ln are:

- satellite catalog which is the full catalog for the locally existing data and some entries for the remaining data
- transient accounting files
- workspace (unformatted space which the system and the user can fill with provisional data)
- local files: they cannot be accessed from other nodes and maintenance is completely local
- global files: this is the local component of the DDB itself.

At the end of a working session a Ln node can send to C the accounting information, and then delete them.

FUNCTIONS OF C. Besides having all Ln functions there are these particular tasks:

- general utilities (copyfile,logs,statistics,system aids)
- concurrency control (update,locking etc.)

The additional files to the normal Ln set are: master catalog - accounting files. It can be pointed out that the local interface interacts with many processes of Ln

CONCURRENCY CONTROL. If node C is disconnected the system cannot allow any update. Nevertheless non local files can be read by remote users , if a communication link is available. As a first step it is possible to identify the term global file with the term relation (4)(5).

2- OPTIMIZATION IN RELATIONAL DISTRIBUTED DATA BASE

It can be easily asserted that there is certainly a cost reduction minimizing the traffic among the nodes even if the processing in the nodes somehow increases. Here the attention is given to the transmission cost reduction which is the most relevant part of the network hardware cost. Distinguish between a generic optimization of the traffic and a forced optimization of particular queries.

a)-Generic optimization: this problem has been faced many times from a general point of view (6), and also with different mathematical models. (7) (8) (9) (10) (11)
But these kinds of approach cannot easily take into account the nature itself of the DB. In fact only optimizing some kinds of queries, one can try to introduce in some measure the relational structure of the DB.

b)-Optimization of a particular set of queries: In practice in every data base one can easily find out a set of operations,namely,in this case, queries, which are very frequently used. Moreover it is possible that the total request rate of these queries is of the same order of magnitude of the whole set of possible extemporary queries. The problem was faced investigating the basic components of a relational query , namely the relational algebra operators. The following philosophy is similar to the one proposed by INGRES (2) (3). A complex query is decomposed firstly by detaching from it every possible "one variable" query. These one-variable queries

can immediately be processed in parallel.
The traffic involved in this process is only matter of answer size and distance,therefore it falls down into the direct request from node i of an object j. Secondly, there are the irreducible two-variable queries which , potentially, involve the maximum amount of traffic if the two relations are splitted in different locations.

Consequently consider briefly the relational algebra dyadic operators. It is possible to demonstrate that cartesian product and division can be implemented by using other operators (12).
On the other side set-theoretic operators (union, intersection, difference) could be taken into account in a model such as the one presented here, but in practice they are not frequently used.
The join operation is very common, especially together with the projection operation. A feasible implementation of the "distributed join" is proposed, in which only the distribution aspects of the join operation are considered.

1-the node where the first relation is allocated sends to the second node (where there is the second relation) the theta-operator concerned and also the domain involved.

2-a task starts in the second node which extracts the tuples satisfying the query,if any, and sends them back.

3- the first relation is scanned against the tuples obtained and finally the join is completed.

The process can take place symmetrically. Obviously if the two relations considered are both in the same node the join operation will be performed without any transmission of data.

3- THE ROUTING PROBLEM DERIVING FROM THE DISTRIBUTED JOIN OPERATION

In this section some definitions are introduced which will also be used in the mathematical formalization of the model.
Let be:

i,j =any couple of nodes in the network
f_1,f_2 =two relations allocated at i and j respectively
Df_1 =a subset of f_1 containing the satisfying tuples extracted after a search in the interested domain pursued by the theta-operator
SL_{f_2} =the size of satisfying tuples found in f_2
$SR_{f_1f_2}$ =size of the result of the join

As was pointed out the process can be symmetric, then the indices must be interchanged.

Consider the following situation : a join $f_1 \cdot f_2$ is performed on the common domain D and the result is needed in i

Let c_{ij} be the transmission cost per byte and unit of time between i and j The following transmission costs, due to two different strategies of routing, have to be compared.

$$C_1 = c_{ij} \underset{1}{(D_{f_1} + SL_{f_2})} \qquad C_2 = c_{ij} \underset{2}{(D_{f_2} + SL_{f_1} + SR_{f_1 f_2})}$$

Now it is possible to demonstrate that in some instancies $C_1 > C_2$ and therefore routing 2 should be selected by an appropriate optimized routing technique.

Remark that, if the result of the join operation is desired in a node in which neither f_1 nor f_2 are present, four different routing strategies are obtained, which the mathematical formulation must be able to distinguish. (see fig. 1)

In this way it is possible to manage the distributed join. Moreover some one-variable subqueries contained in the query with the join, could be taken into account. In fact, from a communication point of view, only the choice of appropriate values of D_f or SL_f is sufficient.

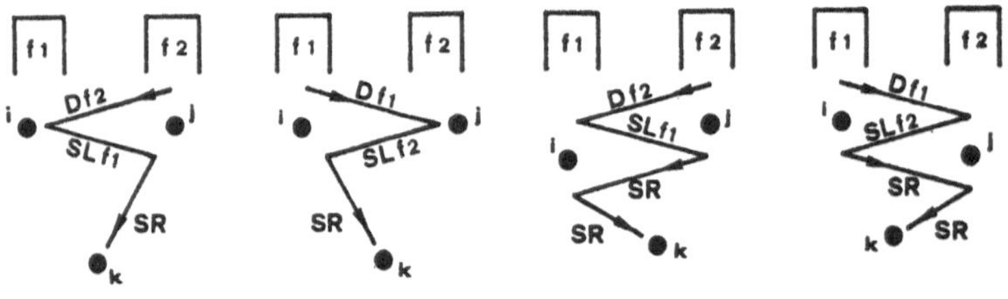

Fig. 1 — "Distributed join" mechanism. General case

4- FRAMEWORK OF THE PROBLEM

This is the area of reference that is considered:

-static behaviour: we are in a statistical equilibrium situation . The network is in busy state.

-data communication: point to point full duplex among nodes: the minimal cost routing is predetermined and fixed. Further it should be obvious that the formalization is accomplished only on the "backbone part" of the network, thus neglecting terminals and their access lines to the data base.

-incomplete information: of course it is impossible to have a complete forecast for the requesting nodes. The model could conceivably work several times using as input data, the information collected by the accounting file located at node C.

-program sharing: as a matter of fact there is only the "query server" component of the DDBMS which is present in every node. As the traffic owed to the transactions flow and system messages can be considered neglegible compared to the data (answer) flow, it can be assumed that the traffic main streams are routed following the minimal distance whithout particular program-file routing . Thus the specialized presence of C node and its "update facility" can be neglected.

5- THE MATHEMATICAL FORMULATION OF THE MODEL

The computer network considered has N nodes and a DB consisting of $F_1 = \left\{ f_1, f_2, \dots f_F \right\}$ relations. For some of these, let they be f_1 f_2 the "distributed join operator" is requested. The problem is to find an optimal file allocation minimizing the operating costs. Two kinds of traffic are taken into account: query traffic and update traffic. Copies of each relations are available. For any query at a certain node a transaction is generated to the "nearest" copy of relations requested or to the "nearest couple" if the query consists in a join between f_1 and f_2 .

As was pointed out in section (3) " nearest couple " means that the routing connecting the three nodes considered must be a minimal one. The same can be said for the "nearest couple" available. For any update demand, of course, the message is transmitted to every copy of the relation. Let the subscripts f,i,j,k, mean respectively relation f and nodes i,j,k. The following notations are introduced:

Let be

M_i =size of the external storage available in i

S_f =size of f

S_f^* =size of the intermediate and temporary result of the join

SU_f =update size for f

E'_i =size of a switching catalog reduced entry in i

E_i =size of a switching catalog full entry in i

k_i =size of the relations owned by a single node and working area

α_{if} =storage-data base mapping factor of f in i (depending on the dishomogeneities in the network)

r_f =number of copies of f

u_{if} =query traffic from i to f

u'_{if} =update traffic from i to f

$U_{if_rf_s}$ =query traffic from i to f_r and f_s implementing the join

$AJ_{f_rf_s}$ =symmetric binary matrix defined in the following way:

$AJ_{f_rf_s}=1$ if $\exists(U_{if_rf_s}, i=1,\dots,N \mid U_{if_rf_s}\neq 0)$

$F_1=\left\{f_1,f_2,\dots,f_F\right\}$ =set of all the relations available in the network (copies are not considered) the cardinality of F_1 is F

$F_2=\left\{f_k,f_1,\dots,f_g\right\}$=subset of F_1 with cardinality F_p i.e. the subset of the relations for which the join operation is requested

$F_3=F_1\cup F_2=\left\{f_1,f_2,\dots,f_F,f_{F+1},\dots,f_{F+F_p}\right\}$=set of cardinality $F'=F+F_p$ .In this set the relations for which the join operation is requested are enumerated twice

c_i =unitary external storage cost in i

c_{ik} =trans. cost per byte and per time unit from i to k

$c_{ijkf_rf_s}$=trans. cost for a minimal routing in the set of the nodes provided that f_r is at node i and f_s is at node j.

$c_{ijkf_rf_s}$ can be easily precalculated (see fig.1)

The variables which control the allocation and the routing discipline are defined this way Let be

$$x_{if}=\begin{cases}1 & \text{if a copy of f is stored at i}\\ 0 & \text{otherwise}\end{cases}$$

$$x_{ijf}=\begin{cases}1 & \text{if a routing of minimal cost exists from i to j provided that f is stored at j}\\ 0 & \text{otherwise (where otherwise means that either the routing exists but it is not a minimal one or } x_{jf}=0)\end{cases}$$

The model can be formulated as a $\{0,1\}$ integer programming problem in the following way:

Let X be a vector whose components are x_{if} $(i=1,\dots,N, f=1,F')$ and $x_{ijf}(i,j=1,\dots,N, f=1,\dots,F')$.

Then the problem is:

$$\min g(X) =$$

$$1) \quad \sum_{i=1}^{N} \sum_{j=1}^{N} \sum_{f=1}^{F} c_{ij} \, SR_f \, u_{if} \, x_{ijf}$$

$$2) \quad + \sum_{k=1}^{N} \sum_{j=1}^{N} \sum_{i=1}^{N} \sum_{\substack{f_r, f_s = F+1 \\ f_r \neq f_s}}^{F'} c_{ijkf_rf_s} \, U_{kf_rf_s} \, AJ_{f_rf_s} \, (x_{kif_r^*} \, x_{kjf_s})$$

$$3) \quad + \sum_{i=1}^{N} \left(\sum_{f=1}^{F} c_i \, S_f \, x_{if} + \sum_{f=F+1}^{F'} c_i \, S_f^* \, x_{if} \right)$$

$$4) \quad + \sum_{i=1}^{N} \sum_{f=1}^{F} (c_i \, E_i' \, (1 - x_{if}) \, r_f + c_i \, E_i \, x_{if})$$

$$5) \quad + \sum_{i=1}^{N} \sum_{k=1}^{N} \sum_{f=1}^{F} c_{ik} \, u_{if}' \, SU_f \, x_{kf}$$

with the following constraints

$$6) \quad \sum_{i=1}^{N} x_{if} = r_f \qquad \forall f = 1, \ldots, F$$

$$7) \quad \sum_{f=1}^{F} (\alpha_{if} \, S_f - r_f \, E_i' + E_i) x_{if} + \sum_{f=1}^{F} r_f \, E_i' + \sum_{f=F+1}^{f=F'} \alpha_{if} S_f^* \, x_{if} +$$

$$+ \quad k_i \leqslant M_i \qquad \forall (i = 1, \ldots, N)$$

$$8) \quad \sum_{j=1}^{N} x_{ijf} = 1 \qquad \forall (i = 1, \ldots, N, \; f = 1, \ldots, F')$$

$$9) \quad \sum_{i=1}^{N} x_{ijf} \leqslant N x_{jf} \qquad \forall (j = 1, \ldots, N, \; f = 1, \ldots, F')$$

$$10) \quad x_{if} = x_{if'} \qquad \forall (i = 1, \ldots, N, \; f' = F+1, \ldots, F', \; f=1, \ldots, F_p)$$

$$11) \quad \begin{array}{ll} x_{ijf} = 0 \text{ or } 1 & \forall (i, j = 1, \ldots, N, \; f = 1, \ldots, F') \\ x_{if} = 0 \text{ or } 1 & \forall (i = 1, \ldots, N, \; f = 1, \ldots, F') \end{array}$$

The different terms of the objective function have the following meaning:

1. Transmission cost of the query traffic which does not require the distributed join operation

2. Transmission cost of the query traffic involving the join operation

3. Storage cost which consists of two parts, the first referring to the storage of the relations themselves, the second referring to the storage of the temporary results of the join operation

4. Storage cost of the " switching catalog "

5. Transmission cost of the update traffic

The quadratic term 2) in the objective function can be easily linearized introducing new variables suitably defined . Then a linear integer $\{0-1\}$ programming problem is obtained.

If the storage constraints 7) are neglected, it can be easily proved that the problem can be decomposed in K subproblems where $K=F-(F_p/2)$. Each subproblem corresponds to an optimal allocation problem of one single file, or a couple of files requiring the join operation.

6- NUMERICAL RESULTS

To point out that the parallelism can effectively change the optimal allocation of the files, i.e. can change the optimal solution of the problem considered, the following simple example was run. Consider a three nodes network in which two files have to be allocated, such that the join operation is requested for them. The optimal allocation given by the two different models is the following:

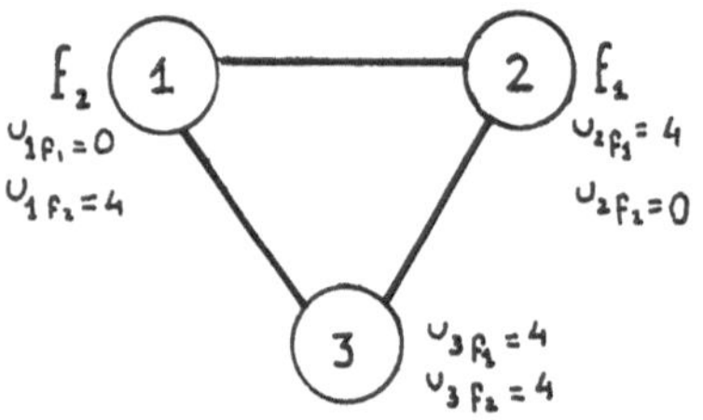

1) model whithout parallelism

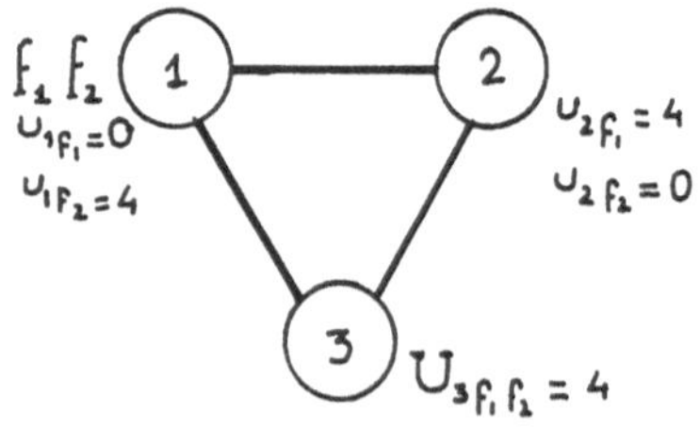

2) model with parallelism

In order to respect the consistency of the proof the size of the information transmitted to node 3 in case 1 is a subset of f_1 and f_2 suitably calculated. Hereinafter a set of proof was run on a 5-nodes network in which 7 files are to be allocated; the last 2 of them coupled by the join operation. The whole set of constants was suitably chosen. For instance the size (Mbytes) of the 7 relations is : S_1=.2 S_2=.5 S_3=1 S_4 =2 S_5=15 S_6 =15 S_7=15. This can be a real data base. The following results were obtained varying the number of copies of the last 3 files which are the most significant. In the table only a subset of the proofs run is considered.

	Allocation					Routing 1)					Obj 2)
node	1	2	3	4	5	from 1	2	3	4	5	
files	34	12567			12345	to 2	1	21	12	2	.1360
files	34	12567			1234	to 2	1	21	12	2	.1571
files	34	12567			123456	to 2	1	12	12	2	.1150
files	34	12567		34	12567	to 2	1	12	2	4	.1035
files	34	12567			123467	to 2	1	12	12	2	.1055
files	34	12567			12346	to 2	1	12	12	2	.1360
files	34	12567	67	34	12567	to 2	1	12	2	4	.1110
files	34	12567	57	34	12567	to 2	1	12	2	4	.1117
files	34	12567	6		12346	to 2	1	12	12	2	.1664
files	34	12567	567	34	12567	to 2	1	12	2	4	.1127

1) routing from node to node(s) where there are the files required
2) Obj = value of the objective function in the optimal solution

From the above table one can argue that in many cases some computers, where there is no file allocated, can be considered as simple switching nodes. Another interesting aspect is that, knowing the routing from each node, one can think to dimension the capacities of the network. This way the model can be used to design the network itself, since allocation allows to assign storage to the nodes and routing allows to assign capacity to the links.

7- REFERENCES

1. D.D.Chamberlin et al.: Sequel 2. A unified approach to Data

Definition,Manipulation and Control - IBM J.of Research and Development,20,6(Nov.1976),560-575

2. M.Stonebraker et al.: The design and implementation of INGRES - ACM Transact. on Database Systems 1,3(Sept.1976),189-222
3. E.Wong and K.Youssefi: Decomposition. A strategy for Query Processing- ACM Transact. on Database Systems 1,3(Sept.1976),223,241
4. E.F.Codd: Relational completeness of data base sublanguage Courant Computer Science Symposium 6- Prentice-Hall(1972),65-99
5. C.J.Date: An Introduction to Database Systems-Addison-Wesley(Jan.1976),42-47
6. M.E. Deppe and J.P. Fry: Distributed data bases:a summary of research- Computer Networks 1,1(1976),130-138
7. W.W. Chu: Optimal File Allocation in a Multiple Computer System - IEEE Transact. on Computers, C-18,10(Oct.1969),885-889
8. K.D. Levin and H.L. Morgan: Optimizing distributed data bases - A framework for research - Proc.of NCC Conf.(1975),473-478
9. H.L. Morgan and K.D. Levin: Optimal program and data location in computer networks - Comm. of ACM 20,5 (May1977), 315-322
10. F.A.Schreiber: Problemi posti dal progetto di database distribuiti : i problemi di allocazione - Rivista di informatica 5,2(jun.1974),118-131
11. G.Bucci and S.Golinelli: A distributed approach to database decomposition in hierarchical computer systems-Proc.on communications of European Conf. on Electronics (May1977),61-70
12. E.F.Codd: Normalized data base structure:a Brief Tutorial-Proc.ACM SIGFIDET workshop on Data Description,Access and Control(1971)

GENERAL PURPOSE NONLINEAR PROGRAMMING PACKAGE

D.F. Rufer
Department of Automatic Control
Swiss Federal Institute of Technology
CH - 8092 Zurich, Switzerland

Abstract

This paper tries to make three steps towards more efficient applicability of mathematical programming algorithms : *First*, a program package is presented which easily allows comparisons of different algorithms applied to a specific minimization problem. All the algorithms use similar, problem-dependent termination criteria. *Secondly*, an efficient standard algorithm for unconstrained optimization without using derivatives is described. It is a quasi-Newton method in which the termination criterion of the unidimensional searches becomes refined according to the convergence of the multidimensional optimization. The required gradients are evaluated by numerical differences using step-lengths which iteratively are optimized to get maximum accuracy. *Thirdly*, the user of the nonlinear programming package has the option to leave the selection of the optimization method up to the program. Numerical examples are presented.

1. INTRODUCTION

During the last decade many algorithms have been developed for solving mathematical programming problems [1], [2], [3]. Using existing publications it is difficult to compare the efficiency of these methods because most numerical experiences are based on different termination criteria, codes, computers or on different test problems. Therefore, before solving numerical optimization problems, one has to solve the problem how to select an algorithm well suited to the actual type of problem. Not only should the reliability and the execution time of the chosen algorithm be satisfactory, but also the user's effort in selecting and applying this algorithm should be minimized.

With this in mind it seems that there is a practical need to develop a program package for nonlinear programming which fulfills the following requirements :

a) Algorithms for solving general nonlinear programming problems are easily applicable.

b) Different algorithms for unidimensional, multidimensional and constrained problems can be combined with each other (modularity of the program package).

c) All algorithms are using the same, problem-oriented termination criteria. All algorithms can be used together with the same problem-scaling. For all algorithms the same coding is used (minimizing the execution time rather than the storage requirements).

d) If the user has to solve several similar optimization problems *(application type A)* it may be worth to compare different algorithms in order to select the one which solves that class of problems most efficiently. Such a comparison is easily possible if the above-mentioned requirements are fulfilled.

e) If the user has to solve just one optimization problem of the same kind *(application type B)*, it is, of course, not worth to compare different algorithms in order to find out the fastest one. For such applications, the program package itself selects a reliable, problem-suited standard algorithm.

2. ORGANIZATION OF THE PROGRAM PACKAGE

A nonlinear programming package, called NLP has been coded in FORTRAN [4]. It is hierarchically (stage I to stage V) structured according to figure 1 :

Stage I : For *applications of type A,* the user must give information on the problem to be solved (number fo parameters and constraints, initial paramter values, required tolerances), and he must define the desired minimization method and its necessary parameters, e.g. factors for constraint weighting and step enlargements, initial step-lengths, maximum number of iterations.
For *applications of type B,* the user only needs to give some basic information about the optimization problem. Then the subprogram NLP itself selects an appropriate standard algorithm.

Stage II : Algorithms to reduce minimization problems with equality and/or inequality constraints to unconstrained optimizations.

Stage III : Algorithms for solving unconstrained, multidimensional problems with or without using derivatives supplied by the user.

Stage IV : This stage contains different algorithms to solve unconstrained unidimensional problems (Block UNIOP), different algorithms for numerical gradient evaluations (Blocks GUNC and GCON) and an algorithm for automatic problem-scaling (Block SCALER).

Stage V : In subprogram FUN the user must define the function to be minimized, the functions describing constraints and, if available, the appropriate derivatives. The subprograms FCON and FUNC are used to count the number of function evaluations and eventually to scale the problem. PENAL calculates the value of the augmented penalty function in case of a constrained minimization.

Up to now the following algorithms have been implemented into the program package :

a) Transforming constrained into unconstrained problems :
  . Exterior point penalty function method of Courant [5] ,
  . Accelerated penalty function method [5], [6].

b) Solving multidimensional, unconstrained problems :
  . Search directions according to method of Davidon-Fletcher-Powell [7] ,
  . Search directions according to method of Fletcher [9] ,
  . Search directions according to rank-one formula of Broyden [8] ,
  . Search directions according to different formulas of Pearson [8] ,
  . Search directions according to method of Powell [2] ,
  . Method of Rosenbrock [2] .

All these algorithms have been adapted to the same type of termination criteria (described in section 3a).

c) Solving unidimensional, unconstrained problems :
  . Golden section search with algorithm to bracket a starting interval [4] ,
  . Iterative quadratic fitting [2] ,
  . Iterative cubic fitting [10] ,
  . Combination of golden section search and cubic fitting [10] ,
  . Approximative search of Fletcher [9] .

d) Evaluation of derivatives with respect to the parameters :
  . Derivatives are supplied by the user in subroutine FUN (analytical derivatives),
  . Numerical evaluations by foreward-differences, step-lengths supplied by the user,
  . Numerical evaluations by foreward-differences, step-lengths calculated by method of Stewart [11] ,
  . Numerical evaluations by foreward-differences, direct optimization of the step-lengths as described in section 3c.

It should be noted that some of the mentioned multidimensional methods do not require derivatives at all.

e) In connection with any algorithm for multidimensional minimization a special parameter scaling is possible : Scaling factors can be calculated such that the sensitives of the performance function with respect to each parameter is the same at the initial point.

On a CDC 6500 computer the program package NLP requires a storage space of about

45k (octal) for optimization problems with up to 40 parameters, 10 equality and 10 inequality constraints.

3. STANDARD ALGORITHM FOR UNCONSTRAINED OPTIMIZATION WITHOUT USING DERIVATIVES

The unconstrained minimization of a nonlinear function without explicitely using its derivatives is an important problem (e.g. for system identification, see [12]). Many different algorithms for solving such problems can be obtained with the program package NLP by combining the available methods for multidimensional optimizations (2b), for unidimensional searches (2c) and for gradient evaluations (2d). In the following an algorithm is described which proved to be particularly reliable and fast in hundreds of practical applications. Basically it combines the quasi-Newton formula of Fletcher [9] with a golden section search and a numerical gradient approximation using optimized step-lengths.

a) Multidimensional search

The problem to be solved can be formulated as

$$M(\underline{p}) \overset{!}{=} \text{minimum}, \quad \underline{p} \in \mathbb{R}^{\ell}.$$

$\underline{p}_0$ are given starting values ; $\xi_1,\ldots,\xi_\ell$ and $\varrho_1,\ldots,\varrho_\ell$ are given absoluate and relative tolerances for the solution point $\underline{p}^*$; ξ_0 and ϱ_0 are given absolute and relative tolerances for $M(\underline{p}^*)$. The parameters i_{res} and i_{max} can be used to reinitialize and to stop the algorithm after a certain number of iterations. $\alpha(\underline{p},\underline{s})$ is a unidimensional search step and $\underline{\gamma}(\underline{p},\cdot)$ a gradient approximation. These functions are defined in the sections 3b and 3c.

Step 1 : $i:=0$; $\underline{\gamma}_0:=\underline{\gamma}(\underline{p}_0,1)$; $H_0:=I$;

Step 2 : $\underline{s} := -H_i\cdot\underline{\gamma}_i/\|H_i\cdot\underline{\gamma}_i\|$; $\alpha^*:=\alpha(\underline{p}_i, \underline{s})$; $\underline{p}_{i+1}:=\underline{p}_i+\alpha^*\cdot\underline{s}$; $\underline{\gamma}_{i+1}:=\underline{\gamma}(\underline{p}_{i+1}, 0)$;

Step 3 : $\underline{x}:=\underline{p}_{i+1}-\underline{p}_i$; $\underline{y}:=\underline{\gamma}_{i+1}-\underline{\gamma}_i$; $H_{i+1}:=H_i+\underline{x}\cdot\underline{x}^T/(\underline{x}^T\cdot\underline{y})-H_i\cdot\underline{y}\cdot\underline{y}^T\cdot H_i/(\underline{y}^T\cdot H_i\cdot\underline{y})$;

Step 4 : If $(\underline{x}^T\cdot\underline{y}\geq\underline{y}^T\cdot H_i\cdot\underline{y})$ go to step 5 ; $\underline{z}:=\underline{x}/(\underline{x}^T\cdot\underline{y})-H_i\cdot\underline{y}/(\underline{y}^T\cdot H_i\cdot\underline{y})$;

$H_{i+1}:=H_{i+1}+\underline{y}^T\cdot H_i\cdot\underline{y}\cdot\underline{z}\cdot\underline{z}^T$;

Step 5 : If $(i\geq i_{max})$ stop the algorithm ; $i:=i+1$;

Step 6 : If one of the conditions $\underline{\gamma}_i^T\cdot H_i\cdot\underline{\gamma}_i\leq 0$ or $i\geq i_{res}$ is satisfied, go to step 8 ;

Step 7 : If not all of the $\ell+1$ conditions

$$|p_{i,k}-p_{i-1,k}| \leq \max\{\xi_k, \varrho_k \cdot |p_{i,k}|\} \quad , \quad k=1,..,\ell$$

$$|M(\underline{p}_i)-M(\underline{p}_{i-1})| \leq \max\{\xi_0, \varrho_0 \cdot |M(\underline{p}_i)|\}$$

are satisfied go to step 2 ;

$\underline{s}:=-\underline{\gamma}_i/\|\underline{\gamma}_i\|$; $\alpha^*:=\alpha(\underline{p}_i,\underline{s})$; $\underline{p}^*:=\underline{p}_i+\alpha^*\cdot\underline{s}$;

If all the $\ell+1$ conditions

$$|p_k^*-p_{i,k}| \leq 0.5 \max\{\xi_k, \varrho_k \cdot |p_{i,k}|\} , \quad k=1,..,\ell$$

$$|M(\underline{p}^*)-M(\underline{p}_i)| \leq 0.5 \max\{\xi_0, \varrho_0 \cdot |M(\underline{p}_i)|\}$$

are satisfied stop the algorithm ;

$H_i:=I$; $\underline{p}_{i+1}:=\underline{p}^*$; $\underline{\gamma}_{i+1}:=\underline{\gamma}(\underline{p}^*,1)$; go to step 3 ;

Step 8 : $H_i:=I$; $i_{res}:=i+i_{res}$; $\underline{s}:=-\underline{\gamma}_i/\|\underline{\gamma}_i\|$; $\alpha^*:=(\underline{p}_i , \underline{s})$;

$\underline{p}_{i+1}:=\underline{p}_i+\alpha^*\cdot\underline{s}$; $\underline{\gamma}_{i+1}:=\underline{\gamma}(\underline{p}_{i+1} , 1)$; go to step 3 ;

This algorithm is similar to the method of Davidon-Fletcher-Powell [7] if step 4 is omitted and $\underline{\gamma}(\underline{p},\cdot)$ is the exact gradient of $M(\underline{p})$. If step 4 is added and $\alpha(\underline{p},\underline{s})$ is a rough quadratic approximation of the unidimensional search, one obtains the method of Fletcher [9]. The convergence properties are greatly improved due to the special termination criteria (in step 7) and the reinitializations (step 6 and 8). In particular this algorithm satisfies Wolfe's general convergence conditions for descent algorithms [2]. Hence, for continuously differentiable functions, sufficiently small values ξ_0, ϱ_0 and sufficiently accurate gradient evaluations, either the algorithm converges to a stationary point of $M(\underline{p})$ or $M(\underline{p})\to-\infty$.

b) Unidimensional search

The previously used function values $\alpha(\underline{p},\underline{s})$ are defined as an approximative solution of the unidimensional optimization

$$\overline{M}(\alpha) \overset{!}{=} \text{minimum}, \quad \overline{M}(\nu) := M(\underline{p}+\nu\cdot\underline{s}) .$$

In any case the condition $\overline{M}(\alpha(\underline{p},\underline{s})) \leq \overline{M}(0)$ should be satisfied. The unidimensional search can be performed by the golden section method [2]. In order to start these iteration, the function $\overline{M}(\cdot)$ must be evaluated at 4 points $a_0=0<b_0<c_0<d_0$ (forming

a golden section of the interval $[0,d_0]$) such that $\bar{M}(a_0) > \bar{M}(b_0)$ and $\bar{M}(c_0) < \bar{M}(d_0)$. This condition can be satisfied by iteratively enlarging or reducing the starting interval $[0,d_0]$ by the ratio of the golden section. Thus, for each enlargement- or reduction-step only one more function evaluation is needed [4].

Executing the golden section iterations one obtains values a_i, b_i, c_i, d_i, i=1, 2, 3,..., converging to the unidimensional optimum. If all $\ell+1$ conditions

$$\left|(d_i-a_i)\cdot s_k\right| \leq \max\left\{\gamma_k,\ \varrho_k\cdot\left|p_k+b_i\cdot s_k\right|,\ \varkappa\cdot\left|b_i\cdot s_k\right|\right\},\quad k=1,..,\ell$$

$$\left|\bar{M}(b_i)-\bar{M}(b_{i-1})\right| \leq \max\left\{\gamma_0,\ \varrho_0\cdot\left|\bar{M}(b_i)\right|,\ \varkappa\cdot\left|\bar{M}(b_i)-\bar{M}(0)\right|\right\}$$

are satisfied, the unidimensional search ends with the statements

$$d_0:=b_i\ ;\quad \alpha(\underline{p}\ ,\ \underline{s}):=b_i\ .$$

The factor $\varkappa$ plays an important role in the termination criterion : If $\varkappa=0$, all unidimensional searches are carried out with the accuracy required for the multi-dimensional problem. If $\varkappa>0$, the termination criterion becomes refined according to the convergence of the multidimensional optimization. Practical experiences demonstrate that for "flat" performance functions best results are obtained with $\varkappa=0.1$ while for functions with "steep valleys" (e.g. resulting from penalty functions to handle constraints) $\varkappa=0.01$ is appropriate. The introduction of such an adaptive termination criterion reduces the overall computation time up to 50 % in many practical applications.

c) Gradient approximation

The function $\underline{\gamma}(\underline{p}\ ,\ n)$ used in the multidimensional search algorithm is defined as

$$\gamma_k(\underline{p},n) := \frac{M(\underline{p}+\delta_k\cdot\underline{e}_k) - M(\underline{p})}{\delta_k}\quad,\quad k=1,\ldots,\ell\ .$$

$\underline{e}_k$ is the kth column of the ℓ-dimensional unit matrix I. If n=1 (at the beginning and after reinitializations of the multidimensional search) the step-lengths δ_k are iteratively evaluated in order to obtain maximum accuracy of γ_k. If n=0 the values δ_k remain unchanged.

The evaluation of correct step-lengths δ_k is of greatest importance when using such a numerical gradient approximation : Small δ_k cause large rounding errors, while large δ_k cause large truncation errors. The maximum rounding errors r_k and the truncation errors t_k can be approximated [4] as

$$r_k \approx \frac{2}{\delta_k}\left|\eta\cdot M(\underline{p})\right|,\quad t_k \approx 2\cdot\left|\gamma_k - \hat{\gamma}_k\right|,\quad k=1,\ldots,\ell\ .$$

where η is a given constant indicating the maximum relative error made in the evaluation of the function $M(\underline{p})$ and $\hat{\gamma}_k$ is a derivative approximation calculated similarly to γ_k but with a step-length of $\delta_k/2$. Therefore, in order to look for a step-length δ_k which minimizes the total error r_k+t_k , the condition $r_k \approx t_k$ must be satisfied. Hence, the optimum value δ_k can be found by iterative doubling of δ_k as long as $r_k > t_k$ or by iterative halving of δ_k as long as $r_k < t_k$. If the initial value of δ_k is much too small, the inequality

$$\frac{1}{\delta_k} \cdot |\eta \cdot M(\underline{p})| > |\gamma_k - \hat{\gamma}_k|$$

may not be true due to very large rounding errors in γ_k and $\hat{\gamma}_k$. Therefore, during the mentioned iterative procedure, it must be checked that the value of $|\gamma_k - \hat{\gamma}_k|$ is decreasing if δ_k is reduced.

In many practical applications the algorithm described in this section proved to be much more reliable and efficient than other known algorithms for function minimization without explicit use of derivatives. Some comparisons with other methods are shown in section 5.

4. SELECTION OF STANDARD ALGORITHMS

As mentioned in the introduction, it is not worth to compare the efficiency of different nonlinear programming algorithms for certain types of applications. In this case, the user has the option to leave the selection of the optimization algorithm up to the program package NLP. One out of nine standard algorithms is automatically selected according to some basic information about the problem, namely the number of parameters to be optimized, the presence of equality or inequality constraints, the availability of analytical derivatives and the availability of an estimate for the minimum function value (see table 1).

The accelerated penalty function method [5] is used to generate unconstrained minimization problems in any case in which constraints must be regarded. The quasi-Newton method described in section 3a is implemented in all standard algorithms for the optimization of more than one parameter ($\ell>1$). The program package selects a reinitialization of these iterations as follows : $i_{res}=3\ell$ if $\ell \geq 20$, $i_{res}=4\ell$ if $3<\ell<20$ and $i_{res}=i_{max}$ (no reinitialization) if $\ell \leq 3$. Unidimensional searches are carried out by the golden section method mentioned in section 3b or by the method of iterative quadratic fitting (using the same termination criterion). The appropriate values of the parameter $\varkappa$ used in the termination criteria are also indicated in table 1. The golden section method has very safe convergence properties but the quadratic fitting method may converge much faster (specially for nearly quadratic functions and if a good estimate of the minimum value is available). In case of a bad convergence of quadratic fit iterations (in standard algorithms 3, 5 or 7) the user can still force the program package to switch over to the safer algorithms 2, 4 or 6 by not indicating any estimate for the minimum function value (see table 1).

5. NUMERICAL APPLICATIONS AND COMPARISON OF DIFFERENT ALGORITHMS

Three examples for the application of the program package NLP are given in this section. Two of them are test functions (with a relatively large number of parameters) for which the automatically selected standard algorithm is compared with other algorithms incorporated in the package. In the presented examples the standard algorithm is the most efficient one. Of course, one could construct problems where one of the other methods is faster, but nevertheless, these results seem to be representative. The third example is a trajectory optimization for a spacecraft maneuver. Other applications of NLP are reported in [12], [13] and [14].

a) Example 1

The function

$$M(\underline{p}) = \sum_{i=1}^{10} 100\cdot(p_{2i} - p_{2i-1}^2)^2 + (1 - p_{2i-1})^2,$$

depending on the variables $p_1,..,p_{20}$, shall be minimized by starting with the values $p_{2i}=1$, $p_{2i-1}=-1.2$, $i=1,...,10$. Analytical derivatives are available ; the required tolerances are $\xi_0=10^{-10}$, $\varsigma_0=10^{-5}$, $\xi_i=10^{-4}$, $\varsigma_i=10^{-2}$, $i=1,..,20$. The optimum values are $p_i^*=1$, $i=1,..,20$ and $M(\underline{p}^*)=0$.

Table 2 shows the numerical results (minimum function value, number of function and gradient evaluations and computation time for a CDC-6500 machine) obtained with different algorithms.

b) Example 2

The same function as in example 1 shall be minimized, but it is assumed that no analytical derivatives are available. In table 3 the results obtained with the selected standard algorithm and with the methods of Stewart, Powell and Rosenbrock can be compared.

c) Example 3

The problem of fuel optimal transfer maneuvers in a central gravitational field is considered in [15]. By using analytical results and a parametrization of the control functions, the original optimal control problem can be solved by a sequence of mathematical programming problems. After introducing transformed variables and piecewise constant thrust-accelerations, all necessary trajectory integrations, as well as the evaluation of the performance function and the constraining functions are performed in closed form. The considered example is a transfer maneuver between two elliptic orbits inclined by 5^0(initial orbit: $I_0=5^0$, $\Omega_0=0^0$, $\omega_0=-90^0$, $e_0=0.2$, $q_0=9266$ km ; final orbit: $I_f=0^0$, $\Omega_f=0^0$, $\omega_f=30^0$, $e_f=0.2$, $q_f=11120$ km). The maneuver starts with the true anormaly $\phi_0=90^0$. The spacecraft is equipped with a thrust-limited propulsion system (c=3924 m/s, $d_{max}=0.001\ m_0/s$) which shall twice be switched on. The duration of the maneuver as well as the parametrized functions of thrust-direction shall be optimized to get minimum fuel consumption. To solve this problem with a realistic accuracy, a nonlinear function of 15 variables must be minimized and simultaneously 5 nonlinear equality constraints must be fulfilled. This parameter optimization task has been solved with standard algorithm 1 of the NLP-package (no analytical derivatives can be used because of the complexity of the nonlinear functions) within 189 seconds of computation time. The resulting trajectory is shown in figure 2 ; the maneuver duration is 9855 seconds and the fuel consumption 0.318 times the initial spacecraft mass m_0.

REFERENCES

[1] L.C.W. Dixon : Optimization in Action. Academic Press, London, 1976.

[2] S.L. Jacoby, J.S. Kowalik, J.T. Pizzo : Iterative Methods for Nonlinear Optimization Problems. Prentice-Hall, 1972.

[3] D. Rufer : Mathematische Programmierung - eine Uebersicht. Angewandte Informatik, 3, 1974.

[4] D. Rufer : Optimale Steuerung des Zweikörperproblems. Dissertation No.5519, Eidgenössische Technische Hochschule Zürich, 1975.

[5] H.J. Kelley, W.F. Denham, I.L. Johnson, P.O. Wheatley : An Accelerated Gradient Method for Parameter Optimization with Nonlinear Constraints. Journal of the Astronautical Sciences, Vol XIII, No 4, 1966.

[6] I.L. Johnson : The Davidon-Fletcher-Powell Penalty Function Method - a Generalized Iterative Technique for Solving Parameter Optimization Problems. Johnson Space Center, Internal Note No 75-FM-34, Houston 1975.

[7] R. Fletcher, M. Powell : A Rapidly Convergent Descent Method for Minimization. Computer Journal, Vol 6, 1963.

[8] J.D. Pearson : Variable Metric Methods of Minimization. Computer Journal, Vol 12, 1969.

[9] R. Fletcher : A New Approach to Variable Metric Algorithms. Computer Journal, Vol 13, 1970.

[10] G. Guilfoyle, I. Johnson, P. Wheatley : One-dimensional Search Combining Golden Section and Cubic Fit Techniques. NASA CR-65994, 1967.

[11] G.W. Stewart : A Modification of Davidon's Minimization Method to Accept Difference Approximations of Derivatives. Journal of ACM, Vol 14, 1967.

[12] D. Rufer : Implementation and Properties of a Method for the Identification of Nonlinear Continuous Time Models. Bericht Nr. 77-01, Fachgruppe für Automatik, Eidgenössische Technische Hochschule, Zürich, 1977.

[13] H. Mäder, P. Wehrli : Modelling of a Reservoir. Proceedings of the IFIP Working Conference on Modeling and Simulation of Land, Air, Water Ressources Systems, Ghent, Belgium, August 1977.

[14] J. Leimgruber ; Stationary and Dynamic Behaviour of a Speed Controlled Synchronous Motor with cosφ - or Commutation Limit Line Control. IFAC Symposium on Control in Power Electronics and Electrical Drives, Düsseldorf, Oct. 3- 5, 1977.

[15] D. Rufer : Trajectory Optimization by Making Use of the Closed Solution of Constant Thrust-Acceleration Motion. Celestial Mechanics, Vol 14, 1976.

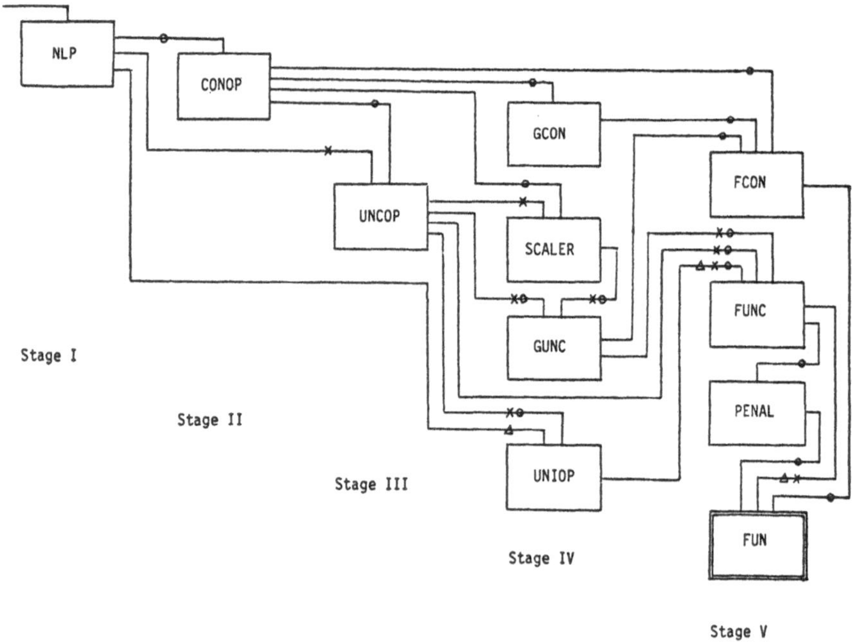

Figure 1 : Structure of the program package ;
o Program flow for constrained problems,
x Program flow for unconstrained multidim. problems,
Δ Program flow for unconstrained unidim. problems.

NONLINEAR PROGRAMMING PROBLEM				SELECTED STANDARD ALGORITHM		
MORE THAN ONE PARAMETER TO OPTIMIZE	EQUALITY OR INEQUALITY CONSTRAINTS	ANALYTICAL DERIVATIVES AVAILABLE	ESTIMATE OF MIN. FUNCTION VALUE AVAILABLE	No.	UNIDIMENSIONAL SEARCH	$\varkappa$
YES	YES	NO	YES/NO	1	GOLDEN SECTION	0.01
YES	YES	YES	NO	2	GOLDEN SECTION	0.01
YES	YES	YES	YES	3	QUADRATIC FIT	0.1
YES	NO	NO	NO	4	GOLDEN SECTION	0.01
YES	NO	NO	YES	5	QUADRATIC FIT	0.1
YES	NO	YES	NO	6	GOLDEN SECTION	0.01
YES	NO	YES	YES	7	QUADRATIC FIT	0.1
NO	YES	YES/NO	YES/NO	8	GOLDEN SECTION	0
NO	NO	YES/NO	YES/NO	9	GOLDEN SECTION	0

Table 1 : Selection of standard algorithms.

METHOD	MIN. FUNCTION VALUE	No.OF FUNCTION EVALUATIONS	No.OF GRADIENT EVALUATIONS	COMPUTATION TIME sec
STANDARD ALGORITHM (No.7)	$2.3 \; 10^{-13}$	393	72	3.7
DAVIDON-FLETCHER-POWELL METHOD (Quadr.fit, $\varkappa$=0)	$2.6 \; 10^{-11}$	614	95	5.1
FLETCHER METHOD (Unidim.search of [9])	13.4	220	201	9.0
MODIF.DAVIDON-FLETCHER-POWELL M. (Quadr.fit,$\varkappa$=0.1)	$4.2 \; 10^{-10}$	433	84	4.3
MODIF.FLETCHER METHOD (Standard algorithm No.6)	$1.5 \; 10^{-12}$	1537	80	5.8

Table 2 : Numerical results of example 1.

METHOD	MIN. FUNCTION VALUE	No.OF FUNCTION EVALUATIONS	COMPUTATION TIME sec
STANDARD ALGORITHM (No.5)	$3.2 \; 10^{-10}$	2799	6.3
STEWART METHOD (Quadr. fit, $\varkappa$=0)	$4.9 \; 10^{-5}$	6031	17.9
POWELL METHOD (Quadr. fit, $\varkappa$=0)	$6.1 \; 10^{-13}$	17745	31.3
ROSENBROCK METHOD	$7.8 \; 10^{-10}$	13205	15.1

Table 3 : Numerical results of example 2.

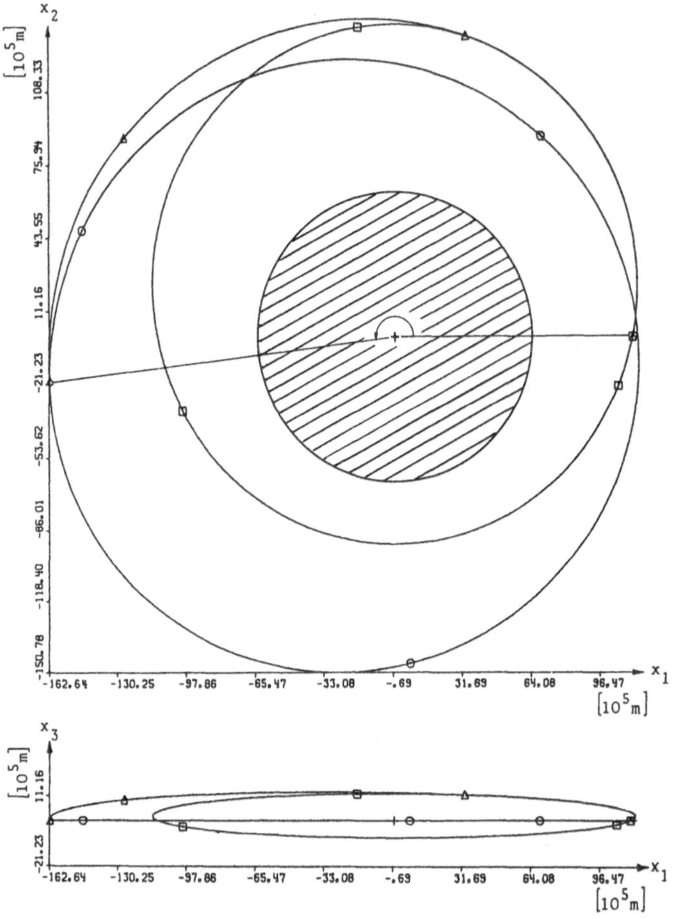

Figure 2 : Fuel optimal transfer maneuver ;
□ Initial orbit,
△ Transfer orbit,
○ Final orbit.

LIST OF AUTHORS

*paper not received

Lecture Notes in Economics and Mathematical Systems

For information about Vols. 1–99 please contact your bookseller or Springer-Verlag

Vol. 100: B. S. Duran and P. L. Odell, Cluster Analysis. A Survey. VI, 137 pages. 1974.

Vol. 101: W. M. Wonham, Linear Multivariable Control. A Geometric Approach. X, 344 pages. 1974.

Vol. 102: Analyse Convexe et Ses Applications. Comptes Rendus, Janvier 1974. Edited by J.-P. Aubin. IV, 244 pages. 1974.

Vol. 103: D. E. Boyce, A. Farhi, R. Weischedel, Optimal Subset Selection. Multiple Regression, Interdependence and Optimal Network Algorithms. XIII, 187 pages. 1974.

Vol. 104: S. Fujino, A Neo-Keynesian Theory of Inflation and Economic Growth. V, 96 pages. 1974.

Vol. 105: Optimal Control Theory and its Applications. Part I. Proceedings 1973. Edited by B. J. Kirby. VI, 425 pages. 1974.

Vol. 106: Optimal Control Theory and its Applications. Part II. Proceedings 1973. Edited by B. J. Kirby. VI, 403 pages. 1974.

Vol. 107: Control Theory, Numerical Methods and Computer Systems Modeling. International Symposium, Rocquencourt, June 17–21, 1974. Edited by A. Bensoussan and J. L. Lions. VIII, 757 pages. 1975.

Vol. 108: F. Bauer et al., Supercritical Wing Sections II. A Handbook. V, 296 pages. 1975.

Vol. 109: R. von Randow, Introduction to the Theory of Matroids. IX, 102 pages. 1975.

Vol. 110: C. Striebel, Optimal Control of Discrete Time Stochastic Systems. III. 208 pages. 1975.

Vol. 111: Variable Structure Systems with Application to Economics and Biology. Proceedings 1974. Edited by A. Ruberti and R. R. Mohler. VI, 321 pages. 1975.

Vol. 112: J. Wilhlem, Objectives and Multi-Objective Decision Making Under Uncertainty. IV, 111 pages. 1975.

Vol. 113: G. A. Aschinger, Stabilitätsaussagen über Klassen von Matrizen mit verschwindenden Zeilensummen. V, 102 Seiten. 1975.

Vol. 114: G. Uebe, Produktionstheorie. XVII, 301 Seiten. 1976.

Vol. 115: Anderson et al., Foundations of System Theory: Finitary and Infinitary Conditions. VII, 93 pages. 1976

Vol. 116: K. Miyazawa, Input-Output Analysis and the Structure of Income Distribution. IX, 135 pages. 1976.

Vol. 117: Optimization and Operations Research. Proceedings 1975. Edited by W. Oettli and K. Ritter. IV, 316 pages. 1976.

Vol. 118: Traffic Equilibrium Methods, Proceedings 1974. Edited by M. A. Florian. XXIII, 432 pages. 1976.

Vol. 119: Inflation in Small Countries. Proceedings 1974. Edited by H. Frisch. VI, 356 pages. 1976.

Vol. 120: G. Hasenkamp, Specification and Estimation of Multiple-Output Production Functions. VII, 151 pages. 1976.

Vol. 121: J. W. Cohen, On Regenerative Processes in Queueing Theory. IX, 93 pages. 1976.

Vol. 122: M. S. Bazaraa, and C. M. Shetty, Foundations of Optimization VI. 193 pages. 1976

Vol. 123: Multiple Criteria Decision Making. Kyoto 1975. Edited by M. Zeleny. XXVII, 345 pages. 1976.

Vol. 124: M. J. Todd. The Computation of Fixed Points and Applications. VII, 129 pages. 1976.

Vol. 125: Karl C. Mosler. Optimale Transportnetze. Zur Bestimmung ihres kostengünstigsten Standorts bei gegebener Nachfrage. VI, 142 Seiten. 1976.

Vol. 126: Energy, Regional Science and Public Policy. Energy and Environment I. Proceedings 1975. Edited by M. Chatterji and P. Van Rompuy. VIII, 316 pages. 1976.

Vol. 127: Environment, Regional Science and Interregional Modeling. Energy and Environment II. Proceedings 1975. Edited by M. Chatterji and P. Van Rompuy. IX, 211 pages. 1976.

Vol. 128: Integer Programming and Related Areas. A Classified Bibliography. Edited by C. Kastning. XII, 495 pages. 1976.

Vol. 129: H.-J. Lüthi, Komplementaritäts- und Fixpunktalgorithmen in der mathematischen Programmierung. Spieltheorie und Ökonomie. VII, 145 Seiten. 1976.

Vol. 130: Multiple Criteria Decision Making, Jouy-en-Josas, France. Proceedings 1975. Edited by H. Thiriez and S. Zionts. VI, 409 pages. 1976.

Vol. 131: Mathematical Systems Theory. Proccedings 1975. Edited by G. Marchesini and S. K. Mitter. X, 408 pages. 1976.

Vol. 132: U. H. Funke, Mathematical Models in Marketing. A Collection of Abstracts. XX, 514 pages. 1976.

Vol. 133: Warsaw Fall Seminars in Mathematical Economics 1975. Edited by M. W. Loś, J. Loś, and A. Wieczorek. V. 159 pages. 1976.

Vol. 134: Computing Methods in Applied Sciences and Engineering. Proceedings 1975. VIII, 390 pages. 1976.

Vol. 135: H. Haga, A Disequilibrium – Equilibrium Model with Money and Bonds. A Keynesian – Walrasian Synthesis. VI, 119 pages. 1976.

Vol. 136: E. Kofler und G. Menges, Entscheidungen bei unvollständiger Information. XII, 357 Seiten. 1976.

Vol. 137: R. Wets, Grundlagen Konvexer Optimierung. VI, 146 Seiten. 1976.

Vol. 138: K. Okuguchi, Expectations and Stability in Oligopoly Models. VI, 103 pages. 1976.

Vol. 139: Production Theory and Its Applications. Proceedings. Edited by H. Albach and G. Bergendahl. VIII, 193 pages. 1977.

Vol. 140: W. Eichhorn and J. Voeller, Theory of the Price Index. Fisher's Test Approach and Generalizations. VII, 95 pages. 1976.

Vol. 141: Mathematical Economics and Game Theory. Essays in Honor of Oskar Morgenstern. Edited by R. Henn and O. Moeschlin. XIV, 703 pages. 1977.

Vol. 142: J. S. Lane, On Optimal Population Paths. V, 123 pages. 1977.

Vol. 143: B. Näslund, An Analysis of Economic Size Distributions. XV, 100 pages. 1977.

Vol. 144: Convex Analysis and Its Applications. Proceedings 1976. Edited by A. Auslender. VI, 219 pages. 1977.

Vol. 145: J. Rosenmüller, Extreme Games and Their Solutions. IV, 126 pages. 1977

Vol. 146: In Search of Economic Indicators. Edited by W. H. Strigel. XVI, 198 pages. 1977.

Vol. 147: Resource Allocation and Division of Space. Proceedings. Edited by T. Fujii and R. Sato. VIII, 184 pages. 1977.

Vol. 148: C. E. Mandl, Simulationstechnik und Simulationsmodelle in den Sozial- und Wirtschaftswissenschaften. IX, 173 Seiten. 1977.

Vol. 149: Stationäre und schrumpfende Bevölkerungen: Demographisches Null- und Negativwachstum in Österreich. Herausgegeben von G. Feichtinger. VI, 262 Seiten. 1977.

Vol. 150: Bauer et al., Supercritical Wing Sections III. VI, 179 pages. 1977.

Vol. 151: C. A. Schneeweiß, Inventory-Production Theory. VI, 116 pages. 1977.

Vol. 152: Kirsch et al., Notwendige Optimalitätsbedingungen und ihre Anwendung. VI, 157 Seiten. 1978.

Vol. 153: Kombinatorische Entscheidungsprobleme: Methoden und Anwendungen. Herausgegeben von T. M. Liebling und M. Rössler. VIII, 206 Seiten. 1978.